MOLECULAR MECHANISMS OF METAL TOXICITY AND CARCINOGENESIS

Molecular Mechanisms of Metal Toxicity and Carcinogenesis

Edited by

XIANGLIN SHI

Pathology and Physiology Research Branch
National Institute for Occupational Safety and Health
1095 Willowdale Road
Mall Stop B167
26505, Morgantown WV
USA
Phone: 304-285-6158
E-mail: xas0@cdc.gov
Fax: 304-285-5938

VINCE CASTRANOVA

Pathology and Physiology Research Branch
National Institute for Occupational Safety and Health
1095 Willowdale Road
Mall Stop B167
26505, Morgantown WV
USA

VAL VALLYATHAN
Pathology and Physiology Research Branch
National Institute for Occupational Safety and Health
1095 Willowdale Road
Mall Stop B167
26505, Morgantown WV
USA

WILLIAM G PERRY
Occupational Safety and Health Administration
Washington DC
USA

Reprinted from *Molecular and Cellular Biochemistry*, Volume 222 (2001)

Kluwer Academic Publishers
Dordrecht / Boston / London

Distributors for North, Central and South America:
Kluwer Academic Publishers
101 Philip Drive
Assinippi Park
Norwell, Massachusetts 02061 USA
Telephone (781) 871-6600
Fax (781) 681-9045
E-mail <kluwer@wkap.com>

Distributors for all other countries:
Kluwer Academic Publishers Group
Distribution Centre
Post Office Box 322
3300 AH Dordrecht, THE NETHERLANDS
Telephone 31 78 6392 392
Fax 31 78 6546 474
E-mail <services@wkap.nl>

Electronic Services <http://www.wkap.nl>

Library of Congress Cataloging-in-Publication Data

Molecular mechanisms of metal toxicity and
carcinogenesis / edited by Xianglin Shi.
p.cm. –(Developments in molecular and cellular
biochemistry; v. 34) "Reprinted from Molecular and
cellular biochemistry, Volume 222 (2001)."
ISBN 0-7923-7498-3 (alk. paper)
1. Metals–Carcinogenicity–Congresses. 2. Metals–
Toxicology–Congress. 3. Molecular toxicology–
Congresses. 1. Shi, Xianglin, Ph.D. II. Series.

RC268.7.M45.M65 2001
616.99'4071–dc21 2001046196

Printed on acid-free paper

Molecular and Cellular Biochemistry:

An International Journal for Chemical Biology in Health and Disease

CONTENTS VOLUME 222, Nos. 1 & 2, June 2001

MOLECULAR MECHANISMS OF METAL TOXICITY AND CARCINOGENESIS
Drs. Xianglin Shi, Vince Castranova, Val Vallyathan and William G. Perry

Molecular and Cellular Biochemistry **222:** 1, 2001.

Preface

Epidemiologic and animal studies have identified several metals and metal-containing compounds as potent mutagens and carcinogens. The metals associated with these biological effects include As, Cd, Cr, and Ni. During the last two decades, chemical and cellular studies have contributed enormously to our understanding of metal-induced carcinogenesis. Many hypotheses have been extensively investigated to understand the role of metals in pathophysiological processes. One of the major breakthroughs has been the elucidation of the role of reactive oxygen/nitrogen species in metal-induced carcinogenesis. Carcinogenic metals induce genotoxicity in a multiplicity of ways, either alone or by enhancing the effects of other agents by additive or synergistic action. Although each of these metals is unique in its mechanism of action, several common pathways, for example oxidative stress, may be shared by many of these carcinogenic metals. Recently, the sub-discipline of molecular toxicology and carcinogenesis has been developed. New techniques are available now to unravel the mechanism of carcinogenesis in precise molecular terms so that intricate biological interrelationships can be elucidated.

The conference on Molecular Mechanisms of Metal Toxicity and Carcinogenesis was held at the National Institute for Occupational Safety and Health, Morgantown, West Virginia, from September 10–12, 2000. The conference focused on the latest developments concerning the molecular mechanisms of metal-induced toxicity and carcinogenesis. During this conference, international experts provided in-depth state-of-the-art information pertinent to the field. The conference promoted a multidisciplinary investigative approach among attendees from academia, federal agencies and industries in pursing new avenues of research in metal toxicity and carcinogenesis.

We express our thanks to those who cooperated in the conference and in the publication of this special issue.

Xianglin Shi
Vince Castranova
Val Vallyathan
National Institute for Occupational Safety and Health,
Morgantown, WV, USA

William G. Perry
Occupational Safety and Health Administration,
Washington DC, USA

Guest Editors

Molecular and Cellular Biochemistry **222**: 3–9, 2001.

Molecular mechanisms of metal toxicity and carcinogenesis

Suwei Wang and Xianglin Shi

Pathology and Physiology Research Branch, Health Effects Laboratory Division, National Institute for Occupational Safety and Health, Morgantown, WV; Department of Basic Pharmaceutical Sciences, West Virginia University, Morgantown, WV, USA

Abstract

Many metals and metal-containing compounds have been identified to be potent mutagens and carcinogens. Recently, a new sub-discipline of molecular toxicology and carcinogenesis has been developed. The combination of newly developed molecular techniques and free radical approach makes it possible to insightfully examine metal-induced carcinogenesis in precise molecular terms so that intricate biological interrelationships can be elucidated. In consideration of the increased amount of new findings deciphered by utilizing these new methods, the 1st Conference on Molecular Mechanisms of Metal Toxicity and Carcinogenesis was held. In this conference, more than 50 scientists from nine countries presented their novel discoveries concerning metal-induced carcinogenesis, delineated molecular mechanism of metal carcinogenesis, and proposed novel therapeutic intervention and prevention strategies. This article reviews some of the state-of-the-art information presented at the meeting regarding the molecular mechanisms of metal cytotoxicity and carcinogenesis. (Mol Cell Biochem **222**: 3–9, 2001)

Key words: ROS, cell growth regulation, transcription factor, signal transduction, cytotoxicity and carcinogenesis

Introduction

Both environmental and occupational exposures to chromium (Cr), nickel (Ni), and arsenic (As) compounds have been reported to cause human tumors in various systems [1, 2]. Although a variety of metals, including Cr, Ni, As, vanadium (V), cadmium (Cd), and cobalt (Co), have been reported to induce carcinogenesis both in animals and in humans, the underlying mechanisms are not well understood. Since carcinogenesis induced by metals involves a variety of facets, it is difficult to identify one common mechanism. It is very likely that each metal has its own unique molecular mechanisms which contribute to the cancer development.

During the last two decades, chemical and cellular studies have provided a variety of contributions to extend our understanding for metal-induced carcinogenesis. Generally, carcinogenesis is considered to have the following four stages: initiation, promotion, progression, and metastasis. Although mutations on genome DNA which are capable of activating oncogenes or inactivating tumor suppressors, are traditionally considered to be the crucial factor for cancer initiation, other events, such as transcription activation, recombination, and oncogene amplification, which may or may not require DNA damage, also contribute to the tumor initiation process [3–7]. Altered gene expression and cell signaling are considered to be related to tumor promotion and progression [8, 9]. The mechanisms of metal-induced carcinogenesis are believed to be involved in all stages of cancer development.

Metal compounds are found to have effects on the cellular organelles and components, such as cell membrane, mitochondrial, lysosome, endoplasmic reticulum, nuclei, and some enzymes involved in metabolism, detoxification, and damage repair [10]. All these systems are considered to influence metal-induced cellular responses. Metal ions have been found to directly bind to some cell components such as DNA and nuclear proteins through ionic and coordination bonds, causing

Address for offprints: X. Shi, Pathology and Physiology Research Branch, National Institute for Occupational Safety and Health, Morgantown, WV 26505, USA

DNA damage and conformation changes [11, 12]. Since the binding of metal ions to DNA and proteins are reversible, it appears that this direct binding can not be fully responsible for carcinogenic metal-induced DNA lesions, including DNA strand breaks and base modifications, which result from the breakage or formation of covalent bonds [13–16]. Based on these observations, in addition to direct binding, other indirect effects of metals must be involved in metal-induced cell responses.

Accumulating evidence suggests that reactive oxygen species (ROS) play a major role in mediating metal-induced cellular responses and carcinogenesis [16–19]. ROS mainly include superoxide radical ($O_2^{\cdot-}$), hydrogen peroxide (H_2O_2), hypochloride (HOCl), hydroxyl radical (·OH), peroxyl radical (ROO·), alkoxyl radical (RO·), thiyl radical (RS·), and nitric oxide (NO). Several groups have reported that carcinogenic metal-induced chromatin damage can be mimicked by ROS, suggesting the important role of ROS in these processes [16–19]. In addition, various metal ions, such as Cr (VI), V (V), and Co (II), have been reported to produce ROS in cellular systems through a Fenton-like reaction and Haber–Weiss cycle [20–23]. The generated ROS are believed to mediate metal-induced cell injuries.

Recently, a new sub-discipline of molecular toxicology and carcinogenesis has been developed. The combination of newly developed molecular techniques and free radical approach makes it possible to insightfully examine metal-induced carcinogenesis in precise molecular terms so that intricate biological interrelationships can be elucidated. In view of the increased amount of new findings deciphered by utilizing these new methods, the 1st Conference on Molecular Mechanisms of Metal Toxicity and Carcinogenesis was held at National Institute for Occupational Safety and Health (NIOSH), Morgantown, WV, in September, 2000. In this conference, more than 50 scientists from nine countries presented their novel discoveries concerning metal-induced carcinogenesis, delineated molecular mechanism of metal carcinogenesis, and proposed novel therapeutic intervention and prevention strategies. The following categories have been emphasized during the meeting: (a) effects of metal on apoptosis and cell growth regulation; (b) mechanism of metal-induced regulation of nuclear transcription factors, such as NF-κB, AP-1, p53, and NFAT; (c) activation of signal transduction pathway, MAP kinase pathway; (d) effects of metals on gene expression; (e) mechanism of metal-induced cytotoxicity, cellular responses, mutagenesis, and carcinogenesis; (f) detection and mechanism of metal-induced free radical generation.

Effects of metal on apoptosis and cell growth regulation

Apoptosis, or programmed cell death, is a genetically controlled response for cells to commit suicide. It provides a major form of cell death in normal development of vertebrate embryos, appropriate functions of the immune system, and maintenance of tissue homeostasis [24–29]. Apoptosis can be triggered by a variety of extrinsic and intrinsic signals, including UV and ion radiations, viral infection, bacterial toxins, nutrient deprivation, oncogene expressions, oxidants, etc. [30–32]. Apoptosis that results from these stimuli regulates the elimination of excess, genetically damaged, and improperly developed cells from tissues. Various toxic metals, including Cr, V, Cd, Ni, and As, have been reported to be able to induce apoptosis [33–37]. However, the mechanisms and the effects of apoptosis are still not clear.

In this conference, Ye *et al.* presented the evidence indicating that both ROS and p53 play very important roles in Cr (VI)-induced apoptosis. ROS were generated within a few minutes after Cr (VI) treatment, while p53 induction took at least 5 h. The observations that p53 positive cells have a higher apoptosis level compared to p53 negative cells only at later stage (3–24 h) but not at early stage (0–3 h) indicate that ROS generation through Cr (VI) reduction was responsible for the early stage of apoptosis, whereas p53 contributed to the later enhanced apoptosis induced by Cr (VI). Meanwhile, Patierno *et al.* and Bagchi *et al.* also reported that Cr (VI) was able to induce apoptosis in several different cell lines at different levels in a dose and time dependent manner both *in vitro* and *in vivo*. Induction of p53, disruption of the mitochondrial membrane permeability transition (MMPT), release of mitochondrial cytochrome C into the cytoplasm, and activation of caspase 3 have been implicated in the process of apoptosis induced by Cr (VI). More interestingly, they found that Cr-induced DNA–DNA crosslinks were crucial for apoptosis resulting from Cr exposure. These findings indicate that both ROS and Cr metabolites from Cr (VI) reduction play important roles in the mechanism of Cr (VI)-induced apoptosis.

Ong *et al.* investigated the role of calcium (Ca^{2+}) in Cd-induced apoptosis in primary cultured mouse thymocytes. Exposure of thymocytes to Cd resulted in a rapid and sustained intracellular Ca^{2+} elevation, followed by caspase-3 activation, PARP cleavage, and DNA fragmentation. They believe that intracellular Ca^{2+} elevation may trigger caspase-3 activation, contributing to Cd-induced apoptosis in thymocytes. This study has provided new information for a better understanding of the immunotoxic and immunomodulatory effects of Cd.

Under normal conditions, cell cycle proceeds from G_1 phase to S phase, then to G_2 and M phase, without interruption. These processes are monitored at several checkpoints for the proper DNA duplication, regulated by complexes of cyclins and cyclin-dependent kinases. Two important checkpoints are in charge of G_1/S and G_2/M transitions. When damage occurs to the cells, most of them are capable of arresting the cell cycle, repairing the damage, and resuming the proliferation when the damages are fixed. By this way, the cells take control of the precision of DNA replication and maintain the integrity of the genome [38–41]. However, for the damage that cannot be repaired by arresting the cell cycle, the cells would undergo apoptosis. Therefore, cell cycle arrest is a passive defense mechanism of cells in response to external toxicants, and improper cell growth regulation may result in cancer development.

At present, there are only a few reports available on the effects of metal on cell growth regulation. In this meeting, two articles from Dr. Xianglin Shi's laboratory presented their novel findings on mechanisms of metal-induced cell growth arrest. Zhang *et al.* found that V (V) was able to induce G_2/M phase arrest in A549 cells. During this process, several regulatory pathways are involved, including activation of p53 and p21 and inactivation of cdc2/cyclin B complex resulting from both increased chk1 expression and inhibition of cdc25C. ROS generated in V (V)-stimulated cells, especially H_2O_2, are involved in the regulatory mechanisms. In another study, Chen *et al.* utilized Cr (VI) and As (III), focusing on GADD45, a cell cycle checkpoint protein whose activation arrests cells at G_2/M phase. Inhibition of NF-κB activation caused an increased and prolonged induction of GADD45, resulting in G_2/M arrest induced by As (III). On the other hand, abrogation of c-*jun* N-terminal kinase (JNK) activation decreased GADD45 induction by As (III), suggesting that JNK and NF-κB may have distinct contribution to cell growth regulation through their opposite effects on GADD45 in response to toxic metals. The results indicate that toxic metals indeed can cause cell cycle arrest, and several different pathways and regulatory proteins are involved in these cell growth arrest process. The investigations on metal-induced cell cycle arrest opened a new window to study the mechanisms of metal-induced cancer development.

Mechanism of metal-induced regulation of nuclear transcription factors

One of the characteristics of nuclear transcription factors is that they are able to recognize and bind to specific sequences of DNA, regulating the expression of their target genes. Due to this unique property, the activation of transcription factors act as a key switch in mediating cell responses upon stimulation. Accumulating evidence indicate that some metals such as Cr, V, and lead (Pb), are able to activate various transcription factors including NF-κB, AP-1, and p53 protein [42–52]. While it has been suggested that ROS may be involved in the process of transcription factors activation, the underlying mechanisms are still need to be investigated.

In this conference, Wang *et al.* provided direct evidence, indicating that ROS do play an important role in mediating p53 activation induced by Cr (VI). Among all the ROS generated during the Cr (VI) reduction, ·OH radical is the one that is responsible for Cr (VI)-induced p53 activation. Thorough investigation on transcription factor activation in response to vanadate exposure has been carried out by Dr. Xianglin Shi's group. In addition to the findings that vanadate is able to induce the activation of NF-κB, AP-1, and p53, they performed studies to uncover the mechanisms of those activation from different angles. Huang *et al.* reported an increase in p53 transactivation induced by vanadate. Furthermore, they pointed out that this activation of p53 was mainly through H_2O_2, which is generated through vanadate reduction.

Chen *et al.* presented evidence that vanadate also was able to induce the activation of both NF-κB and JNK in macrophages. In their study, they also investigated the possible cross talk between MAP kinase pathway and NF-κB upstream signal transduction. Blockage of JNK activation resulted in partial inhibition of vanadate-induced IκBα degradation, but inactivation of IKKβ, which results in decreased IκBα degradation, did not have any effect on the activation of JNK induced by vanadate. These results suggest that the involvement of JNK in regulating the upstream signal transduction of NF-κB activation. Furthermore, they found that both vanadate-induced IκBα degradation and JNK activation could be inhibited by antioxidant, N-acetylcysteine (NAC), indicating that regulatory role of ROS in the upstream of NF-κB activation. Transcription factor AP-1, which is important in mediating cell proliferation, also could be induced by vanadate. Ding *et al.* observed AP-1 activation induced by vanadate both in JB6 cells and rat lung epithelia cells, and the activation was mediated by $O_2^{\cdot-}$ and H_2O_2, but not by ·OH radical.

NFAT (nuclear factor of activated T cells) family proteins, which are expressed in most immune-system cells, play a pivotal role in the transcription of cytokine genes and other genes critical for the immune response. Huang *et al.* reported that NFAT could be activated upon vanadate exposure. Blockage of both calcium and calcineurin inhibited the activation of NFAT by vanadate. Similar inhibition was observed when specific antioxidants for H_2O_2 was added, suggesting the potential role of H_2O_2 in mediating calcium/calcineurin pathway-dependent activation of NFAT induced by vanadate. In addition, another heavy metal, nickel was also found to be able to induce NFAT activation. The activation of NFAT by nickel is involved in regulating the expression of Cap43 gene, a novel gene induced by nickel and hypoxia [53, 54]. These findings provide evidence for the activation of NFAT by metal for the first time, and help to understand the signal transduction pathways involved in carcinogenic effects of nickel compounds, as well as other transition metals.

Effects of metals on signal transduction pathway

Since metals have been found to activate transcription factor AP-1, it is not surprising to find them affecting the AP-1 upstream regulatory signal transduction pathway, namely the MAP kinase pathway. Different effects of Cr (VI) and Cd on the activation of three major MAP kinases, ERK, JNK, and p38 kinase were presented by Dr. Jia-LingYang's group. Cr (VI) is able to activate all of the three MAP kinases in dose- and time-dependent manner. In contrast, Cd decreased the ERK activity with a concomitant enhancement in JNK at low cytotoxic dosages, but persistently activated the three MAP kinases at high cytotoxic dosages. Cross talk of the MAP kinase signals was observed in cells exposed to Cr (VI) but not to Cd. Increased JNK and p38 kinase activity, as well as reduced ERK activity, contribute to Cd-induced apoptosis. In contrast, the activation of these three MAP kinases was not involved in apoptosis induced by Cr (VI). These results imply that although ROS are considered to play important roles in the cell responses induced by both Cd and Cr (VI), these two metals may induce MAP kinase activation and regulate cell growth arrest and apoptosis through different mechanisms.

Another interesting and novel finding presented in this conference is that ROS were reported to induce the phosphorylation of tyrosine kinase. Cr (VI) was utilized as a source for whole spectrum of ROS generation. A dose- and time-dependent increase of tyrosine phosphorylation was observed in human lung epithelial cells after Cr (VI) exposure. Among all the ROS generated by Cr (VI) reduction, H_2O_2 and ·OH radical are the major species to mediate the increased tyrosine phosphorylation. This study provides direct evidence demonstrating the involvement of ROS, especially H_2O_2 and ·OH radical, in tyrosine phosphorylation for the first time, providing a new direction for the further mechanism studies on the effects of ROS on regulating signal transduction pathway.

Effects of metals on gene expression

Most of the toxic heavy metals are also reported to be carcinogens and mutagens due to their ability to cause gene mutation. Water-insoluble nickel has been reported by Costa *et al.* to silence genes in mammalian cells [55]. In this meeting, Dr. Costa presented the results of their recent investigations on the molecular mechanisms of nickel-induced gene silencing. The inhibition of the acetylation of histone H4 at specific N-terminal lysine sites by nickel was observed to contribute to nickel induced gene silencing. Trichostatin A, which is an inhibitor of histone deacetylase, is able to inhibit the ability of nickel to silence genes by reducing the inhibition of histone H4 acetylation. The results indicated that as a major component of the transcriptional machinery, the acetylation of histone plays an important role in nickel-induced silencing, which is presumed to be a crucial step in nickel-induced carcinogenesis.

In addition to inducing gene silencing, insoluble nickel compounds can also cause morphological transformation by positively altering gene expression as presented by Landolph *et al.* By using RAP-PCR mRNA Differential Display to examine cells transformed by carcinogenic insoluble nickel compounds, they found that 130 genes were differentially expressed compared to nontransformed cells. These results indicate that the global gene expression in the transformed cells was substantially modified. Among these altered 130 genes, 10 genes which are involved in various functions have been identified. Further work to determine how these alterations contribute to nickel-induced transformation will help in understanding the tumor development caused by nickel.

Similarly, by using microarray technique, Ye *et al.* found that Cr (VI) was able to alter the expression of about 220 genes, of which 150 genes was up-regulated and 70 were down-regulated by Cr (VI). Functional analysis indicates that these genes are involved in several categories, including redox stress, calcium mobilization, fat degradation, protein metabolism, cell cycle arrest, and carcinogenesis. The results provide critical information and directions for understanding the molecular mechanisms involved in Cr (VI)-induced carcinogenesis.

To further understand the precise mechanisms of As and Cr induced toxicity and carcinogenesis, Hamilton *et al.* examined the effects of these two metals on gene expression. Their previous studies indicate that both As (III) and Cr (VI) have profound but preferential effects on expression of several inducible genes, including the hormone-regulated phosphoenolpyruvate carboxykinase (PEPCK) gene, whose expression is associated with the glucocorticoid receptor (GR)-mediated regulatory pathway [56, 57]. Their novel study presented at this conference indicated that As (III) significantly suppressed both basal and inducible expression of PEPCK through specific suppression of GR as a transcription factor within the nucleus following hormone binding and nuclear translocation. In contrast, Cr (VI) enhanced both basal and inducible gene expression of PEPCK via Cr (III)-involved cAMP signaling pathway.

Metallothionein (MT) is a low-molecular-weight protein with approximately one-third of its amino acids being cysteine residues. This protein is inducible by various metals, exerting its role in essential trace metal homeostasis and heavy metal detoxification. MTF-1, a protein which specific binds to metal response element (MRE), plays a major role in regulating metal-inducible MT transcription [58]. At present, there are two models that describe MTF-1-mediated regulation of MT transcription. However, neither of them is able to explain how metals other than zinc can activate MT transcription. Freedman *et al.* presented a novel mechanism for the activation of MT induced by metal: MTF-1 phosphorylation. They found that MTF-1 was phosphorylated following zinc exposure but not by Cd treatment. The results confirmed that MTF-1 can be phosphorylated *in vivo*, and this phosphorylation might be the mechanism for metal-induced MT transcription. Although both zinc and Cd are able to up-regulate MT transcription, the mechanisms for their effects might be different.

Mechanism of metal-induced cytotoxicity, cellular responses, mutagenesis, and carcinogenesis

Cr (VI)-containing compounds are well established mutagens and carcinogens. Although the mechanisms for their effects are not clear yet, it is generally believed that Cr (VI)-induced cellular responses are mediated by the reactive intermediates generated directly from Cr (VI) reduction, including ROS, Cr (V), Cr (IV), and Cr (III). Previous studies and a variety investigations presented in this meeting have shown that ROS generated from Cr (VI) reduction play crucial roles in DNA damage, lipid peroxidation, and transcription factor activation [42–44, 46]. At the meeting, Pritchard reported his recent finding, implicating nitrogen species in the mechanisms of Cr (VI)-induced endothelial cell inflammation. Cr (VI) was observed to be capable of inducing inflammation in human umbilical vein endothelial cells (HUVEC). During this process, ICAM-1 gene expression was increased at mRNA level, suggesting the possible involvement of this gene in the inflammation. Cr (VI) increased the generation of $O_2^{\cdot-}$ and decreased the intracellular NO without affecting the NO generation. At the same time, an increased formation of peroxynitrite was observed, indicating that generated $O_2^{\cdot-}$ scavenged NO to produce peroxynitrite. Through this mechanism, decreased NO activity and increased formation of peroxynitrite appear to play a critical role in enhancing the expression of ICAM-1, which is crucial in Cr (VI)-induced endothelial cell injury. The other project carried out by the same group using bovine coronary endothelial cells (BCEC) investigated the mechanisms of effects of Cr (VI) on eNOS (endothelial NO synthase) activity. Even at very low concentration, Cr (VI) perturbed eNOS function in BCEC cells. This perturbation is possibly due to the disruption of the interaction between eNOS and heat shock protein 90 (Hsp90), whose association to eNOS ensures the proper functions of this synthase. This disruption may be an important step in the mechanisms by which endothelial cells can be induced to generate $O_2^{\cdot-}$ in an eNOS-dependent manner.

In addition to ROS and reactive nitrogen species, the importance of chromates in Cr (VI)-induced cellular responses have been re-emphasized during this meeting. Several groups presented their recent findings on Cr (VI)-induced DNA damage. Cystine may act as a reductant when mixing with Cr (VI), and reduce Cr (VI) to Cr (III). The generated Cr (III) is able to form ternary Cr (III)-DNA adducts, which contribute to a significant increase in mutation frequencies. The study by Zhitkovich *et al.* provided strong evidence for the importance of Cr (III)-dependent pathway in the genotoxicity of Cr (VI) compounds. In addition, Krepkiy *et al.* reported that Cr (III)-Cys might be the intermediate for the formation of Cr (III)-DNA complex. Cr (V), another reactive metabolite of Cr (VI), is also reported to cause DNA damage [59]. To investigate the oxidative DNA damage by Cr (V), Sugden *et al.* established two model Cr (V) complexes: Cr (V)-ehba and Cr (V)-salen, which have similar oxidative mechanisms in terms of their abilities to serve as direct-acting oxidants but oxidizing different sites on DNA. Cr (V)-ehba complex has a preference for oxidation at the deoxyribose sugar moiety resulting in sequence-neutral, frank- and alkaline-labile strand breaks, while Cr (V)-salen complex with has shown a preference for oxidation at the nucleic acid base guanine to yield a piperidine-labile cleavage site. The use of these tow model Cr (V) complexes has shown an excellent correlation with the DNA damage associated with cellular exposure to chromate, facilitating further understanding for the mechanisms of Cr (VI)-induced DNA damage and other responses.

Cr (VI) has been shown to be mutagenic in a variety of assay systems. Using Big Blue transgenic mouse, Dixon *et al.* reported that manipulating glutathione (GSH) level could alter the mutant frequency of Cr (VI): reduced tissue GSH levels resulted in a lower mutant frequency, suggesting the important role of GSH in regulating Cr (VI) mutagenesis. Most interestingly, As, a co-mutagen, was found to suppress the Cr (VI) mutagenesis in this system. The possible explanation is that As reduced the availability of free GSH required for Cr reduction and mutagenesis.

Although As has a much lower potency to induce mutagenesis compared to Cr (VI), it is able to induce significant cytotoxicity and carcinogenesis. Increasing evidence indicates that epigenetic mechanisms, including modulation of cell growth signaling, may play a role in As carcinogenesis. Simeonova *et al.* have demonstrated that As exposure activates cell proliferation of urinary bladder epithelium by using both *in vitro* and *in vivo* models. A persistent activation of AP-1 transcription factor, as well as the altered expression of a variety of genes associated with cell growth and cell arrest were observed following exposure of the cells to As. These molecular alterations might be involved in As-induced carcinogenesis. Moreover, the toxicity and carcinogenesis induced by As can be significantly enhanced by atrazine.

Detection and mechanism of free radical generation

It is generally believed that Cr (VI) compounds exert their functions via their reduction metabolites. However, there is no systematic investigation on how Cr (VI) compounds are converted to the metabolites. To solve this question, Leonard *et al.* thoroughly investigated the mechanisms of the generation of ROS and chromate metabolites. They have shown that Cr (VI) can be reduced to Cr (V) by glutathione reductase. During this process, $O_2^{\cdot-}$ radical is generated from molecular oxygen, followed by the generation of H_2O_2 via $O_2^{\cdot-}$ dismutation catalyzed by superoxide dismutase (SOD). Furthermore, H_2O_2 reacts with Cr (V) through a Fenton-like reaction to produce ·OH radical and regenerate Cr (VI). ·OH radicals generated by the enzymatic Cr (VI) reduction are capable of causing DNA strand breaks. This study provides a clear picture of ROS and reactive Cr intermediates generated from Cr (VI), facilitating further biological studies on Cr (VI)-induced carcinogenesis. The other study carried out by Liu *et al.* confirmed the generation of Cr (V) in the biological system. By using a newly developed *in vivo* electron paramagnetic resonance (EPR) spectrometer, they observed the formation of Cr (V) directly from intact mice after intravenous injection of Cr (VI). The intensity of Cr (V) signals decreased when the animal was pretreated with antioxidants and metal chelators. The study verified the importance of Cr (V) in mediating Cr (VI)-induced responses, and illustrated the possibility of using EPR to study paramagnetic metal ions in intact animals.

Discussion

During the meeting, a number of groups presented their state-of-the-art information in their research with metals. The presentations cover a wide range, including metal-induced cytotoxicity, cellular responses, mutagenesis, apoptosis, cell growth regulation, regulation of nuclear transcription factors, activation of signal transduction pathway, and gene expression. It should be noted that impressive progress has been achieved concerning the molecular mechanisms of metal-induced toxicity and carcinogenesis. However, there are still a lot of questions to be solved. A great number of investigations have been conducted concerning the roles of metals in regulating cell signaling. However, some of the findings are contradictory, and a link is needed to connect all the observations and findings. The question of how ROS and metal metabolites affect the signal transduction still remains to be answered. Metal-induced cell growth control has got more attention during this meeting. Three presentations from Dr. Xianglin Shi's group reported the effects of V on cell growth regulation. These findings provide a new direction for the study of mechanisms of metal-induced carcinogenesis. Furthermore, some new emerging techniques provide more convenient and powerful tools for the investigation of metal-induced carcinogenesis. For example, the microarray technique makes it possible to screen a large amount of genes which are possibly induced by certain metals, providing the hints and directions necessary for more insightful studies.

Acknowledgements

The authors thank Drs Vince Castranova, Murali Rao, and Val Vallyathan for a critical reading of the manuscript. Research funded under Interagency Agreement number 98-18-00m2 between the Occupational Safety and Health Administration (OSHA) and the National Institute for Occupational Safety and Health (NIOSH). The views expressed in the paper are those of the authors and do not necessarily reflect the official position of OSHA.

References

1. IARC: Chromium, nickel, and welding. In: Monographs on the Evaluation of the Carcinogenic Risks to Humans, Vol. 49. International Agency for Research on Cancer, Lyon, France, 1990, pp 49–256
2. NRC: National Research Council Report: Arsenic in the Drinking Water. National Academy Press, Washington, DC, 1999
3. Solomon E, Borrow J, Goddard AD: Chromosome aberrations and cancer. Science 254: 1153–1160, 1991
4. Bos JL: Ras oncogenes in human cancer: A review. Cancer Res 49: 4682–4689, 1989
5. Hollstein M, Sidransky D, Vogelstein B, Harris CC: p53 mutations in human cancers. Science 253: 49–53, 1991
6. Koberle B, Speit G: Molecular characterization of mutations at the hprt locus in V79 Chinese hamster cells induced by bleomycin in the presence of inhibitors of DNA repair. Mutat Res 249: 161–167, 1991
7. Schwab M: Oncogene amplification in neoplastic development and progression of human cancers. Crit Rev Oncog 2: 35–51, 1990
8. Flamigni F, Marmiroli S, Caldarera CM, Guarnieri C: Effect of sodium arsenite on the induction and turnover of ornithine decarboxylase activity in erythroleukemia cells. Cell Biochem Funct 7: 213–217, 1989
9. Mikalsen SO: Effects of heavy metal ions on intercellular communication in Syrian hamster embryo cells. Carcinogenesis 11: 1621–1626, 1990
10. Squibb KS, Fowler BA: Relationship between metal toxicity to subcellular systems and the carcinogenic response. Environ Health Perspect 40: 181–188, 1981
11. Berg JM: Potential metal-binding domains in nucleic acid binding proteins. Science 232: 485–487, 1986
12. Leonard A: Chromosome damage in individuals exposed to heavy metals. In: H. Sigel (ed). Metal Ions in Biological Systems, Vol. 20. Marcel Dekker, New York, 1986, pp 229–258
13. Hamilton-Koch W, Snyder RD, Lavelle JM: Metal-induced DNA damage and repair in human diploid fibroblasts and Chinese hamster ovary cells. Chem Biol Interact 59: 17–28, 1986
14. Oleinick NL, Chiu SM, Ramakrishnan N, Xue LY: The formation, identification, and significance of DNA-protein cross-links in mammalian cells. Br J Cancer (suppl 8): 135–140, 1987
15. Kasprzak KS, Diwan BA, Rice JM, Misra M, Riggs CW, Olinski R, Dizdaroglu M: Nickel (II)-mediated oxidative DNA base damage in renal and hepatic chromatin of pregnant rats and their fetuses. Possible relevance to carcinogenesis. Chem Res Toxicol 5: 809–815, 1992
16. Kasprzak KS: Possible role of oxidative damage in metal-induced carcinogenesis. Cancer Invest 13: 411–430, 1995
17. Breimer LH: Molecular mechanisms of oxygen radical carcinogenesis and mutagenesis: The role of DNA base damage. Mol Carcinog 3: 188–197, 1990
18. Dizdaroglu M: Chemical determination of free radical-induced damage to DNA. Free Radic Biol Med 10: 225–242, 1991

19. Vallyathan V, Shi X, Castranova V: Reactive oxygen species: Their relation to pneumoconiosis and carcinogenesis. Environ Health Perspect 106 (suppl 5): 1151–1155, 1998
20. Shi X, Dalal NS: Evidence for a Fenton-type mechanism for the generation of .OH radicals in the reduction of Cr (VI) in cellular media. Arch Biochem Biophys 281: 90–95, 1990
21. Shi X, Dalal NS: Superoxide-independent reduction of vanadate by rat liver microsomes/NAD(P)H: Vanadate reductase activity. Arch Biochem Biophys 295: 70–75,1992
22. Wang G, Hazra TK, Mitra S, Lee HM, Englander EW: Mitochondrial DNA damage and a hypoxic response are induced by CoCl(2) in rat neuronal PC12 cells. Nucleic Acids Res 28: 2135–2140, 2000
23. Vallyathan V, Shi X: The role of oxygen free radicals in occupational and environmental lung diseases. Environ Health Perspect 105 (suppl 1): 165–177, 1997
24. Kerr JFR, Harmon B, Searle J: An electron-microscope study of cell deletion in the anuran tadpole tail during spontaneous metamorphosis with special reference to apoptosis of striated muscle fibers. J Cell Sci 14: 571–585, 1974
25. Kerr JFR, Serale J, Harmon BV, Bishop CJ: Apoptosis. In: C.S. Potten (ed). Perspectives on Mammalian Cell Death. Oxford University Press, Oxford, 1987, pp 93–128
26. Cohen JJ: Programmed cell death in the immune system. Adv Immunol 50: 55–85, 1991
27. Golstein P, Ojcius DM, Young JD-E: Cell death mechanisms and the immune system. Immunol Rev 121: 29–65, 1991
28. Savill JS, Wyllie AH, Henson JE, Walport MJ, Henson PM, Haslett C: Macrophage phagocytosis of aging neutrophils in inflammation: Programmed cell death in the neutrophil leads to its recognition by macrophages. J Clin Invest 83: 865–875, 1989
29. Schwartzman RA, Cidlowski JA: Apoptosis: The biochemistry and molecular biology of programmed cell death. Endocr Rev 14: 133–150, 1993
30. Sachs L, Lotem J: Control of programmed cell death in normal and leukemia cells: New implications for therapy. Blood 82: 15–21, 1993
31. Reed JC: Bcl-2 and the regulation of programmed cell death. J Cell Biol 124: 1–6, 1994
32. Vaux DL, Weissman IL, Kim SK: Prevention of programmed cell death in *Caenorhabditis elegans* by human bcl-2. Science 258: 1955–1957, 1992
33. Blankenship LJ, Manning FC, Orenstein JM, Patierno SR: Apoptosis is the mode of cell death caused by carcinogenic chromium. Toxicol Appl Pharmacol 126: 75–83, 1994
34. Schieven GL, Wahl AF, Myrdal S, Grosmaire L, Ledbetter JA: Lineage-specific induction of B cell apoptosis and altered signal transduction by the phosphotyrosine phosphatase inhibitor bis(maltolato)oxovanadium (IV). J Biol Chem 270: 20824–20831, 1995
35. el Azzouzi B, Tsangaris GT, Pellegrini O, Manuel Y, Benveniste J, Thomas Y: Cadmium induces apoptosis in a human T cell line. Toxicology 88: 127–139, 1994
36. Shiao YH, Lee SH, Kasprzak KS: Cell cycle arrest, apoptosis and p53 expression in nickel (II) acetate-treated Chinese hamster ovary cells. Carcinogenesis 19: 1203–1207, 1998
37. Ochi T, Nakajima F, Sakurai T, Kaise T, Oya-Ohta Y: Dimethylarsinic acid causes apoptosis in HL-60 cells via interaction with glutathione. Arch Toxicol 70: 815–821, 1996
38. Shackelford RE, Kaufmann WK, Paules RS: Cell cycle control, checkpoint mechanisms, and genotoxic stress. Environ Health Perspect 107 (suppl 1): 5–24, 1999
39. Huang LC, Clarkin KC, Wahl GM: Sensitivity and selectivity of the DNA damage sensor responsible for activating p53-dependent G1 arrest. Proc Natl Acad Sci USA 93: 4827–4832, 1996
40. Hartwell LH, Kastan MB: Cell cycle control and cancer. Science 266: 1821–1828, 1994
41. Mercer WE: Checking on the cell cycle. J Cell Biochem Suppl 30–31: 50–54, 1998
42. Ye J, Zhang X, Young HA, Mao Y, Shi X: Chromium (VI)-induced nuclear factor-kappa B activation in intact cells via free radical reactions. Carcinogenesis 16: 2401–2405, 1995
43. Shi X, Ding M, Ye J, Wang S, Leonard SS, Zang L, Castranova V, Vallyathan V, Chiu A, Dalal N, Liu K: Cr (IV) causes activation of nuclear transcription factor-kappa B, DNA strand breaks and dG hydroxylation via free radical reactions. J Inorg Biochem 75: 37–44, 1999
44. Shi X, Chiu A, Chen CT, Halliwell B, Castranova V, Vallyathan V: Reduction of chromium (VI) and its relationship to carcinogenesis. J Toxicol Environ Health B Crit Rev 2: 87–104, 1999
45. Chen F, Ding M, Lu Y, Leonard SS, Vallyathan V, Castranova V, Shi X: Participation of MAP kinase p38 and IkappaB kinase in chromium (VI)-induced NF-kappaB and AP-1 activation. J Environ Pathol Toxicol Oncol 19: 231–238, 2000
46. Wang S, Leonard SS, Ye J, Ding M, Shi X: The role of hydroxyl radical as a messenger in Cr (VI)-induced p53 activation. Am J Physiol Cell Physiol 279: C868–875, 2000
47. Huang C, Chen N, Ma WY, Dong Z: Vanadium induces AP-1- and NFkappB-dependent transcription activity. Int J Oncol 13: 711–715, 1998
48. Ding M, Li JJ, Leonard SS, Ye JP, Shi X, Colburn NH, Castranova V, Vallyathan V: Vanadate-induced activation of activator protein-1: Role of reactive oxygen species. Carcinogenesis 20: 663–638, 1999
49. Hart BA, Lee CH, Shukla GS, Shukla A, Osier M, Eneman JD, Chiu JF: Characterization of cadmium-induced apoptosis in rat lung epithelial cells: Evidence for the participation of oxidant stress. Toxicology 133: 43–58, 1999
50. Barchowsky A, Dudek EJ, Treadwell MD, Wetterhahn KE: Arsenic induces oxidant stress and NF-kappa B activation in cultured aortic endothelial cells. Free Radic Biol Med 21: 783–790, 1996
51. Cavigelli M, Li WW, Lin A, Su B, Yoshioka K, Karin M: The tumor promoter arsenite stimulates AP-1 activity by inhibiting a JNK phosphatase. EMBO J 15: 6269–6279, 1996
52. Ramesh GT, Manna SK, Aggarwal BB, Jadhav AL: Lead activates nuclear transcription factor-kappaB, activator protein-1, and amino-terminal c-Jun kinase in pheochromocytoma cells. Toxicol Appl Pharmacol 155: 280–286, 1999
53. Zhou D, Salnikow K, Costa M: Cap43, a novel gene specifically induced by Ni^{2+} compounds. Cancer Res 58: 2182–2189, 1998
54. Salnikow K, Blagosklonny MV, Ryan H, Johnson R, Costa M: Carcinogenic nickel induces genes involved with hypoxic stress. Cancer Res 60: 38–41, 2000
55. Broday L, Cai J, Costa M: Nickel enhances telomeric silencing in *Saccharomyces cerevisiae*. Mutat Res 440: 121–130, 1999
56. McCaffrey J, Wolf CM, Hamilton JW: Effects of the genotoxic carcinogen chromium (VI) on basal and hormone-inducible phosphoenolpyruvate carboxykinase gene expression *in vivo*: Correlation with glucocorticoid- and developmentally regulated expression. Mol Carcinog 10: 189–198, 1994
57. Hamilton JW, Kaltreider RC, Bajenova OV, Ihnat MA, McCaffrey J, Turpie BW, Rowell EE, Oh J, Nemeth MJ, Pesce CA, Lariviere JP: Molecular basis for effects of carcinogenic heavy metals on inducible gene expression. Environ Health Perspect 106 Suppl 4: 1005–1015, 1998
58. Murphy BJ, Andrews GK, Bittel D, Discher DJ, McCue J, Green CJ, Yanovsky M, Giaccia A, Sutherland RM, Laderoute KR, Webster KA: Activation of metallothionein gene expression by hypoxia involves metal response elements and metal transcription factor-1. Cancer Res 59: 1315–1322, 1999
59. Bose RN, Fonkeng BS, Moghaddas S, Stroup D: Mechanisms of DNA damage by chromium (V) carcinogens. Nucleic Acids Res 26: 1588–1596, 1998

Molecular and Cellular Biochemistry **222**: 11–20, 2001.

Cadmium-induced apoptosis and phenotypic changes in mouse thymocytes

Suyun Dong,[1] Han-Ming Shen[2] and Choon-Nam Ong[2]

[1]*Department of Environmental Health, Sun Yat-sen University of Medical Sciences, Guangzhou, P.R. of China;* [2]*Centre for Environmental and Occupational Health, Department of Community, Occupational and Family Medicine, MD3, Faculty of Medicine, National University of Singapore, Singapore*

Abstract

At present cadmium (Cd)-induced immunotoxicity and the mechanisms involved have not been fully elucidated. The main objective of the present study is to explore the apoptogenic property of Cd in primary cultured mouse thymocytes and its effect on cell surface marker expression and phenotypic changes. Cd-induced thymocyte apoptosis was determined by TdT-mediated dUTP nick end labeling (TUNEL) assay, DNA content/cell cycle analysis and DNA gel electrophoresis. The results showed that Cd was able to cause apoptosis in mouse thymocytes in a time- and dose-dependent manner. Moreover, different subsets of thymocytes possessed different susceptibility to the apoptotic effect of Cd, in the order of $CD8^+ > CD4^-CD8^-$ (double negative cells, DN) $> CD4^+CD8^+$ (double positive cells, DP) $> CD4^+$. Cd treatment also altered thymocyte surface marker expression, leading to evident phenotypic changes. Such changes were characterized by a decline in DP cells and a marked decrease in $CD4^+/CD8^+$ ratio, mainly due to a significant increase in $CD8^+$ subsets. These observations help to obtain a better understanding of the immunotoxic and immunomodulatory effects of Cd. (Mol Cell Biochem **222**: 11–20, 2001)

Key words: cadmium, thymocyte, apoptosis, susceptibility, development, $CD4^+/CD8^+$

Introduction

Cadmium (Cd), a common environmental contaminant in food, water and air, has a complex toxic mechanism because of its long biological half-life and multi-organ toxicity. In addition to its well document toxic effects on the lung, kidney and liver [1], Cd is a potent immunotoxicant in a variety of mammalian and nonmammalian animal species [2].

Thymus, an important primary lymphoid organ, the place where successive stages of cell development and selection generate functionally competent T cells from immature precursor cells [3], is a target organ of Cd-induced immunotoxicity. For instance, Cd exposure resulted in thymic damage and modified the proliferation rate of thymocytes in rats [4]. A number of *in vivo* experiments have demonstrated that Cd is able to cause marked weight decrease of thymus or thymus atrophy [5–7]. A recent report further demonstrated that Cd is capable of inducing thymocyte apoptosis *in vitro* [8]. One of our earlier studies also revealed that Cd-induced thymocyte apoptosis is mediated by intracellular calcium surge and caspase activation [9]. In addition, Cd has been shown to induce apoptotic cell death in some immune cell lines, such as T-cell line CEM-C12 and CCRF-CEM, the B-cell line Raji and the lymphoblastoid cell line Molt-3 [10, 11]. It is known that apoptosis is an important mechanism of thymus atrophy induced by some other xenobiotics such as tributyltin and dioxin [12, 13]. However, to our knowledge, no in-depth study has been conducted on the implication of Cd-induced thymocyte apoptosis in the immunotoxic effects of Cd. In order to have a better understanding on how Cd may affect the function of T lymphocytes, a systematic investigation on Cd-induced thymocyte apoptosis and its effect on thymocyte development is obviously needed.

Address for offprints: C.N. Ong, Centre for Environmental and Occupational Health, Department of Community, Occupational and Family Medicine, MD3, Faculty of Medicine, National University of Singapore, 16 Medical Drive, Singapore 117597

In this study, we explored the apoptotic response of primary cultured mouse thymocytes treated with different concentrations of $CdCl_2$. In addition, the susceptibility of different thymocyte subpopulations to Cd-induced apoptosis was also examined, together with the phenotypic changes of thymocytes, for evaluating the possible effect of Cd exposure on thymocyte development and differentiation.

Materials and methods

Chemicals and reagents

Phycoerythrin (PE) conjugated-anti CD4, Quantum Red™-conjugated-anti CD8 monoclonal antibodies, trypan blue, RNase A, penicillin and streptomycin, paraformaldehyde, sodium dodecyl sulfate (SDS), dexamethasone (Dex), and 2-mercaptoethonal were all purchased from Sigma (St. Louis, MO, USA). RPMI 1640 medium and fetal bovine serum (FBS) were from Gibco BRL (Gaithersburg, MD, USA). Propidium iodide (PI) was purchased from Molecular Probes (Eugene, OR, USA). TdT-mediated dUTP nick end labeling (TUNEL) assay kit was from Roche (Mannheim, Germany).

Preparation of cell suspension

Male Balb/C mice (3–5 week old) were provided by the Animal Center at the National University of Singapore and handled following the guidelines of animal study by the University. Thymocytes were isolated as previously reported [9]. Briefly, after animals were killed by cervical dislocation, thymuses were excised under aseptic conditions and carefully freed from adjacent lymph nodes. The thymuses were ground in cold PBS with frost glass slides, then filtered with a 100 µm mesh stainless steel filter and centrifuged. Red blood cells in the cell preparation were lysed by incubating in 0.017 M Tris, 0.75% NH_4Cl (pH 7.2) at room temperature for 8 min, and followed by centrifugation and washing with FBS-free medium. Single cell suspensions were prepared at 4°C in RPMI 1640 medium supplemented with 10% FBS, penicillin (100 U/ml), streptomycin (100 µg/ml) and 50 µM 2-mercaptoethonal, referred to as complete cell culture medium. Cell concentration was adjusted (1×10^6/ml) and cells were incubated in culture flasks at 37°C in a humidified incubator under an atmosphere of 5% CO_2 in air. The viability of freshly isolated cells was always over 95% (trypan blue exclusion test).

Thymocytes treatment

Cells were exposed to various Cd concentrations (5, 10, 20, and 50 µM) for 3, 6, 12, and 24 h, respectively. Dex with a final concentration of 0.1 µM was used as a positive control.

Assessment of cell viability

Cell viability was determined by the trypan blue exclusion test. After treatment, cells were stained with 2% trypan blue and counted under microscope. Viable cells were able to exclude the dye and appear bright, while dead cells appeared to be in blue color.

TUNEL assay

DNA fragmentation in apoptosis was determined by the terminal deoxynucleotidyl transferase-mediated dUTP nick end-labeling method (TUNEL assay) [14]. Briefly, after being treated with Cd or Dex for the designated time, thymocytes were collected from the culture flasks and washed with cold PBS once, followed by PBS with 1% BSA once, then fixed with 2% paraformaldehyde for 30 min at room temperature. The fixed cells were then incubated with permeabilization solution (0.1% Triton X-100, 0.1% sodium citrate) for 5 min on ice. Enzyme solution was added and the mixtures were incubated for 60 min at 37°C in the dark. Finally, cells were pelleted and resuspended in 250 µl PBS for flow cytometry analysis (Coulter Epics Elite ESP, Miami, FL, USA). Analyzing conditions were as follows: the exciting at 488 and 525 nm band pass filter and 550 nm dichroic mirror were taken as detectors. At least 20,000 cells were collected for each sample, and a negative control without the addition of TdT enzyme was always included in each test. Data obtained were analyzed using WinMDI 2.7 software (Scripps Research Institute, La Jolla, CA, USA) for subtracting the histogram and calculating the percentage of apoptotic cells in each group.

DNA content/cell cycle analysis

The distinct sub-G1 peak in apoptosis was examined by PI staining and flow cytometry, as described earlier [15]. Cells (0.5×10^6 cells) were first washed twice with cold PBS and fixed in precold (–20°C) 70% ethanol for at least 2 h. Cells were then washed with PBS again and incubated in 0.5 ml staining buffer (0.1% Triton X-100, 200 µg/ml RNase and 20 µg/ml PI in PBS) for 30 min at room temperature. The fluorochrome was excited at 488 nm and the red fluorescence measured using a 620 nm long pass filter. Relative cellular

DNA content was assessed after acquisition of gating imposition to ensure counting of single cells. A minimum of 10,000 counting events was analyzed for each sample. The percentage of apoptotic sub-G1 cells was calculated from the resultant histogram using WinMDI 2.7 software.

Phenotyping and detection of apoptosis in different subsets

Multi-parameter analysis of treated thymocytes was conducted to analyze the apoptosis in subpopulations based on CD4 and CD8 surface molecules. Briefly, 2×10^6 treated thymocytes were centrifuged and resuspended in 100 µl PBS. PE-conjugated anti-CD4 monoclonal antibodies and Quantum Red™-conjugated anti-CD8 monoclonal antibodies were added to thymocytes (4 µg/10^6 cells). The phenotypic staining mixture was incubated at 4°C on ice in dark for 30 min. The cells were washed twice in PBS-1% BSA, then fixed with 2% paraformaldehyde in PBS for 30 min in room temperature, followed by the TUNEL assays as above described. The PE and Quantum Red™ fluorescence was collected after a 575 and a 670 nm ban pass filter respectively, together with the FITC fluorescence which was collected after a 525 nm band pass filter simultaneously. Appropriate electronic compensation was adapted among the three fluorescence channels to overcome residual spectral overlap. Data acquisition and analysis were performed using WinMDI 2.7 software and presented in dotplots for phenotypic analysis and in histograms for calculating the percentage of apoptotic cells in each subsets of thymocytes.

Statistical analysis

Data were presented as mean ± S.D. from at least 3 independent experiments and analyzed by One-way ANOVA with Scheffe's test. The histograms from flow cytometry analysis were presented from one representative set of 3 independent experiments. A p-value below 0.05 was considered statistically significant.

Results

Cell viability

As shown in Fig. 1, there were no significant changes of cell viability within the first 6 h of treatment even for Cd concentration up to 50 µM. The cell viability only started to decline after 12 h of treatment with Cd concentrations over 10 µM. An evident dose-dependent decrease of cell viability was also noted. When cells were treated with 50 µM of Cd for 24 h, only 20% of thymocytes were still alive,

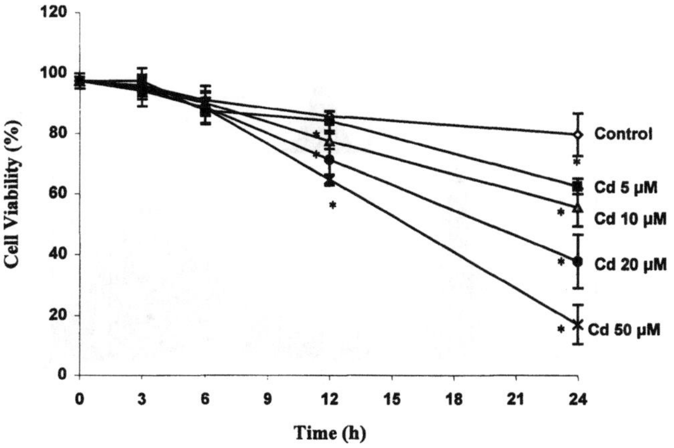

Fig. 1. Changes of cell viability in Cd-treated mouse thymocytes. After thymocytes were incubated with the indicated Cd concentrations for designated periods of time, the percentages of viable cells were measured by trypan blue exclusion test. Results are presented as mean ± S.D. (n = 6). $^*p < 0.05$ compared with controls at respective time points (one-way ANOVA with Scheffe's test).

while the viability was over 80% for untreated control cells. The LC50 for $CdCl_2$ is estimated to be about 17.6 µM at 24 h.

Cd-induced apoptosis in thymocytes

In the present study, Cd-induced apoptosis in thymocytes was studied using the following two assays: (i) TUNEL, and (ii) DNA content/cell cycle analysis.

Figure 2A shows the typical histograms in TUNEL assay. Significant increase of TUNEL-positive cells was observed with Cd treatment, while basically no apoptotic cells were found in the control group (the inserted). The time-course changes are shown in Fig. 2B; the percentage of apoptotic cells increased with incubation time in Cd-treated thymocytes. Although there was no statistical difference between the control and treated cells at 3 and 6 h, the percentage of apoptotic cells in Cd-treat groups was significantly higher than that of the untreated control cells from 12 h onwards; and it was about 5-fold higher at 12 h and 6-fold higher at 24 h. Over 90% of cells were undergoing apoptosis when thymocytes were treated with 50 µM of Cd for 24 h. Meanwhile, a certain degree of spontaneous apoptosis was also observed in untreated control cells as about 20% of the control cells were apoptotic when cultured for 24 h. Moreover, Cd-induced apoptosis was found to be dose-dependent when treated for 12 and 24 h (Fig. 2C).

Figure 3A illustrates the typical histogram showing a distinct sub-G1 peak in Cd-treated thymocytes, while no such peak was observed in the control cell as shown in the inserted histogram. Similar to the results of TUNEL assay (Fig. 2), a relatively low level of spontaneous apoptosis was also ob-

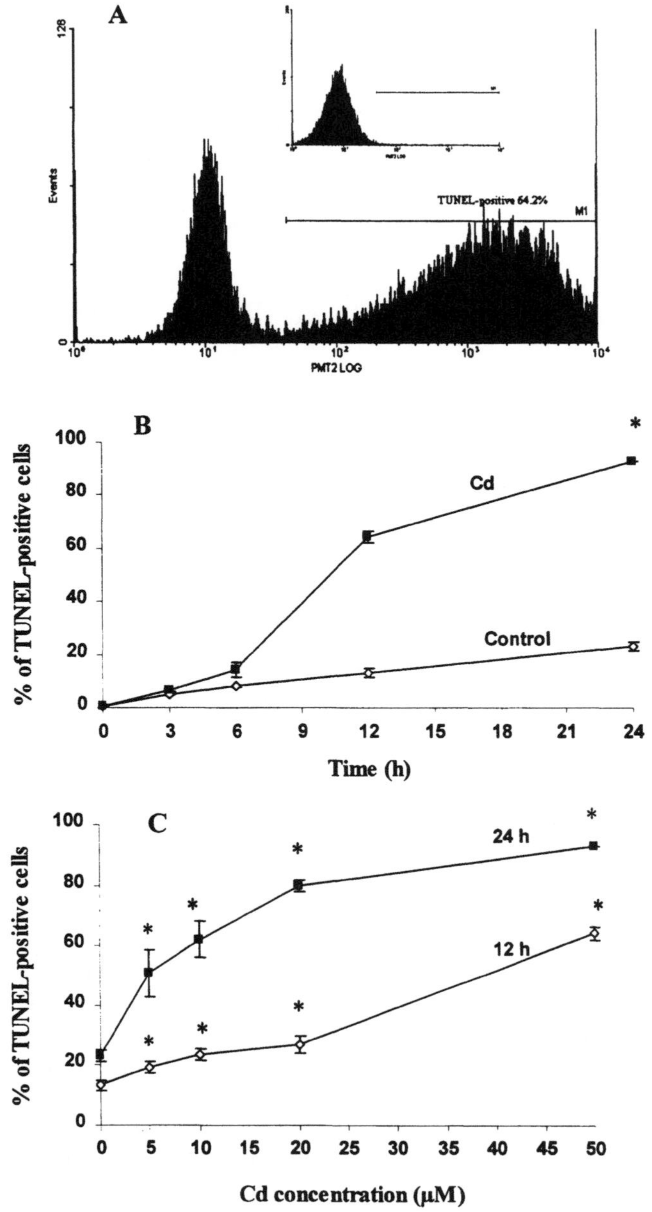

Fig. 2. Cd-induced apoptosis in mouse thymocytes measured by TUNEL assay using flow cytometry. (A) typical histogram showing a significant increase of TUNEL-positive cells in Cd-treated thymocytes, while the inserted histogram was from control cells. (B) time course: cells were treated with 50 μM Cd for up to 24 h. (C) dose-response: thymocytes were treated for 12 and 24 h. Results are presented as mean ± S.D. (n = 6). *p < 0.05 compared with their respective controls (one-way ANOVA with Scheffe's test).

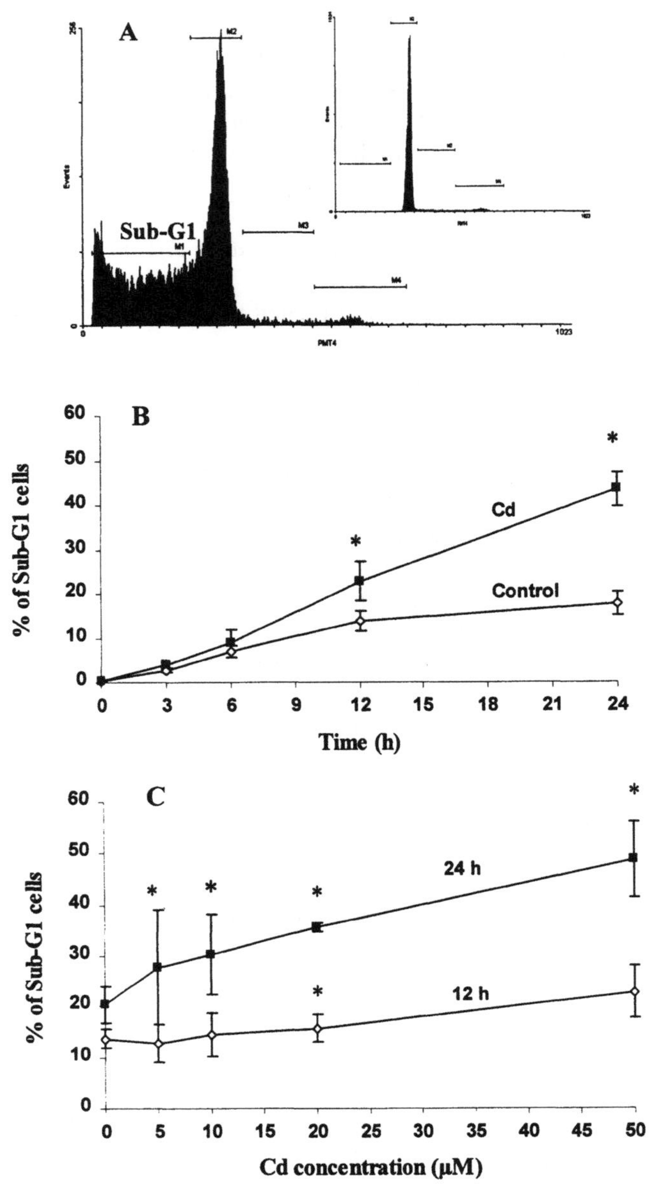

Fig. 3. Cd-induced apoptosis in mouse thymocytes measured using DNA content/cell cycle analysis. (A) typical histogram of Cd-treated thymocytes showing the presence of sub-G1 cells, while no similar changes were observed in control cells (as shown in the inserted histogram). (B) time-dependent of changes sub-G1 cells when thymocytes were treated with 50 μM $CdCl_2$. (C) dose-dependent increase of sub-G1 in Cd-treated thymocytes for 12 and 24 h. Results are presented as mean ± S.D. (n = 6). *p < 0.05 compared with their respective controls (one-way ANOVA with Scheffe's test).

served in thymocytes without the presence of Cd, while Cd-treatment markedly enhanced the percentage of sub-G1 cells, led to more than two folds increase of sub-G1 cells (44%) compared with the control (17%) at 24 h (Fig. 3B). The progressive rise of Cd induced thymocyte apoptosis at 12 and 24 h with the increasing dose of Cd is depicted in Fig. 3C. At 12 h, only cells treated with higher concentrations (20 or 50 μM) of Cd exhibited significant higher percentage of sub-G1 cells, and with 24 h incubation, all Cd-treated groups were found to have significantly higher percentage of sub-G1 cells, as compared to the control.

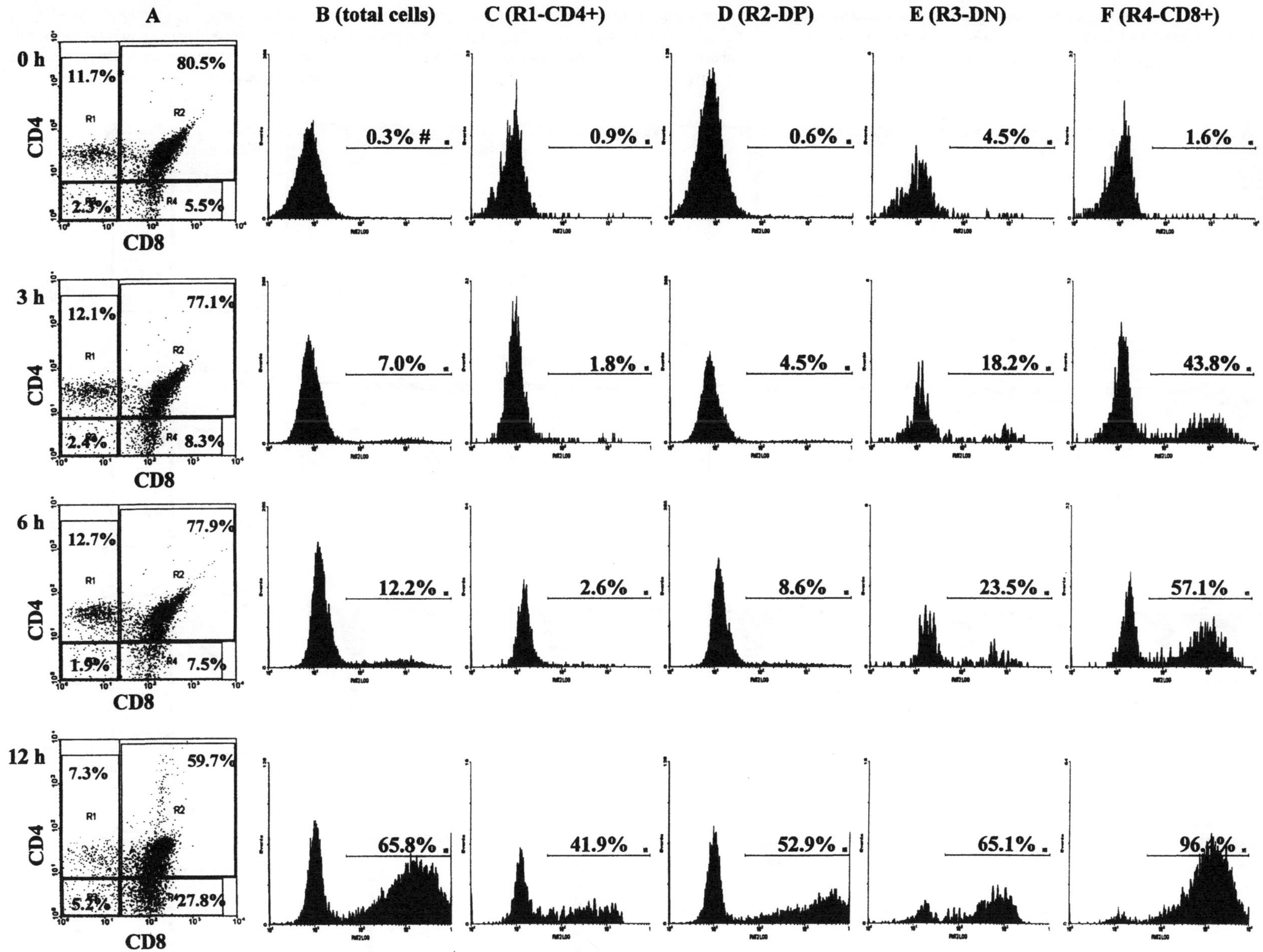

Fig. 4. Time-course of Cd-induced phenotypic changes and apoptosis in thymocyte sub-populations. After designated treatment, cells were first stained with PE-conjugated anti-CD4 and Quantum Red-conjugated anti-CD8 antibody, followed by TUNEL assay. Column A: dotplots showing the relative frequencies of the 4 sub-populations. Column B: histograms of TUNEL assay for total cells. Columns C–F present the histograms of TUNEL assay for subsets of $CD4^+$, DP, DN and $CD8^+$ cells, respectively. The histograms in column C–F were subtracted from each gated quadrant as shown in the dotplots (column A). [#]Indicates the percentage of TUNEL-positive cells in each specified cell populations Results were representatives of 3 independent experiments.

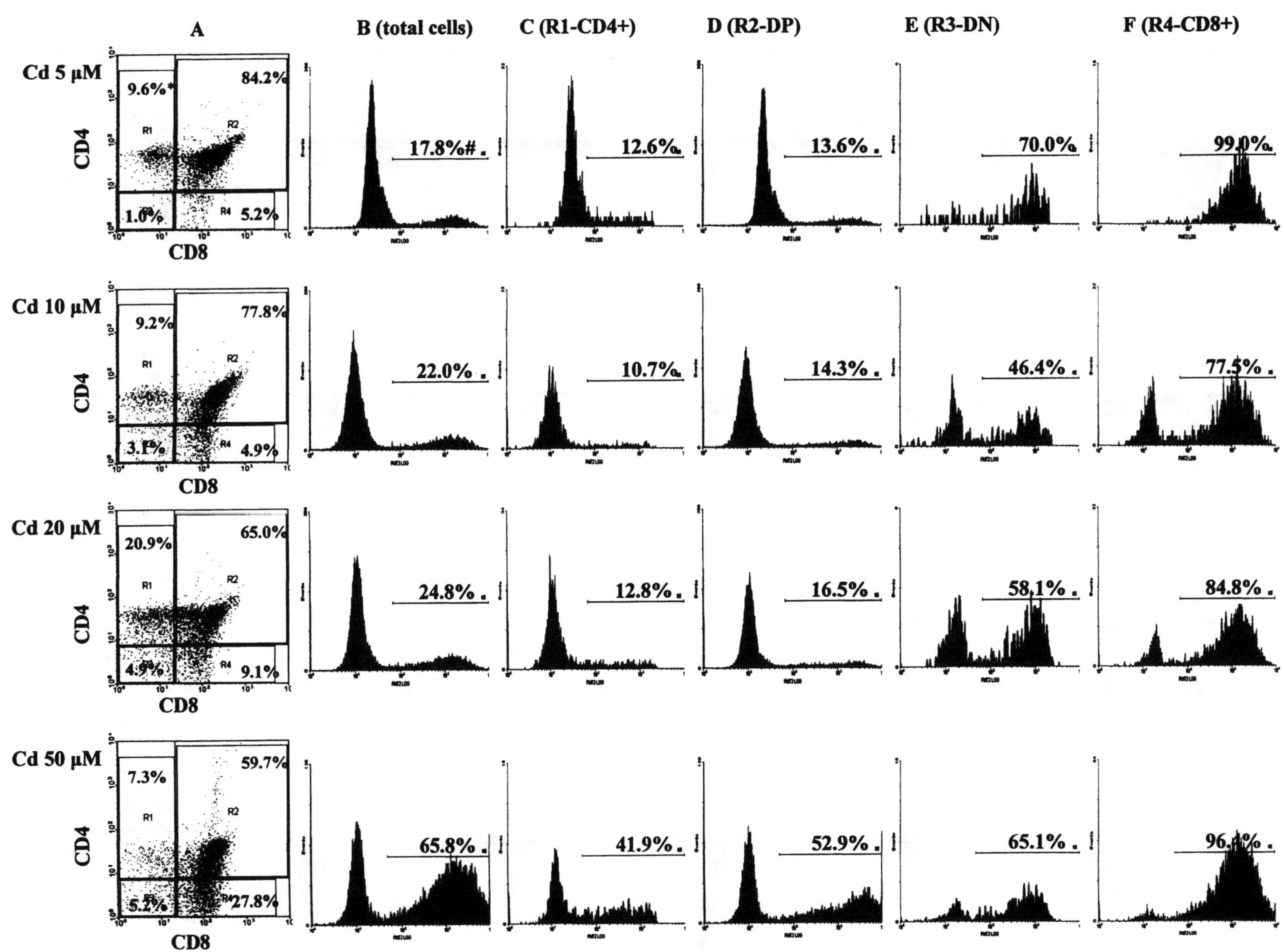

Fig. 5. Dose-response of Cd-induced phenotypic changes and apoptosis in thymocyte sub-populations. After designated treatment, cells were first stained with PE-conjugated anti-CD4 and Quantum Red-conjugated anti-CD8 antibody, followed by TUNEL assay. Column A: dotplots showing the relative frequencies of the 4 sub-populations. Column B: histograms of TUNEL assay for total cells. Columns C–F present the histograms of TUNEL assay for subsets of $CD4^+$, DP, DN and $CD8^+$ cells, respectively. The histograms in columns C–F were subtracted from each gated quadrant as shown in the dotplots (column A). #Indicates the percentage of TUNEL-positive cells in each specified cell populations Results were representatives of 3 independent experiments.

Different susceptibility of thymocyte subsets to Cd-induced apoptosis

In this study, the phenotyping and the TUNEL assay were conducted simultaneously in order to examine the different susceptibility of different thymocyte subsets to Cd-induced apoptosis. The time and dose-dependent patterns of Cd-induced apoptosis in total thymocytes and their four sub-populations are illustrated in Figs 4 and 5 (columns B–F), respectively. The histograms in each column present the percentage of TUNEL positive cells in a specific cell population. In the time-course study (Fig. 4), all subgroups of thymocytes show increased apoptosis with prolonged exposure to Cd. In the dose-response study (Fig. 5), apoptosis was also observed in each subset of thymocytes with a dose-dependent manner. It was interesting to observe the different susceptibilities of different subpopulations to Cd-induced apoptosis. Among the 4 subsets studied, $CD8^+$ was the most susceptible and $CD4^+$ was the most resistant with an order of $CD8^+ > DN > DP > CD4^+$. This trend was consistently found in both time-course and dose-response studies. On the other hand, although $CD8^+$ was the most susceptible cells, the high incidence of apoptosis in $CD8^+$ did not contribute much to the overall percentage of apoptosis in the whole population of thymocytes due to the relative small proportion of $CD8^+$ cells. The unusual high percentage of apoptosis in DN and $CD8^+$ cells with a low Cd dosage (5 μM) (Fig. 5, columns E and F) may be caused by some technical factors in this particular experiment. Such discrepancy was no longer found in repeated experiments (data not shown).

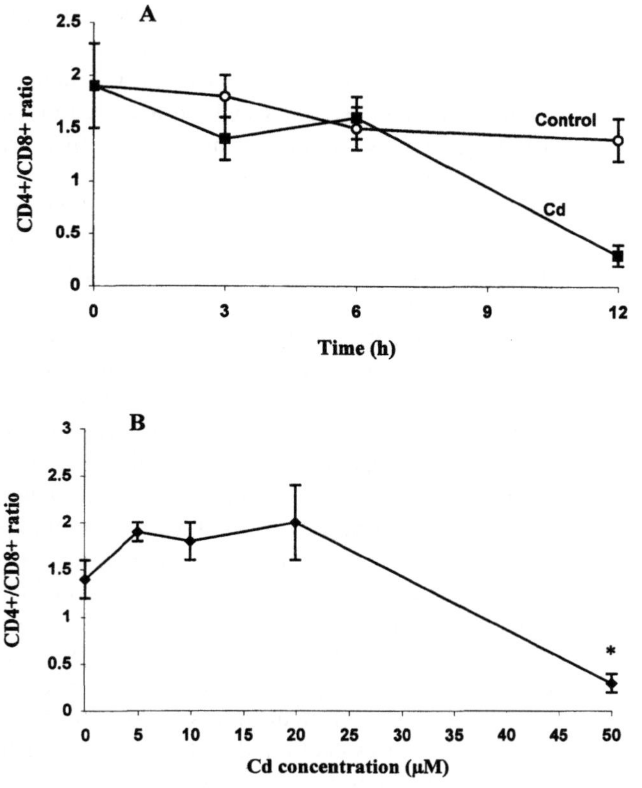

Fig. 6. Increase of $CD4^+/Cd8^+$ ratio induced by Cd treatment. (A) time-course: cells were treated with 50 μM $CdCl_2$ up to 12 h. (B) dose-response: cells were treated with various concentrations of $CdCl_2$ for 12 h. Results are presented as mean ± S.D. (n = 3). *$p < 0.05$ indicates significant difference compared with the control at respective time points (one-way ANOVA with Scheffe's test).

Cd-induced thymocyte phenotypic changes

In order to evaluate the possible effects of Cd exposure on thymocyte differentiation and development, the effect of Cd on thymocytes surface phenotypes was measured by the following parameters: (i) the relative frequencies of each subpopulation, and (ii) the ratio of $CD4^+/CD8^+$. As demonstrated in Figs 4 and 5 (column A, dotplots), the proportion of each subpopulation in control cells remained unchanged up to 12 h. Similarly, there were no significant phenotypic changes at 3 and 6 h for cells treated with 50 μM of Cd (Fig. 4, column A). However, there was a marked decrease of DP cells when treated with Cd for 12 h, accompanying by the increase of the proportion of $CD8^+$ cells (Fig. 4, column A). Such changes were also noted to be dose-dependent. As demonstrated in Fig. 5 (column A, dotplots), cells treated with higher concentration of Cd (20 and 50 μM) showed obvious phenotypic alterations.

Another important phenotypic change caused by Cd is the decrease of $CD4^+/CD8^+$ ratio. These data are summarized in Figs 6A and 6B for time-dose and dose-dependent response, respectively. Significant reduction of $CD4^+/CD8^+$ ratio was observed when cells were treated for 12 h (Fig. 6A) or with the highest concentration of Cd (50 μM) (Fig. 6B). As the proportion of $CD4^+$ cells did not change significantly, it is believed that the decrease $CD4^+/CD8^+$ ratio is largely due to the increase of $CD8^+$ cells (Figs 4 and 5, dotplots in column A).

Discussion

Cd-induced apoptosis in thymocytes

Numerous studies have demonstrated that Cd exposure, under either *in vivo* or *in vitro* conditions, induces apoptosis in various target tissues or cells including liver, lung, kidney, prostate, testes, and hematocytes [16–20]. Moreover, a couple of earlier studies also revealed that Cd is capable of causing apoptosis in immune cells including both T cells and B cells [10, 11]. In the present study, we provided convincing evidence showing that exposure to micro-molar concentrations of $CdCl_2$ resulted in time- and dose-dependent apoptosis in mouse thymocytes, based on the data from both TUNEL assay and DNA content/cell cycle analysis. Such finding is generally consistent with a recent study showing Cd-induced apoptosis in thymocytes [8]. The $CdCl_2$ dosage used in the present study is generally within the range reported by other *in vitro* studies for the induction of apoptosis in its target cells [8, 10, 20, 21]. The LC50 in the present study was found to be about 17.6 μM, which is consistent with an early report that the LC50 of Cd ranged from 13.5 ± 2.2 to 25 ± 20 μM in a number of immune cells [11].

Thymus is an important primary lymphoid organ, responsible for generating functionally competent T cells from immature precursor cells. It has been well studied that thymocyte apoptosis has two important aspects of biological significance. First, thymocyte apoptosis plays a critical role in T cell development or differentiation through the so-called 'negative selection' mechanism [22]. Second, various immunotoxins act through induction of thymocyte apoptosis via different signaling pathways. Such immunotoxins include dioxin, organotin compounds (tributyltin) and fusarenon-X [13, 23, 24]. Based on the results from the present study, it is believed that induction of apoptosis in thymocytes by Cd may constitute an underlying mechanism responsible for to its immunotoxic effects. Results from this study may also help to explain the results from some *in vivo* experiments that Cd treatment causes T cell depletion and thymus atrophy [5–7].

Different susceptibility of thymocyte subpopulations to Cd-induced apoptosis

In the present study, thymocytes were categorized into 4 subsets according to the expression of their surface markers: double negative (DN), double positive (DP), $CD4^+$, and $CD8^+$. As different subpopulations of thymocytes possess different immuno-activities, it would be of interest to find out whether there are differences in susceptibility of these four subpopulations of thymocytes to Cd-induced apoptosis. The results, based on flow cytometric analysis of TUNEL assay, clearly demonstrated that apoptosis occurred in each subsets of thymocytes; however, different thymocytes subsets possess different susceptibility, in the order of $CD8^+ > DN > DP > CD4^+$ (Figs 4 and 5). Such characteristic selectivity of Cd is indeed different from the effects of some other immunotoxins such as dioxins, fusarenon-X, linomide, and Dex [24–27]. For example, fusarenon-X was able to almost completely deplete DP thymocytes in mouse thymus with *in vivo* treatment, while the other subpopulations were not affected at a similar extent, suggesting the DP thymocytes are especially susceptible to the cytotoxic effect of fusarenon-X [24]. At present, there is no mechanistic explanation for such discrepancy. Since mature $CD8^+$ cells mainly function as cytotoxic or suppressor cells and are closely related to anti-tumor defense mechanisms [28], the high susceptibility of $CD8^+$ thymocytes to Cd-induced apoptosis may be associated, at least in part, with the carcinogenic effect of Cd. The exact biological implications of the high susceptibility of $CD8^+$ thymocytes to Cd-induced apoptosis remain to be further established.

Changes of phenotypes and cell development in Cd-treated thymocytes

Within the thymus, the T cell development starts from T lymphocyte precursors to $CD4^-CD8^-$ double negative immature T cells, subsequently to $CD4^+CD8^+$ double positive thymocytes, and eventually to mature T cells expressing either $CD4^+$ or $CD8^+$ single positive cells prior to their release into the circulation [29]. The normal distribution of these four thymocyte subpopulations in mouse is as follows: 5–7% of DN, 72% of DP, 11–16% of $CD4^+$, and 7–8% of $CD8^+$ [30]. In the present study, we found similar values for each subset in the control group (Fig. 4). However, it is noted that Cd treatment was able to induce evident phenotypic changes in thymocytes, featured by (i) the significant decrease of DP cell proportion, and (ii) increase of the percentage of $CD8^+$ cells and DN cells. Such phenotypic alterations were especially significant when thymocytes were treated with higher Cd concentration or prolonged exposure time (Figs 4 and 5).

So far, a number of *in vitro* studies have demonstrated that there are significant changes of expression of surface molecules during the course of thymocyte apoptosis, resulting in thymocyte phenotypic alterations [27, 31]. $CD4^+$ and $CD8^+$ are among a number of surface molecules that are active in positive and negative selection of thymocytes [32]. At present, the nature of such changes that occur in thymocytes undergoing apoptosis is not well understood; and it is still controversial whether the phenotypic changes occur as a result of apoptosis or whether they precede the apoptotic death of thymocytes. In the present study, evident apoptosis was observed in Cd-treated thymocytes when no significant phenotypic changes took place, suggesting that such phenotypic alterations result from apoptosis, and not *vice versa*. Cd exposure significantly reduced the number of DP cells and increased the proportions of $CD8^+$ and DN cells. Similar results were also found in thymocytes treated with some other immunotoxins such as dioxin [27, 33, 34]. It is believed that the reduction of DP cells in Cd-treated thymocytes might be caused by either the direct apoptotic cell death or the down-regulation of their surface markers, most probably due to the degradation of cell components as a consequence of apoptosis [31]. Therefore, results from this study suggest that Cd is capable of altering thymocyte development and change cell phenotypes through the induction of apoptosis in thymocytes.

Another important finding of the present study is the altered $CD4^+/CD8^+$ ratio in Cd-exposed thymocytes. Such ratio in peripheral blood is usually used for measurement of immune function and response [35]. The changes of $CD4^+/CD8^+$ ratio may reflect the immunotoxic or immunomodulatory effect of xenobiotic agents, characterized by changes in the phenotypic expression patterns or alterations in the quantity of lymphocyte subpopulations [36]. In the present study, the decrease of this ratio is mainly due to the increase of $CD8^+$ cells. Based on the observations that (i) $CD8^+$ subset is the most susceptible one towards Cd-induced apoptosis, and (ii) there was concurrent decrease of DP cells, it is believed that increased percentage of $CD8^+$ cells can either be the result of the $CD8^+$ surface marker up-regulation occurred in apoptotic cells, or the enhanced conversion rate of DP cells to $CD8^+$ cells. Moreover, the altered $CD4^+/CD8^+$ ratio could also be the result of differential interactions of the T cell receptor (TCR) and $CD8^+$ co-receptor complex with the respective MHC II haplotypes caused by the toxic effect on thymocytes [37].

In summary, the present study showed that Cd was capable of inducing apoptosis in cultured mouse thymocytes in a dose and time-dependent manner. Different subsets of thymocytes respond differently to the apoptotic effect of Cd, in the order of $CD8^+ > DN > DP > CD4^+$. Cd treatment also affected cell surface marker expression and led to phenotypic changes, characterized by a dose-dependent decline in DP cells and a marked decrease in the $CD4^+/CD8^+$ ratio. Results from this study may help to obtain a better understanding of the immunotoxic mechanisms of Cd.

Acknowledgements

This research was partially supported by the China Medical Board (New York) and The Centre for Environmental and Occupational Health, National University of Singapore. Dong S was a recipient of a CMB Fellowship on Environmental Health. We thank Ms. B.L. Ng of Clinical Research Centre, NUS for her excellent technical support in flow cytometry analysis.

References

1. Nath R, Prasad R, Palinal VK, Chopra RK: Molecular basis of cadmium toxicity. Prog Food Nutr Sci 8: 109–163, 1984
2. Descotes J: Immunotoxicology of cadmium. IARC Sci Publ 118: 385–390, 1992
3. Kisielow P, Von Boehmer H: Development and selection of T cells: Facts and puzzles. Adv Immunol 58: 87–209, 1995
4. Morselt AF, Leene W, De Groot C, Kipp JB, Evers M, Roelofsen AM, Bosch KS: Differences in immunological susceptibility to cadmium toxicity between two rat strains as demonstrated with cell biological methods. Effect of cadmium on DNA synthesis of thymus lymphocytes. Toxicology 48: 127–139, 1988
5. Borgman RF, Au B, Chandra RK: Immunopathology of chronic cadmium administration in mice. Int J Immunopharmacol 8: 813–817, 1986
6. Mackova NO, Lenikova S, Fedorocko P, Brezani P, Fedorockova A: Effects of cadmium on haemopoiesis in irradiated and non-irradiated mice: 2. Relationship to the number of circulating blood cells and haemopoiesis. Physiol Res 45: 101–106, 1996
7. Liu J, Liu Y, Habeebu SS, Klaassen CD: Metallothionein-null mice are highly susceptible to the hematotoxic and immunotoxic effects of chronic $CdCl_2$ exposure. Toxicol Appl Pharmacol 159: 98–108, 1999
8. Fujimaki H, Ishido M, Nohara K: Induction of apoptosis in mouse thymocytes by cadmium. Toxicol Lett 115: 99–105, 2000
9. Shen HM, Dong SY, Ong CN: Critical role of calcium overloading in cadmium-induced apoptosis in mouse thymocytes. Toxicol Appl Pharmacol (in press)
10. El Azzouzi B, Tsangaris GT, Pellegrini O, Manuel Y, Benveniste J, Thomas Y: Cadmium induced apoptosis in a human T cell line. Toxicology 88: 127–139, 1994
11. Tsangaris GT, Tzortzatou-Stathopoulou F: Cadmium induces apoptosis differentially on immune system cell lines. Toxicology 128: 143–150, 1998
12. Raffray M, McCarthy D, Snowden RT, Cohen GM: Apoptosis as a mechanism of tributyltin cytotoxicity to thymocytes: Relationship of apoptotic markers to biochemical and cellular effects. Toxicol Appl Pharmacol 119: 122–130, 1993
13. Kamath AB, Nagarkatti PS, Nagarkatti M: Characterization of phenotypic alterations induced by 2,3,7,8-tetrachlorodibenzo-p-dioxin on thymocytes *in vivo* and its effect on apoptosis. Toxicol Appl Pharmacol 150: 117–124, 1998
14. Shen HM, Yang CF, Ong CN: Sodium selenite-induced oxidative stress apoptosis in human hepatoma $HepG_2$ cells. Int J Cancer 81: 820–828, 1999
15. Yang CF, Shen HM, Ong CN: Ebselen induces apoptosis in $HepG_2$ cells through rapid depletion of intracellular thiols. Arch Biochem Biophys 374: 142–152, 2000
16. Habeebu SS, Liu J, Klaassen CD: Cadmium-induced apoptosis in mouse liver. Toxicol Appl Pharmacol 149: 203–209, 1998
17. Ishido M, Homma-Takeda S, Tohyama C, Suzuki T: Apoptosis in rat renal proximal tubular cells induced by cadmium. J Toxicol Environ Health 55: 1–12, 1998
18. Hart BA, Lee CH, Shukla GS, Shukla A, Osier M, Eneman JD, Chiu JF: Characterization of cadmium-induced apoptosis in rat lung epithelia cells: Evidence for the participation of oxidant stress. Toxicology 133: 43–58, 1999
19. Zhou T, Zhou G, Song W, Eguchi N, Lu W, Lundin E, Jin T, Nordberg G: Cadmium-induced apoptosis and changes in expression of p53, c-jun and MT-I genes in testes and ventral prostate of rats. Toxicology 142: 1–13, 1999
20. Bagchi D, Joshi SS, Bagchi M, Balmoori J, Benner EJ, Kuszynski CA, Stohs SJ: Cadmium- and chromium-induced oxidative stress, DNA damage, and apoptotic cell death in cultured human chronic myelogenous leukemic K562 cells, promyelocytic leukemic HL-60 cells, and normal human peripheral blood mononuclear cells. J Biochem Mol Toxicol 14: 33–41, 2000
21. Galan A, Garcia-Bermejo ML, Troyano A, Vilaboa NE, de Blas E, Kazanietz MG, Aller P: Stimulation of p38 mitogen-activated protein kinase is an early regulatory event for the cadmium-induced apoptosis in human promonocytic cells. J Biol Chem 275: 11418–11424, 2000
22. Kisielow P: Apoptosis in intrathymic T-cell development. In: C.D. Gregory (ed). Apoptosis and the Immune Response. Wiley-Liss Inc., New York, 1995, pp 13–53
23. Pieters RH, Bol M, Penninks AH: Immunotoxic organotins as possible model compounds in studying apoptosis and thymocyte differentiation. Toxicology 91: 189–202, 1994

24. Miura K, Nakajima Y, Yamanaka N, Terao K, Shibato T, Ishino S: Induction of apoptosis with fusarenon-X in mouse thymocytes. Toxicology 127: 195–206, 1998
25. Baumann CA, Badamchian M, Goldstein AL: Thymosin α1 antagonizes dexamethasone and CD3-induced apoptosis of $CD4^+CD8^+$ thymocytes through the activation of cAMP and protein kinase C dependent second messenger pathways. Mech Ageing Dev 94: 85–101, 1997
26. Harring AC, Xu Z, Andersson G, Hedlund G: Linomide enhances apoptosis in $CD4^+CD8^+$ thymocytes. Scand J Immunol 46: 488–494, 1997
27. Lai ZW, Fiore NC, Gasiewicz TA, Silverstone AE: 2,3,7,8-Tetrachlorodibenzo-p-dioxin and diethylstilbestrol affect thymocytes at different stages of development in fetal thymus organ culture. Toxicol Appl Pharmacol 149: 167–177, 1998
28. Bartik MM, Baumgartel-Scofield BA, Mokyr MB: Enhanced expansion of the thymic $CD8^+$ cell subset as a potential mechanism for the generation of enhanced antitumor cytotoxicity by thymocytes from low-dose melphalan-treated MOPC-315 tumor bearers. Cancer Immunol Immunother 34: 79–89, 1991
29. Rothenberg EV: The development of functionally responsive T cells. Adv Immunol 51: 85–214, 1992
30. Thomas TJ, Gunnia UB, Thomas T: Reversal of the abnormal development of T cell subpopulations in the thymus of autoimmune MRL-lpr/lpr mice by a polyamine biosynthesis inhibitor. Autoimmunity 13: 275–283, 1992
31. Kishimoto H, Surh CD, Sprent J: Upregulation of surface markers on dying thymocytes. J Exp Med 181: 649–655, 1995
32. Conroy LA, Alexander DR: The role of intracellular signalling pathways regulating thymocyte and leukemic T cell apoptosis. Leukemia 10: 1422–1435, 1996
33. Holladay SD, Lindstrom P, Blaylock BL, Comment CE, Germolec DR, Heindell JJ, Luster MI: Perinatal thymocyte antigen expression and postnatal immune development altered by gestational exposure to tetrachlorodibenzo-p-dioxin (TCDD). Teratology 44: 385–393, 1991
34. Blaylock BL, Holladay SD, Comment CE, Heindel JJ, Luster MI: Exposure to tetrachlorodibenzo-p-dioxin (TCDD) alters fetal thymocyte maturation. Toxicol Appl Pharmacol 112: 207–213, 1992
35. Chakravarti A: The CD4/CD8 ratio: Message in a bottle? Nat Med 1: 1240–1241, 1995
36. Morris DL, Komocsar WJ: Immunophenotyping analysis of peripheral blood, splenic, and thymic lymphocytes in male and female rats. J Pharmacol Toxicol Meth 37: 37–46, 1997
37. Damoiseaux JG, Cautain B, Berard I, Mas M, van Breda Vriesman PJ, Druet P, Fournie G, Saodi A: A dominant role for the thymus and MHC genes in determining the peripheral CD4/CD8 T cell ratio in the rat. J Immunol 163: 2983–2989, 1999

Molecular and Cellular Biochemistry **222:** 21–28, 2001.

Cytotoxicity and transcriptional activation of stress genes in human liver carcinoma cells ($HepG_2$) exposed to cadmium chloride

Paul Bernard Tchounwou,[1] Ali Baba Ishaque[1] and John Schneider[2]

[1]*Molecular Toxicology Research Laboratory, NIH-Center for Environmental Health, School of Science and Technology, Jackson State University, Jackson, MS;* [2]*Client Research Laboratory, Xenometrix Inc., Boulder, CO, USA*

Abstract

Epidemiological studies have shown that there exists some correlation between cadmium exposure and human cancers. The evidence that cadmium and cadmium compounds are probable human carcinogens is also supported by experimental studies reporting induction of malignant tumors formation in multiple species of laboratory animals exposed to these compounds. *In vitro* studies with mammalian cells have also shown that cadmium is clastogenic, but its mutagenic potential is rather weak. In this research, we performed the MTT assay for cell viability to assess the cytotoxicity of cadmium chloride ($CdCl_2$), and the CAT-Tox (L) assay to measure the induction of stress genes in thirteen different recombinant cell lines generated from human liver carcinoma cells ($HepG_2$), by creating stable transfectants of different mammalian promoter – chloramphenicol acetyl-transferase (CAT) gene fusions. Cytotoxicity experiments with the parental cell line yielded a LC_{50} of 6.1 ± 0.8 µg/mL, upon 48 h of exposure. Four (metallothionein – HMTIIA, 70-kDa heat shock protein – HSP70, xenobitic response element – XRE, and cyclic adenosine monophosphate response element – CRE) out of the 13 constructs evaluated showed statistically significant inductions ($p < 0.05$). The induction of these genes was concentration-dependent. Marginal inductions were also recorded for the *c-fos*, and 153-kDa growth arrest DNA damage (GADD153) promoters, indicating a potential for $CdCl_2$ to damage DNA. However, no significant inductions ($p > 0.05$) of gene expression were recorded for cytochrome P450 1A1 – CYP1A1, glutathion-*S*-transferase Ya subunit – GST Ya, nuclear factor kappa (B site) response element – NFκBRE, tumor suppressor protein response element – *p53*RE, 45-kDa growth arrest DNA damage – GADD45, 78-kDa glucose regulated protein – GRP78, and retinoic acid response element – RARE. As expected, these results indicate that metallothioneins and heat shock proteins appear to be excellent candidates for biomarkers for detecting cadmium-induced proteotoxic effects at the molecular and cellular levels. Induction of XRE indicates the potential involvement of $CdCl_2$ in the biotransformation process in the liver, while activation of CRE indicates stimulation of cellular signaling through the protein kinases pathway. (Mol Cell Biochem **222**: 21–28, 2001)

Key words: cadmium, cytotoxicity, stress genes, transcriptional activation, human liver carcinoma cells

Introduction

Cadmium, a relatively rare heavy metal, is biologically non-essential. It is mainly used in the battery, alloy manufacturing and the electroplating industries. Roughly about 50,000 workers are occupationally exposed to cadmium in the USA [1]. Cadmium, apart from occupational concerns also poses environmental problems due to its extremely long biological half-life. It has been implicated as the cause of severe deleterious effects on fish and wildlife.

Address for offprints: P.B. Tchounwou, Molecular Toxicology Research Laboratory, NIH-Center for Environmental Health, School of Science and Technology, Jackson State University, 1400 Lynch Street, P.O. Box 18540, Jackson, Mississippi 39217, USA

Food is recognized as the major source of cadmium in humans, except in comparatively rare cases of occupational air exposure. Chronic exposure of cadmium leads to biomagnification in specific tissues. Ingested cadmium is translocated mainly to the kidney. This cadmium build-up causes renal damage. Once renal damage occurs, cadmium excretion increases and the kidneys lose their cadmium burden: liver and kidney burdens increase until a 40 ppm concentration is reached in the liver, after which kidney levels decrease while liver burden continues to rise [2]. In human tissues cadmium content in the renal cortex portion of the kidney increased by a factor of 47 and whole body burden increased by a factor near 5 in the years 1897–1939 vs. 1980 [3]. The significance of this increase is not fully clear; however, it has been suggested that cadmium and lead were associated with increased risk of heart-related death, even in the light of known conventional causes of such fatalities [4].

Cadmium is a known teratogen. Its teratogenic effect on animals appears to be greater than for other metals, including lead, mercury, copper, indium, and arsenic [5]. Though, cadmium is a known teratogen in laboratory animals yet no conclusive evidence is available when it comes to human teratogenesis. From a variety of studies in which mice and bacteria were used as models, it appears likely that cadmium has mutagenic effects. Mice injected with 3 or 6 mg $CdCl_2$/kg body weight showed changes in chromosome number 12 h later; similar changes were observed in hamsters at 1.5–3.0 mg/kg [5].

The available epidemiological evidence among humans before 1993 was not sufficient enough to conclude that cadmium was a definite carcinogen [6, 7]. The International Agency for Research on Cancer (IARC), after updating its assessments, reported limited evidence for cadmium carcinogenicity to humans and sufficient evidence for carcinogenicity to animals [8]. Hence in 1993, IARC classified cadmium as a human carcinogen [9].

Elevated levels of respiratory, prostate, and other cancers have been reported in cadmium-exposed workers. The evidence is strongest for lung cancer [10–15]. Tumors of the kidney and the prostrate have been associated with cadmium exposure [16, 17]. Cadmium has also been associated with stomach cancer in exposed workers [16]. Effects of cadmium on the liver have been investigated in animals and humans [18, 19]. Although effects have been seen in some studies, the evidence is not strong enough to link exposure to cadmium with specific effects.

It has been reported that the liver plays a very active and pivotal role in rapidly clearing cadmium ions from the blood. This rapid clearance of cadmium ions from blood to the liver constitutes an important risk factor for hepatocellular toxicity. Cadmium is retained in the liver with a long biological half life, leading to a variety of toxic responses by the liver cells. Classically, hepatocyte damage both *in vivo* and *in vitro* has been identified by cellular release of liver enzymes, and potassium ion, as well as ultrastructural changes in mitochondria and endoplasmic reticulum [20]. Several other toxic end-points have also been identified under acute exposure to cadmium including: induction of genes associated with carcinogenesis, oxidative damage, changes in glutathion content, Kuffer cell activation, endothelial cell damage, and cytoplasmic acidification leading to decreased mitochondrial respiration [21].

Although cadmium is a systemic toxicant that has been found to cause significant hepatotoxicity in experimental animals, the biochemical and molecular mechanisms by which this metal exerts its toxic action in humans remains to be elucidated. Therefore, the objectives of this study were: (a) to determine the cytotoxicity of cadmium chloride to human liver carcinoma ($HepG_2$) cells, and (b) to assess the transcriptional activation of stress genes associated with cadmium exposure in these hepatocytes. Establishing the degree of cytotoxicity, and the profile of stress genes induced by cadmium will provide a scientific basis for elucidating the potential mechanisms involved in cadmium-induced hepatotoxicity.

Materials and methods

Chemicals and media

Reference solution (1000 ± 10 ppm) of cadmium chloride ($CdCl_2$, CAS No. 7440-43-9, Lot No. 981734-24) with a purity of 98% was purchased from Fisher Scientific in Fair Lawn, NJ, USA. Dulbecco's Modified Eagle's Minimal Essential Medium (DMEM, Lot. 1016511) was purchased from life Technologies in Grand Island, NY, USA.

Cytotoxicity assay

Parental $HepG_2$ cells stored in liquid nitrogen were thawed by gentle agitation of their containers (vials) for 2 min in a water bath at 37°C. After thawing, the content of each vial was transferred to a 75 cm^2 tissue culture flask, diluted with DMEM supplemented with 10% fetal bovine serum (FBS) and 1% streptomycin and penicillin, and incubated for 24 h at 37°C in a 5% CO_2 incubator to allow the cells to grow, and form a monolayer in the flask. Cells grown to 80–95% confluence were washed with phosphate buffer saline (PBS), trypsinized with 3 mL of 0.25% (v) trypsin-0.0.3%/v) EDTA, diluted, counted, and seeded (5×10^5 cells/well) in two sets of 96-well microtiter tissue culture plates.

Seeded plates were incubated for 24 h at 37°C in a 5% CO_2 incubator. The old medium was replaced by 180 µl of fresh medium. Twenty µl of serial dilutions of cadmium chloride (0, 1.9, 3.9, 7.8, 15.62, 31.25, 62.5, 125, 250, 500, and 1000 μg/mL) were added column wise to the 96-well microtiter tissue culture plates and incubated for 48 h. Cell viability assay was performed using the MTT {3-(4,5-dimethylthiazol-2-yl)-2,5-diphenyltetrazolium bromide} method. The absorbance was read at a wavelength of 550 nm [20] using microtiter plate reader (Bio-Tek Instruments Inc.). The absorbance values obtained per concentration were converted to percentage cell viability. Regression analysis was conducted on the percentage cell viability data and the resulted equation was used to compute the lethal concentration needed to kill 50% of the cells (LC_{50}).

Gene profile assay

The mammalian gene profile assay (CAT-Tox) for measuring differential gene expression in immortalized human liver cells was performed. Thirteen different recombinant constructs generated by creating stable transfectants of different mammalian promoter – chloramphenicol acetyltransferase (CAT) were obtained from Xenometrix, Inc. (Boulder, CO, USA). Each construct contained a unique stress gene promoter or response element fused to the CAT reporter gene. The promoter/response element-fusion constructs and their respective biological functions have been published elsewhere in literature [21].

Seeded plates were incubated for 24 h at 37°C in a 5% CO_2 incubator. Cells dosing was done as described above and they were done in triplicates. For quality assurance/quality control purposes, positive control plates were also made using known inducers including 3-methyl cholanthrene (3-MC-10 μM) for CYP 1A1, CRE, GADD45, *p53*RE, and XRE; methyl methane sulphonate (MMS-100 μg/mL) for GST Ya, HMTIIA, *c-fos*, HSP70, NFκBRE, GADD153, and GRP78; and all-trans retinoic acid (RA-10 μM) for RARE. All chemical exposures involved the use of polypropylene 96-well microtiter plates for the purpose of chemical dilutions.

Following chemical exposure, the cells were re-incubated for 48 h at 37°C, and 5% CO_2. After the incubation period, the total protein was measured by the Bradford method, at 600 nm using a microtiter plate reader (Bio-Tek Instruments Inc.). A standard sandwich ELISA was performed and in the final step, horseradish peroxidase catalyzed a color change reaction that was measured at 405 nm [21, 22].

Statistical analysis

The average fold inductions of the CAT-gene for each recombinant cell line at each cadmium chloride concentration were calculated using the CAT-Tox computer software based on the optical density readings at 600 nm and 405 nm. The software also converted the 550 nm readings to cell viability percentages. Standard deviations were determined, and the Student's *t*-test values were computed to determine if there were significant differences in cell viability and gene induction in treated cells compared to the control cells. Graphs were made to illustrate the dose-response relationship with respect to cytotoxicity and gene expression.

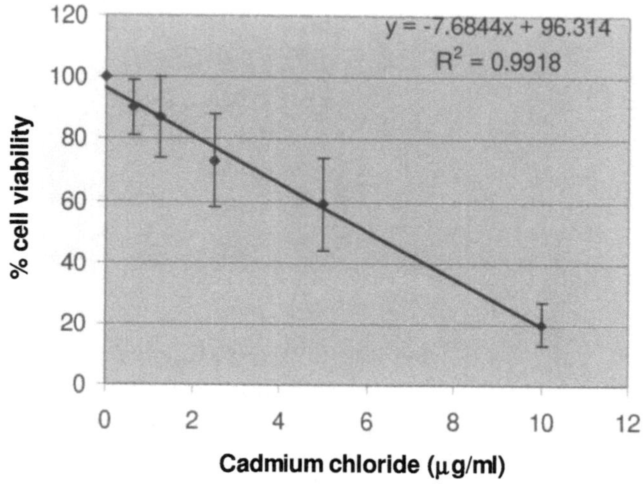

Fig. 1. Cell viability of parental $HepG_2$ cells exposed to cadmium chloride. Dose-dependence of $CdCl_2$-induced cell death. The cells were treated with serial dilutions (0–10 μg/mL) of $CdCl_2$. Cell viability was measured by MTT assay as stated in Materials and methods. Absorbance readings taken from survival cells were converted to percentage cell viability using the CAT-Tox software. Each point represents a mean and standard deviation of results from 8 individual measurements.

Results

Cytotoxicity assay

The result of the cytotoxic effect of cadmium chloride ($CdCl_2$) on $HepG_2$ cell line after 48 h exposure by MTT assay is shown in Fig. 1. This figure depicts a strong dose-response relationship with a correlation coefficient of 1. The percentage cell viabilities were 100.0 ± 0.0%, 90.0 ± 8.6%, 87.0 ± 12.8%, 73.0 ± 15.2%, 59.0 ± 14.7% and 20.0 ± 6.8% for the control, 0.6, 1.25, 2.5, 5 and 10 μg/mL of cadmium chloride, respectively. The concentration of cadmium chloride required to produce 50% reduction in the viability of HepG2 cells (LC_{50}) was computed to be 6.1 ± 0.8 μg/mL.

Gene profile (CAT-Tox) assay

Four (HMTIIA, HSP70, XRE and CRE) out of the 13 constructs evaluated showed a statistically significant level ($p < 0.05$) of induction/activation associated with cadmium

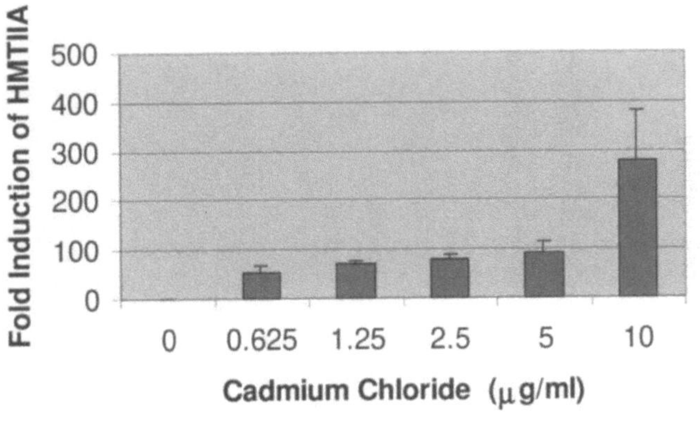

Fig. 2. Fold induction of metallothionein (HMTIIA) in human liver carcinoma ($HepG_2$) cells exposed to cadmium chloride. Recombinant $HepG_2$ cells transfected with a HMTIIA gene promoter fused to the CAT protein were treated with serial dilutions of cadmium chloride (0–10 μg/mL). HMTIIA induction was based on the expression of the CAT-protein using a sandwich ELISA, as indicated in Materials and methods. Each point represents a mean value and standard deviation of 3 experiments.

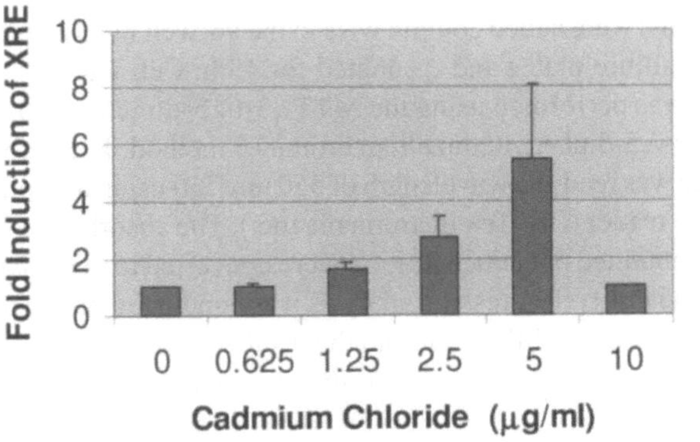

Fig. 3. Fold induction of xenobiotic response element (XRE) in human liver carcinoma ($HepG_2$) cells exposed to cadmium chloride. Recombinant $HepG_2$ cells transfected with a XRE fused to the CAT protein were treated with serial dilutions of cadmium chloride (0–10 μg/mL). XRE induction was based on the expression of the CAT-protein using a sandwich ELISA, as indicated in Materials and methods. Each point represents a mean value and standard deviation of 3 experiments.

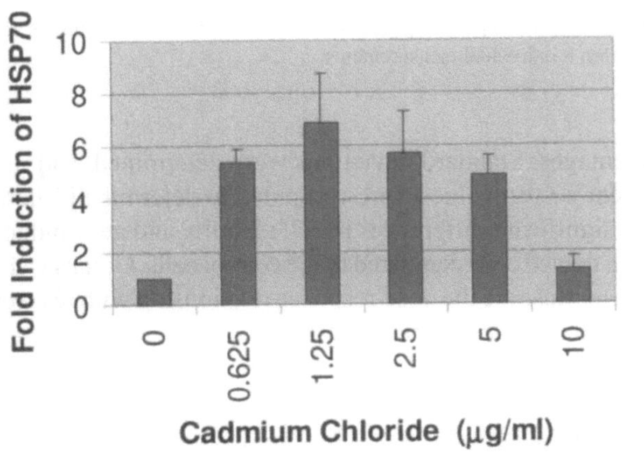

Fig. 4. Fold induction of 70-kDa heat shock protein (HSP70) in human liver carcinoma ($HepG_2$) cells exposed to cadmium chloride. Recombinant $HepG_2$ cells transfected with a HSP70 gene promoter fused to the CAT protein were treated with serial dilutions of cadmium chloride (0–10 μg/mL). HSP70 induction was based on the expression of the CAT-protein using a sandwich ELISA, as indicated in Materials and methods. Each point represents a mean value and standard deviation of 3 experiments.

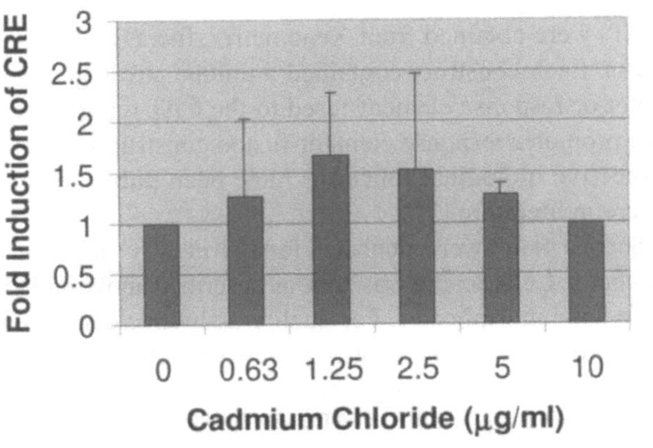

Fig. 5. Fold induction of cyclic adenosine monophosphate response element (CRE) in human liver carcinoma ($HepG_2$) cells exposed to cadmium chloride. Recombinant $HepG_2$ cells transfected with a CRE fused to the CAT protein were treated with serial dilutions of cadmium chloride (0–10 μg/mL). CRE induction was based on the expression of the CAT-protein using a sandwich ELISA, as indicated in Materials and methods. Each point represents a mean value and standard deviation of 3 experiments.

chloride exposure compared to the negative control. The inductions of HMTIIA and HSP70 genes and the activation of the XRE and CRE by cadmium were concentration dependent (Figs 2–5). The fold inductions were 54.3 ± 25.0, 70.2 ± 10.0, 80.8 ± 11.0, 91.3 ± 26.0, and 279.5 ± 37.0 for HMTIIA; 5.4 ± 0.5, 6.9 ± 2.0, 5.7 ± 2.0, 4.9 ± 0.7, and 1.3 ± 0.5 for HSP70; 1.0 ± 0.1, 1.6 ± 0.3, 2.7 ± 0.7, 5.5 ± 2.6 and 1.0 ± 0.0 for XRE; and 1.3 ± 0.8, 1.7 ± 0.6, 1.5 ± 1.0, 1.3 ± 0.1, and 1.0 ± 0.1 for CRE, at 0.6, 1.25, 2.5, 5 and 10 μg/mL of cadmium chloride, respectively. Though the fold inductions were concentration-dependent, each follows a specific pattern.

Detectable but not significant inductions of *c-fos* and GADD153 genes were observed. No significant inductions ($p > 0.05$) were recorded for the cytochrome P450 1A1 – CYP 1A1, glutathione-*S*-transferase Ya subunit – GST Ya, nuclear factor kappa (B site) response element – NFκBRE, tumor suppressor protein response element – *p53*RE, 45-kDa growth arrest DNA damage – GADD45, 78-kDa glucose regulated protein – GRP78, and retinoic acid response element – RARE compared to the negative control. All QA/QC tests with positive control chemicals (3-MC, MMS, and RA) resulted in significant inductions ($p < 0.05$) of stress genes in all recombinant cell lines (data not shown).

Discussion

Cytotoxicity of cadmium

Data obtained from the present study clearly indicate that cadmium chloride is highly cytotoxic to human liver carcinoma cells. The LC_{50} was computed to be 6.1 ± 0.8 µg/mL. It has been suggested that hepatocyte injury following exposure to cadmium may be due to binding to the inner membrane and accumulation of cadmium in the mitochondria, leading to the collapse of the mitochondrial membrane potential, followed by plasma membrane depolarization and cell death. Although the mechanism causing collapse of the mitochondrial membrane potential is unclear, it has been shown that inhibition of sulfhydryl-containing enzymes in the citric acid cycle and in the electron transport chain may be involved [25]. Cadmium has also been demonstrated to cause programmed cell death (apoptosis) in the liver [26].

Metallothionein-HMTIIA induction by cadmium

The fold induction of metallothionein (HMTIIA) gene was concentration-dependent even at concentrations where cytotoxicity was observed (Fig. 2). The highest fold induction was observed at 10 µg/mL of cadmium chloride though only 20% of the cell survived at that concentration. This construct is induced either through metal response elements (MRE) by heavy metals and/or through glucocorticoids response elements (GRE) by glucocorticoids [27, 28]. The induction of HMTIIA-gene by Cd was through the MRE since MRE acts as a metal-inducible enhancer, hence, all genes that have MRE in the upstream region could be induced by heavy metals.

Metallothioneins (MT) are low molecular weight cysteine-rich, heavy metal binding proteins. MT are said to be an important storage site for copper (Cu), zinc (Zn), cadmium (Cd) and mercury (Hg). Metal (Cd, Hg and excess Cu and Zn) binding to MT serves as a protective mechanism since they are no longer available to exert any toxic action. MT gene is known to be involved in the cellular defense system against heavy metal toxicity. The expression of this gene is investigated as a model for elucidating the molecular mechanism of cellular responses to heavy metals. It has been reported that the induction of MT through cadmium exposure protects organisms against acute Cd-induced lethality and hepatotoxicity in animals [29, 30]. Also, newborn animals are resistant to Cd-induced hepatotoxicity due to their high level of hepatic MT [31, 32]. Hence, the induction of MT by $HepG_2$ cells due to Cd exposure could be viewed as a protection mechanism against Cd hepatotoxicity. On the other hand the induction of MT by the liver through cadmium exposure has been implicated in nephrotoxicity. This is because in the liver where Cd-MT complex are stored, some leak out into the plasma and then are taken up by the kidney. This leakage is said to be responsible for nephrotoxicity since Cd-MT complex circulating in the plasma is known to be a potent nephrotoxicant [29, 30]. Hence, whether MT is beneficial or detrimental for chronic Cd-induced nephrotoxicity is debatable [35, 36].

Xenobiotic response element-XRE induction by cadmium

The highest fold induction of XRE was observed at 5 µg/mL of cadmium chloride (Fig. 3). This fold induction corresponds to 59% cell viability. At cadmium chloride concentration of 10 µg/mL, no XRE was induced though we had 20% cell survival. For XRE to respond to an exogenous compound (ligand), the ligand must first enter the cell, presumably by passive diffusion and bind to the Aryl hydrocarbon Receptor (AhR) in the cytosol. The AhR-ligand complex is then translocated to the nucleus where they undergo transformation leading to its high affinity DNA binding form [38, 39]. Ligands for AhR are mainly halogenated aromatic hydrocarbons (HAH) and to a less extent polycyclic aromatic hydrocarbons (PAHs). Available evidence suggests that, the differences in potency between HAHs and PAHs are likely due to a combination of the higher AhR binding affinity of HAHs and the increased resistance of HAHs to metabolic degradation, which results in sustained AhR occupancy by HAHs and persistent activation of gene expression [40–43].

CAT expression by the XRE construct resulting from cadmium chloride exposure suggests that this chemical is indirectly acting as a ligand to AhR, perhaps through breakdown or release of other cell products as a consequence of cadmium toxicity, leading to the production of non classical ligand(s) of AhR, as suggested by recent studies [44–46]. Hence, an AhR-mediated response such as CYP1A1 induction of should be expected. Above all, the CYP1A1 construct has a XRE within its promoter [47, 48], indicating that its induction should involve an Ah-dependent mechanism. On the contrary, there was no induction of CYP1A1-gene. This could be due to the inability of the cadmium chloride-induced ligand-AhR complex to go through the entire process necessary for the increase in synthesis of microsomal cytochrome P4501A1. These processes are: DNA bending, disruption of the chromatin and the nucleosome, increase promoter accessibility, increase rates of transcription initiation of the CYP1A1 gene, the subsequent accumulation of CYP1A1-specific mRNA and increase synthesis of microsomal cytochrome P4501A1 [49, 50].

Hence, it could be said that ligand-AhR complex binding to XRE *in vitro* does not mean *a priori* that the compound will confer 3,4,7,8-tetrachlorodibenzo p-dioxin (TCDD) responsiveness. A similar result has been reported by Safe [40] using PCBs, which are known inducers of CYP1A1. Our laboratory has also reported similar results (induction of XRE and lack of subsequent induction of CYP1A1) following exposure of $HepG_2$ cells to other metal-containing compounds such as arsenic trioxide [51] and lead nitrate (in preparation). Further studies are needed to ascertain and to elucidate the real biochemical mechanism(s) involved in the activation of XRE by these metals.

70 kDa heat shock protein-HSP70 induction by cadmium

The induction of HSP70 increases from 0.625 μg/mL and peaks at 1.25 μg/mL cadmium chloride. It starts leveling off from 2.5 μg/mL and reaches its minimum induction at 10 μg/mL, probably due to cell death at higher cadmium chloride concentrations. The genes responsible for heat shock proteins are expressed in response to a wide range of physiologically and chemically induced stress conditions. Protein perturbations either by direct protein damage or disruption of nascent chain elongation or folding are the necessary conditions needed to up-regulate HSP70 transcription [52, 53]. Hence the induction of HSP70 gene by Cd is to facilitate proper folding of proteins, signifying protein protection from Cd toxicity or repair of damaged protein. The heat shock elements (HSE) located up-stream region of the HSP genes as well as the heat shock factor (HSF) is required for the transcriptional activation of the heat shock genes [54]. The binding ability of HSF to DNA increases in association with heavy metal treatment [53, 55–57]. Since Cd is a classical inducer of HSP70, it could be said that, the induction of HSP70 by Cd may be due to the activation of HSF through some indirect effects of Cd, for example generating of abnormal proteins; a form of protein perturbation. This perturbation could be due to binding of Cd to the sulfur in proteins since Cd has a strong affinity for sulfur. Hence, the induction of HSP70 gene by $HepG_2$ cells could be interpreted as cellular defense mechanism against cadmium-induced toxicity.

Cyclic AMP response element-CRE induction by cadmium

CRE activation increases from 0.625 μg/mL and reaches the highest activation at 1.25 μg/mL cadmium chloride. It starts leveling off from 2.5 μg/mL and reaches an induction level comparable to the control at 10 μg/mL, most probably due to cell death at higher chemical concentrations. Receptor-mediated events, which activate adenylate cyclase, can be measured by cyclic AMP levels. These cyclic AMP levels can be monitored through activation of the cAMP response element binding (CREB) protein. The CREB protein binds to several genes. The increased in intracellular levels of cAMP specifically activates CRE fusion construct [58, 59].

The activation of CRE through Cd exposure makes this metal an endocrine receptor (ER) agonist. There are two pathways that depict the mechanism of action for activation of ER. First an ER ligand (agonist) must enter the cell and bind to the non-ligand bound form of ER, which is a part of a complex with associated proteins [60] and is predominantly localized in the nucleus [61]. The ER-ligand complex then binds to estrogen-responsive elements (EREs) in the promoter region of ER-responsive genes.

The second pathway is through ligand-independent activation of ER. This is done through phosphorlylation of ER, binding of phosphorlylated ER to EREs, which then activates transcription [62–64]. This pathway is known to be initiated by growth factors, e.g. epidermal growth factors (EGF), insulin-like growth factor 1 and platelet-derived growth factor or agents that increase cAMP levels. Since the activation of CRE fusion construct is increased in intracellular levels of cAMP specific, it is obvious that cadmium interaction with ER is through the ligand-independent pathway. This pathway elucidates the mechanism behind the endocrine-disrupting potential of cadmium and it also confirms cadmium as an endocrine-disrupting metal.

C-fos and GADD153 induction by cadmium

Statistically, marginal inductions (1<fold inductions<2) were recorded for the *c-fos*, and 153-kDa growth arrest DNA damage (GADD153) promoters. A study by Tang and Enger [65] showed that cadmium has the ability to stimulate the expression of proto-oncogenes, and genes related to cell proliferation including *c-fos, c-jun*, and *c-myc*. The induction of the 153-kDa growth arrest and DNA damage (GADD153) gene by cadmium chloride is an indication of its potential to cause DNA damage. Studies with HeLa cells have also shown a strong induction of GADD153 following cadmium exposure [66].

Conclusions

The results from this study indicates that: (1) cadmium chloride causes significant cytotoxicity to human liver carcinoma ($HepG_2$) cells, with a LC_{50} of about 6 ug/mL; (2) cadmium chloride induces the expression of XRE, indicating its potential involvement in the biotransformation process in the liver; (3) cadmium chloride significantly induces the expression of CRE, indicating its ability to disrupt endocrine functions; (4) cadmium chloride marginally induces the GADD153 and *c-fos* promoters, indicating that it may have some potential for DNA damage and activation of cell proliferation; and (5) as expected, cadmium chloride strongly induces the HMTIIA and HSP70 gene promoters, indicating its ability to cause proteotoxic effects at the molecular and cellular levels.

Acknowledgments

This research was financially supported in part by a grant from the US Department of Energy through the HBCU/MI-Environmental Technology Consortium under Cooperative Agreement No. DE-FC04-90AL66158, and in part by a grant from the US Department of Army (Grant No. DE-W-7505-ENG48) to Lawrence Berkeley National Laboratory, and Subcontract No. 648515 to Jackson State University.

References

1. Chettle DR, Ellis KJ: Further scientific issues in determining an occupational standard for cadmium. Am J Ind 22: 117–124, 1992
2. Mason HJ, Davison AG, Wright AL, Guthrie CJG, Fayers PM, Venables KM, Smith NJ, Chettle DR, Franklin DM, Scott MC, Holden H, Gompertz D, Newman-Taylor AJ: Relations between liver cadmium, cumulative exposure, and renal function in cadmium alloy workers. Br J Ind Med 45: 793–805, 1988
3. Drasch GA: An increase of cadmium body burden for this century: An investigation on human tissues. Sci Total Environ 26: 111–120, 1983
4. Voors AW, Shuman MS, Johnson WD: Additive statistical effects of cadmium and lead on heart related disease in a North Carolina USA autopsy series. Arch Environ Health 37: 98–102, 1982
5. Ferm VH, Layton WM Jr.: Teratogenic and mutagenic effects of cadmium. In: J.O. Nriagu (ed). Cadmium in the Environment. Part 2, Health Effects. John Wiley, New York, 1981, pp 743–756
6. EPA: Ambient Water Quality Criteria for Cadmium. Rep. 440/5-80-025. US Environmental Protection Agency, Washington, DC, 1980, 183 pp
7. Nomiyama K: Carcinogenicity of cadmium. Jpn J Indust Health 24: 13–23, 1982
8. International Agency for Research on Cancer: Cadmium and cadmium compounds. IARC Monographs on the Evaluation of Carcinogenic Risks to Humans: Supplement 7. WHO, Lyons, France, 1987, pp 140–142
9. International Agency for Research on Cancer: IARC Monographs on the Evaluation of Carcinogenic Risks to Humans: Beryllium, Cadmium Mercury and Exposures in The Glass Industry. IARC Scientific Publications no. 58. IARC, Lyons, France, 1993, pp 119–238
10. Elinder CG, Kjellstrom T, Hogstedt C, Andersson K, Spang G: Cancer mortality of cadmium workers. Br J Ind Med 42: 651–655, 1985
11. Occupational Safety and Health Administration, US Department of Labor: Occupational exposure to cadmium; final rules. Fed Reg 57: 42102–42463, 1992
12. Kazantzis G, Lam TH, Sullivan KR: Mortality of cadmium-exposed workers: A five-year update. Scand J Work Environ Health 14: 220–223, 1988
13. Oberdorster G: Airborne cadmium and carcinogenesis of the respiratory tract. Scand J Work Environ Health 12: 523–537, 1986
14. Sullivan K, Waterman L: Cadmium and cancer: The current position. Report of an international meeting in London, September 1988. Ann Occup Hyg 32: 557–560, 1988
15. Thun MJ, Elinder CG, Friberg L: Scientific basis for an occupational standard for cadmium. Am J Ind Med 20: 629–642, 1991
16. Waalkes MP, Oberdorster G: Cadmium Carcinogenesis. Biological Effects of Heavy Metals, Vol. II. CRC Press, Boca Raton, FL, 1990
17. Waalkes MP, Coogan TP, Barter RA: Toxicological principles of metal carcinogenesis with special emphasis on cadmium. Crit Rev Toxicol 22: 175–201, 1994
18. de Kort WLAM, Verschoor MA, Wibowo AAE, van Hemmen JJ: Am J Ind Med 11: 145–156, 1987
19. Elinder CG: Other toxic effects. In: K.J. Ellis (ed). Cadmium and Health, Vol. 2. CRC Press, Boca Raton, FL, 1986, pp 160–197
20. Bucio L, Souza V, Albores A, Sierra A, Chavez E, Carabez A, Gutierrez-Ruiz MC: Cadmium and mercury toxicity in a human fetal hepatic cell line (WRL-68 cells). Toxicology 102: 285–299, 1995
21. Bhattacharyya MH, Wilson AK, Rajan SS, Jonah M: Biochemical pathways in cadmium toxicity. In: R.K. Zalups, J. Koropatnick (eds). Molecular Biology and Toxicology of Metals. Taylor and Francis, New York, 2000, pp 34–74
22. Mosmann T: Rapid colorimetric assay for cellular growth and survival: Applications to proliferation and cytotoxicity assays. J Immunol Meth 65: 55–63, 1983
23. Todd MD, Lee MJ, Williams JL, Nalenzy JM, Gee P, Benjamin MB, Farr SB: The CAT-Tox assay: A sensitive and specific measure of stress induced transcription in transformed human liver cells. Fundament Appl Toxicol 28: 118–128, 1995
24. Tchounwou PB, Wilson BA, Schneider J, Ishaque A: Cytogenetic assessment of arsenic trioxide toxicity in the Mutatox, Ames II, and CAT-Tox assays. Metal Ions Biol Med 6: 89–91, 2000
25. Chang LW: An introduction to metal effects on other organ systems. In: L.W. Chang, L. Mgs, T. Suzuki (eds). Toxicology of Metals. CRC Lewis Publishers, Boca Raton, FL, 1996, pp 885–899
26. Habeebu SS, Liu J, Klaassen CD: Cadmium-induced apoptosis in mouse liver. Toxicol Appl Pharmacol 149: 203–209, 1998
27. Cherian MG, Howell SB, Imura N, Klaassen CD, Koropatnick J, Lazo JS, Waalkes MP: Contemporary issues in toxicology: Role of metallothionein in carcinogenesis. Toxicol Appl Pharmacol 126: 1–5, 1994
28. Richards RI, Heguy A, Karin M: Structural and functional analysis. The human metallothionein-I gene: Differential induction by metal ions and glucocorticoids. Cell 7: 263–272, 1984
29. Leber AP, Miya TS: A mechanism for cadmium- and zinc-induced tolerance to cadmium toxicity: Involvement of metallothionein. Toxicol Appl Pharmacol 37: 403–414, 1976
30. Goering PL, Klaassen CD: Tolerance to cadmium-induced hepatotoxicity following cadmium pretreatment. Toxicol Appl Pharmacol 74: 308–313, 1984
31. Wong KL, Cachia R, Klaassen CD: Comparison of the toxicity and tissue distribution of cadmium in newborn and adult rats after repeated administration. Toxicol Appl Pharmacol 56: 317–325, 1980
32. Goering PL, Klaassemn CD: Resistance to cadmium-induced hepatotoxicity in immature rats. Toxicol Appl Pharmacol 74: 321–329, 1984
33. Nordberg GF, Goyer R, Nordberg M: Comparative toxicity of cadmium chloride on mouse kidney. Arch Pathol 99: 192–197, 1975
34. Dudley RE, Gammal LM, Klaassen CD: Cadmium-induced hepatic and renal injury in chronically exposed rats: Likely role of hepatic cadmium-metallothionein in nephrotoxicity. Toxicol Appl Pharmacol 77: 414–426, 1985
35. Goering PL, Waalkes MP, Klaassen CD: Toxicology of cadmium. In: R.A. Goyer, M.G. Cherian (eds). Handbook of Experimental Pharmacology, Vol. 115. Springer-Verlag, Heidelberg, New York, 1995, pp 189–213
36. Petering DH, Fowler BA: Roles of metallothionein and related proteins in metal metabolism and toxicity: Problems and perspectives. Environ Health Perspect 65: 217–224, 1986
37. Cherian MG: Metallothionein and its interaction with metals. In: R.A. Goyer, M.G. Cherian (eds). Handbook of Experimental Pharmacology, Vol. 115. Springer-Verlag, Heidelberg, New York, 1995, pp 121–137
38. Yao EG, Denison MS: DNA sequence determinants for binding of transformed Ah receptor to a dixion-responsive enhancer. Biochemistry 31: 5060–5067, 1992

39. Pimental RA, Liang B, Yee GK, Wilhelmsson A, Poellinger L, Paulson KE: Dioxin receptor and C/EBP regulate the function of the glutathione-*S*-transferase Ya gene xenobiotic element. Mol Cell Biol 13: 4265–4373, 1993
40. Safe S: Polychlorinated biphenyls (BCBs), dibenzo-p-dioxins (PCDDs), dibenzofurans (PCDFs), and related compounds: Environmental and mechanistic considerations, which support the development of toxic equivalency factors (TEFs). Crit Rev Toxicol 21: 51–88, 1990
41. Poland A, Knutson JC: 2,3,7,8-tetrachlorodibenzo-p-dioxin and related halogenated aromatic hydrocarbons: Examination of the mechanism of toxicity. Annu Rev Pharmacol Toxicol 22: 517–542, 1982
42. Schmidt JV, Bradfield CA: Ah receptor signaling pathways. Annu Rev Cell Dev Biol 12: 55–89, 1996
43. Safe S: Comparative toxicology and mechanism of action of polychlorinated dibenzo-p-dioxins and dibenzofurans. Annu Rev Pharmacol Toxicol 26: 371–399, 1986
44. Sadek CM, Allen-Hoffman BL: Suspension-mediated induction of hepa 1c1c7 CYP1a-1 expression is dependent on the Ah receptor signal transduction pathway. J Biol Chem 269: 31505–31509, 1994
45. Weiss C, Kolluri SK, Kiefer F, Gottlicher M: Complementation of Ah receptor deficiency in hepatoma cells: Negative feedback regulation and cell cycle control by the Ah receptor. Exp Cell Res 226: 154–163, 1996
46. Ma Q, Whitlock JP Jr: The aromatic hydrocarbon receptor modulates the Hepa 1c1c7 cell cycle and differentiated state independently of dioxin. Mol Cell Biol 16: 2144–2150, 1996
47. Fujii-Kuriyama Y, Imataka H, Sogawa K, Yasumoto KI, Kikuchi Y: Regulation of CYP1A1 expression. FASEB J 6: 706–710, 1992
48. Gonzales GJ, Nebert DW: Autoregulation plus upstream positive and negative control regions associated with transcriptional activation of the mouse P1 (450) gene. Nucleic Acids Res 13: 7269–7288, 1985
49. Hankinson O: The aryl hydrocarbon receptor complex. Annu Rev Pharmacol Toxicol 35: 307–340, 1995
50. Whitlock JP Jr, Okino S, Dong L, Ko H, Clarke-Ktzenberg R, Ma Q, Li H: Induction of cytochrome P4501A1: A model for analyzing mammalian gene transcription. FASEB J 10: 809–818, 1996
51. Tchouwnou PB, Wilson BA, Schneider J, Ishaque AB: Cytogenetic assessment of arsenic trioxide in the Mutatox, Ames II and CAT-Tox (L) assays. In: J.A. Centeno, Ph. Collery, G. Vernet, R.B. Finkelman, H. Gibb, J.C. Etienne (eds). Metal Ions in Biology and Medicine, Vol. 6. Paris, France, 2000, pp 89–91
52. Morimoto RI, Tissieres A, Georgopoulos C (eds): Stress Proteins in Biology and Medicine. Cold Spring Harbor Laboratory Press, New York, 1990
53. Mosser DD, Theodorakis NG, Morimoto RI: Coordinate changes in heat shock element-binding activity and HSP70 gene transcription rates in human cells. Mol Cell Biol 8: 4736–4744, 1988
54. Morimoto RI, Sarge KD, Abravaya K: Transcriptional regulation of heat shock genes: A paradigm for inducible genomic responses. J Biol Chem 267: 2987–21990, 1992
55. Sarge KD, Murphy SP, Morimoto RI: Activation of heat shock gene transcription by heat shock factor1 involves oligomerization, acquisition of DNA-binding activity, and nuclear localization and can occur in the absence of stress. Mol Cell Biol 13: 1392–1407, 1993
56. Hatayama T, Asai Y, Wakatsuki T, Kitamura T, Imahara H: Regulation of hsp 70 synthesis induced by cupric sulfate and zinc sulfate in thermotolerant HeLa cells. J Biochem (Tokyo) 114: 592–597, 1993
57. Abe T, Konishi T, Katoh T, Hirano H, Matsukuma K, Kashimura M, Higashi K: Inductiion of heat shock 70 mRNA by cadmium is mediated by gluthathione suppressive and non-suppressive triggers. Biochem Biophys Acta 201: 29–36, 1994
58. Borrelli E, Montmayeur JP, Foulkes NS, Sassone-Corsi P: Signal transduction and gene control: The camp pathway. Crit Rev Oncol 3: 321–338, 1992
59. Lee KAW, Masson N: Transcriptional regulation by CREB and its relatives. Biochim Biophys Acta 1174: 221–233, 1993
60. Smith DR, Toft DO: Steroid receptors and their associated proteins. Mol Endocrinol 7: 4–11, 1993
61. Gorski J, Furlow JD, Murdoch FE, Fritsch M, Kanenko K, Ying C, Malayer JR: Perturbations in the model of estrogen receptor regulation of gene expression. Biol Reprod 48: 8–14, 1993
62. Ignar-Trowbridge DM, Pimentel M, Parker MG, McLachlan JA, Korach KS: Peptide growth factor cross-talk with the estrogen requires the A/B domain and occurs independently of protein kinase C or estradiol. Endocrinol 137: 1735–1744, 1996
63. Ignar-Trowbridge DM, Nelson KG, Bidwell MC, Curtis SW, Washborn TF, McLachlan JA, Korach KS: Coupling of dual signaling pathways: Epidermal growth factor action involves the estrogen receptor. Proc Natl Acad Sci 89: 4658–4662, 1992
64. Cho H, Katzenellenbogen BS: Synergistic activation of estrogen receptor-mediated transcription by estradiol and protein kinase activators. Mol Endocrinol 7: 441–452, 1993
65. Tang N, Enger MD: Cd^{2+}-induced c-myc mRNA accumulation in NRK-49F cells is blocked by the protein kinase inhibitor H7 but not HA 1004, indicating that protein kinase C is a mediator of the response. Toxicology 81: 155–164, 1993
66. Luethy JD, Holbrook NJ: The pathway regulating GADD153 induction in response to DNA damage is independent of protein kinase C and tyrosine kinases. Cancer Res 54: 1902–1906, 1994

Molecular and Cellular Biochemistry **222**: 29–34, 2001.

Arsenic-induced NFκB transactivation through Erks- and JNKs-dependent pathways in mouse epidermal JB6 cells

Chuanshu Huang,[1] Jingxia Li,[1] Min Ding,[2] Liying Wang,[2] Xianglin Shi,[2] Vincent Castranova,[2] Val Vallyathan,[2] Gong Ju[3] and Max Costa[1]

[1]*Nelson Institute of Environmental Medicine, New York University School of Medicine, NY;* [2]*Health Effects Laboratory Division, National Institute for Occupational Safety and Health, Morgantown, WV, USA;* [3]*The Institute of Neuroscience, The Fourth Military Medical University, Xi'an, P.R. China*

Abstract

Tumor promoting effects of arsenic are believed to be associated with its transactivation activity on transcription factors, such as AP-1 and NFκB. However, the results from different groups studying the effects of arsenic on NFκB activation are contradictory in different cell models. Since arsenic is a strong skin carcinogen, we have investigated the activation of NFκB by arsenic in a mouse skin epidermal cell line, JB6 cells. Exposure of cells to arsenite or arsenate led to NFκB transactivation in mouse epidermal JB6 NFκB-luciferase reporter stable transfectants, C141 NFκB $mass_1$. This induction of NFκB activity by arsenic was dose- and time-dependent. The transactivation of NFκB by arsenic appeared to be through activation of Erks and JNKs pathways because increased NFκB activity by arsenic could be dramatically inhibited by either pre-treatment of cells with PD98059 or overexpression of dominant negative JNK_1. That Erks activation is required for arsenic-induced NFκB transactivation was further supported by the findings that arsenic-induced NFκB transactivation was impaired in JB6 30.7b cells, which were deficient in Erks. (Mol Cell Biochem **222**: 29–34, 2001)

Key words: arsenic, NFκB, MAP kinase

Abbreviations: NFκB – nuclear factor-κB; AP-1 – activator protein-1; Erks – extracellular signal-regulated protein kinases; MAPK – mitogen-activated protein kinase; P^+ – tumor promoter-sensitive; P^- – tumor promoter-resistant; JNKs – c-Jun N-terminal kinases; FBS – fetal bovine serum; MEM – minimal essential medium

Introduction

Arsenic is a well-recognized compound widely distributed in food, water, air and soil [1–3]. Arsenic exposure is accompanied by many diseases, such as cancers of the lung, skin and liver [2–6]. Arsenic is released to the atmosphere from both natural and anthropogenic sources. Global natural emissions of arsenic and arsenic compounds have been estimated to be 8000 tons each year, whereas anthropogenic emissions are about three times higher, 23,600 tons per year [7]. NIOSH estimated that 1.5 million industrial workers are potentially exposed to arsenic and its compounds during manufacturing and processing operations [8]. Arsenic can exist in many chemical forms. Inorganic trivalent arsenite and pentavalent arsenate are the most important forms in causing diseases [8, 9].

One of the sensitive targets for arsenic toxicity is the skin. Both inorganic and organic forms are absorbed by human and animal skin [8, 9]. Animal experiments show that following chronic exposure, arsenic accumulates in the skin and hair

Address for offprints: C. Huang, Nelson Institute of Environmental Medicine, New York University School of Medicine, NY 10016, USA

[8, 9]. Although arsenic is ingested and absorbed from the diet, the skin is a major target organ. Some previous studies suggested that the accumulation of arsenic in skin increased sensitivity of skin to ultraviolet (UV) light and elevated the risks of UV carcinogenic effects [10, 11]. Many cases of skin cancer have been recorded among peoples with occupational or medical exposure to arsenic [8, 9]. Epidemiological evidence has shown an excess of skin cancers following arsenic exposure. Although arsenic is acknowledged to be highly carcinogenic to human skin, its mechanism of action at the cellular and molecular level in skin is not well understood. Arsenite was demonstrated to promote altered gene expression and modify intra- or inter-cellular communications [12, 13]. Arsenite was found to be a potent stimulator of AP-1 activity, but not NFκB activity in Hela cells [14]. Ludwig *et al.* [15] reported that arsenite induces Erks activation through MAPK kinase 6/p38-dependent pathways. Recently, we demonstrated that low concentrations of arsenite induced Erks activation and this Erks activation was required for arsenite-induced cell transformation [13]. High concentrations of arsenic induced cell apoptosis through JNKs pathways [16]. Here we investigated the effects of arsenic on NFκB activation and the signal transduction pathways involved in NFκB activation.

Materials and methods

Cell culture and agents

Mouse epidermal JB6, C141 and 30.7b cells as well as their NFκB-luciferase reporter stable transfectants were cultured in monolayers in Eagle's minimal essential medium containing 5% fetal bovine serum (FBS), 2 mM L-glutamine, and 25 μg of gentamicin/ml [17]. Fetal bovine serum was from Life Technologies, Inc. (Rockville, MD, USA); Eagle's minimal essential medium (MEM) was from Calbiochem (San Diego, CA, USA); luciferase assay substrate was from Promega (Madison, WI, USA); both arsenite and arsenate were from Sigma (St. Louis, MO, USA).

Assay for NFκB activation in JB6 C141 cells

Confluent monolayers of JB6 C141 NFκB $mass_1$ cells were trypsinized, and 8×10^3 viable cells suspended in 100 μ l of 5% FBS MEM were added to each well of a 96-well plate. Plates were incubated at 37°C in a humidified atmosphere of 5% CO_2. Twelve to 24 h later, cells were starved by culturing them in 0.1% FBS MEM for 24 h. Then, the cells were exposed to arsenic (diluted in 0.1% FBS MEM at the concentration indicated; storage concentration of arsenic was 50 mM dissolved in PBS) for NFκB induction and maintained in culture. After various time periods, the cells were extracted with lysis buffer, and luciferase activity was measured using a luminometer (Monolight 2010). The results are expressed as relative NFκB activity or relative luciferase activity [18].

Statistical analysis

The significance of the difference in the NFκB activity was determined with the Student's *t*-test.

Results

Induction of NFκB in mouse epidermal JB6 cells by arsenite and arsenate

Arsenite and arsenate were reported to have no effect on NFκB activity in Hela cells [14]. To determine the effects of arsenic on NFκB activation in mouse epidermal cells, we incubated C141 NFκB $mass_1$ cells, a stable JB6 C141 transfectant with the NFκB-luciferase reporter, with either arsenite or arsenate. The results show that treatment of C141 cells with either arsenite or arsenate led to significant NFκB induction (Fig. 1). A dose-response study indicates a concentration dependent induction of NFκB activation by arsenic (Fig. 2a). A time-course study suggests that treatment of cells with arsenic resulted in maximal induction of NFκB activity (~ 70-fold) at 24–48 h of-arsenic treatment of cells (Fig. 2b). Then the NFκB activity returned to near basal levels (Fig. 2b). These results demonstrate that both arsenite and arsenate induce very strong activation of NFκB activity in mouse epidermal cells.

Inhibition of arsenic-induced NFκB activation by PD98059

Arsenic was reported to activate JNKs and P38 kinase as well as Erks in previous studies [13–15]. Our recent studies also demonstrate that arsenite could induce activation of Erks and JNKs in JB6 cells [13, 16]. To test the possible role of Erks activation in arsenic-induced NFκB activation in JB6 cells, we pretreated C141 NFκB $mass_1$ cells with PD98059, an MEK1-Erks pathway inhibitor, before exposure of cells to arsenic. The results show that pretreatment of cells with PD98059 results in marked inhibition of arsenic-induced NFκB activation (Fig. 3). These data suggest that Erks may play a role in arsenic-induced NFκB activation.

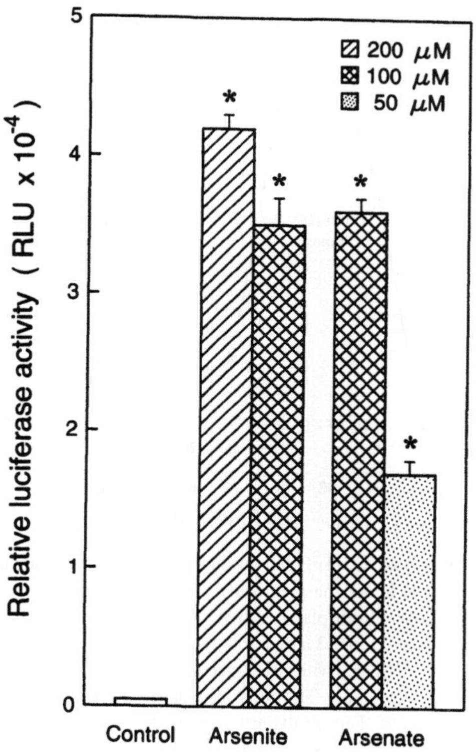

Fig. 1. Induction of NFκB activity by arsenic in mouse epidermal JB6 cells. 8 × 10^3 Cl41 NF B $mass_1$ cells were seeded into each well of 96-well plates. After being cultured at 37°C overnight, the cells were starved for 12 h by replacing medium with 0.1% FBS MEM. Then the cells were treated with either arsenite or arsenate at the concentrations indicated. After being cultured for 24 h, the luciferase activity was measured. The results are presented as relative luciferase activity. Each bar indicates the mean and standard deviation of 4 identically treated assay wells. *Indicates a significant increase from control ($p < 0.05$).

Arsenic-induced NFκB activation was impaired in JB6 30.7b cells

The JB6 mouse epidermal cell system includes tumor promotion-sensitive (P^+, C141) and tumor promotion-resistant (P^-, 30.7b) cells [18–22]. Our recent studies demonstrated that the lack of AP-1 activation and cell transformation responses to TPA and EGF in P^- cells is due to the shortage of Erks in this cell line. To further determine the role of Erks activation in arsenic-induced activation of NFκB, we compared NFκB activation of C141 NFκB $mass_4$ cells to 30.7b NFκB $mass_1$ cells. Results show that arsenic-induced NFκB activity was dramatically impaired in 30.7b NFκB $mass_1$ cells, while marked NFκB activation was observed in C141 NFκB $mass_4$ cells (Fig. 4). This result provides additional evidence that Erks activation is required for arsenic-induced NFκB activation.

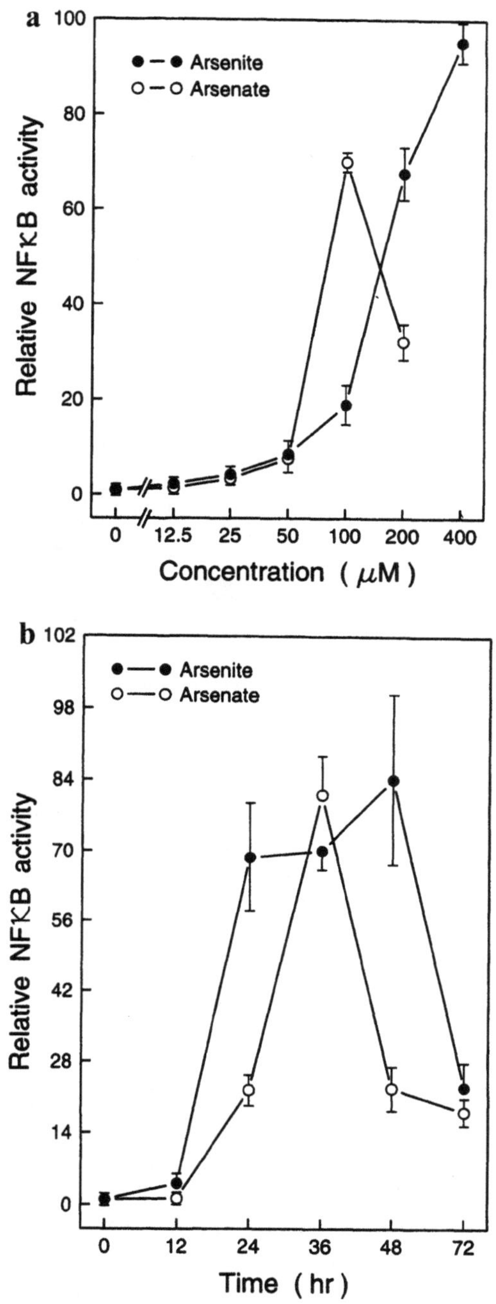

Fig. 2. Dose-response and time course-study of NFκB activation by arsenic. 8 × 10^3 C141 NFκB $mass_1$ cells were seeded into each well of 96-well plates. After being cultured at 37°C overnight, the cells were starved for 12 h by replacing medium with 0.1% FBS MEM. (A) For dose response study, the cells were treated with different concentrations of either arsenite or arsenate as indicated. After being cultured for 24 h, the luciferase activity was measured. The results are presented as relative NFκB activity. (B) For time course study, the cells were treated with either arsenite (100 μM) or arsenate (100 μM) for various times indicated and the luciferase activity was measured. The results are presented as relative NFκB activity. Each value indicates the mean and standard deviation of 4 identically treated assay wells.

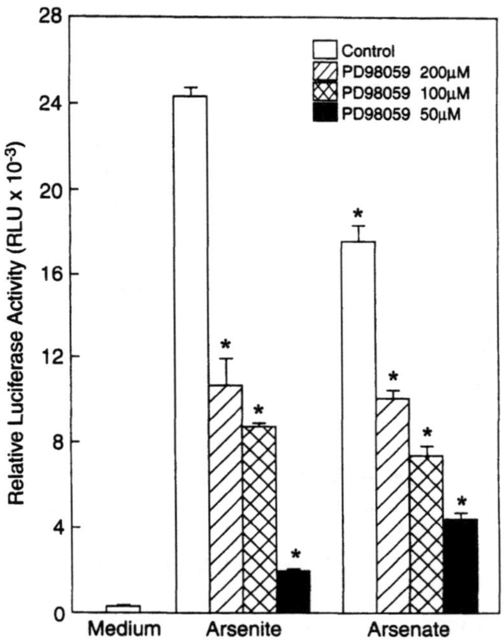

Fig. 3. Inhibition of arsenic-induced NFκB activity by pretreatment of cells with PD98059. 8×10^3 C141 NFκB $mass_1$ cells were seeded into each well of 96-well plates. After being cultured at 37°C overnight, the cells were starved for 12 h by replacing medium with 0.1% FBS MEM. Then the cells were first incubated with different concentrations of PD98059 for 30 min and then treated with either arsenite (100 μM) or arsenate (100 μM). After being cultured for 24 h, the luciferase activity was measured and the results are presented as relative luciferase activity. Each bar indicates the mean and standard deviation of 4 identically treated assay wells. *Indicates a significant increase from control ($p < 0.05$).

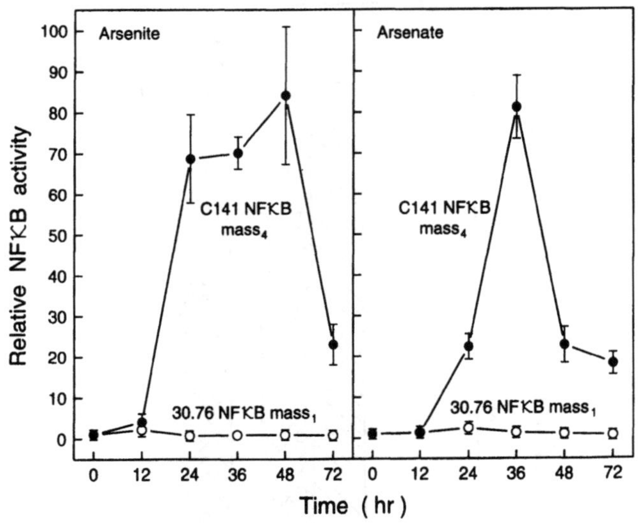

Fig. 4. Arsenic induces NFκB activation in JB6 C141 cells, but not in 30.7b cells. 8×10^3 C141 NFκB $mass_4$ cells or 30.7b NFκB $mass_1$ cells were seeded into each well of 96-well plates. After being cultured at 37°C overnight, the cells were starved for 12 h by replacing medium with 0.1% FBS MEM. Then the cells were treated with either arsenite (100 μM) or arsenate (100 μM) for various times indicated. The cells were harvested and the luciferase activity was measured. The results are presented as relative NFκB activity. Each value indicates the mean and standard deviation of 4 identically treated assay wells.

Blocking of arsenic-induced NFκB activation by overexpression of dominant negative JNK_1

Our previous results suggest that arsenite induces JNκB activation [13, 16]. To test possible involvement of JNKs in arsenic-induced NFκB activation, we used a well-characterized JB6 cell stable co-transfectants, C141 DN JNK_1 $mass_2$ and C141 DN JNK_1 $mass_4$ [13, 16, 23]. Results showed that arsenic-induced NFκB activity was dramatically blocked by overexpression of dominant negative mutant of JNK_1. This result demonstrates that JNK_1 was involved in the signal transduction pathway leading to NFκB activation in response to arsenic.

Discussion

In the present study, we investigated the effects of arsenite and arsenate on NFκB transactivation as well as its signal transduction pathways in mouse epidermal JB6 cell lines. The results show that treatment of NFκB-luciferase reporter transfected mouse epidermal C141 cells with either arsenite or arsenate leads to an increase in NFκB activity. This NFκB activation appears to be time- and dose-dependent. The induction of NFκB could be significantly inhibited by pretreatment of cells with PD98059, a specific inhibitor of MEK_1-Erks pathway. The important role of Erks in arsenic-induced NFκB activation was further supported by the finding that NFκB activation was impaired in 30.7b cell, a cell deficient in Erks, as compared with C141 cells that exhibit normal expression of Erks. Furthermore, the overexpression of the dominant negative mutant of JNK_1 also inhibited arsenic-induced NFκB activation. These results demonstrate that both arsenite and arsenate could induce NFκB transactivation in mouse epidermal cells, and that this induction appears to be mediated by activation of MAP kinase family members, including Erks and JNκB.

Environmental and occupational exposure to arsenic is associated with increased risk of skin, urinary bladder and respiratory tract cancers. A growing amount of evidence demonstrates that arsenic acts at the level of tumor promotion by activating the signal transduction pathways leading to activation of transcription factors, which in turn modulate the gene expression [12–15]. It has been reported that arsenite is a potent stimulator for induction of c-fos and c-jun gene expression, AP-1 transactivational activity and JNK activity, but not NFκB activity in Hela cells [14]. Ludwig *et al.* [15] reported that arsenite induces Erks activation through MAPK Kinase 6/p38-dependent pathways. Because the skin is one

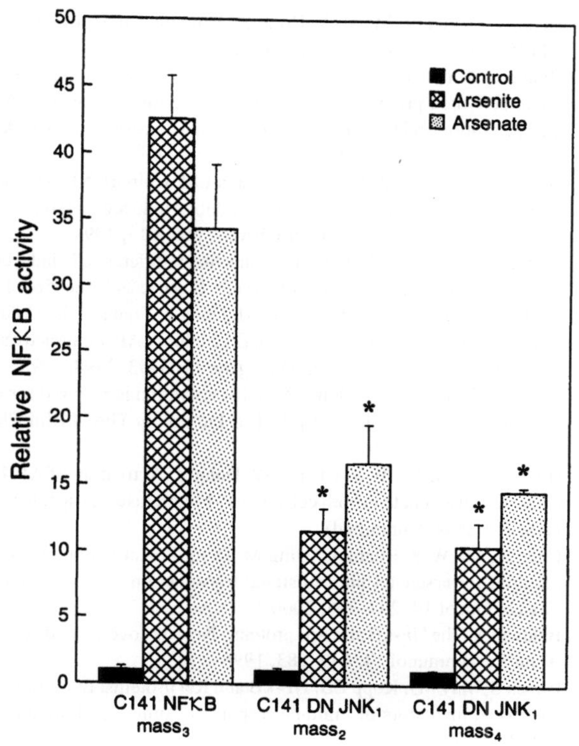

Fig. 5. Blockade of arsenic-induced NFκB activity by overexpression of dominant negative mutant JNK_1. 8×10^3 C141 NF B $mass_3$ cells, C141 DN JNK_1 $mass_2$ cells and C141 DN JNK_1 $mass_4$ cells were seeded into each well of 96-well plates. After being cultured at 37°C overnight, the cells were starved for 12 h by replacing medium with 0.1% FBS MEM. Then the cells were treated with either arsenite (100 μM) or arsenate (100 μM). After being cultured for 24 h, the cells were harvested and the luciferase activity was measured. The results are presented as relative NFκB activity. Each bar indicates the mean and standard deviation of 4 identically treated assay wells. *Indicates a significant increase from control ($p < 0.05$).

of the major targets of arsenic, we recently addressed the effects of arsenic on signal transduction pathways leading to activation of transcription factors and MAP kinases in mouse epidermal cell line, JB6 cells, and mouse skin [13, 24]. We demonstrate that low concentrations of arsenite induced Erks activation and that Erks activation was required for arsenite-induced cell transformation [13]. In contrast high concentrations of arsenite caused activation of JNKs, which mediated cell apoptosis in mouse epidermal JB6 cells [16]. Very recently, we also determined arsenic-induced activation of AP-1 *in vivo* using AP-1-luciferase reporter transgenic mice [24]. It was found that arsenic could induce activation of AP-1 *in vitro* and *in vivo* [24]. Arsenic was also found to induce activation of Erks, JNKs, and P38 kinase [13, 16, 24]. The induction of AP-1 activity by arsenic appears to be mediated by activation of PKC and MAP kinase family members, because induction of AP-1 could be dramatically blocked by either pretreatment of cells with PD98059 or overexpression of dominant negative mutant of PKC [24, 25]. The results from this study show that both arsenite and arsenate markedly induce activation of NFκB in mouse epidermal JB6 C141 cells, but not 30.7b cells, which had low levels of Erks [19], suggesting that Erks activation is involved in NFκB activation by arsenic. This notion was further supported by data that specific inhibition of arsenic-induced Erks activation by PD98059 dramatically inhibited NFκB activation induced by arsenic. Furthermore, it was also found that over-expression of dominant negative mutant JNK_1 also inhibited arsenic-induced NFκB activation, indicating that JNKs activation is also required for arsenic-induced NFκB activation. Because blockade of activation of either Erks or JNKs led to inhibition of arsenic-induced NFκB activation, we speculate that there may be cross-talk between Erks and JNKs in the JB6 cells response to arsenic stimulation, and that this cross-talk is required for NFκB activation.

NFκB has attracted widespread attention among researchers in many fields in the last decade [26, 27]. NFκB consists of two major subunits, p50 and p65 (Rel-A) [26, 27]. It exists in cells as an inactive cytoplasmic precursor by forming a complex with the IκB, an inhibitor for NFκB [26, 27]. Signaling leading to IκB phosphorylation triggers the release of NFκB from IκB, resulting in the activation and translocation of NFκB from the cytoplasm to nucleus where NFκB binds to the promoter region of its specific targeting genes [26, 27]. There is increasing evidence that activation of NFκB is a part of mechanisms involved in tumor promotion that has been demonstrated in different cell models by different groups [28–30]. NFκB was first considered to be a mediator of tumor promotion because of its ability to alter gene expression in response to tumor promoters and oncogenes, including bacterial lipopolysaccharide, TPA, TNF-α, UV radiation, metals, reactive oxygen species and Her2/Neu [18, 26, 30–32]. Two members of the NFκB family, v-rel and p52/lyt-10, and the IκB family member Bcl-3 are potentially oncogenic [26]. H-Ras and Raf-1 can also activate NFκB [33, 34]. Overexpression of the NFκB inhibitor, IκBa, blocked the ability of oncogenic Ras alleles to induce focus formation in 3T3 cells [35]. It was reported that v-Rel expressed in transgenic mice under T cell specific control caused early death of animals from multicentric aggressive T cell lymphomas, and that v-Rel was predominantly associated with p50 and partially with IκBα [30]. Further, it was found that crossing v-Rel transgenic mice with mice overexpressing IκBα caused a delay in death from leukemia [30]. In light of the important role of NFκB activation in tumor development, we investigated the signal transduction pathways leading to NFκB activation by arsenic in mouse the epidermal cell line, JB6 C141 cells. We found that both arsenite and arsenate induced activation of NFκB in JB6 cells, revealing that arsenic-induce NFκB activation may play a role in arsenic-NFκB induced carcinogenesis. In addition, results in this study show that arsenic-induced activation was inhibited by pre-treatment of cells with PD98059 or overexpression of dominant negative mutant JNK_1, suggesting chemical agents or molecular approaches that target MAP kinase-NFκB pathways may block the carcinogenic effects of arsenic.

References

1. Knowles FC, Benson AA: The enzyme inhibitory form of inorganic arsenic. Z Gesamte Hyg 30: 625–626, 1984
2. Bettley FR, O'Shea JA: The absorption of arsenic and its relation to carcinoma. Br J Dermatol 92: 563–568, 1975
3. Landolph JR: Molecular mechanisms of transformation of C3/H10T1/2 C1 8 mouse embryo cells and diploid human fibroblasts by carcinogenic metal compounds. Environ Health Perspect 102(suppl 3): 119–125, 1994
4. Evans S: Arsenic and cancer. Br J Dermatol 97(suppl 15): 13, 1977
5. Goyer RA: Toxic effects of metals. In: C.D. Klaassen, M.O. Amdur, J. Doull (eds). Toxicology. The Basic Science of Poisons. Macmillan Publishing Company, New York, 1986, pp 582–635
6. Chen C-J, Wang C-J: Ecological correlation between arsenic level in well water and age-adjusted mortality from malignant neoplasms. Cancer Res 50: 5470–5474, 1990
7. Chan PC, Huff J: Arsenic carcinogenesis in animals and in humans: Mechanistic, experimental, and epidemiological evidence. Environ Carcinogen Ecotoxicol Rev C15: 83–122, 1997
8. National Institute for Occupational Safety and Health: National Occupational Exposure Survey (1980–1983). Department of Health and Human Services, Cincinnati, OH, 1984
9. Lansdown ABG: Physiological and toxicological changes in the skin resulting from the action and interaction of metal ions. Crit Rev Toxicol 25: 397–462, 1995
10. Luchtrath H: The consequences of chronic arsenic poisoning among Moselle wine growers. Pathoanatomical investigations of post-mortem examinations performed between 1960 and 1977. J Cancer Res Clin Oncol 105: 173–182, 1983
11. Philipp R: Arsenic exposure: Health effects and the risk of cancer. Rev Environ Health 5: 27–57, 1985
12. Snow ET: Metal carcinogenesis: mechanistic implications. Pharmac Ther 53: 31–65, 1992
13. Huang C, Ma W-Y, Li J, Goranson A, Dong Z: Requirement of Erks, but not JNKs, for arsenite-induced cell transformation. J Biol Chem 274: 14595–14601, 1999
14. Cavigelli M, Li WW, Lin A, Su B, Yoshioka K, Karin M: The tumor promoter arsenite stimulates AP-1 activity by inhibiting a JNK phosphatase. EMBO J 15: 6269–6279, 1996
15. Ludwig S, Hoffmeyer A, Goebeler M, Kilian K, Häfner H, Neufeld B, Han J, Rapp UR: The stress inducer arsenite activates mitogen-activated protein kinases extracellular signal-regulated kinases 1 and 2 via a MAPK kinase 6/p38-dependent pathway. J Biol Chem 273: 1917–1922, 1998
16. Huang C, Ma WY, Li J, Dong Z: Arsenic induces apoptosis through a c-Jun NH_2-terminal kinase-dependent, p53-independent pathway. Cancer Res 59: 3053–3058, 1999
17. Huang C, Ma W-Y, Bowden GT, Dong Z: Ultraviolet B-induced activated protein-1 activation does not require epidermal growth factor receptor but is blocked by a dominant negative PKCλ/ι. J Biol Chem 49: 31262–31268, 1996
18. Huang C, Chen N, Ma W-Y, Dong Z: Vanadium induces AP-1 and NFκB-dependent transcription activity. Int J Oncol 13: 711–715.
19. Huang C, Ma W-Y, Young MR, Colburn N, Dong Z: Shortage of mitogen-activated protein kinase is responsible for resistance to AP-1 transactivation and transformation in mouse JB6 cells. Proc Natl Acad Sci USA 95: 156–161, 1998
20. Dong Z, Birrer MJ, Watts RG, Matrisian LM, Colburn NH: Blocking tumor promoter induced AP-1 activity inhibits induced transformation in JB6 cells. Proc Natl Acad Sci USA 91: 609–613, 1994
21. Bernstein LR, Colburn NH: AP-1/jun function is differentially induced in promotion-sensitive and resistant JB6 cells. Science 244: 566–569, 1989
22. Watts RG, Huang C, Young MR, Li JJ, Dong Z, Pennie WD, Colburn NH: Expression of dominant negative Erk2 inhibits AP-1 transactivation and neoplastic transformation. Oncogene 17: 3493–3498, 1998
23. Huang C, Li J, Ma W-Y, Dong, Z: JNKs activation is required for JB6 cell transformation induced by TNF-α but not by TPA. J Biol Chem 274: 29672–29676, 1999
24. Huang, C, Li J, Ma W-Y, Chen N-Y, Dong Z: Activation of AP-1 by arsenate and arsenate in both cell and AP-1 luciferase-transgenic mice. FASEB J: 2000 (submitted)
25. Chen N, Ma W-Y, Huang C, Ding M, Dong Z: Activation of PKC is required for arsenate-induced signal transduction. J Environ Pathol Toxicol Oncol 19: 297–305, 2000
26. Baldwin A: The NF-κB and IκB proteins: New discoveries and insights. Annu Rev Immunol 14: 649–683, 1996
27. Ghosh S, May MJ, Kopp EB: NF-κB and Rel proteins: Evolutionarily conserved mediators of immune responses. Annu Rev Immunol 16: 225–260, 1998
28. Li JJ, Westergaard C, Ghosh P, Colburn NH: Inhibitors of both nuclear factor-κB and activator protein-1 activation block the neoplastic transformation response. Cancer Res 57: 3569–3576, 1997
29. Gilmore TD, Koedood M, Piffat KA, White DW: Rel/NF-κB/κB proteins and cancer. Oncogene 13: 1367–1378, 1996
30. Baeuerle PA, Baltimore D: NF-κB: Ten years after. Cell 87: 13–20, 1996
31. Verma IM, Stevenson JK, Schwarz EM, Van Antwerp D, Miyamoto S: Rel/NF-κB/IκB family: Intimate tales of association and dissociation. Genes Dev 9: 2723–2735, 1995
32. Galang CK, Garcia-Ramirez J, Solski PA, Westwick JK, Der CJ, Neznanov NN, Oshima RG, Hauser CA: Oncogenic Neu/ErbB-2 increases Ets, AP-1, and NF-κB-dependent gene expression, and inhibiting Ets activation blocks Neu-mediated cellular transformation. J Biol Chem 271: 7992–7998, 1996
33. Finco TS, Baldwin AS Jr: Kappa B site-dependent induction of gene expression by diverse inducers of nuclear factor κB requires Raf-1. J Biol Chem 268: 17676–17679, 1993
34. Bruder JT, Heidecker G, Rapp UR: Serum-, TPA-, and Ras-induced expression from Ap-1/Ets-driven promoters requires Raf-1 kinase. Genes Dev 6: 545–556, 1992
35. Finco TS, Westwick JK, Norris JL, Beg AA, Der CJ, Baldwin AS Jr: Oncogenic Ha-Ras-induced signaling activates NF-κB transcriptional activity, which is required for cellular transformation. J Biol Chem 272: 24113–24116, 1997
36. Simeonova PP, Luster MI: Mechanisms of arsenic carcinogenicity: genetic or epigenetic mechanisms? J Environ Pathol Toxicol Oncol 19: 281–286, 2000
37. Simeonova PP, Wang S, Toriuma W, Kommineni V, Matheson J, Unimye N, Kayama F, Harki D, Ding M, Vallyathan V, Luster MI: Arsenic mediates cell proliferation and gene expression in the bladder epithelium: Association with activating protein-1 transactivation. Cancer Res 60: 3445–3453, 2000

Molecular and Cellular Biochemistry **222**: 35–40, 2001.

Lead-related effects on rat fibroblasts

Ivo Iavicoli,[1] Alessandro Sgambato,[2] Giovanni Carelli,[1] Raffaele Ardito,[2] Achille Cittadini[2] and Nicolò Castellino[1]

[1]*Institute of Occupational Health, School of Medicine, Università Cattolica del Sacro Cuore, Rome;* [2]*Centro di Ricerche Oncologiche 'Giovanni XXIII', Institute of General Pathology, School of Medicine, Università Cattolica del Sacro Cuore, Rome, Italy*

Abstract

Lead (Pb) is an environmental toxicant that can induce structural and functional abnormalities of multiple organ systems, including the central nervous and the immune systems. The aim of this study was to evaluate the effects of extracellular Pb supplementation on the cellular content of the metal and on the proliferation and the survival of normal rat fibroblasts.

We found that the concentration of Pb in the culture medium was 0.060 μM and the normal Pb concentration in rat fibroblasts was 3.1 ± 0.1 ng/10^7 cells. Then we exposed the cells to increasing concentration of Pb (as Pb acetate) from 0.078–320 μM. We observed a dose-dependent inhibition of cell proliferation after 48 h, which was already apparent at a concentration of 0.312 μM ($p = 0.122$) and became statistically significant for concentration higher than 0.625 μM ($p = 0.0003$ at 5 μM). Cell proliferation was completely compromised at 320 μM Pb total inhibition of cell proliferation.

To investigate the mechanisms of Pb-mediated inhibition of cell proliferation, we evaluated the occurrence of apoptosis in the same cells and found that cytosolic DNA fragments, hallmark of apoptotic cell death, increased significantly at Pb concentrations from 2.5–10.0 μM. The occurrence of apoptosis was also confirmed by FACS analysis which showed the appearance of a subdiploid peak at Pb concentrations from 5–20 μM. The distribution of cells in the cell cycle showed a dose-dependent accumulation of cells in the G_0/G_1 phase mainly compensated by a decrease in the percentage of cells in the S phase. In conclusion, our results demonstrate that induction of apoptosis contributes to the Pb-induced inhibition of cell proliferation in rat fibroblasts. (Mol Cell Biochem **222**: 35–40, 2001)

Key words: lead, rat fibroblasts, cell proliferation, apoptosis

Introduction

In the 1960s, radioactive tracers (^{203}Pb and ^{210}Pb) were used to study *in vivo* the biological distribution of lead (Pb, as Pb^{2+}). Results demonstrated that in rats this metal rapidly diffuses into extracellular spaces and penetrates the cell membrane [1]. The amount of Pb content in the subcellular fraction varies with the kind of tissue examined. However, nuclei (kidney) and mitochondria (liver) contain the highest percentage Pb in cells [1–3]. Pb in the cytosol seems to be in equilibrium both with Pb present in the intracellular organelles and also with Pb in the extracellular fluids. In fact, this metal is present in all the various subcellular structures that can be obtained by ultracentrifugation of tissue homogenate like liver, kidney, spleen, heart and brain [1–3]. The metal is known to form complexes with various nucleophilic functional groups (COOH, NH_2 and SH) whose stability progressively increases and it reaches a maximum for those containing S (SH). Pb is the cation that forms the most stable complexes with the sulphydryl groups (SH) found in dithiol compounds (S-S). Interaction with the latter involves the initial breaking down of the S-S bridge and the formation of the S-Pb complex after which S-Pb-S is formed. Therefore, in theory, an unlimited number of cell functions may be affected [4–7]. It is generally accepted that Pb is toxic at a protoplasmatic level [8, 9].

Address for offprints: I. Iavicoli, Università Cattolica del Sacro Cuore, Istituto di Medicina del Lavoro, Centro di Igiene Industriale, Largo Francesco Vito, 1-00168 Roma, Italy

Although scientists discovered centuries ago the harmful effects on health related to occupational and non-occupational Pb exposure, only in the last 10 years they have focused epidemiological and experimental studies on the effects of low doses of this metal [10]. This new trend has come about partly on account of a reduction in Pb exposure following the application of international laws on risk prevention in the workplace and the general environment, and partly due to the evidence that low dose Pb exposure can have serious effects on children, damaging in particular the central nervous system [10–13].

Both *in vivo* and *in vitro* experiments can be conducted to investigate the mechanisms underlying the toxic effects of Pb exposure. However, *in vitro* models have considerable advantages over *in vivo* studies since they are relatively easy to perform, can be repeated at reasonable costs and offer the possibility of obtaining rapid results, even when experimental conditions vary.

The aim of the present study was the evaluation of the effects of Pb supplementation to the medium of normal rat fibroblasts at doses ranging from 0.078–320 μM (Pb levels progressively increased by a factor of 2) by means of: (1) The observation of cell proliferation and survival after 48 h incubation of normal rat fibroblasts; (2) The evaluation of apoptosis after exposure to the aforementioned doses; (3) A comparison of the effects obtained in cells exposed to the aforementioned doses with those grown in normal medium without Pb supplementation.

Materials and methods

Analytical methods

Cellular Pb analysis

Parallel exponentially growing cultures of Rat-1 fibroblasts were rinsed twice with Eagle's Minimum Essential Medium (EMEM) (Gibco, Merelbeke, Belgium) and exposed to Pb supplemented EMEM plus 10% (v/v) FBS. After 48 h exposure, cells were scraped, and the Pb content of cell pellets was determined. Concomitantly, parallel treated cells were counted in duplicate.

Double deionized water, obtained from two deionizers connected in series (Nuovo RD 30 Elettracqua, Italy and Milli-Q, Millipore, USA), was used throughout this study.

Cell pellets (10^7 cells) were resuspended with the aid of 3 aliquots of 200 μl water in 3 ml teflon vessels (CEM Corporation, Matthews, NC, USA) and evaporated to dryness on a heated plate at 150°C. Blanks were obtained by evaporating the reagents used for the preparation of pellets. After cooling, 50 ml HNO_3 (Aristar, BDH, UK) and 10 μl H_2O_2 30% (w/w) (Farmitalia, Carlo Erba S.p.A., Milano, Italy) were added into the vessels which were closed with a torque wrench and placed into 120 ml teflon vessels. Digestion was carried out in a MDS – microwave oven mod. 2000 (CEM Corporation, Matthews, NC, USA) under the conditions described in Table 1. After digestion and cooling, the vessels were opened and 200 μl deionized water was added to the residue.

Pb standard stock solution (Plasma Chem Corp, USA) contained 1000 mg Pb /l as $Pb(NO_3)_2$ in 2% (v/v) HNO_3 and working standards at 20, 40 and 60 μg/l were obtained by serial dilutions with 0.2% (v/v) HNO_3 (69–70% Baker Instra, Pb content lower than 0.5 μg/l). The determination of Pb was performed with the PerkinElmer Model 4100 ZL atomic absorption spectrophotometer equipped with longitudinal Zeeman-effect background correction and transversely heathed atomizer. A PerkinElmer Lumina™ hollow cathode lamp was operated at 12 mA current. The primary Pb resonance line at 283.3 nm was separated with a low spectral bandpass of 0.7 nm. The read delay and the baseline offset correction were set to 0 and 3 sec respectively. Samples were automatically injected in triplicate into the graphite tube using the PerkinElmer Model AS-70 autosampler. When necessary the sample was diluted accordingly; calculations were carried out with the external standard method on the basis of peak area measurements. Recoveries were measured by diluting samples with 0.2% (v/v) HNO_3 or with the Pb standard solution at 20 μg/l. The recovery of added Pb ranged from 91.0 to 109.0, mean ± S.D. = 97.4 ± 5.0, (n = 9).

Aliquots of each 10 μl of sample and matrix modifier were analyzed. The matrix modifier aliquot contained 0.050 mg $NH_4H_2PO_4$ and 0.003 mg $Mg(NO_3)_2$, and was added to standards, samples and blank. Pyrolysis was carried out at 850°C and the atomization temperature of 1500°C was chosen as it offered the best compromise between atomic peak shape and analytical sensitivity. Instrumental conditions are described in Table 2.

Pb analysis in EMEM and phosphate buffer saline (PBS)

Analysis of EMEM Pb was carried out using a method similar to that adopted for the determination of Pb in cells. However, instead of evaporating cell suspensions, aliquots of 0.5 ml of EMEM or PBS were dried and digested after addition of 0.2 ml of HNO_3 to the residue, according to the procedure described above. After completion of digestion, 0.3 ml

Table 1. Microwave programme variables

Parameter	Stage 1	Stage 2
Power	100%	100%
Pressure	35 psi	60 psi
Run time	5.00 min	15.00 min
Time at p	5.00 min	15.00 min
Temperature	90°C	140°C
Fan speed	100%	100%

Table 2. Graphite furnace programme

Step	Temp (°C)	Ramp (sec)	Hold (sec)	Gas flow (ml/min)	Read	Gas type
1	130	1	30	250		Argon
2	150	5	20	250		Argon
3	850	5	20	250		Argon
4	1500	0	4	0	*	Argon
5	2400	1	3	250		Argon

of water were added to the samples. Pb recovery was evaluated as described above for cellular Pb analysis.

Pb release from plastic dishes

Pb was determined after filling the culture polystyrene dishes with 5% (v/v) HNO_3 at different contact times up to 72 h. Analytical determination was carried out using the instrumental parameters described in the subsection 'Pb cell analysis', the only variation being that the samples were determined calibrating against Pb standards prepared in 5%(v/v) HNO_3.

Cell culture

The Rat-1 normal rat fibroblasts were grown and maintained in EMEM (Gibco, Merelbeke, Belgium) supplemented with 10% heat inactivated fetal bovine serum (FBS), and were cultured at 37°C in a 5% CO_2 atmosphere. Pb acetate trihydrate, $Pb(CH_3COO)_2\ 3H_2O$, was purchased from Merck, (Darmstadt, Germany) and the 10 mM stock solution was prepared by dissolving the salt in water and filtering through a Millex-GV 0.22-μm syringe filter (Millipore, Bedford, MA, USA).

Cell proliferation assay

Cells were plated in triplicate in 24-well plates at a density of 2×10^4 cells per well and incubated 24–36 h. After incubation the cells were rinsed twice with EMEM plus 10% (v/v) FBS supplemented with different concentrations of Pb. After 48 h-incubations, the medium was removed and cultures were rinsed twice with EMEM and then incubated with medium containing 1 mg/ml 3-(4,5-dimethylthiazol-2-yl)-2,5-diphenyl-tetrazolium bromide (MTT, Sigma, Milan, Italy) for 2 h at 37°C. The medium was then aspirated and 500 μl 0.04 N HCl in isopropanol was added to each well in order to stop the enzymatic reaction. After 1 h 50 μl of solution from each well were transferred in each well of a 96-well plate and the level of the coloured formazan derivative was determined on a multiscan reader at the wavelength of 540 nm (reference wavelength 630 nm).

Detection of DNA fragmentation

Cytosolic DNA fragments were quantified with a cell death detection ELISA assay (Boehringer Mannheim GmbH, Germany). Briefly, cells were plated in triplicate in 24-well plates at a density of 2×10^4 cells per well and were exposed to different concentrations of Pb for 48 h, as previously described. Floating and attached cells were then harvested and assayed with a quantitative sandwich-enzyme-immunoassay using mouse monoclonal antibodies directed against DNA and histones. This allows the specific determination of mono- and oligonucleosomes in the cytoplasmatic fraction of lysates. The occurrence of apoptosis in the cultures is expressed as enrichment of nucleosomes in the cytoplasm of treated cells compared with control cells using the following formula:

$$\frac{\text{absorbance of the sample}}{\text{absorbance of the corresponding control}} = \text{enrichment factor}$$

Determination of DNA content by FACS analysis

Cells were plated in duplicate in 6-cm dishes at a density of 5×10^5 cells per dish and incubated 24–36 h. They were then rinsed twice with EMEM and grown in EMEM plus 10 % (v/v) FBS supplemented with different concentrations of Pb for 48 h. Cells were then trypsinized, collected and washed twice with PBS. Cell pellets were resuspended in 1 ml PBS, fixed in 5 ml of 70% (v/v) ethanol and stored at 4°C. For the analysis, cells were collected by centrifugation and the pellets were resuspended in 0.2 mg/ml of propidium iodide in Hank's balanced salt solution (HBSS) containing 0.6% (w/v) NP-40. RNAase A (1 mg/ml) was added and the suspension was incubated in the dark at room temperature for 30 min. The cell suspension was then filtered and analyzed for DNA content on a Coulter EPICS 753 flow cytometer. The percent of cells in different phases of the cell cycle and the subdiploid peak (< 2N DNA content), which indicates apoptotic cells, were determined using the Multicycle software version 2.53.

Cell viability assay

Cell viability assay was evaluated on parallel treated cells, for each of the above-described assays by the Trypan blue exclusion test. Briefly, cells were detached with trypsin-EDTA (Gibco), rinsed with PBS and 0.04% (w/v) trypan blue dye was added. Within 5 min of incubation, the number of cells excluding dye was expressed as a percentage of total cells counted from five random squares of a Burger chamber.

Results

Pb concentration in the medium does not increase during cell culturing

EMEM medium supplemented with 10% FBS was analyzed fresh and after incubation in the absence of cells in cell culture dishes kept in the incubator for up to 168 h, as illustrated in Table 3. Pb was detected at the trace levels of 0.060 ± 0.006 μM (mean ± S.D.; n = 6) in fresh EMEM and no detectable increase in the amount of Pb was observed following a contact period up to 168 h, thus suggesting that no Pb is released from the plastic dishes used for cell growth. A similar concentration of Pb was also observed in the PBS buffer used for cell culturing (data not shown).

Effect of extracellular Pb supplementation on the cellular Pb level

We evaluated cellular Pb uptake following exposure of cultures to increasing amounts of extracellular Pb for 48 h. As shown in Fig. 1 cellular Pb level was affected by changes in extracellular Pb content (from 0 up to 10 μM) and this effect was evident at all exposure levels.

Extracellular Pb supplementation and cell proliferation

The antiproliferative activity of extracellular Pb supplementation was determined by exposing exponentially growing cultures of Rat-1 cells to increasing concentrations of Pb: 0.00, 0.078, 0.156, 0.312, 0.625, 1.25, 2.5, 5.0, 10.0, 20.0, 40.0, 80.0, 160 and 320 μM. Cell survival was assessed after 48 h incubation by the MTT test and viability was quantified on parallel cultures by the trypan blue dye exclusion test. The antiproliferative activity of Pb which was already visible at a concentration of 0.312 μM, was statistically significant at concentrations of 0.625 μM ($p = 0.014$) and over ($p < 0.0001$ at 20 μM). As shown in Fig. 2, a dose-dependent decrease in cell number was observed with a 50% inhibitory concentration (IC_{50}) of 62.1 ± 3.7 μM (mean ± S.D., range 59–64, n = 6). It is of interest that exposure to these conditions did not produce a dramatic increase in the percentage of cells capable of excluding dye (data not shown) thus suggesting that, rather than inducing cell death via necrosis, Pb supplementation mainly inhibits cell proliferation.

Table 3. Pb levels in EMEM at different culture times

Time (h)	Pb (μM) mean ± S.D. (n = 3)	p*
0	0.060 ± 0.005	
24	0.062 ± 0.006	0.616
48	0.054 ± 0.005	0.216
72	0.065 ± 0.005	0.288
120	0.058 ± 0.006	0.718
168	0.055 ± 0.004	0.248

*No statistical differences were observed in the Pb levels between 0 h and all the intervals indicated up to 168 h.

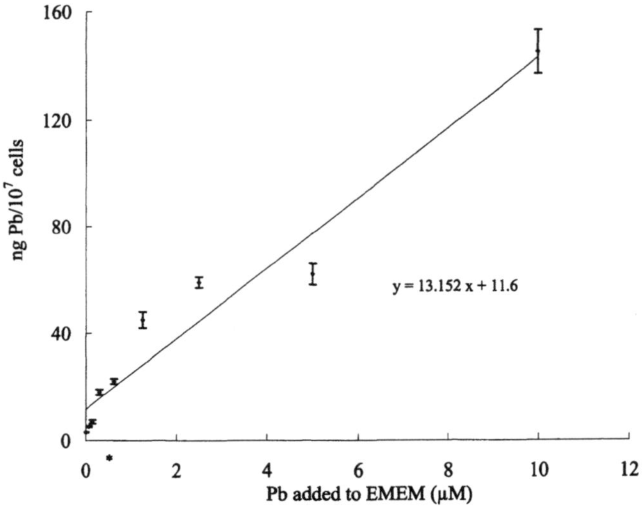

Fig. 1. Pb cellular uptake after exposure (48 h) to increasing amounts of extracellular Pb.

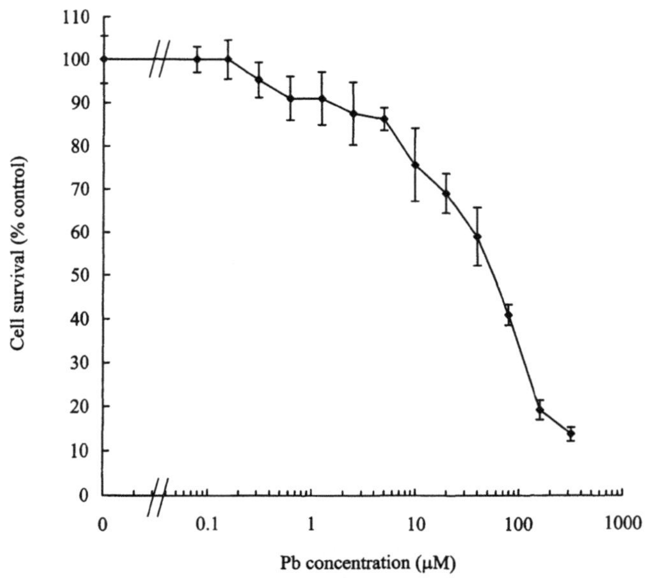

Fig. 2. Effects of extracellular lead supplementation on the growth of normal rat fibroblasts. Exponentially growing cultures of Rat-1 cells were incubated in media supplemented with the indicated amount of Pb and cell number was determined in triplicate after 48 h by the MMT test. The results are shown as % of the control. The data reported are the mean ± S.D. (n = 6).

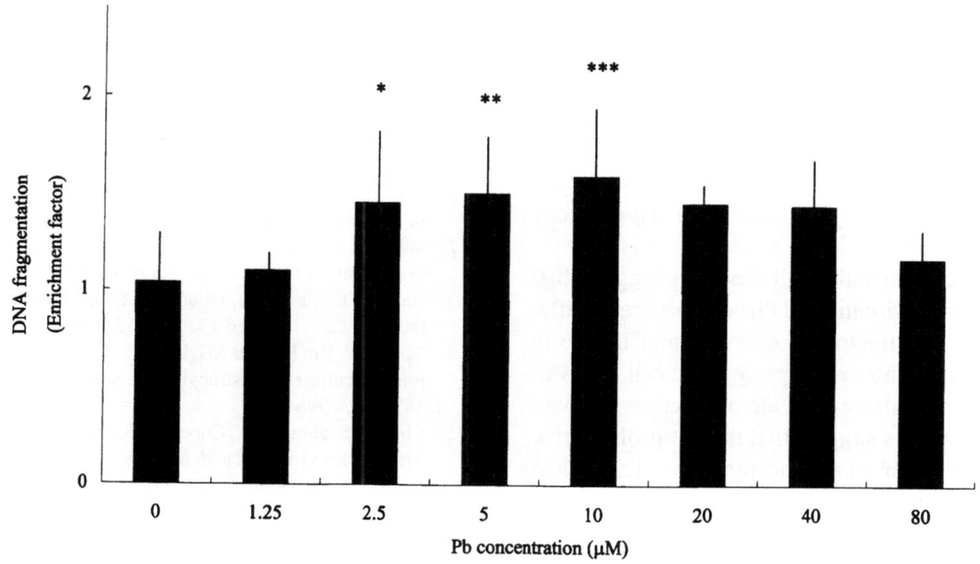

Fig. 3. Dose dependency of the Pb-induced apoptosis in Rat-1 cells. The cells were cultured for 48 h in the presence or absence (control) of increasing concentration of Pb, and then the quantitation of apoptosis was performed. The values represent enrichment of nucleosomes in the cytoplasm of cells exposed to the indicated concentration of Pb for 48 h, as compared to control (mean ± S.D., n = 6). *Indicates significant differences as compared to the control.

Extracellular Pb supplementation and apoptosis

Exponentially growing cultures were exposed to increasing concentrations of Pb for 48 h and apoptosis was evaluated by quantifying cytosolic DNA fragments. As shown in Fig. 3, a significant increase in cytosolic DNA fragments ($p < 0.05$) was observed at Pb concentrations ranging from 2.5–10.0 μM. Further increases in extracellular Pb concentration did not augment apoptosis.

To further investigate the antiproliferative effects of Pb supplementation on rat fibroblasts, we evaluated the distribution in the cell cycle and the occurrence of a subdiploid peak in Pb-exposed cells. Exponentially growing cultures of Rat-1 fibroblasts were exposed to increasing concentrations of Pb and the DNA content was analyzed after 48 h in both treated and parallel untreated cultures. As shown in Table 4, Pb supplementation caused a dose-dependent accumulation of cells in the G_0/G_1 phase of the cell cycle compensated by a decrease in the percentage of cells in the S phase. Moreover, FACS analysis confirmed that apoptosis occurred more frequently for Pb concentrations ranging from 5–20 μM and its occurrence decreased thereafter (Table 4).

Discussion

In vitro models are suitable for the study of toxic effects of Pb exposure because, in comparison with *in vivo* experiments, they are relatively easy to perform, can be repeated at acceptable costs and offer the advantage of obtaining rapid results, even under different experimental conditions.

This study investigated the effects of extracellular Pb supplementation on the cell content of the metal and on the proliferation and survival of normal rat fibroblasts.

We found that Pb is normally present in the medium and PBS buffer used for cell culturing in trace amounts (0.060 μM) and in non-exposed cells (3.1 ng/10^7 cells). This background Pb level must be taken into consideration particularly at low Pb supplementation, a condition where the endogenous Pb in the medium and added Pb are of the same order of magnitude.

In order to reduce this Pb background we reproduced a medium containing the lowest possible Pb level by using ultrapure components under normal laboratory environmental conditions. This attempt was not however successful since the resulting medium contained approximately the same Pb

Table 4. Effects of Pb supplementation on cell cycle distribution and apoptosis in Rat-1 fibroblasts*

Pb (μM)	G_0/G_1	S	G_2/M	Sub G_1
0	74.6	15.9	9.4	1.2
5	78.9	8.6	12.6	11.3
20	83.7	4.9	11.4	21.6
80	85.5	4.3	9.8	5.1

*Flow cytometry analyses of lead exposed cells. Parallel exponentially growing cultures of Rat-1 cells were incubated with the indicated concentration of lead for 48 h before analysis. The values represent the percentage of the total cell population in each phase of the cell cycle. The Sub G1 peak corresponds to the percent of apoptotic cells. All assays were performed in triplicate, and all experiments were repeated at least twice. The data reported are the results of a typical experiment. Similar results were obtained in replicate experiments.

concentration of the EMEM. We are therefore aware that each of our experimental findings has been evaluated on the basis of a control value that cannot considered a zero, even if the Pb level was very low. Consequently it seems feasible to claim a No Observed Adverse Effect Level (NOAEL) at concentrations ranging from 0.078–0.312 μM, but we cannot affirm this finding with certainty on account of the presence of a very low background Pb level.

No release occurred from culture dishes. This suggests that no contamination from environmental Pb takes place even after long exposure of the medium to the environment. Changes in the extracellular Pb concentration rapidly affect cell Pb level. Although Pb uptake in culture cells already occurs at lower concentrations, our results suggest that the antiproliferative activity of Pb is only apparent at concentrations greater than 0.312 μM and is statistically significant at concentrations of 0.625 μM ($p = 0.014$) and over. These results enable us to identify a NOAEL, in terms of proliferation inhibition, at extracellular Pb levels of approximately 0.3 μM.

The mechanism assessment of Pb-related inhibition of cell growth is important for our understanding of the final effects of low dose Pb toxicity on biological systems. To our knowledge, this is the first study to investigate the effects of Pb supplementation on cell cycle distribution in normal fibroblasts. Our results suggest that Pb-induced inhibition of cell growth is probably due to a direct inhibition of cell proliferation rather than to a direct cytotoxic effect of the metal. In fact, we found that Pb supplementation causes a dose-dependent accumulation of cells in the G_0/G_1 phase of the cell cycle with subsequent reduction of cells in the S phase. Moreover, Pb exposure from 2.5–10 μM was associated with apoptotic cell death (up to 20%). Apoptosis decreased at higher Pb concentrations, suggesting that at high doses, Pb completely inhibits its enzymatic activities, including the ones needed for the induction of apoptotic cell death.

References

1. Castellino N, Aloj S: Intracellular distribution of lead in the liver and in the kidney of the rat. Br J Ind Med 26: 139–143, 1969
2. Sabbioni E, Marafante E: Identification of lead-binding components in rat liver: *In vivo* study. Chem Biol Interact 15: 1–20, 1976
3. Barltrop D, Barret AJ, Dingle JT: Subcellular distribution of lead in the rat. J Lab Clin Med 77: 705-712, 1971
4. Tang HW, Yan HL, Hu XH, Liang YX, Shen XY: Lead cytotoxicity in primary cultured rat astrocytes and Schwann cells. J Appl Toxicol 16: 187–196, 1996
5. Chetty CS, Stewart TC, Cooper A, Rajanna B, Rajanna S: *In vitro* interaction of heavy metals with ouabain receptors in rat brain microsomes. Drug Chem Toxicol 16: 101–110, 1993
6. Dabrowska Bouta B, Struzynska L, Rafalowska U: Effect of acute and chronic lead exposure on the level of sulphydryl groups in rat brain. Acta Neurobiol Exp 56: 233–236, 1996
7. Sakai T: Biomarkers of lead exposure. Ind Health 38: 127–142, 2000
8. Quinlan GJ, Halliwell B, Moorehous CP, Gutteridge JMC: Action of lead (II) and aluminium (III) ions on iron-stimulated lipid peroxidation in liposomes, erythrocytes and rat liver microsomal fractions. Biochim Biophys Acta 962: 196–200, 1988
9. Raghanan SR, Culver BD, Gonick HC: Erythrocyte lead binding protein after occupational exposure. II. Influence on lead inhibition of membrane Na^+K^+ adenosine triphosphate. J Toxicol Environ Health 7: 561–568, 1981
10. Castellino N, Castellino P, Sannolo N: In: Inorganic Lead Exposure. Lewis Publishers, Boca Raton, 1995, 516 pp
11. Silbergeld EK: Toward the twenty-first century: Lessons from lead and lessons yet to learn. Environ Health Perspect 86: 191–196, 1990
12. Stiles KM, Bellinger DC: Neuropsychological correlates of low-level lead exposure in school-age children: A prospective study. Neurotoxicol Teratol 15: 27–35, 1993
13. Winneke G: Endpoints of developmental neurotoxicity in environmentally exposed children. Toxicol Lett 77: 127–136, 1995

Molecular and Cellular Biochemistry **222:** 41–47, 2001.

In vivo reduction of chromium (VI) and its related free radical generation

Ke Jian Liu[1] and Xianglin Shi[2]

[1]*College of Pharmacy, University of New Mexico, Albuquerque, NM;* [2]*Pathology and Physiology Research Branch, National Institute For Occupational Safety and Health, Morgantown, WV, USA*

Abstract

Chromium (VI) compounds are widely recognized as human carcinogens. Extensive studies *in vitro* and in model systems indicate that the reactive intermediate, Cr (V), generated by cellular reduction of Cr (VI), is likely the candidate for the ultimate carcinogenic form of chromium compounds. Here we review our current understanding of the in vivo reduction of Cr (VI) and its related free radical generation. Our results demonstrate that Cr (V) is indeed generated from the reduction of Cr (VI) *in vivo*, and that Cr (V) thus formed can mediate the generation of free radicals. Cr (V) and its related free radicals are very likely to be involved in the mechanism of Cr (VI)-induced toxicity and carcinogenesis. These studies also illustrate that *in vivo* EPR spectroscopy and magnetic resonance imaging can be very useful and powerful tools for studying paramagnetic metal ions in chemical and biochemical reactions occurring in intact animals. (Mol Cell Biochem **222**: 41–47, 2001)

Key words: EPR, chromium, free radical, MRI, reactive oxygen species

Introduction

Chromium is a naturally occurring element found in rocks, animals, plants, soil, and in volcanic dust and gases. Chromium can exist in several oxidative forms, most commonly as Cr (0), Cr (III), and Cr (VI). Cr (III) occurs naturally in the environment and is an essential nutrient in humans. Chromate (Cr (VI)) compounds, widely used in industry, have been shown to have serious toxic and carcinogenic effects on humans. Intensive epidemiological evidence has been published on the high incidence of respiratory tract cancers in workers occupationally exposed to Cr (VI) compounds [1, 2]. The mechanism of Cr (VI)-induced toxicity and carcinogenesis is not well understood, although recent results suggest the involvement of transient chromium intermediate states, such as Cr (V) and Cr (IV), and their related free radicals formation.

Cr (VI) has been demonstrated to induce a variety of DNA lesions such as single-strand breaks, alkali-labile sites, and DNA protein cross-links [3, 4]. In contrast, most Cr (III) compounds, the final product in the reduction of Cr (VI), are relatively nontoxic, noncarcinogenic, and nonmutagenic [3, 5]. Since Cr (VI) does not react with isolated DNA, the reduction of Cr (VI) to lower oxidation states after entering the human system has been considered an important step in Cr (VI)-induced carcinogenesis [5, 6]. While several studies have shown that the chromium-induced DNA damage is strongly dependent on the formation of Cr (V) intermediates, recent studies have suggested that free radicals generated by Cr (V) may also play an important role [7–9].

In this paper, we will review our current understanding of the *in vivo* reduction of Cr (VI), based mostly on our previously published results. New experiments have since been conducted, and are also included here. The goal of these studies is to demonstrate that upon administration into animals, Cr (VI) is reduced to Cr (V) that can mediate the formation of free radicals. These processes are likely important steps in the mechanism of Cr (VI)-induced carcinogenesis.

Address for offprints: K.J. Liu, College of Pharmacy, University of New Mexico, Albuquerque, NM 87131, USA

Materials and methods

Animal preparation

BALB/c mice weighing 18–20 g were obtained from Charles River Laboratories (Wilminton, MA, USA). After the animals were anesthetized with ketamine/xylazine (100/10 mg/kg body wt, im), 100 μl of sodium dichromate or other chemical solutions were introduced into the animal either intravenously, intraperitoneally, or intratracheally. Electron magnetic resonance (EPR) or magnetic resonance imaging (MRI) measurements followed immediately.

For the measurement of the reduction of Cr (VI) on the skin, Wistar rats from Charles River Laboratories were used. The hair on the back of the anesthetized rat was shaved, and aqueous sodium dichromate solution was applied topically to the exposed skin. The EPR spectra were collected using an extended loop resonator positioned over the area of interest.

In vivo *EPR measurement of Cr (V) in living mice*

In vivo EPR measurement were obtained using a custom-designed EPR spectrometer equipped with a low frequency (1.2 GHz, L-band) microwave bridge and a cylinder-shaped loop gap resonator. After administration of sodium dichromate, the animal was immediately inserted into the resonator, which was placed between the poles of the magnet. The thorax cavity of the mouse was placed in the center of the resonator. For kinetics studies, the EPR spectra were collected every 60 sec to record the formation and decay of the Cr (V) signal. Typical settings for the spectrometer included magnetic field, 425 G; incident microwave power, 50 mW; modulation frequency, 27 kHz; modulation amplitude, 1.6 G.

Magnetic resonance imaging of Cr (V) in mice

The MRI measurements were undertaken using a S.M.I.S.7.0 Tesla MR imager. Anesthetized mice were placed inside the 6 cm bird cage coil, and the tail vein was cannulated so that chromium compounds could be administered without moving the animal out of the magnet. A series of pre- and post-injection T1 weighted gradient echo MR image were collected with TR/TE/α = 100 msec/7 msec/70°, FOV = 4 cm, a slice thickness of 2 mm and a pixel resolution of 312 μm. There were four averages, resulting in an acquisition time of 51 sec.

Results and discussion

Detection of in vivo *formation of Cr (V) intermediate species by EPR*

When a mouse was intravenously injected with sodium dichromate (Cr (VI)), the *in vivo* EPR spectrometer recorded a signal from the thorax cavity of the mouse a few minutes after the administration (Fig. 1). Based on the spectral characteristics of the EPR spectra, including g-value (g = 1.9798), linewidth, splitting, and satellite peaks of Cr^{53} (Fig. 1D), the EPR signal in Fig. 1A was assigned to CR (V) [10, 11]. Administration of CR (VI) to mice by intraperitoneal or intratracheal injection produced identical EPR signal (Figs 1B and 1C), albeit at reduced signal intensity, due to delayed uptake and redistribution process. Since Cr (VI) is EPR silent, the observation of Cr (V) signal indicates that Cr (VI) is rapidly reduced to Cr (V) upon administration.

The formation of Cr (V) through the reduction of Cr (VI) is supported by conventional EPR experiments using isolated tissues after treatment *in vivo* [12]. In agreement with our *in vivo* experiments, liver and kidney have been reported to be the organs with the highest chromium accumulation in treated rats [13], and therefore is the target organ of many studies. Rats were intraperitoneally injected with Cr (VI), the liver and kidney tissue was collected immediately after sacrificing the animal, and then the EPR signals were recorded at liquid nitrogen temperature using conventional X-band EPR spectrometer. Again, Cr (V) species was found in both liver and kidney (Fig. 2). Here both of the in vivo EPR (L-band, 1.2 GHz) and conventional EPR (X-band, 9.6 GHz) are utilized to their full advantage. The *in vivo* EPR provides real time, localized formation of Cr (V) in living animals, while the conventional EPR, with higher detection sensitivity, can afford much lower administered concentrations of Cr (VI) to the animal, as well as more detailed spectral information about the Cr (V) species. Combination of the results from both types of EPR measurements shows unequivocal evidence that Cr (V) is formed through the reduction of Cr (VI) *in vivo*.

In addition to oral ingestion and inhalation as the main routes for occupational and accidental exposure, skin represents another possible route for chromium to enter into humans and exert its carcinogenecity. Application of an aqueous dichromate solution to the skin of a living rat resulted in an EPR signal which was identical to that of Cr (V) observed *in vivo* [14]. Removing the stratum corneum increased the rates of both formation and decay of Cr (V) (Fig. 3), indicating its important role as a protective skin barrier. Together with result from administration of Cr (VI) through intravenous, in-

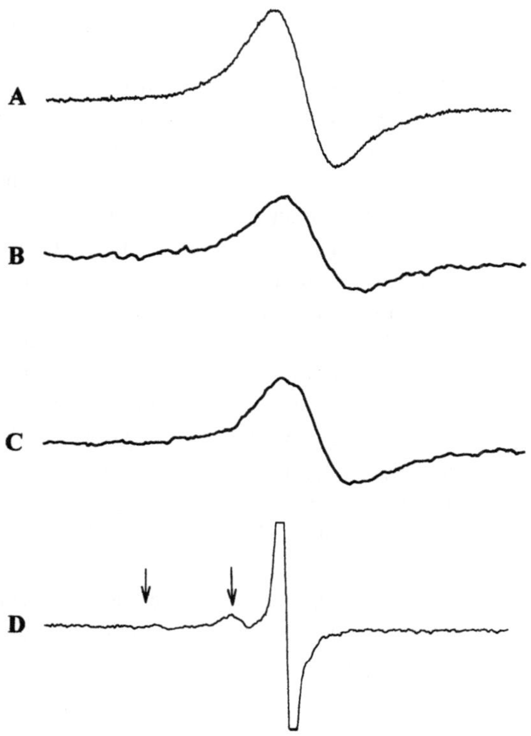

Fig. 1. *In vivo* EPR spectra of Cr (V) from a mouse. The L-band EPR spectra were obtained from mice 8 min after treatment with 100 μl of 100 mM sodium dichromate: (a) intravenous injection; (b) intraperitoneal injection; (c) intratracheal injection; and (d) 9.6 GHz X-band EPR spectrum of Cr (V) from the blood of a mouse. The blood sample was drawn from the tail vein of the same mouse as in (a). The arrows indicate Cr^{53} satellite peaks.

traperitoneal, and intratracheal injection, these studies lead to the conclusion that regardless of the route of administration, Cr (VI) will be reduced to Cr (V) once absorbed in the animal.

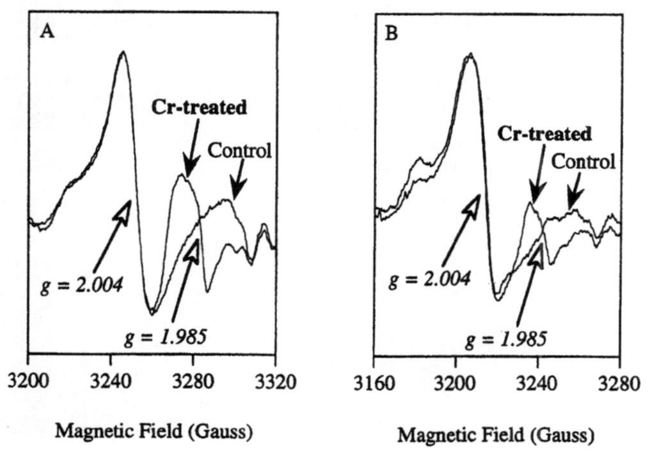

Fig. 2. X-band EPR spectra from liver and kidney of rats treated with Cr (VI). (A) Liver and (B) kidney was removed 30 min after intraperitoneal injection of sodium dichromate (10 mg/kg). Tissue Cr (V) was measured at 77 K in a finger Dewar fill with liquid nitrogen using a conventional X-band EPR spectrometer. From ref. [12] with permission.

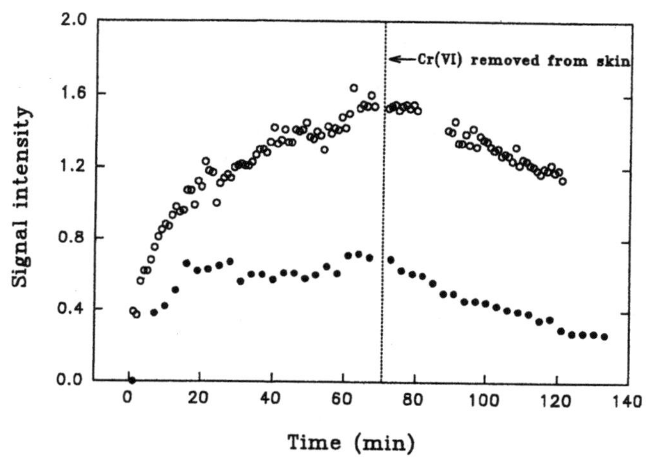

Fig. 3. The time course of the formation and decay of Cr (V) in the skin of a mouse after topical treatment with Cr (VI). Time zero is the time of application. (O) Skin was shaved and stripped 10 times with surgical tapes; (●) shaved skin without stripping. From ref. [14] with permission.

In all above studies, the intermediate species Cr (V) is relatively short lived, as shown in the case of skin (Fig. 3). The kinetics of the formation and decay of Cr (V) are well demonstrated in Fig. 4, where the signal intensity of Cr (V) reached a maximum approximately 10 min after intravenous injection of Cr (VI), and then decayed slowly with an observed lifetime of about 30 min. It is expected that Cr (V) will be further reduced to Cr (IV), and eventually to the stable Cr (III). Therefore, the time course in Fig. 4 represents the steady state concentration of Cr (V) over time, due to its continu-

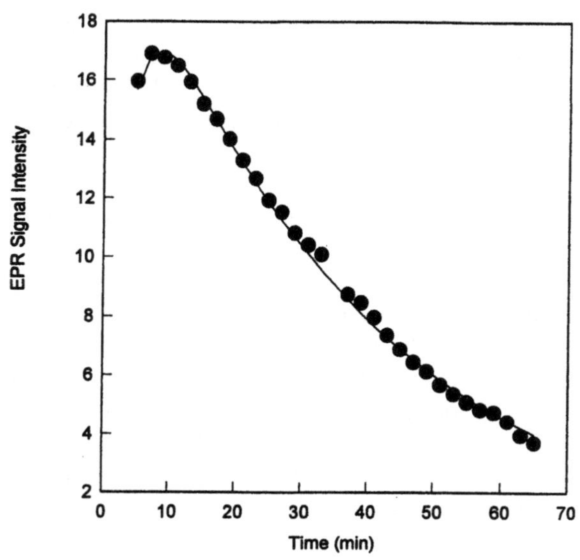

Fig. 4. The time course of the formation and decay of Cr (V). The EPR signal of Cr (V) was scanned every minute immediately following intravenous injection of 100 μl of 100 mM sodium dichromate into a mouse. Time zero is the time of injection. From ref. [10] with permission.

ous formation and decay. The actual lifetime of Cr (V) could be significantly shorter than shown in the figure. In order to estimate the real value, a Cr (V) model compound, K_3CrO_8, was synthesized [15, 16], and was intravenously injected to a mouse. The lifetime of the EPR signal of Cr (V) was found to be less than 1 min, suggesting that under in vivo conditions, Cr (V) is indeed very short lived [11].

Detection of in vivo *formation of Cr (V) and Cr (III) by MRI*

In spite of the transient nature of Cr (V), EPR is the technique of choice for characterizing the formation of Cr (V) species, and has been proved to be a valuable tool in chromium studies. On the other hand, the noninvasive magnetic resonance imaging (MRI) technique provides a much needed complimentary method on the *in vivo* localization and distribution of Cr (V). Because of their paramagnetic properties, Cr (V) and its final product Cr (III), can alter relaxation time of protons, hence Cr (V) and Cr (III) formed *in vivo* in the reduction of the non-paramagnetic Cr (VI) could act as contrast agents for proton MRI [17]. To test this idea, feasibility studies were carried out to record the MRI images of the liver and kidney of a mouse before and after administration of Cr (VI). High contrast was observed in the kidney cortex and in the liver (Fig. 5), indicating that Cr (VI) was rapidly reduced to Cr (V) and Cr (III) upon administration to a mouse, and that chromium accumulated at high concentrations in the liver and kidney. These results are consistent with, and further confirm, those findings obtained by EPR, suggesting that MRI could potentially be a very useful tool for chromium studies and produce certain information that are not obtainable by any existing methods. Experiments are currently underway to distinguish the contribution of contrast enhancement from each of the two chromium species, i.e. Cr (V) and Cr (III).

Effect of antioxidants and chelators on Cr (V) formation

Results presented above, and other evidence reported in the literature [18–25], indicate that the intermediate species Cr (V) is involved in the mechanism of chromium induced toxicity and carcinogenesis. Some of the approaches to attenuate the deleterious effects of Cr (V) include the treatment of the animal with agents that either prevent the reduction of Cr (VI) to Cr (V), or that would accelerate the reduction of Cr (VI) to the final product, Cr (III). Metal chelators are the first type of these agents [10, 23, 26, 27]. When the animals were pretreated with chelators EDTA, DETAPAC, or 1,10-phenanthroline before the administration of Cr (VI), a significant reduction in the yield of Cr (V) was observed (Fig. 6). The second type of agents are the antioxidants, or reductants, which can affect the reduction process of Cr (VI) metabolism [11, 20, 26–31]. Figure 7 shows the effect of glutathione, ascorbate, and NADH on the signal intensity of Cr (V). Pretreatment of the animal with ascorbate or glutathione significantly reduced the Cr (V) signal, while NADH increased the signal. Ascorbate and glutathione results can be easily understood because at higher concentrations, these cellular antioxidants would reduce Cr (VI) rapidly to the final product Cr (III), yielding lower steady state concentration of Cr (V). The increased EPR signal of Cr (V) when treated with NADH, however, is likely because NADH serves as a cofactor for the enzymes responsible for the one-electron reduction of Cr (VI), and the formation of a Cr (V)-NAD(P)H complex that enhance the stability of Cr (V) *in vivo* [11, 32–34]. Treatment with NADH therefore would increase production of Cr (V) and enhance Cr (V) stability, leading to higher steady state concentration of Cr (V)-NAD(P)H complex.

These results indicate that to prevent or attenuate Cr (VI) induced cellular damage, either proper chelation to inhibit the formation of Cr (V), or treatment with antioxidants, could be used as viable strategies.

A. Kidney

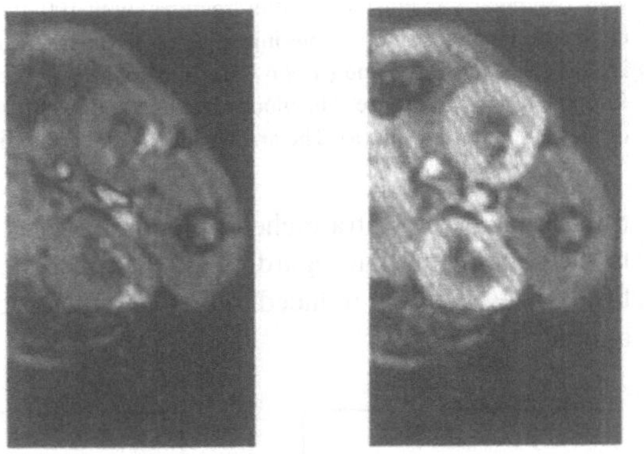

B. Liver

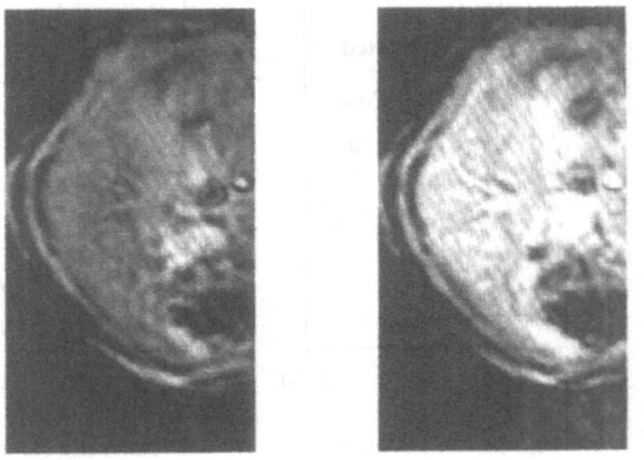

Fig. 5. Tissue localization of chromium based on T1 weighted MR imaging. Gradient echo images were collected before (left) and 10 min post (right) injection of 100 μl of 100 mM sodium dichromate. Increased intensity indicates an increased concentration of Cr (V) and Cr (III). Acquisition parameters are described in the Materials and methods.

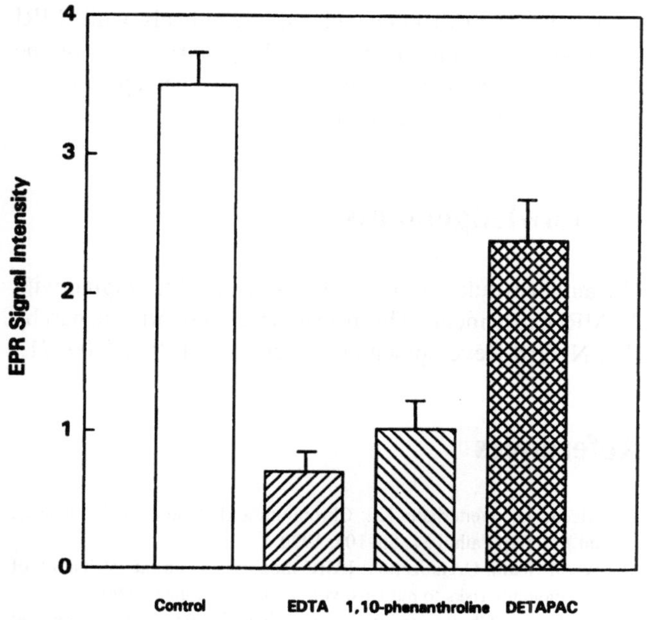

Fig. 6. Effect of metal ion chelators on the intensity of Cr (V) EPR signal. Three minutes before the intravenous injection of 100 μl of 100 mM sodium dichromate, the animals were injected intravenously with 200 μl of 200 mM EDTA, 100 μl of 200 mM DETAPAC, or 50 μl of 200 mM 1,10-phenanthroline, respectively. The values represent means ± S.E. (n = 3). From ref. [10] with permission.

Formation of Cr (V) and free radical in cells

The intermediate Cr (V) has been demonstrated in chemical solutions to be capable of generating free radicals, such as superoxide and hydroxyl radicals, commonly referred to reactive oxygen species (ROS) [7, 8, 20, 24, 26, 27, 35–41]. It is believed that these free radicals, mediated by Cr (V), are largely responsible for the observed Cr (VI) toxicity and carcinogenicity. Although attempts have been made to detect these Cr (V)-mediated free radicals directly in living animals by EPR techniques, to date, the effort has largely failed, due to the very short lifetime of the radicals, and low concentration of the radicals being formed. On the other hand, a significant amount of evidence, both direct and indirect, has been obtained in cellular studies, unambiguously demonstrating that free radicals are indeed formed through Cr (V)-mediated reactions. Figure 8 is a typical example of these studies in which characteristic EPR signal of Cr (V) was recorded when Jurkat cells were incubated with Cr (VI). Addition of glutathione reductase, which catalyzed the Cr (VI) reduction, enhanced the Cr (V) formation. If the Cr (VI) was incubated with the cells in the presence of the spin trap DMPO, the EPR signal of DMPO-OH is formed. At the same time, a reduc-

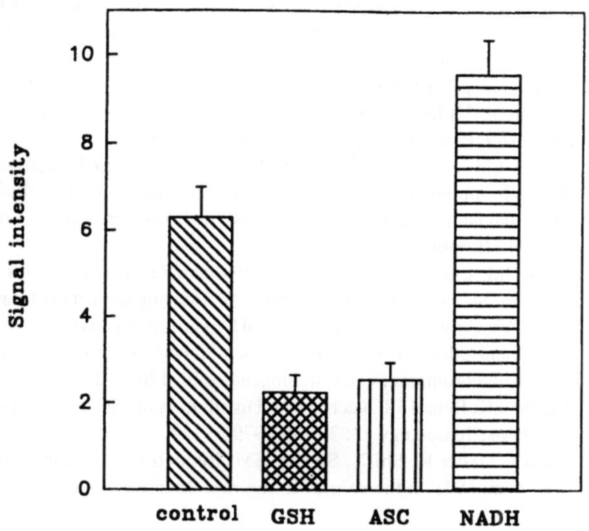

Fig. 7. Effect of glutathione, ascorbate, and NADH on the signal intensity of Cr (V). Three minutes before the intravenous injection of 100 μl of 100 mM sodium dichromate, the animals were injected intravenously with 100 μl of 250 mM glutathione, 250 mM ascorbate, or 500 mM NADH. The values represent means ± S.E. (n = 3). From ref. [11] with permission.

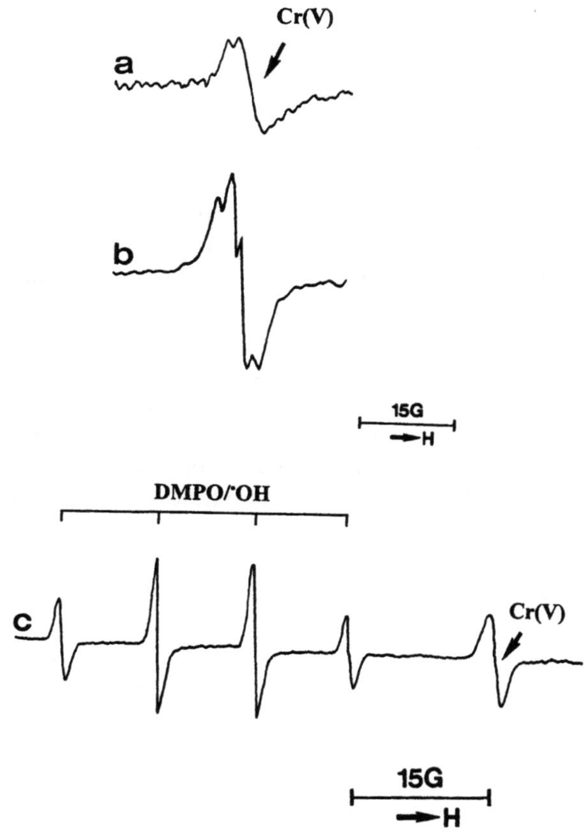

Fig. 8. Generation of hydroxyl radical through Fenton-like reaction. (a) An EPR spectrum recorded from a phosphate buffer solution containing 2 mM $K_2Cr_2O_7$ and 2×10^6/ml Jurkat cells. The spectrum was recorded 4 min after reaction initiation. (b) Same as (a) but with 2.5 units/ml glutathione reductase added. (c) Same as (a) but with 100 mM DMPO added. From ref. [26] with permission.

tion in the signal intensity of Cr (V) is noticed, as shown at the right side of Fig. 8c. These studies firmly establish that Cr (V) can generate hydroxyl radicals through a Fenton-like reaction, as in the following:

$$Cr(V) + H_2O_2 \rightarrow {}^{\bullet}OH + Cr(VI) + OH^-$$

The great advantage of EPR to study free radicals and Cr (V) is well demonstrated here. Both of these species are paramagnetic (i.e. EPR active) and can be recorded simultaneously in a single scan, making it possible to study the causal relationship of the two types of species. Perhaps because of the unique capability of EPR to both measure chromium intermediates and related free radicals, EPR has been one of the popular methods for chromium studies.

Extensive studies conducted recently have shown that the free radicals mediated by Cr (V) can cause DNA damages [8, 20, 24, 27, 39, 42–45], and play an important role in Cr (VI)-induced cell responses, including apoptosis, cell growth arrest, and activation of nuclear transcription factors, NF-κB, AP-1 and p53 [26, 46–49]. It has become increasingly clear that Cr (V) and its related free radicals are involved in the mechanism of Cr (VI)-induced toxicity and carcinogenesis.

Conclusion

The results presented above, and those from the literature lead to the following conclusions: (a) When entered into biological systems, Cr (VI) is reduced to the transient intermediate species Cr (V). Cr (V) is further reduced to the final stable product Cr (III); (b) Cr (V) is capable of generating hydroxyl radicals through a Fenton-like reaction; (c) The free radicals mediated by Cr (V) can cause DNA damage, as well as activate a variety of signal transduction pathways, leading to toxicity and carcinogenicity. These overall mechanism of Cr (VI) is summarized in Fig. 9; and (d) EPR is a valuable tool

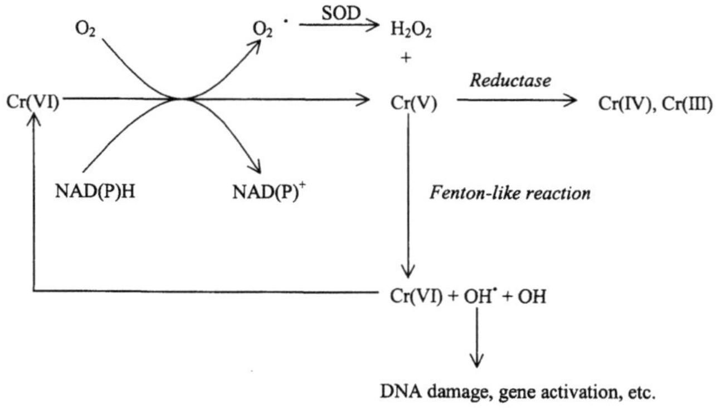

Fig. 9. The overall scheme for the reduction of Cr (VI) and its related free radical generation.

to study both Cr (V) and its related free radicals, while MRI could potentially provide the complementary results on the localization of chromium intermediates, and perhaps, with the right spin trap, even free radicals.

Acknowledgements

The author would like to Dr. Goran Bacic for helping with the MRI experiments. This research was supported in part by NIH/NIEHS Developmental Center Grant PZO ES 09871.

References

1. Alcedo JA, Wetterhahn KE: Chromium toxicity and carcinogenesis. Int Rev Exp Pathol 31: 85–108, 1990
2. World Health Organization: IARC monographs on the evaluation of carcinogenic risks to humans, vol. 49. Lyon, France, 1990
3. Bartlett RJ: Chromium cycling in soils and water: Links, gaps, and methods. Environ Health Perspect 92: 17–24, 1991
4. De Flora S, Serra D, Basso C, Zanacchi P: Mechanistic aspects of chromium carcinogenicity. Arch Toxicol 13(suppl): 28–39, 1989
5. Connett PH, Wetterhahn KE: Metabolism of the carcinogen chromate by cellular constituents. Struct Bonding 54: 93–124, 1983
6. Jennette KE: Microsomal reduction of the carcinogen chromate produces chromium (V). J Am Chem Soc 104: 874–875, 1982
7. Shi X, Dalal NS: Generation of hydroxyl radical by chromate in biologically relevant systems: Role of Cr (V) complexes vs. etraperoxochromate (V). Environ Health Perspect 102(suppl 3): 231–236, 1994
8. Shi X, Mao Y, Knapton AD, Ding M, Rojanasakul Y, Gannett PM, Dalal N, Liu K: Reaction of Cr (VI) with ascorbate and hydrogen peroxide generates hydroxyl radicals and causes DNA damage: Role of a Cr (IV)-mediated Fenton-like reaction. Carcinogenesis 15: 2475–2478, 1994
9. Shi X, Dalal NS: The role of superoxide radical in chromium (VI)-generated hydroxyl radical: The Cr (VI) Haber-Weiss cycle. Arch Biochem Biophys 292: 323–327, 1992
10. Liu KJ, Jiang J, Swartz HM, Shi X. Low-frequency EPR detection of chromium (V) formation by chromium (VI) reduction in whole live mice. Arch Biochem Biophys 313: 248–252, 1994
11. Liu KJ, Shi X, Jiang JJ, Goda F, Dalal N, Swartz HM: Chromate-induced chromium (V) formation in live mice and its control by cellular antioxidants: An L-band electron paramagnetic resonance study. Arch Biochem Biophys 323: 33–9, 1995
12. Yuann JM, Liu KJ, Hamilton JW, Wetterhahn KE: *In vivo* effects of ascorbate and glutathione on the uptake of chromium, formation of chromium (V), chromium-DNA binding and 8-hydroxy-2′-deoxyguanosine in liver and kidney of osteogenic disorder shionogi rats following treatment with chromium (VI). Carcinogenesis 20: 1267–1275, 1999
13. Tandon SK, Behari JR, Kachru DN: Distribution of chromium in poisoned rats. Toxicology 13: 29–34, 1979
14. Liu KJ, Mader K, Shi X, Swartz HM: Reduction of carcinogenic chromium (VI) on the skin of living rats. Mag Res Med 38: 524–526, 1997
15. Dalal NS, Suryan MM, Seehra MS: Potassium perchromate standard for determination of paramagnetic spin concentration, g values, and magnetic moments of fossil fuels. Anal Chem 53: 938–940, 1981
16. Liebross RH, Wetterhahn KE: *In vivo* formation of chromium (V) in chick embryo liver and red blood cells. Carcinogenesis 13: 2113–2120, 1992

17. Fujii H, Wan X, Zhong J, Berliner LJ, Yoshikawa K: *In vivo* imaging of spin-trapped nitric oxide in rats with septic shock: MRI spin trapping. Mag Res Med 42: 235–239, 1999
18. Ueno S, Susa N, Furukawa Y, Sugiyama M: Formation of paramagnetic chromium in liver of mice treated with dichromate (VI). Toxicol Appl Pharmacol 135: 165–171, 1995
19. Sugiyama M, Tsuzuki K: Effect of glutathione depletion on formation of paramagnetic chromium in Chinese hamster V-79 cells. FEBS Lett 341: 273–276, 1994
20. Stearns DM, Kennedy LJ, Courtney KD, Giangrande PH, Phieffer LS, Wetterhahn KE: Reduction of chromium (VI) by ascorbate leads to chromium-DNA binding and DNA strand breaks *in vitro*. Biochemistry 34: 910–919, 1995
21. Rossi SC, Gorman N, Wetterhahn KE: Mitochondrial reduction of the carcinogen chromate: Formation of Cr (V), Chem Res Toxicol 1: 101–107, 1988
22. Micera G, Dessi A: Chromium adsorption by plant roots and formation of long-lived Cr (V) species: An ecological hazard? J Inorg Biochem 34: 157–166, 1988
23. Mao Y, Zang L, Shi X: Generation of free radicals by Cr (IV) from lipid hydroperoxides and its inhibition by chelators. Biochem Mol Biol Int 36: 327–337, 1995
24. Faux SP, Gao M, Chipman JK, Levy LS: Production of 8-hydroxydeoxyguanosine in isolated DNA by chromium (VI) and chromium (V). Carcinogenesis 13: 1667–1669, 1992
25. Barr-David G, Charara M, Codd R, Farrell R, Irwin J, Lay P, Bramley R, Brumby S, Ji JY, Hanson G: EPR characterization of the Cr (V) intermediates in the Cr (V/VI) oxidations of organic substrates and of relevance to Cr-induced cancers. J Chem Soc Faraday Trans 91: 1207–1216, 1995
26. Shi X, Dong Z, Huang C, Ma W, Liu K, Ye J, Chen F, Leonard SS, Ding M, Castranova V, Vallyathan V: The role of hydroxyl radical as a messenger in the activation of nuclear transcription factor NF-kappa B. Mol Cell Biochem 194: 63–70, 1999
27. Shi X, Ding M, Ye J, Wang S, Leonard SS, Zang L, Castranova V, Vallyathan V, Chiu A, Dalal N, Liu K: Cr (IV) causes activation of nuclear transcription factor-kappa B, DNA strand breaks and dG hydroxylation via free radical reactions. J Inorg Biochem 75: 37–44, 1999
28. Sugiyama M: Role of cellular antioxidants in metal-induced damage. Cell Biol Toxicol 10: 1–22, 1994
29. Standeven AM, Wetterhahn KE: Ascorbate is the principal reductant of chromium (VI) in rat lung ultrafiltrates and cytosols, and mediates chromium-DNA binding *in vitro*. Carcinogenesis 13: 1319–1324, 1992
30. Moghaddas S, Gelerinter E, Bose RN: Mechanisms of formation and decomposition of hypervalent chromium metabolites in the glutathione-chromium (VI) reaction. J Inorg Biochem 57: 135–146, 1995
31. Gannett PM, Dalal NS, Shi XL, Toth B: 8-Hydroxy-2′-deoxyguanosine formation during the catalytic oxidation of hydrazines in the presence of 2′-deoxyguanosine. Chem Biol Interact 80: 57–72, 1991
32. Shi XL, Dalal NS: NADPH-dependent flavoenzymes catalyze one electron reduction of metal ions and molecular oxygen and generate hydroxyl radicals. FEBS Lett 276: 189–91, 1990
33. Liu KJ, Shi X, Jiang J, Goda F, Dalal N, Swartz HM: Low frequency electron paramagnetic resonance investigation on metabolism of chromium (VI) by whole live mice. Ann Clinical Lab Sci 26: 176–184, 1996
34. Shi XL, Dalal NS: Chromium (V) and hydroxyl radical formation during the glutathione reductase-catalyzed reduction of chromium (VI). Biochem Biophys Res Commun 163: 627–634, 1989
35. Ueno S, Sugiyama M, Susa N, Furukawa Y: Effect of dimethylthiourea on chromium (VI)-induced DNA single-strand breaks in Chinese hamster V-79 cells. Mutat Res 346: 247–253, 1995
36. Sugiyama M, Tsuzuki K, Haramaki N: Influence of o-phenanthroline on DNA single-strand breaks, alkali-labile sites, glutathione reductase, and formation of chromium (V) in Chinese hamster V-79 cells treated with sodium chromate (VI). Arch Biochem Biophys 305: 261–266, 1993
37. Sugiyama M: Role of physiological antioxidants in chromium (VI)-induced cellular injury. Free Rad Biol Med 12: 397–407, 1992
38. Sugiyama M: Effects of vitamins on chromium (VI)-induced damage. Environ Health Perspect 92: 63–70, 1991
39. Kawanishi S, Inoue S, Yamamoto K: Active oxygen species in DNA damage induced by carcinogenic metal compounds. Environ Health Perspect 102(suppl 3): 17–20, 1994
40. Kadiiska MB, Xiang QH, Mason RP: *In vivo* free radical generation by chromium (VI): An electron spin resonance spin-trapping investigation. Chem Res Toxicol 7: 800–805, 1994
41. Aiyar J, Berkovits HJ, Floyd RA, Wetterhahn KE: Reaction of chromium (VI) with glutathione or with hydrogen peroxide: Identification of reactive intermediates and their role in chromium (VI)-induced DNA damage. Environ Health Perspect 92: 53–62, 1991
42. Aiyar J, Berkovits HJ, Floyd RA, Wetterhahn KE: Reaction of chromium (VI) with hydrogen peroxide in the presence of glutathione: Reactive intermediates and resulting DNA damage. Chem Res Toxicol 3: 595–603, 1990
43. Bagchi D, Hassoun EA, Bagchi M, Stohs SJ: Chromium-induced excretion of urinary lipid metabolites, DNA damage, nitric oxide production, and generation of reactive oxygen species in Sprague-Dawley rats. Comp Biochem Physiol C Pharmacol Toxicol Endocrinol 110: 177–187, 1995
44. Sugiyama M, Lin XH, Costa M: Protective effect of vitamin E against chromosomal aberrations and mutation induced by sodium chromate in Chinese hamster V79 cells. Mutat Res 260: 19–23, 1991
45. Tsuzuki K, Sugiyama M, Haramaki N: DNA single-strand breaks and cytotoxicity induced by chromate (VI), cadmium (II), and mercury (II) in hydrogen peroxide-resistant cell lines. Environ Health Perspect 102(suppl 3): 341–342, 1994
46. Chen F, Ding M, Lu Y, Leonard SS, Vallyathan V, Castranova V, Shi X: Participation of MAP kinase p38 and I-kappaB kinase in chromium (VI)-induced NF-kappaB and AP-1 activation. J Environ Pathol Toxicol Oncol 19: 231–238, 2000
47. Wang S, Leonard SS, Ye J, Ding M, Shi X: The role of hydroxyl radical as a messenger in Cr (VI)-induced p53 activation. Am J Physiol – Cell Physiol 279: C868–C875, 2000
48. Ye J, Wang S, Leonard SS, Sun Y, Butterworth L, Antonini J, Ding M, Rojanasakul Y, Vallyathan V, Castranova V, Shi X: Role of reactive oxygen species and p53 in chromium (VI)-induced apoptosis. J Biol Chem 274: 34974–34980, 1999
49. Ye J, Zhang X, Young HA, Mao Y, Shi X: Chromium (VI)-induced nuclear factor-kappa B activation in intact cells via free radical reactions. Carcinogenesis 16: 2401–2405, 1995

Molecular and Cellular Biochemistry **222**: 49–59, 2001.

Atrazine potentiation of arsenic trioxide-induced cytotoxicity and gene expression in human liver carcinoma cells ($HepG_2$)

Paul B. Tchounwou,[1] Barbara A. Wilson,[1] Ali B. Ishaque[1] and John Schneider[2]

[1]*Molecular Toxicology Research Laboratory, NIH-Center for Environmental Health, School of Science and Technology, Jackson State University, Jackson, MS;* [2]*Client Research Laboratory, Xenometrix Inc., Boulder, CO, USA*

Abstract

Recent studies in our laboratory indicated that arsenic trioxide has the ability to cause significant cytotoxicity, and induction of a significant number of stress genes in human liver carcinoma cells, HepG2 [1]. However, similar investigations with atrazine did not show any significant effects of this chemical on $HepG_2$ cells, even at its maximum solubility of 100 μg/mL in 1% dimethyl sulfoxide (DMSO). Further cytogenetic studies were therefore carried out to investigate the combined effects of arsenic trioxide and atrazine on cell viability and gene expression in immortalized human hepatocytes. Cytotoxicity was evaluated using the MTT-assay for cell viability, while the CAT-Tox (L) assay was performed to measure the induction of stress genes in thirteen different recombinant cell lines generated from human liver carcinoma cells ($HepG_2$), by creating stable transfectants of different mammalian promoter – chloramphenicol acetyltransferase (CAT) gene fusions. Cytotoxicity experiments yielded LC_{50} values of 11.9 ± 2.6 μg/mL for arsenic trioxide in de-ionized water, and 3.6 ± 0.4 μg/mL for arsenic trioxide in 100 μg/mL atrazine; indicating a 3 fold increase in arsenic toxicity associated with the atrazine exposure. Co-exposure of $HepG_2$ cells to atrazine also resulted in a significant increase in the potency of arsenic trioxide to upregulate a number of stress genes including those of the glutathione-*S*-transferase Ya subunit-GST Ya, metallothionein$_{IIa}$-HMTIIA, 70-kDa heat shock protein-HSP70, *c-fos*, 153-kDa growth arrest and DNA damage (GADD153), 45-kDa growth arrest and DNA damage (GADD45), and 78-kDa glucose regulated protein-GRP78 promoters, as well as the xenobiotic response element-XRE, tumor suppressor protein response element-*p53*RE, cyclic adenosine monophosphate response element-CRE, and retinoic acid response element-RARE. No significant changes were observed with respect to the influence of atrazine on the modulation of cytochrome P450 1A1-CYP 1A1, and nuclear factor kappa (B site) response element-NFκBRE by arsenic trioxide. These results indicate that co-exposure to atrazine strongly potentiates arsenic trioxide-induced cytotoxicity and transcriptional activation of stress genes in transformed human hepatocytes. (Mol Cell Biochem **222**: 49–59, 2001)

Key words: atrazine, potentiation, arsenic, cytotoxicity, gene expression, $HepG_2$ cells

Introduction

Arsenic is a metalloid and exists in the earth's crust. Its sources as an environmental contaminant are mainly from natural phenomena such as erosion and volcanoes. Natural levels in soil are usually between 1–40 mg/kg. Anthropogenic activities such as metal smelting, coal combustion, and industrial production of arsenic-containing pesticides, pharmaceuticals, and other byproducts could significantly augment these levels. More than 80% of arsenic compounds are used to manufacture products with agricultural applications such as insecticides, herbicides, fungicides, algicides, sheep dips, wood preservatives, dyestuffs, and medicines for the eradication of tapeworms in sheep

Address for offprints: P.B. Tchounwou, Molecular Toxicology Research Laboratory, NIH-Center for Environmental Health, School of Science and Technology, Jackson State University, 1400 Lynch Street, P.O. Box 18540, Jackson, MS 39217, USA and cattle [2]. Arsenic compounds have been used for at least a century in the treatment of syphilis, yaws, amoebic dysentery, and trypanosomiasis. Arsenical drugs are still used in treating certain tropical diseases such as African sleeping sickness and amoebic dysentery, and in veterinary medicine to treat parasitic diseases, including filariasis in dogs and black head in turkeys and chickens [2].

In the environment, arsenic may be transported by wind or water erosion of small particles, or by leaching into rainfall or snowmelt. Transport and partitioning of arsenic in water depends upon the chemical form of arsenic and on interactions with other materials present. Soluble forms move with the water and may be carried long distances through rivers, while insoluble forms may be absorbed from water onto soils or sediments, especially clays, irons, aluminum hydroxides, manganese compounds, and organic materials [3]. Bioconcentration of arsenic occurs in aquatic organisms, primarily algae and lower invertebrates. In water, arsenic can undergo a series of transformations, including oxido-reduction reactions, ligand exchange, and biotransformation [3]. In a contaminated river, sediments can contain substantial amounts of arsenic that are predominantly mobile during water-sediments interactions. Factors influencing the fate of arsenicals in water include the oxido-reduction potential, pH, metal sulfide and sulfide ion concentration, iron concentrations, temperature, salinity, distribution and composition of the biota. The mobility of arsenic in these sediments is affected by the physical and chemical forms of the arsenic species as well as by environmental conditions [4].

In the United States, the primary source of arsenic in humans is through food, mainly ingested from fish and shellfish which amounts to about 50 µg/d on the average [5]. Several epidemiological studies have demonstrated that arsenic is a known human carcinogen [6, 7]. It has been implicated in the etiology of skin cancer, as well as several internal cancers involving the lung, the bladder and the liver [8–10]. The general health effects associated with arsenic exposure include cardiovascular and peripheral vascular disease, developmental abnormalities, neurologic and neurobehavioral disorders, diabetes, hearing loss, portal fibrosis of the liver, lung fibrosis, hematologic disorders, and carcinoma [11, 12].

Atrazine, on the other hand, is a xenobiotic, and it is one of the most heavily used herbicides in Mississippi and the United States as a whole. It is used mainly in corn and sorghum farms in the Mississippi Delta where most of the agricultural activities are performed. Atrazine may be released into the environment through effluent discharge from manufacturing facilities, and through its use as herbicide. In the United States, monitoring studies of surface waters as well as ground waters have shown widespread detection of triazine herbicides including atrazine [13, 14]. Surface waters collected from 149 sites in 122 mid-western river basins indicated that 52% of the sites exceeded the United States Environmental Protection Agency (U.S. EPA)'s maximum contaminant level (MCL) of 3 µg/L for atrazine [15].

Atrazine can be absorbed into the blood stream through oral, dermal, and inhalation exposure. The U.S. EPA considers atrazine as a systemic toxicant that has a potential of causing a variety of adverse health effects including: congestion of the heart, lungs and kidneys, hypotension, antidiuresis, muscle spasms, weight loss, bradycardia, and adrenal degeneration [16]. Death has been reported in studies with laboratory animals following exposure to high doses of atrazine [17]. Upon chronic exposure, atrazine also has the potential of causing weight loss, cardiovascular damage, retinal degeneration, and cancer [18]. Atrazine is considered as a possible human carcinogen (group C) based on evidence of induction of mammary gland tumors in laboratory animals [18].

Arsenic-containing compounds such as monosodium methanearsonate (MSMA) and disodium metahanearsonate (DMSA) have been and are still being used as herbicides in conjunction with atrazine, to control unwanted weeds in several agricultural areas of the Mississippi Delta. MSMA is also currently used by the Mississippi Department of Transportation to control vegetation along highways and rights of way. Therefore, it is expected that occupational co-exposure to arsenic and atrazine is likely to occur in several communities where both chemicals are applied. Various phenyl arsonic acids, especially arsanilic acid, sodium arsanilate and 3-nitro-4-hydroxyphenyl arsonic acid, have been used as feed additives for disease control and for improvement of weight gain in swine for almost 40 years. Most importantly, recent data have renewed the interest for arsenic-containing compounds as anticancer agents. In particular, arsenic trioxide has been demonstrated to be an effective drug in the treatment of acute promeylocytic leukemia by inducing programmed cell death (apoptosis) in leukemia cells, both *in vivo*, and *in vitro* [19]. However, the biochemical mechanisms involved in the therapeutic properties of these arsenic compounds are poorly understood.

Recent studies in our laboratory indicated that arsenic trioxide has the ability to cause significant cytotoxicity, and induction of a significant number of stress genes in human liver carcinoma cells, HepG2 [1]. However, similar investigations with atrazine did not show any significant effects of this chemical on HepG2 cells, even at its maximum solubility of 100 µg/mL in 1% dimethyl sulfoxide (DMSO). Therefore, the present research was designed to investigate the combined effects of arsenic trioxide and atrazine on cell viability and gene expression in immortalized human hepatocytes.

Materials and methods

Chemicals and test media

Arsenic trioxide (As_2O_3) – lot no. 97313-24, with an active ingredient of 100% arsenic in 10% nitric acid was purchased from Fisher Scientific, Houston, TX, USA. Atrazine (2-choro-4-ethylamino-6-isopropylamino-s-triazine) – lot no. 207-62B, with an active ingredient of 98% was purchased from Chem Service Inc., West Chester, PA, USA. Dimethyl sulfoxide (DMSO) – lot no. 9224-01, were purchased from J.T. Baker, Phillipsburg, PA, USA. Dulbecco's Modified Eagle's Minimal Essential Medium (DMEM) – lot no. 1016511 was purchased from Life Technologies, Grand Island, NY, USA.

Gene profile and cytotoxicity assays

The mammalian gene profile assay (CAT-Tox) was performed for measuring differential gene expression in immortalized human liver cells. The parental cell line ($HepG_2$), and thirteen different recombinant constructs generated by creating stable transfectants of different mammalian promoter - chloramphenicol acetyltransferase (CAT) were obtained from Xenometrix, Inc. (Boulder, CO, USA). Each construct contained a unique stress gene promoter or response element fused to the CAT reporter gene. Table 1 presents the promoter/response element-fusion constructs and their respective biological functions [20].

In the laboratory, cells were stored in liquid nitrogen until use. They were next thawed by gentle agitation of their containers (vials) for 2 min in a water bath at 37°C. After thawing, the content of each vial was transferred to a 75 cm^2 tissue culture flask, diluted with DMEM supplemented with 10% fetal bovine serum (FBS) and 1% streptomicin and penicillin, and incubated for 24 h at 37°C in a 5% CO_2 incubator to allow the cells to grow, and form a monolayer in the flask. Cells grown to 80–95% confluence were washed with phosphate buffered saline (PBS), trypsinized with 10 mL of 0.25% (w/v) trypsin – 0.03% (w/v) EDTA, diluted, counted, and seeded (5×10^5 cells/well) in two sets of 96-well microtiter tissue culture plates.

Seeded plates were incubated for 24 h at 37°C in the presence of 5% CO_2/95% air, followed by a replacement of the old medium by a fresh one containing the appropriate amount of each of the test chemicals: 0, 0.3, 0.6, 1.2, 2.5, and 5 μg/mL arsenic trioxide, using deionized water or atrazine (100 μg/mL in 1% DMSO) as solvent. For quality assurance/quality control purposes, positive control plates were also made using known inducers including 3-methyl cholanthrene (3-MC-10 μM) for CYP 1A1, CRE, GADD45, p53RE, and XRE; methyl methane sulphonate (MMS – 100 μg/mL) for GST Ya, HMTIIA, *c-fos*, HSP70, NFκBRE, GADD153, and GRP78; and all-*trans* retinoic acid (RA – 10 μM) for RARE. All chemical exposures involved the use of polypropylene 96-well microtiter plates for the purpose of chemical dilutions. A specific, constant volume (20 μL) was transferred from each well of the chemical dilution plate to the plate containing the cells to give each cell line five chemical doses and a zero control dose, each in triplicate.

Following chemical exposure, the cells were re-incubated for 48 h at 37°C, and 5% CO_2. After the incubation period, the total protein was measured by the Bradford method, at 600 nm using a microtiter plate reader (Bio-Tek Instruments Inc.). A standard sandwich ELISA was performed and in the final step, horse radish peroxidase catalyzed a color change reaction that was measured at 405 nm [1, 20].

The parental $HepG_2$ cell line was dosed in the same manner as the recombinant cell lines, and was used to perform the MTT [3-(4,5-dimethylthiazol-2-yl)-2,5-diphenyltetrazolium bromide]-based cell viability assay using a microtiter plate reader (Bio-Tek Instruments Inc.) with the wavelength set at 550 nm [21].

Statistical analysis

The transcriptional fold inductions for each recombinant cell line at each arsenic trioxide concentration were calculated

Table 1. Stress gene promoter/response element – CAT fusion constructs and their biologic functions

Promoter	Name/endogenous gene product	Biologic function
CYP 1A1	Cytochrome P450 1A1	Phase I biotransformation enzyme
GST Ya	Glutathion-*S*-transferase Ya subunit	Phase II biotransformation enzyme
HMTIIA	$Metallothionein_{IIA}$	Sequestration of heavy metals
FOS	*c-fos*	Member of AP-1 transcription factor complex
HSP70	70-kDa heat shock protein	Protein chaperone – helps to refold or sequester damaged protein
GADD153	153-kDa growth arrest and DNA damage protein	Involves in cell cycle regulation and response to genotoxic agents
GADD45	45-kDa growth arrest and DNA damage protein	Involves in cell cycle regulation and response to genotoxic agents
GRP78	78-kDa glucose-regulated protein	Endoplasmic reticulum protein chaperone Response to DNA damage and calcium ionophores
XRE	Xenobiotic response element	Binding site for Ah-receptor-planar aromatic hydrocarbon complexes
NFκBRE	Nuclear factor kappa (B site) response element	Binding site for NFκB transcription factor (oxidative stress)
CRE	cAMP response element	Binding site for CREB protein
*p53*RE	53-kDa protein tumor suppressor repsonse element	Binding site for *p53* transcription factor (response to DNA damage)
RARE	Retinoic acid response element	Binding site for retinoic acid – retinoic acid receptor complexes

using the CAT-Tox computer software based on the optical density readings at 600 and 405 nm. The software also converted the 550 nm readings to cell viability percentages. Standard deviations were determined, and the Student's *t*-test values were computed to determine if there were significant differences in cell viability and gene induction in treated cells compared to the control cells. Graphs were made to illustrate the dose-response relationship with respect to cytotoxicity and gene expression. Only five (0, 0.3, 0.6, 1.2, and 2.5 µg/mL) instead of 6 chemical concentrations were shown in the graphs related to gene expression because of the very low percentage of cell viability (only about 4%) in the 5.0 µg/mL arsenic trioxide-atrazine mixture (Fig. 1).

Results

Cytotoxicity assay

The effects of arsenic trioxide alone, and arsenic trioxide in combination with atrazine (100 µg/mL in 1% DMSO), on the viability of the parental $HepG_2$ cells are shown on Fig. 1. Data presented in this figure indicate a concentration-response relationship with respect to the cytotoxicity of arsenic trioxide. Upon 48 h of exposure, LC_{50} values of 11.9 ± 2.6 µg/mL, and 3.6 ± 0.4 µg/mL were determined for this chemical (arsenic trioxide) alone, and in mixture with atrazine, respectively; indicating a 3 fold increase in toxicity due to the presence of atrazine. Highly significant differences ($p < 0.01$) in cell viability associated with the presence of atrazine were found at arsenic trioxide concentrations of 2.5 µg/mL (94.7 ± 11.5% in arsenic alone vs. 74.1 ± 4.5% in arsenic-atrazine mixture), and 5 µg/mL (69.1 ± 9.6% in arsenic alone vs. 4.1 ± 1.0% in arsenic-atrazine mixture) (Fig. 1).

Gene profile assay

The induction levels of stress genes in human liver carcinoma cells exposed to arsenic trioxide alone, and arsenic trioxide in combination of atrazine are presented in Fig. 2 for GSTYa, Fig. 3 for XRE, Fig. 4 for HMTIIA, Fig. 5 for *c-fos*, Fig. 6 for NFκBRE, Fig. 7 for HSP70, Fig. 8 for CRE, Fig. 9 for *p53*RE, Fig. 10 for GADD153, Fig. 11 for GADD45, Fig. 12 for GRP78, and Fig. 13 for RARE.

Data presented in Fig. 2 indicated a concentration-response relationship with respect to arsenic trioxide induction of the glutathione-*S*-transferase gene promoter. This induction of GSTYa was significantly upregulated by atrazine within the concentration range of 0–2.5 µg/mL arsenic trioxide. No significant difference was found in GSTYa induction at 5 µg/mL arsenic trioxide, with or without co-exposure to atrazine. As expected, exposure to arsenic trioxide alone exerted only marginal induction (fold induction < 2) of xenobiotic response element (XRE). However, co-exposure to atrazine significantly ($p < 0.01$) boosted the activation of XRE in a dose-dependent manner (Fig. 3). As shown in Fig. 4, the induction of metallothionein by arsenic trioxide alone was also

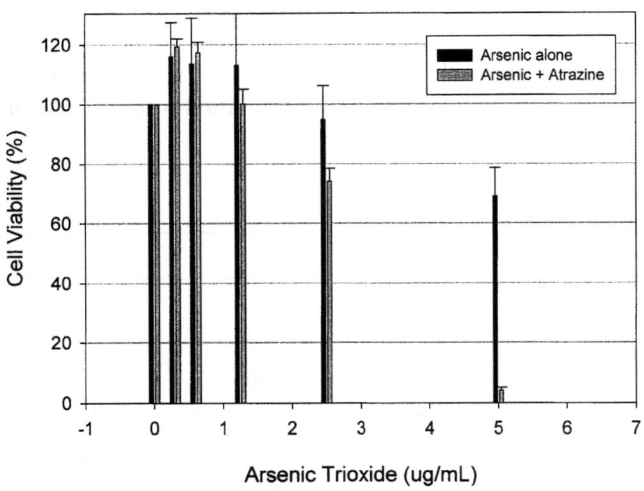

Fig. 1. Cytotoxicity in human liver carcinoma ($HepG_2$) cells exposed to arsenic trioxide alone, and a mixture of arsenic trioxide and atrazine (100 µg/mL). The cells were treated with different doses of arsenic trioxide, with or without atrazine, as indicated. Cell viability was determined based on the MTT assay, as indicated in Materials and methods. Each point represents a mean value and S.D. of 3 experiments.

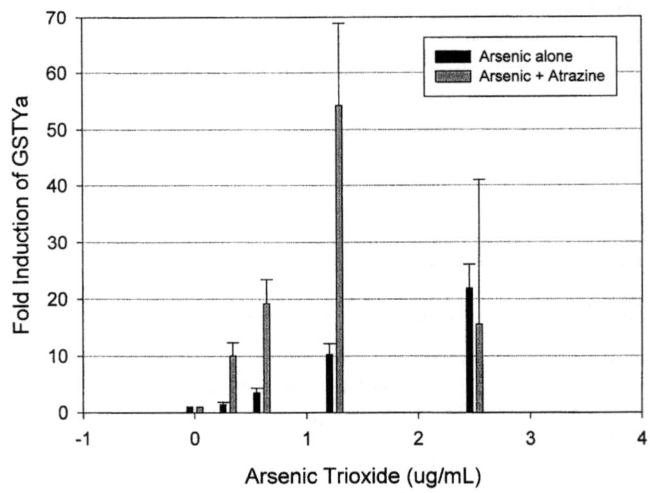

Fig. 2. Fold induction of glutathion-*S*-transferase (GSTYa) in human liver carcinoma ($HepG_2$) cells exposed to arsenic trioxide alone, and a mixture of arsenic trioxide and atrazine (100 µg/mL). Recombinant $HepG_2$ cells transfected with a GSTYa gene promoter fused to the CAT protein were treated with different doses of arsenic trioxide, with or without atrazine, as indicated. GSTYa induction was based on the expression of the CAT-protein using a sandwich ELISA, as indicated in Materials and methods. Each point represents a mean value and S.D. of 3 experiments.

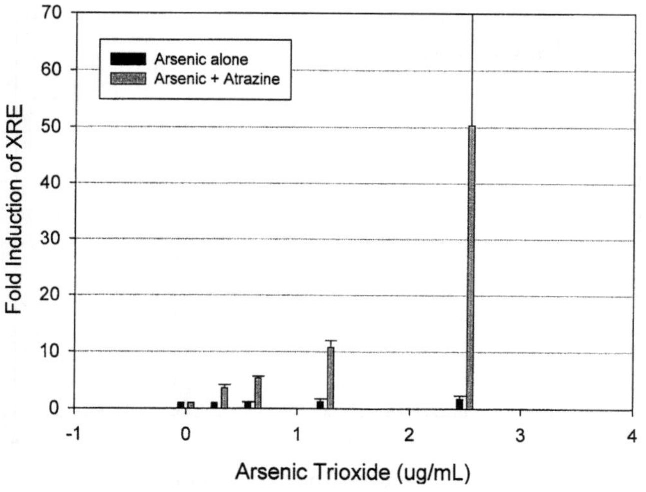

Fig. 3. Fold induction of xenobiotic response element (XRE) in human liver carcinoma ($HepG_2$) cells exposed to arsenic trioxide alone, and a mixture of arsenic trioxide and atrazine (100 μg/mL). Recombinant $HepG_2$ cells transfected with a XRE gene promoter fused to the CAT protein were treated with different doses of arsenic trioxide, with or without atrazine, as indicated. XRE induction was based on the expression of the CAT-protein using a sandwich ELISA, as indicated in Materials and methods. Each point represents a mean value and S.D. of 3 experiments.

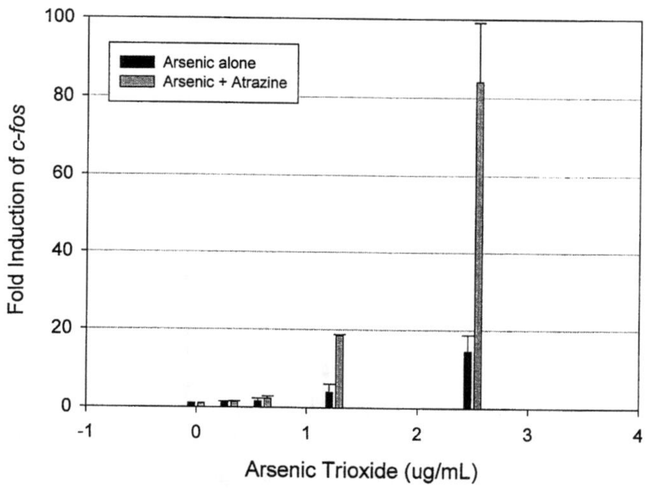

Fig. 5. Fold induction of *c-fos* in human liver carcinoma ($HepG_2$) cells exposed to arsenic trioxide alone, and a mixture of arsenic trioxide and atrazine (100 μg/mL). Recombinant $HepG_2$ cells transfected with a *c-fos* gene promoter fused to the CAT protein were treated with different doses of arsenic trioxide, with or without atrazine, as indicated. *C-fos* induction was based on the expression of the CAT-protein using a sandwich ELISA, as indicated in Materials and methods. Each point represents a mean value and S.D. of 3 experiments.

dose dependent, and co-exposure to atrazine upregulated the induction of this promoter at lower level of arsenic trioxide exposure (< 1 μg/mL). Activation of the *c-fos* gene promoter by arsenic trioxide was also concentration-dependent, and the co-exposure to atrazine positively modulated its induction in transformed human hepatocytes (Fig. 5). Arsenic trioxide significantly induced the nuclear factor kappa (B site) response element (NFκBRE) at concentrations of 1.2 and 2.5 μg/mL, however co-exposure to atrazine did not significantly modulate the induction of this response element (Fig. 6).

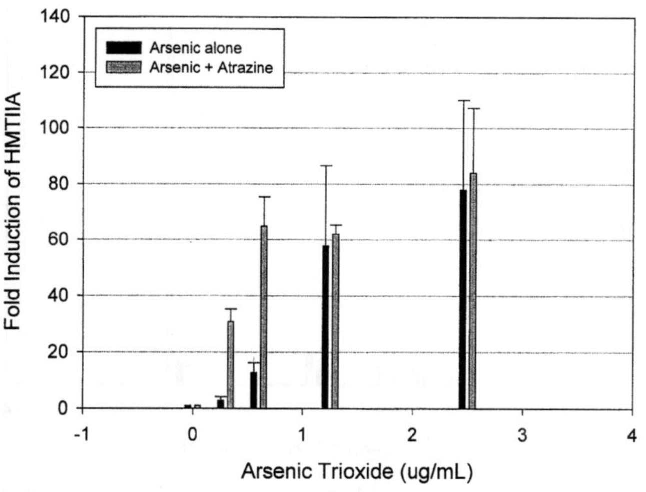

Fig. 4. Fold induction of metallothionein (HMTIIA) in human liver carcinoma ($HepG_2$) cells exposed to arsenic trioxide alone, and a mixture of arsenic trioxide and atrazine (100 μg/mL). Recombinant $HepG_2$ cells transfected with a HMTIIA gene promoter fused to the CAT protein were treated with different doses of arsenic trioxide, with or without atrazine, as indicated. HMTIIA induction was based on the expression of the CAT-protein using a sandwich ELISA, as indicated in Materials and methods. Each point represents a mean value and S.D. of 3 experiments.

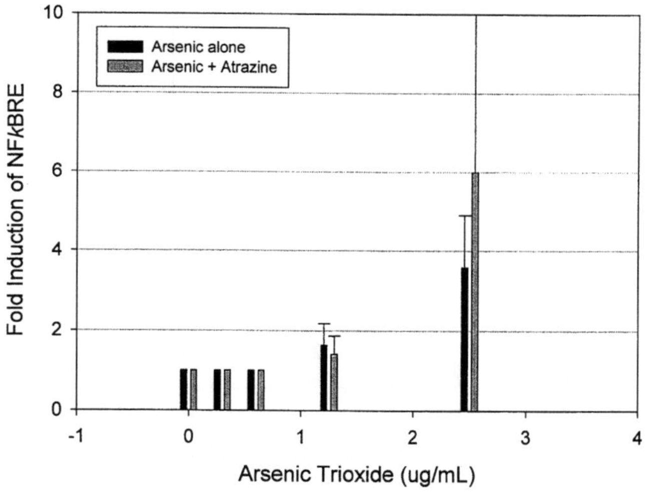

Fig. 6. Fold induction of nuclear factor kappa (B site) response element (NFκBRE) in human liver carcinoma ($HepG_2$) cells exposed to arsenic trioxide alone, and a mixture of arsenic trioxide and atrazine (100 μg/mL). Recombinant $HepG_2$ cells transfected with a NFκB gene promoter fused to the CAT protein were treated with different doses of arsenic trioxide, with or without atrazine, as indicated. NFκB induction was based on the expression of the CAT-protein using a sandwich ELISA, as indicated in Materials and methods. Each point represents a mean value and S.D. of 3 experiments.

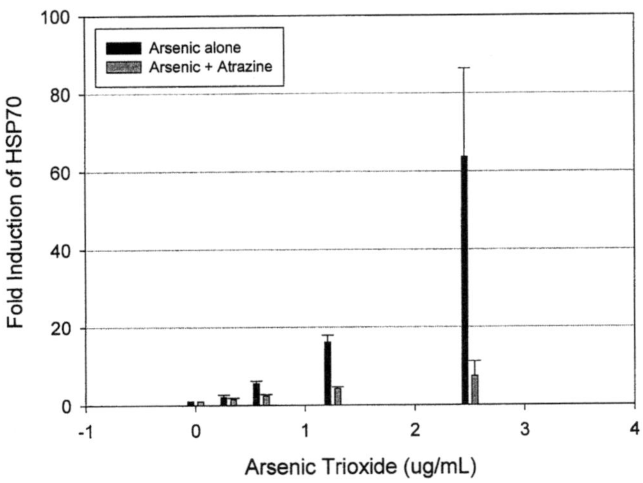

Fig. 7. Fold induction of 70-kDa heat shock protein (HSP70) in human liver carcinoma ($HepG_2$) cells exposed to arsenic trioxide alone, and a mixture of arsenic trioxide and atrazine (100 µg/mL). Recombinant $HepG_2$ cells transfected with a HSP70 gene promoter fused to the CAT protein were treated with different doses of arsenic trioxide, with or without atrazine, as indicated. HSP78 induction was based on the expression of the CAT-protein using a sandwich ELISA, as indicated in Materials and methods. Each point represents a mean value and S.D. of 3 experiments.

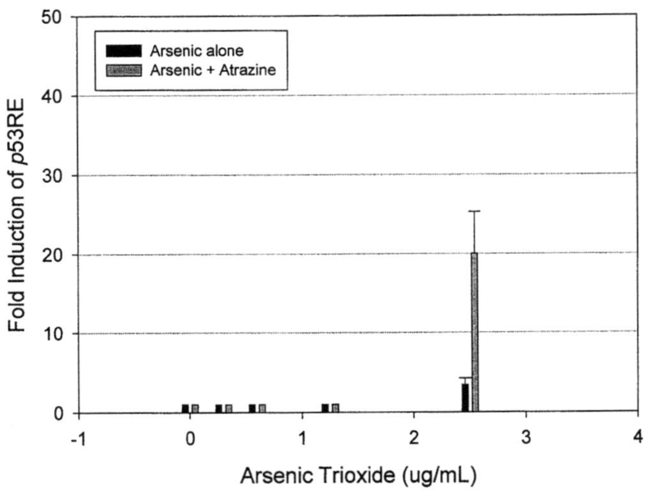

Fig. 9. Fold induction of *p53* tumor suppressor response element (*p53*RE) in human liver carcinoma ($HepG_2$) cells exposed to arsenic trioxide alone, and a mixture of arsenic trioxide and atrazine (100 µg/mL). Recombinant $HepG_2$ cells transfected with a *p53* gene promoter fused to the CAT protein were treated with different doses of arsenic trioxide, with or without atrazine, as indicated. *p53* induction was based on the expression of the CAT-protein using a sandwich ELISA, as indicated in Materials and methods. Each point represents a mean value and S.D. of 3 experiments.

Data presented in Fig. 7 indicated a strong dose-response relationship with respect to arsenic trioxide induction of the 70-kDa heat shock protein (HSP70). However, this induction of HSP70 was significantly down-regulated by atrazine. In contrary, co-exposure to atrazine was found to boost the induction levels of five other stress genes including the cyclic

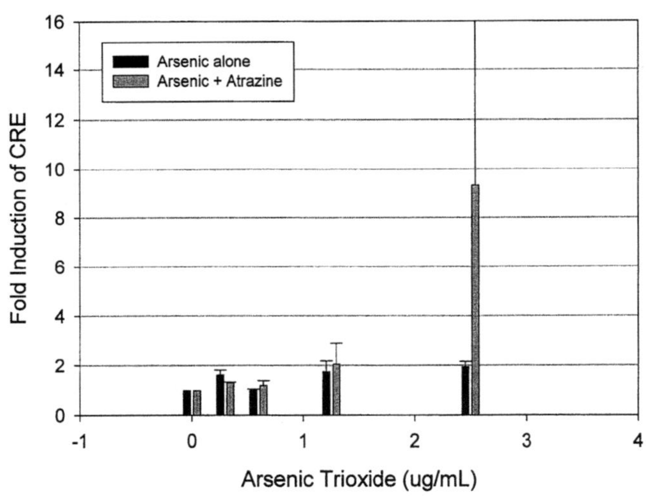

Fig. 8. Fold induction of cyclic adenosine monophosphate response element (CRE) in human liver carcinoma ($HepG_2$) cells exposed to arsenic trioxide alone, and a mixture of arsenic trioxide and atrazine (100 µg/mL). Recombinant $HepG_2$ cells transfected with a cAMP gene promoter fused to the CAT protein were treated with different doses of arsenic trioxide, with or without atrazine, as indicated. CRE induction was based on the expression of the CAT-protein using a sandwich ELISA, as indicated in Materials and methods. Each point represents a mean value and S.D. of 3 experiments.

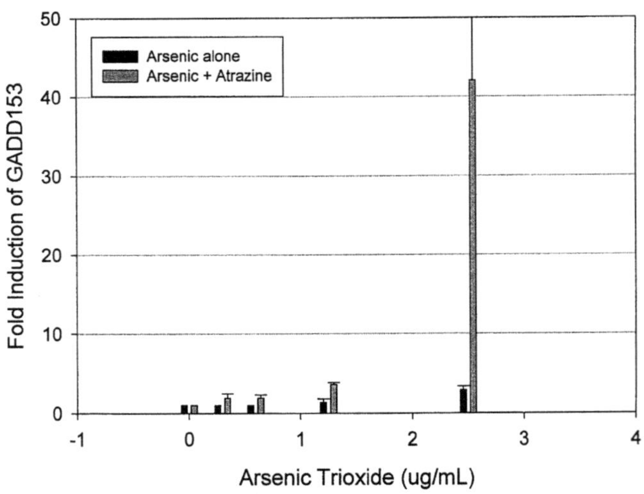

Fig. 10. Fold induction of 153-kDa growth arrest and DNA damage gene (GADD153) in human liver carcinoma ($HepG_2$) cells exposed to arsenic trioxide alone, and a mixture of arsenic trioxide and atrazine (100 µg/mL). Recombinant $HepG_2$ cells transfected with a GADD153 gene promoter fused to the CAT protein were treated with different doses of arsenic trioxide, with or without atrazine, as indicated. GADD153 induction was based on the expression of the CAT-protein using a sandwich ELISA, as indicated in Materials and methods. Each point represents a mean value and S.D. of 3 experiments.

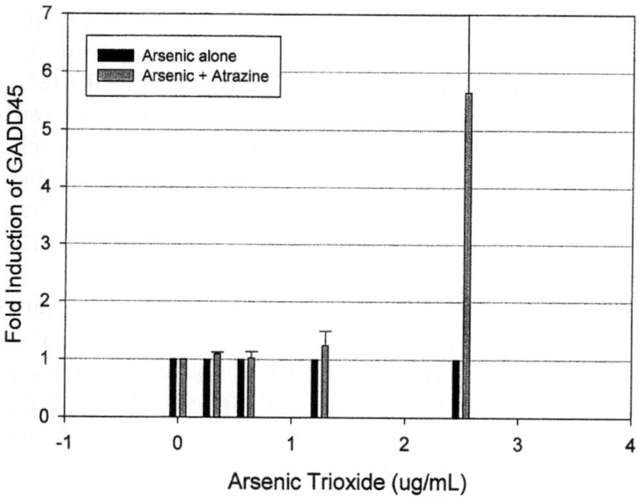

Fig. 11. Fold induction of 45-kDa growth arrest and DNA damage gene (GADD45) in human liver carcinoma ($HepG_2$) cells exposed to arsenic trioxide alone, and a mixture of arsenic trioxide and atrazine (100 μg/mL). Recombinant $HepG_2$ cells transfected with a GADD45 gene promoter fused to the CAT protein were treated with different doses of arsenic trioxide, with or without atrazine, as indicated. GADD45 induction was based on the expression of the CAT-protein using a sandwich ELISA, as indicated in Materials and methods. Each point represents a mean value and S.D. of 3 experiments.

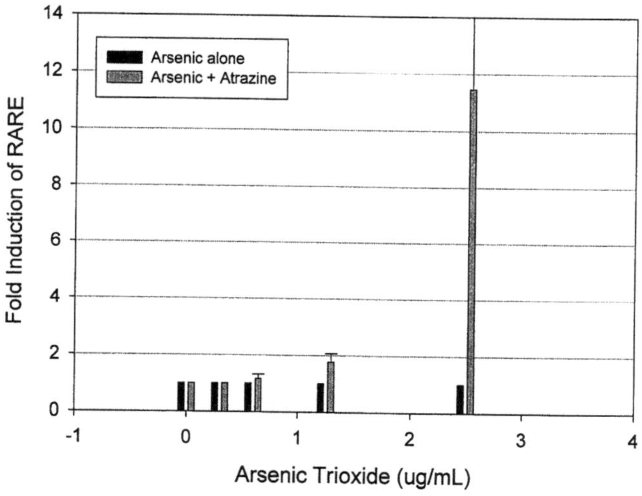

Fig. 13. Fold induction of retinoic acid response element (RARE) in human liver carcinoma ($HepG_2$) cells exposed to arsenic trioxide alone, and a mixture of arsenic trioxide and atrazine (100 μg/mL). Recombinant $HepG_2$ cells transfected with a RARE gene promoter fused to the CAT protein were treated with different doses of arsenic trioxide, with or without atrazine, as indicated. RARE induction was based on the expression of the CAT-protein using a sandwich ELISA, as indicated in Materials and methods. Each point represents a mean value and S.D. of 3 experiments.

AMP response element (Fig. 8), the *p53* tumor suppressor protein (Fig. 9), the 153-kDa growth arrest and DNA damage promoter (Fig. 10), the 45-kDa growth arrest and DNA damage promoter (Fig. 11), the 78-kDa glucose regulated protein (Fig. 12), and the retinoic acid response element (Fig. 13). All QA/QC tests with positive control chemicals (3-MC, MMS, and RA) resulted in significant inductions ($p < 0.05$) of stress genes in all recombinant cell lines (data not shown).

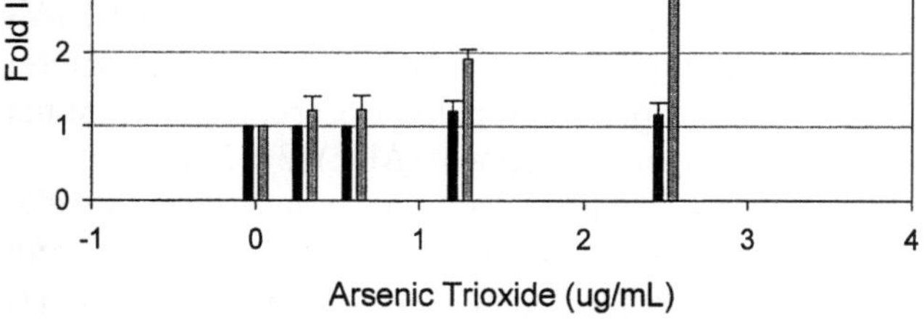

Fig. 12. Fold induction of 78-kDa glucose regulated protein (GRP78) in human liver carcinoma ($HepG_2$) cells exposed to arsenic trioxide alone, and a mixture of arsenic trioxide and atrazine (100 μg/mL). Recombinant $HepG_2$ cells transfected with a GRP78 gene promoter fused to the CAT protein were treated with different doses of arsenic trioxide, with or without atrazine, as indicated. GRP78 induction was based on the expression of the CAT-protein using a sandwich ELISA, as indicated in Materials and methods. Each point represents a mean value and S.D. of 3 experiments.

Discussion

Cytotoxicity assay

This study demonstrated that arsenic trioxide is acutely toxic to human hepatocytes. The toxic potency was concentration-dependent. A 3 fold reduction was observed in the LC_{50} of the chemical (from about 12 to about 4 μg/mL) in the presence of atrazine; indicating a potentiation by atrazine, of arsenic trioxide cytotoxicity. Atrazine itself, was found to be non toxic to human liver carcinoma cells ($HepG_2$), even at its maximum solubility of 100 mg/L in 1% DMSO. These results were in agreement with those of previous studies indicating that the acute toxicity of atrazine to vertebrate animals, and fish in particular, is very low [22].

Many case reports of death in humans from ingestion of high doses of arsenic have been reported. A fatal dose of arsenic trioxide is probably in the 200–300 mg/kg range, yet a dose as low as 10–20 mg/kg has been life threatening, and recovery from 10 g/kg has occurred [24]. Symptoms of acute intoxication usually occur within 30 min of ingestion, but may be delayed if arsenic is taken with food. In nearly all cases, the most immediate effects are severe nausea and vomiting, colicky abdominal pain, profuse diarrhea with rice stools, and gastrointestinal hemorrhage; death may ensue from fluid loss and circulatory collapse. Drowsiness and confusion are also seen along the development of psychosis, which is associated with paranoid delusions, hallucinations, and delirium. Finally, seizures, coma, and death, usually from shock may ensue. If death does not occur in the first 24 h from irreversible circulatory insufficiency, then over the next several days death may result from hepatic or renal failure [24, 25].

Previous research has pointed out that the mechanism by which arsenic exerts its toxic effect is through impairment of cellular respiration by the inhibition of various mitochondrial enzymes, and the uncoupling of oxidative phosphorylation. Most toxicity of arsenic results from its ability to interact with sulphydryl groups of proteins and enzymes, and to substitute phosphorous in a variety of biochemical reactions [26]. Experimentally, arsenic compounds are fetotoxic and teratogenic in mice, rats, and hamsters. The common developmental effects seen include malformations of the brain, urogenital organs, skeleton, ear, as well as small or missing eyes [27]. There are also several epidemiological studies reporting an association between exposure to inorganic arsenic and increased risk of adverse developmental effects such as congenital malformations, low birth weight, and spontaneous abortion [28–30].

Different species of arsenic have different degrees of toxicity, with arsine and the trivalent forms causing the most injury. In addition to the exposure route, and the arsenic species involved, the toxicity also depends on the exposure dose, frequency, duration, and the biological species, as well as on individual susceptibilities, genetic and nutritional factors [12, 26].

Gene profile assay

For most constructs evaluated in the CAT-Tox (L) assay with arsenic trioxide, a dose-response relationship was recorded with respect to gene induction. For example, induction levels of 2.8 ± 1.4, 12.7 ± 3.5, 57.7 ± 29.0, and 78.1 ± 32.3 were recorded for HMTIIA at 0.3, 0.6, 1.25, and 2.5 µg/mL arsenic trioxide, respectively. Overall, 8 out of the 13 tested constructs showed inductions to statistically significant levels ($p < 0.05$). At 2.5 µg/mL arsenic trioxide, the average levels of induction were 21.9 ± 4.2, 78.0 ± 32.3, 14.5 ± 4.2, 3.6 ± 1.3, 63.7 ± 22.7, 1.9 ± 0.2, 3.4 ± 0.8, and 2.9 ± 0.5 for GSTYa, XRE, HMTIIA, *c-fos*, NFκ*B*RE, HSP70, *p53*RE, GADD153, GADD45, and GRP78, respectively. At this arsenic trioxide concentration (2.5 µg/mL), the induction of CYP 1A1 (1.3 ± 0.1), XRE (1.7 ± 0.7), RARE (1.0 ± 0.0), GADD45 (0.8 ± 0.0), and GRP78 (1.2 ± 0.2) were not significant ($p > 0.05$).

Exposure to atrazine alone did not show any significant induction of any of the gene promoters and response elements, even at its maximum solubility of 100 µg/mL in 1% DMSO. However, co-exposure to atrazine significantly modulated arsenic trioxide-induced gene expression in human liver carcinoma cells. In general, atrazine was found to potentiate the ability of arsenic trioxide to upregulate expression of GSTYa, XRE, HMTIIA, *c-fos*, *p53*RE, GADD153, GADD45, GRP78, CRE, and RARE.

Data from this study show a biphasic response with regard to GSTYa induction in the presence of atrazine (Fig. 2). A gradual increase in GSTYa induction was obtained within the concentration range of 0–1.25 µg/mL, followed by a marked reduction in the 2.5 µg/mL concentration, probably due to the relative increase in cytotoxicity at higher chemical concentrations. However, a monophasic response was obtained with XRE stimulation, showing a strong concentration-effect relationship (Fig. 3). Induction of the GSTYa promoter and XRE by the arsenic trioxide-atrazine mixture indicates the potential involvement of the Phase II biotransformation pathway (conjugation) in the liver. Upregulation of GSTYa genes is controlled by the xenobiotic response element – XRE [31–33]. Inducing agents bind to the cytoplasmic ah (aryl hydrocarbon) receptor, which is present not only in the liver, but also in other extra hepatic tissues such as lung, skin and kidney. This gene responds to polycyclic aromatic hydrocarbons (PAHs) such as benzo (a) pyrene, 7- ethoxycoumarin, and 3-mehtylcholanthrene (3-MC). This promoter has also been reported to respond to other chemicals such as 2,3,7,8-tetrachlorlibenzo-dioxin (TCDD), polychlorinated biphenyls (PCBs) and quinone methides [20, 34]. The capacity of liver cells to detoxify active compounds is correlated with the level of bio-transformation enzymes produced. Many of these enzymes including those of cytochrome P450 1A1, and the Ya subunit of glutathione-*S*-transferase are transcriptionally induced when cells are exposed to specific chemicals [20]. The GSTs are a supergene family of dimeric enzymes that catalyze the conjugation of reduced glutathion (GSH) to a variety of electrophiles including arene oxides, unsaturated carbonyls, organic halides, and other substrates [35]. GSTYa inducible transcription is controlled by two distinct response elements, the xenobiotic response element (XRE), and the anti-oxidant responsive element (ARE) [36, 37].

Stimulation of the HMTIIA promoter by arsenic trioxide-atrazine mixture is similar to the response elicited by a variety of toxic metals such as cadmium, zinc, copper and silver [1, 38], as well as arsenic [1]. Alkylating agents such as MMS have also been reported to transcriptionally induce HMTIIA [20]. It has been hypothesized that metallothioneins act as intracellular storage and concentration sites for heavy metals such as cadmium and zinc. Metallothioneins therefore mediate resistance to the toxic effects of heavy metals. They also play a role in cellular resistance to alkylating agents and radiation [39].

The induction of *p53*RE appears to be a very important factor in the pharmacology of arsenic trioxide. This tumor suppressor gene with its *p53* transcription factor shows a strong activation following exposure to arsenic trioxide. This is indicative of the potential mechanism of action of arsenic trioxide at low doses, and hence, its use as a salvage therapy for relapsed and refractory acute promyelocytic leukemia [40]. The *p53* wild type protein is involved in transcriptional activation of genes that negatively control growth and/or invasion through the *p53* response element [42]. Mutant forms of *p53* have been implicated in deregulation of gene expression in many tumor types. The *p53* response element can be upregulated by DNA damaging agents [1, 42, 43].

From a toxicological standpoint, DNA damage as a consequence of exposure to arsenic trioxide and atrazine mixture is reflected by strong inductions of *c-fos*, growth arrest DNA damage (GADD153, GADD45), and *p53*RE. In this study, the *c-fos* promoter was highly responsive to arsenic trioxide-atrazine mixture. *c-fos* forms an integral part of the AP-1 transcriptional complex along with other nuclear protooncogenes including *c-jun* [44]. The *c-fos* promoter is regulated by a variety of response elements including: 12-O-tetradecanoylphorbol 13-acetate response element (TRE), serum response element (SRE) and a cAMP response element (CRE) [45]. The cyclic AMP response element is involved in protein kinase C and protein kinase A signaling pathways [46]. Kinases are enzymes that catalyze the transfer of phosphate groups from ATP, ADP or AMP to a substance. The *c-fos* has also been implicated in the mammalian UV response pathway which appears to have an important role in the response to DNA damaging agents such as methyl methane sulfonate and 4-nitroquinoline N-oxide [47]. Activation of the GADD153 and GADD45 promoters by arsenic trioxide, and the potentiation of this activity by atrazine indicates a potential damage at the genomic level. This damage may be associated with alterations in DNA sequence, as well as conformational changes in its helical structure. Growth arrest and DNA damage promoters have been shown to respond to a variety of DNA damaging agents including UV, DNA-crosslinking, and alkylating agents [48, 49].

Atrazine did not seem to have any significant effect on the potency of arsenic trioxide to induce the nuclear factor kappa (B site) response element (NFκBRE) (Fig. 6). The activation of this promoter by arsenic trioxide, in combination with the *c-fos* activation, is indicative of oxidative damage by arsenic ions. The NFκB response element is a multi-subunit transcription factor that can rapidly activate the expression of genes involved in inflammatory, immune, and acute phase responses [50]. Many different agents can activate the DNA binding activity of NFκB including viruses, bacterial lyopolysaccharides, protein synthesis inhibitors, cytokines including TNF-alpha, TNF-beta and interleukin-1, phorbol esters, lectins, and calcium ionophores [51].

A dose-response relationship was observed with regard to the effect of arsenic trioxide on the transcriptional activation of the 70-kDa heat shock protein (HSP70). However, the co-exposure with atrazine down-regulated its induction level (Fig. 7). Stimulation of this promoter by arsenic trioxide is in line with HSP induction by many other toxicants including alkylating agents [20, 52, 53], toxic metals [1, 54], and quinone methides [34]. It has been suggested that increase in HSP expression may be a direct result of alterations in protein structure through adduct formation; a secondary consequence of decreased non-protein thiol groups and formation of disulfide linkages between proteins [53]; or of release of metal ions due to alkylation of metal-sequesting proteins [54]. The 78-kDa glucose regulated protein (GRP78) is a major endoplasmic reticulum (ER) protein that functions as a chaperone. Upregulation of this promoter is associated with proteins that are malfolded because of mutagenesis, under-glycosylation, or other stress conditions in the ER [55].

None of the two individually tested chemicals seems to induce the transcription of retinoic acid response element (RARE). However, co-exposure of transformed human hepatocytes to the mixture of arsenic trioxide and atrazine significantly induces the expression of this promoter (Fig. 13); indicating a potential for adverse developmental effects associated with exposure to these compounds. RARE is activated by retinoid acid analogs including all-*trans* retinoid acid and 9-*cis* retinoid acid. Retinoids have a broad range of normal biological activities in growth and differentiation [56]. They are vital to normal embryonic development and it is the fine tuning of their concentrations through a milieu of receptors and binding proteins that provides the necessary framework for differentiation to occur. Therefore, it is not surprising that alterations in the normal concentration equilibrium such as those resulting from the lack of, as well as excessive retinoid exposure [57]. Characteristic features of exposure to exogenous retinoids include craniofacial (microtia or anotia; facial bone maldevelopment; micrognathia), cardiac (conotruncal or branchial arch), thymic, and CNS (hydrocephalus) abnormalities [58].

Because of its high potential to cause adverse effects in exposed persons, a number of regulations and guidelines have been established for various inorganic and organic forms of arsenic by international, federal, and state agencies. The International Agency for Research on Cancer classifies arsenic in group 1 – known human carcinogens [7]. The permissible limit for arsenic in drinking water has been fixed at 50 μg/L [59]. The World Health Organization's tolerable daily intake for inorganic arsenic is 2 μg/kg body wt [60]. The action level for arsenic in the air is 5 $\mu g/m^3$. The permissible exposure limit-total weighted average (PEL-TWA) is 10 $\mu g/m^3$ for inorganic arsenicals, and 500 $\mu g/m^3$ for organic arsenicals [61]. The criterion for irrigation waste is 100 μg/L. Tolerances for residues in agricultural products/commodities vary from

0.35–3.5 ppm. The oral reference dose (RfD) is 3×10^{-4} mg/kg/day, and the cancer potency factor is 15 $(mg/kg/day)^{-1}$ for inhalation, and 1.75 $(mg/kg/day)^{-1}$ for ingestion [8, 12, 62].

Conclusions

Findings from this study indicate that: (a) acute exposure to arsenic trioxide significantly reduces the viability of human liver carcinoma ($HepG_2$) cells; (b) acute exposure to atrazine up to a maximum concentration of 100 µg/mL in 1% DMSO for 48 h does not have any significant effect on the viability of $HepG_2$ cells; (c) co-exposure of $HepG_2$ cells to atrazine significantly potentiates the cytotoxicity of arsenic trioxide, resulting in a 3 fold reduction in the mean value of LC_{50}; (d) arsenic trioxide exposure up to 2.5 µg/mL significantly induces a significant number of stress genes in $HepG_2$ cells, including: GSTYa, HMTIIA, HSP70, CRE, GADD153, *p53*RE, *c-fos*, and NFκBRE; (e) atrazine exposure does not show any significant inductions ($p > 0.05$) of any of the thirteen stress genes evaluated; (f) co-exposure to atrazine significantly modulates arsenic trioxide-induced transcription of stress genes in $HepG_2$ cells; with up-regulation of GSTYa, XRE, HMTIIA, CRE, GADD153, GADD45, *p53*RE, *c-fos*, GRP78, and RARE; down-regulation of HSP70; and marginal effect on NFκBRE. The transcriptional activation of these gene promoters and response elements presents significant and profound implications on the pharmacology and toxicology of arsenic trioxide.

Acknowledgements

This research was financially supported by a grant from the National Institutes of Health (Grant No. 1G12RR13459), through the RCMI-Center for Environmental Health at Jackson State University. We thank Dr. Abdul Mohamed, Dean of the School of Science and Technology at Jackson State University, and Dr. William Toscano, Chair of the Environmental Health Sciences Department at the University of Minnesota School of Public Health, for their technical advise on this project.

References

1. Tchounwou PB, Wilson BA, Schneider J, Ishaque A: Cytogenetic assessment of arsenic trioxide toxicity in the Mutatox, Ames II, and CAT-Tox assays. Metal Ions Biol Med 6: 89–91, 2000
2. NAS: Arsenic. National Academy of Science, Washington DC, 1977
3. U.S. EPA: An Exposure and Risk Assessment for Arsenic. EPA 440/4-85-005. EPA Office of Water Regulations and Standards, Washington DC, 1982
4. Mock WM, Wai CM: Mobilization of arsenic in contaminated waters. In: J.O. Nriagu (ed). Arsenic in the Environment, part 1: Cycling and Characterization. John Wiley & Sons, Inc., NY, 1994, pp 99–117
5. Borum DR, Abernathy CO: Human oral exposure to inorganic arsenic. In: W.R. Chappell, C.O. Abernathy, C.R. Cothern (eds). Arsenic Exposure and Health. Science and Technology Press, Northwood, UK, 1994, pp 119–128
6. IARC: Monographs on the Evaluation of Carcinogenic Risks of Chemicals to Humans. Supplement F. Overall Evaluation of Carcinogenicity. International Agency for Research on Cancer. World Health Organization, Lyon, France, 1987, pp 29–57
7. IRIS: Integrated Risk Information System. United States Environmental Protection Agency, Washington DC, 1992
8. Tseng WP, Chu HM, How SW, Fong JM, Lin CS, Yeh S: Prevalence of skin cancer in an endemic area of chronic arsenicism in Taiwan. J Natl Cancer Inst 40: 453–463, 1968
9. Chen CJ, Chuang YC, Lin TM, Wu HY: Malignant neoplasms among residents of a blackfoot disease endemic area in Taiwan: High-arsenic artesian well water and cancers. Cancer Res 45: 5895–5899, 1985
10. Chen CJ, Chen CW, Wu MM, Kuo TL: Cancer potential in liver, lung, bladder and kidney due to ingested inorganic arsenic in drinking water. Brit J Cancer 66: 888–892, 1992
11. Tchounwou PB, Wilson B, Ishaque A: Important considerations in the development of public health advisories for arsenic and arsenic-containing compounds in drinking water. Rev Environ Health 24: 211–229, 1999
12. Tchounwou PB: Health risk assessment and management of arsenic and other toxic and hazardous metals in drinking water. Metal Ions Biol Med 6: 21–23, 2000
13. Coupe RH, Thurman EM, Zimmerman LR: Relation of usage to the occurrence of cotton and rice herbicides in three streams of the Mississippi Delta. Environ Sci Technol 32: 3673–3680, 1998
14. Squillace PJ, Thurman, EM, Furlong ET: Ground water as a non point source for atrazine and deethyl atrazine in a river during base flow conditions. Water Resources Res 29: 1719–1729, 1993
15. Thurman EM, Goolsby DA, Meyer MT, Kolpin DW: Herbicides in surface waters of the midwestern United States: The effects of Spring flush. Environ Sci Technol 25: 1794–1796, 1991
16. Hallenbeck WH, Cummingham B: In: Pesticides and Human Health. Springer-Verlag, New York, 1985
17. U.S. EPA: Atrazine Health Advisory. Office of Drinking Water. United States Environmental Protection Agency, Washington DC, 1988
18. U.S. EPA: Drinking Water and Health: Technical Fact Sheet on Atrazine. United States Environmental Protection Agency, Washington DC, 1999 (on line at: http://www.epa.gov/ OGWDW/dwh/t-soc/atrazine.html)
19. Rousselot P, Laboume S, Marolleau JP, Larghero T, Noguera ML, Brouet JC, Fermand JP: Arsenic trioxide and melarsoprol induce apoptosis in plasma cell lines and in plasma cells from myeloma patients. Cancer Res 59: 1041–1048, 1999
20. Todd MD, Lee MJ, Williams JL, Nalenzny JM, Gee P, Benjamin MB, Farr SB: The CAT-Tox assay: A sensitive and specific measure of stress induced transcription in transformed human liver cells. Fund Appl Toxicol 28: 118–128, 1995
21. Mosmann T: Rapid colorimetric assay for cellular growth and survival: Applications to proliferation and cytotoxicity assays. J Immunol Meth 65: 55–63, 1983
22. Solomon KR, Baker DB, Richards RP, Dixon KR, Klaine SJ, LaPoint TW, Kendall RJ, Weisskopf CP, Giddings JM, Giesy JP, Hall LW, Williams WM: Ecological risk assessment of atrazine in north american surface waters. Environ Toxicol Chem 15: 31–76, 1996
23. Schoolmeester WL, White DR: Arsenic poisoning. Southern Med J 73:198–208, 1980

24. Gordby MS: Arsenic in human medicine. In: J.O. Nriagu (ed). Arsenic in the Environment, part II: Human Health and Ecosystem Effects. John Wiley & Sons, Inc., New York, 1994, pp 1–16
25. Saady JJ, Blanke RV, Pollis A: Estimation of the body burden of arsenic in a child fatally poisoned by arsenic weed killer. J Analyt Toxicol 13: 310–312, 1989
26. ATSDR: Toxicological Profile for Arsenic. TP-92/09. Agency for Toxic Substances and Disease Registry, Center for Disease Control, Atlanta, Georgia, 1993
27. Fan AM: Assessment of metals in drinking water with specific references to lead, copper, arsenic, and selenium. In: L.W. Chan, L. Magos, T. Suzuki (eds). Toxicology of Metals. Lewis Publishers–CRC Press, Boca Raton, Florida, 1996, pp 39–43
28. Nordstorm S, Beckman L, Nordenson I: Occupational and environmental risks in and around a smelter in northern Sweden. V. Spontaneous abortion among female employees and decreased birth weight in their offspring. Hereditas 90: 291–296, 1979a
29. Nordstorm S, Beckman L, Nordenson I: Occupational and environmental risks in and around a smelter in northern Sweden: VI. Congenital malformations. Hereditas 90: 297–302, 1979b
30. Aschengran A, Zierler S, Cohen A: Quality of community drinking water and the occurrence of spontaneous abortion. Arch Environ Health 44: 283–290, 1989
31. Fujii-Kuriyama Y, Imataka H, Sogawa K, Yasumoto KI, Kikuchi Y: Regulation of CYP1A1 expression. Fasc Am Soc Exp Biol 6: 706–710, 1992
32. Neuhold LA, Shirayoshi Y, Ozato K, Jones JE, Nebert DW: Regulation of mouse CYP1A1 gene expression by dioxin: Requirement of two cis-acting elements during induction. Mol Cell Biol 9: 2378–2386, 1989
33. Yao EF, Denison MS: DNA sequence determinants for binding of transformed Ah receptor to a dioxin-responsive enhancer. Biochem 31: 5060–5067, 1992
34. Desjardins JP, Beard SE, Mapoles JE, Gee P, Thompson JA: Transcriptional activity of quinone methides derived from the tumor promoter butylated hydroxytoluene in HepG2 cells. Cancer Lett 131: 201–207, 1998
35. Strange RC, Jones PW, Fryer AA: Glutathion-S-transferase: Genetics and role in toxicology. Toxicol Lett 112–113: 357–363, 2000
36. Rushmore TH, Pickett CB: Xenobiotic responsive elements controlling inducible expression by planar aromatic compounds and phenolic antioxidants. Meth Enzymol 206: 409–420, 1991
37. Rushmore TH, Pickett CB: Glutathion-*S*-transferases, structure, regulation and therapeutic implications. J Biol Chem 268: 11475–11478, 1993
38. Richards RI, Heguy A, Karin M: Structural and functional analysis of the human metallothionein-I gene: Differential induction by metal ions and glucocorticoids. Cell 37: 263– 272, 1984
39. Cherian MB, Howell SB, Imura N, Klaassen CD, Koropatnick K, Lazo J, Waalkes MP: Contemporary issues in toxicology: Role of metallothioneins in carcinogenesis. Toxicol Appl Pharmacol 126: 1–5, 1994
40. Huang SY, Chan CS, Tang JL, Tien HF, Kuo TL, Huang SF, Yoa YT, Chou WC, Chung CY, Wang CH, Shen MC, Chen YC. Acute and chronic poisoning associated with treatment of acute promyelocytic leukemia. Br J Haematol 103: 1092–1095, 1998
41. Vogelstein B, Kinzler KW: p53 function and dysfunction. Cell 70: 523–526, 1992
42. Kastan MB, Onyekwere O, Sidransy D, Vogelstein B, Craig RW: Participation of p53 protein in the cellular response to DNA damage. Cancer Res 51:6304–6311, 1991
43. Zhan Q, Carrier F, Fornace AJ: Induction of cellular p53 activity by DNA-damaging agents and growth arrest. Mol Cell Biol 13: 442–4250, 1993
44. Angel P, Karin M: The role of *jun, fos* and the AP-1 complex in cell-proliferation and transformation. Biochim Biophys Acta 1072: 129–157, 1991
45. Sassone-Corsi P, Sission JC, Verma IM: Transcriptional autoregulation of the proto-oncogene fos. Nature 334: 314–319, 1988
46. Borrelli E, Montmayeur JP, Foulkes NS, Sassone-Corsi P: Signal transduction and gene control: The cAMP pathway. Crit Rev Oncol 3: 321–338, 1992
47. Hollander MC, Fornace AJ: Induction of *fos* RNA by DNA-damaging agents. Cancer Res 29: 1687–1692, 1989
48. Barlett JD, Luethy JD, Carlson SG, Sollott SJ, Holbrook NM: Calcium ionophore A23187 induces expression of the growth arrest and DNA damage inducible CCAAT/enhancer- binding protein (C/EBP)-related gene, gadd153. J Biol Chem 267: 20465–20470, 1992
49. Luethy JD, Holbrook NJ: Activation of gadd153 promoter by genotoxic agents: a rapid and specific response to DNA damage. Cancer Res 52: 5–10, 1992
50. Baeuerle PA: The inducible transcription activator NFκB: Regulation by distinct protein subunits. Biochim Biophys Acta 1072: 63–80, 1991
51. Nelsen B, Hellman L, Shen R: The NF-κB-binding site mediates phorbolester-inducible transcription in non-lymphoid cells. Mol Cell Biol 8: 3526–3531, 1988
52. Hartl FU, Martin J, Neupert W: Protein folding in the cell: The role of molecular chaperones hsp70 and hsp60. Annual Rev Biophys Biomol Struct 21: 293–322, 1992
53. Liu H, Lightfoot R, Stevens JL: Activation of heat shock factor by alkylating agent is triggered by glutathione depletion and oxidation of protein thiols. J Biol Chem 271: 4805–3812, 1996
54. Ovelgonne JH, Souren JEM, Wiegant FAC, Van-Wijk R: Relationship between cadmium-induced expression of heat shock genes, inhibition of protein synthesis and cell death. Toxicol 99: 19–30, 1995
55. Wooden SK, Li LJ, Navarro D, Qadri I, Pereira L: Transactivation of the grp78 promoter by malfolded proteins, glycosylation block, and calcium ionophore is mediated through a proximal region containing a CCAAT motif which interacts with CTF/NF-I. Mol Cell Biol 11: 5612–5623, 1991
56. Allenby G, Bosque MT, Saunders M, Kramer S, Speck J: Retinoic acid receptors and retinoic X receptors: Interactions with endogenous retinoic acids. Proc Natl Acad Sci 90: 30–34, 1993
57. Kirchner SC, Faustman EM: Identifying mechanisms of developmental toxicity: Application of tools from molecular biology. In: A. Puga, K.B. Wallace (eds). Molecular Biology of the Toxic Response. Taylor & Francis, Philadelphia, PA, 1999, pp 411–437
58. Lammer EJ, Chen DT, Hoar RM, Agnish ND, Benke PJ, Braun JT, Curry CJ, Fernhoff PM, Grix AW, Lott IT: Retinoid acid and embryopathology. New Engl J Med 313: 837–841, 1985
59. U.S. EPA: Special Report on Ingested Inorganic Arsenic: Skin Cancer and Nutritional Essentiality. Risk Assessment Forum. United States Environmental Protection Agency, Washington DC, 1987
60. WHO: Guidelines for Drinking Water Quality. World Health Organization, Geneva, Switzerland, 1984
61. OSHA: Occupational Safety and Health Administration. Fed Reg 54: 2332–2335, 1989
62. Tchounwou PB, Abdelghani AA, Pramar YV, Heyer LR, Steward CM: Assessment of potential health risks associated with ingesting heavy metals in fish collected from a hazardous-waste contaminated wetland in Louisiana, USA. Rev Environ Health 11: 191–203, 1996

Molecular and Cellular Biochemistry **222:** 61–68, 2001.

Combining *Drosophila melanogaster* somatic-mutation-recombination and electron-spin-resonance-spectroscopy data to interpret epidemiologic observations on chromium carcinogenicity

Alan J Katz,[1] Arthur Chiu,[2] Jefferson Beaubier[3] and Xianglin Shi[4]

[1]*Department of Biological Sciences, Illinois State University, Normal, IL;* [2]*National Center for Environment Assessment, Office of Research and Development, Washington, DC;* [3]*Office of Pollution Prevention and Toxics, United States Environmental Protection Agency, Washington, DC;* [4]*Pathology and Physiology Research Branch, National Institute for Occupational Safety and Health, Morgantown, WV, USA*

Abstract

Lung cancers are significantly increased among workers exposed to chromate (Cr^{6+}, Cr^{3+}), chromium pigments (Cr^{6+}) and chromium plating (Cr^{6+}). Chromium lung burdens and cancer risk increase proportionately with duration of employment at long latencies. However, this epidemiologic information alone is insufficient in determining whether Cr^{6+} or Cr^{3+} are equally important in causing cancer. We have attempted to combine epidemiologic data with data from the *Drosophila melanogaster* somatic-mutation-recombination-test and from the *in vitro* electron-spin-resonance spectroscopy study to demonstrate that following somatic recombination plays a more important role than somatic mutation in chromium carcinogenesis. Cr^{4+} is more important than Cr^{5+} or Cr^{6+} in inducing somatic recombination while Cr^{6+} produces more and bigger clones than Cr^{4+} in somatic mutation. Cr^{3+} produces negative results in this fruit-fly wing-spot-assay. When the larvae and flies exposed to Cr^{6+} and Cr^{4+} are examined by ESR, only Cr^{5+} and Cr^{3+} are found. Thermodynamic parameters ΔE, ΔH, and ΔS are also estimated from these latter experiments to explain the relative importance of Cr^{6+}, Cr^{4+}, Cr^{3+} in chromium carcinogenesis among exposed industrial workers. (Mol Cell Biochem **222**: 61–68, 2001)

Key words: fruit fly, somatic-mutation-recombination test, electron-spin-resonance-spectroscopy, epidemiolog, chromium carcinogenicity

Introduction

Favorable physicochemical properties of chromate compounds (e.g. color, malleability, ductility, and tensile strength) have led to extensive use of this metal in various industries including chromate production, chromate pigment production, chrome plating, stainless steel welding, ferrochrome alloy production, and leather tanning. Epidemiologic studies of workers in various chromium processing and manufacturing industries have revealed significantly elevated cancer incidence and mortality rates for cancer sites including lung, nasopharyngeal and urinary bladder. Lung cancer mortality rates increase among chromate production workers exposed to Cr^{6+}/Cr^{3+} and among chromate pigment workers exposed to Cr^{6+}. Chrome platers similarly exposed to Cr^{6+} and nickel (Ni^{2+}) likewise have significantly elevated lung cancer risks.

Address for offprints: A. Chiu, National Center for Environment Assessment, Office of Research and Development, Washington, DC 20460, USA

Stainless steel and ferrochrome workers have high Cr^{3+}, Cr^{0} exposures and to a lesser extent Cr^{6+} exposures. However, associations between exposure and lung cancer among these stainless steel and ferrochrome workers are inconclusive if based on surrogate measures of exposure intensity. In other words, thus far it is unlikely that carcinogenic potentials of Cr^{3+} and Cr^{0} can be specifically quantified from existing cohort data sets solely using epidemiologic methods. Likewise, cancer epidemiologic studies of smelter workers with exposures to Cr^{6+}, Cr^{3+}, Cr^{0} and lanthanum, La^{2+} are burdened by multiple confounders. The same can be said for studies of foundry workers exposed to particulate matter and Cr^{3+}, Cr^{0}.

Despite this there seems little doubt that chromium is carcinogenic with increased risk among exposed workers based on duration of employment. Biopsies of tissues from workers and autopsies of workers who have died of lung cancer also show accumulated chromium loads with increasing employment duration in chromium production and processing. Analyses of exposure data by time from first exposure and total exposure duration show that lung cancer increases, but with long latency times.

Reasoning from cell physiology, Cr^{6+} is readily taken up in cells where it is believed to be reduced to lower valence states (Cr^{5+}, Cr^{4+}) prior to exerting its genotoxic effects. Neither pentavalent nor tetravalent chromium interacts directly with DNA, but rather interacts with hydrogen peroxide to yield highly reactive hydroxyl free radicals. While hydroxyl free radicals likely are the ultimate mutagens, it is not known if Cr^{5+} or Cr^{4+} is the penultimate mutagen.

Since hexavalent chromium is rapidly converted in the lung to insoluble trivalent chromium, one questions whether the trivalent compound is equally carcinogenic. Thus, the potential carcinogenicity of hexavalent vs. trivalent chromium of exposed industrial workers is a research concern for industry and public health officials.

This paper reports on the evaluation of Cr^{6+}, Cr^{4+} and Cr^{3+} for genotoxic effects in somatic tissue of the fruit fly, *Drosophila melanogaster*. The results suggest that Cr^{4+} is highly genotoxic in flies via induction of mitotic recombination. Mitotic recombination leads to a *loss-of-heterozygosity* in an organism, which is suspected of playing a major role in the development of human cancers [1–5].

Materials and methods

Somatic mutation and recombination Test of Drosophila melanogaster

The somatic mutation and recombination test of *Drosophila* (i.e. somatic wing spot assay) was developed in the 1980s to provide researchers with a rapid and inexpensive assay for genotoxicity in somatic cells of higher eukaryotes [6–7]. Details of the assay can be found in Graf *et al.* [6]. The assay can detect a variety of genetic alterations, including gene mutations, deletions, aneuploidy/segmental aneuploidy and mitotic recombination. The detection of mitotic recombination is an important feature of the assay since mitotic recombination can lead to *loss-of-heterozygosity* [3–10]. The assay utilizes two loci located on the left arm of chromosome 3 – *mwh* (*multiple wing hair*) and *flr*^{3} (*flare*). Both loci influence development of hair growth in each adult wing blade cell. An induced genetic alteration in a developing wing disc cell of a treated larva can result in a clone of cells possessing malformed hairs (i.e. a spot) on the adult wing blade.

Larvae exposed to a test substance are allowed to develop into adults, then their wings are removed and mounted in microscopic slides for scoring at 400× magnification. Three endpoints are distinguished in the assay: (1) small single spots of either *mwh* or *flr*^{3} phenotype, which consist of just 1–2 cells; (2) large single spots of either *mwh* or *flr*^{3} phenotype, which consist of 3 or more cells; and (3) twin spots, which consist of adjacent *mwh* and *flr*^{3} spots. While small and large single spots can arise from a variety of genetic alterations, twin spots result *solely* from mitotic recombination. The vast majority (90–95%) of spontaneously arising spots on wings of negative control flies are of the small single type while large single spots and twin spots account for the remaining 5–10% of negative control spots.

Experimental and statistical methodology

Virgin females of genotype *mwh* were mated to males of genotype *flr*^{3}*/TM3, Ser*. The *flr*^{3} allele is a homozygotic lethal maintained in a heterozygous state by using the multiply inverted third chromosome balancer *TM3,* which also contains the dominant wing mutation *Ser* (*serrate*). Female parental flies lay eggs for an 8-h interval on standard fly media in glass bottles. The parents were then removed and eggs allowed to continue normal development. Approximately 72 h after beginning egg-laying, the bottled 3rd instar larvae were collected, washed and randomly allocated to different treatments.

All treatments included feeding larvae for 6 h with a neutral cellulose powder or a test substance mixture both wetted with distilled water to form a slurry. Test substances included $K_2Cr_2O_7$ and GSH-Cr^{4+} evaluated at 20 and 40 mM concentrations, and $CrCl_3$ evaluated at 40 mM. This latter acute exposure treatment was used to overcome insolubility of GSH-Cr^{4+} in water. The negative control treatment was plain cellulose powder with water. Larvae were then collected after 6 h, washed and placed in glass vials containing Instant Drosophila Medium (Carolina Biological Supply Co., Burlington, NC, USA) wetted with distilled water. Following appearance in the vials, adult flies were removed and stored in

70% ethanol until needed. Culturing of strains and all experimental procedures were conducted at 25°C.

Experimental matings produced two types of progeny: (1) trans-heterozygous flies (*mwh flr⁺/mwh⁺ flr³*) identified as adults by their wild-type (non-Serrate) wings, and (2) inversion-heterozygous flies (*mwh/TM3, Ser*) identified by their Serrate wings. While all three endpoints (small singles, large singles and twin spots) develop on wings of the trans-heterozygous flies, only small and large *mwh* single spots develop on wings of inversion-heterozygous flies. This is because only *mwh* spots can occur in inversion-heterozygous flies since *flr³* is not present in the strain, and furthermore, mitotic recombination is suppressed by the multiple-inverted *TM3* chromosome. Among trans-heterozygous flies, 40 negative control wings were scored along with 20 wings from each mutagen-treated group. For inversion-heterozygous flies, 20 wings were scored for effects of the 40 mM treatments with $K_2Cr^{6+}{}_2O_7$ and GSH-Cr^{4+}.

Statistical analyses were conducted for tested endpoint variables. Data analysis yielded mean, range and standard deviations of spots observed in treated and negative control trans-heterozygous wings. Then a multiple-decision methodology described by Frei and Würgler [8] was applied to these descriptive statistics determining the test mutagen's capacity to induce specific endpoints as either (1) positive, (2) weakly positive, (3) inconclusive or (4) negative. Statistical significance was assessed at the $p = 0.05$ level.

Electron spin resonance or paramagnetic resonance spectroscopy

Following 6 h feeding of 3rd instar larvae that included test substances Cr^{6+}, Cr^{4+}, Cr^{3+}, some larvae were collected, washed and preserved in chloroform for later ESR analysis. Cr^{3+} signals can be measured on a Varian E-109E EPR spectrometer (9.6 GHz, X-band) since the machine can detect a wide 'scan' range (2000 gauss) with relatively high sensitivity. Spectrometer settings were: magnetic field, 3240 gauss; incident microwave power, 5 mW; modulation frequency, 100 kHz; and modulation amplitude, 2.0 gauss. Cr^{5+} and Cr^{4+} were a flat cell assembly. The settings were: receiver gain, 3.2×10^3; time constant, 0.3 sec; incident microwave power, 20 mW; modulation amplitude, 4.0 G; center field, 3,520 G; sweep width, 400 G for Cr^{4+} and 2,000 G for Cr^{3+}. Settings for the Cr^{5+} measurements were: 1.2 GHz, L-band; 425 gauss, magnetic field; incident microwave power, 30 mW; scan time, 1 min; and modulation frequency, 27 kHz [6–8]. A summary of the ESR study parameters is presented in Table 1.

Table 1. Parameters of the ESR analyses

Ionic species	Centerfield	Range	Modulation frequency	Modulation amplitude
Cr^{3+}	3520 gauss	2000 gauss	100 kHz (9.6 GHz)	2.0 gauss
Cr^{4+}	3520 gauss	400 gauss	100 kHz (9.6 GHz)	4.0 gauss
Cr^{5+}	3520 gauss	425 gauss	27 kHz (1.2 GHz)	0.2 gauss

Results

Somatic mutation and recombination test

A summary of genotoxic effects in trans-heterozygous fruit flies for various valences of chromium is presented in Table 2. Numbers and means of observed spots per wing are listed for endpoints (types of spots) at each test compound concentration. Cr^{6+} was a positive inducer of all endpoints at high and low dose concentrations.

Similarly, Cr^{4+} was found to be a positive inducer of all three endpoints at the higher concentration, but was judged inconclusive for inducing small and large single spots at the lower concentration. However, Cr^{4+} was a positive inducer of twin spots at the lower concentration. Findings that Cr^{6+} and Cr^{4+} induce twin spots demonstrate that chromium in these valence states induce mitotic recombinations. Cr^{3+} was inconclusive for inducing small, large single and twin spots. In other words, there was no Cr^{3+} induced-somatic-mutation or mitotic-recombination.

These above results will now be interpreted in light of established principles of biochemical reaction kinetics. It is well known that in a series of reactions, if the concentration of intermediate chemical species, e.g. in this case Cr^{4+} markedly increases above the concentration of the initial chemical reactant species (Cr^{6+}), there will be a corresponding faster and more concentrated buildup of the terminal product (somatic recombination) compared to the initial reactants.

Table 2. Frequency N and means per wings of induced spots in treated trans-heterozygous (*mwh flr⁺/mwh⁺ flr³*) flies*

Treatment dose	Wings, number	Small spots N	Small spots per wing	Large spots N	Large spots per wing	Twin spots N	Twin spots per wing
H_2O	40	5	0.125	1	0.025	0	0
Cr^{6+}, $K_2Cr_2O_7$							
40 mM	20	20	1.00 (POS)	82	4.10 (POS)	66	3.30 (POS)
20 mM	20	9	0.45 (POS)	39	1.95 (POS)	53	2.65 (POS)
Cr^{4+}, GSH-Cr							
40 mM	20	22	1.10 (POS)	23	1.15 (POS)	26	1.30 (POS)
20 mM	20	5	0.25 (INC)	5	0.20 (INC)	8	0.45 (POS)
Cr^{3+}, $CrCl_3$							
40 mM	20	3	0.15 (INC)	1	0.05 (INC)	0	0

*Comparison of statistical significance between H_2O-negative-control flies *vs* chromium treated flies where POS = positive result or INC = inconclusive result.

Table 3. Total wing spots in trans-heterozygous flies (*mwh flr*$^+$/*mwh*$^+$ *flr*^{3}) treated with Cr^{6+} and Cr^{4+}

Treatment	Total number spots	Mean number spots/wing
40 mM Cr^{6+}	168	8.40
20 mM Cr^{6+}	101	5.05
40 mM Cr^{4+}	71	3.55
20 mM Cr^{4+}	18	0.90

$$Cr^{6+} \Leftrightarrow Cr^{5+} \Leftrightarrow Cr^{4+} \Leftrightarrow \text{somatic recombination}$$
$$\Downarrow \quad\quad \Downarrow$$
$$\text{somatic mutation}$$

The total number of wing spots induced by Cr^{6+} and Cr^{4+} treatments at 20 and 40 mM in trans-heterozygous flies are summarized in Table 3. It shows that Cr^{6+} induced 8.40 spots per wing at 40 mM and 5.05 spots per wing at 20 mM, a 1.66 increase in mean spots per wing (i.e. 8.40/5.05 = 1.66 more for the higher treatment concentration. On the other hand, there was a 3.94 mean increase of Cr^{4+} induced spots (3.55/0.90 = 3.94) comparing the ratio of higher vs. lower treatment concentrations. The higher ratio of response associated with Cr^{4+} over Cr^{6+} with a doubling of mutagen concentration treatments and Cr^{4+} ability to increase twin spots or somatic recombination, therefore loss-of-heterozygosity contrast the results from Cr^{6+} treatment. These two observations suggest that Cr^{4+} *is* the penultimate genotoxic/carcinogenic agent.

A wing scoring summary of inversion-heterozygous flies treated with Cr^{6+} and Cr^{4+} is presented in Table 4. Only small and large single spots of phenotype *mwh* can arise in these flies (*flr*^{3} single spots and twin spots are not possible). For comparison purposes, the number of *mwh* single spots observed in trans-heterozygous flies are also presented, along with the total number of *mwh* single spots observed in the two strains.

The smaller number of *mwh* spots observed in inversion heterozygous flies compared to those in trans-heterozygous flies is due to suppression of mitotic recombination in the inversion-heterozygous strain. This is evidenced by a 91.5% reduction in *mwh* single spots induced by Cr^{6+} between the two strains [i.e. 1 – (7/82) = 0.915] and a similar 85.4% reduction observed for Cr^{4+}-induced spots. Similar results [6] were previously reported for Cr^{6+} in the wing spot assay.

Table 4. Mwh wing spots in inversion-heterozygous and trans-heterozygous flies treated with 40 mM Cr^{6+} and Cr^{4+}.

Strain	Mutagen	No. wings	Small spots	Large spots	Total spots
Inversion-heterozygous	Cr^{6+}	20	7	0	7
(*mwh/TM3, Ser*)	Cr^{4+}	20	5	1	6
Trans-heterozygous	Cr^{6+}	20	16	66	82
(*mwh flr*$^+$/*mwh*$^+$ *flr*^{3})	Cr^{4+}	20	21	20	41

Table 5. Estimated thermodynamic parameters for activation of chromium: $\Delta E = \Delta H + T\Delta S$; or ΔS = 1–200 cal/mole·degree

	ΔE	ΔH	ΔS
cal/mol	3–4000	660	1–200

Hence, one can infer that the majority (~90%) of single spot induction by Cr^{6+} and Cr^{4+} in trans-heterozygous flies arise from mitotic recombination *as opposed to somatic mutation.*

Thermodynamic parameters estimated from larval ESR spectra [11–13]

The estimated energy changes associated with the reduction mechanism of chromium ions ($Cr^{6+} \Rightarrow Cr^{5+} \Rightarrow Cr^{4+} \Rightarrow Cr^{3+}$) within cells is shown in Table 5.

Eight vials of *Drosophila* larvae exposed to 40 mM K_2CrO_4, 40 mM $CrCl_3$, and H_2O for 1, 3, and 6 h. ESR spectra of the larvae are plotted on premium quality bond paper. Peaks above background tracing are cut out and weighed. Density and thickness of the paper cutouts are assumed to be constant throughout. Therefore, the *weight ratios* of peaks become an appropriate estimate of the ratio of integrated areas under the ESR spectra. Larvae exposed to K_2CrO_4 for 1, 3 or 6 h have, respectively, values of 0.1, 0.07 and 0.04 g. Linear regression of decrease in concentrations of Cr^{5+} in larvae exposed to Cr^{6+} gives the following estimates: a = 0.0296, b = 8.93×10^{-4}, $\gamma = 0.95$. For larvae exposed to $CrCl_3$ for 1 and 3 h, the estimates are 0.06 and 0.08 g. Linear regression analysis yields estimates of: a = 0.0098, b = 9.16×10^{-4}, $\gamma = 0.818$.

An estimate of the ratio of kinetic constants of forward and backward reactions leading to the disappearance of paramagnetic radicals ($Cr^{5+} \Rightarrow Cr^{3+}$, $Cr^{3+} \Rightarrow Cr^{5+}$) is obtained by dividing the constant '**a**' of 2 linear regression coefficients 0.0098/0.0296 = 0.3, or 30.2%/h. This estimate is comparable to *in vivo* rodent data [14], where formation-half-life of $Cr^{6+} \Rightarrow Cr^{5+}$ paramagnetic species was reported as 37 min. In other words, Q_{10} for the elimination of paramagnetic radical Cr^{5+} is between 1–2 in living organisms, assuming a rodent body temperature of 30°C and *Drosophila* body temperature of 20°C.

Van't Hoff's equation states: $d\ln K/dT = \mathbf{\Delta H}/RT^2$;

> $\ln K = -\mathbf{\Delta H}/RT$, where R = 2.3; T = 298, and $\mathbf{\Delta H}$ is estimated to be 656.7 cal/mole

From an earlier estimate [9], the formation of the paramagnetic radical Cr^{+5} from Cr^{+6} is $\Delta E = RT \ln K$, where R = 2.3, T = 303, and K = 1.5×10^{-2}, which yields: ΔE = 3–4 Kcal/mole for formation of Cr^{5+} at g = 1.9950, and probably 1.9793. This amount of change in free energy is less than that released from the hydrolysis of high energy bonds in reactions involving ATP/GTP, 7–10 Kcal/mole; GSH, 10 Kcal/mole; or NADH/

NADPH, 12 Kcal/mole. A coupled reaction involving either high energy compounds such as glutathione or NADH would provide sufficient free energy to drive the ion reduction, $Cr^{+6} \Rightarrow Cr^{+4} \Rightarrow Cr^{+5}$ to completion. Now an estimate can be made of how these energy changes relate to the location and reduction mechanisms of chromium ions inside cells, i.e. $Cr^{+6} \Rightarrow Cr^{+5} \Rightarrow Cr^{+4}$. These estimates suggest how xenobiotics such as chromium ions enter cells and become detoxified or activated. The suggested thermodynamic parameters presented here identify the location of such reactions, whether they occur inside the nucleus, an organelle or in the cytoplasm. Our modelling indicates that major proportion of free energy change in mediating chromium ion reduction is provided for by the decrease in *cellular entropy* as chromium is now restricted inside the cell. The decrease in entropy is close to 1–200 cal/mol.deg. The consumption of *free energy* in the chromium reduction is even higher than the 7 kcal/mole of energy used by ATPase in the Na^+-K^+ pump at the plasma membrane. This energy is needed to keep a pentavalent cation Cr^{+5} inside the cell in combination with the coenzyme NAD or a tetravalent Cr^{+4} in combination with glutathione since chromium is no longer in the form of the divalent anion $Cr_2O_4^{2-}$. Moreover, neither our modelling or any evidence from molecular kinetics suggest that a specific enzyme is needed to drive the $Cr^{+6} \Rightarrow Cr^{+5} \Rightarrow Cr^{+4} \Rightarrow Cr^{+3}$ series of reduction reactions. Rather now it appears that activated chromium ions can be cytotoxic in the Cr^{+5} state in the cytoplasm or genotoxic in the Cr^{+4} state.

Discussion

Epidemiology vs. in vitro *experiments*

Evidence from epidemiologic studies of industrial workers strongly suggests chromate is a human carcinogen. Nonetheless, chromate is not mutagenic *in vitro* in the presence of liver P450 nonspecific oxidase preparations and does not form covalent bonds with DNA under physiological conditions [22–24, 26]. This latter idea is relevant since most ambient environmental carcinogens must be activated or metabolized to an ultimate carcinogen [14–18, 20, 21] for carcinogenesis to occur within exposed humans and other organisms.

Hexavalent chromium compounds are, in contrast [30], reduced sequentially and ultimately to a single trivalent chromium compound that has much lower carcinogenic potency. Nevertheless, a cogent theory of chromate-compound-reduction that includes hexavalent chromium Cr^{6+} as a necessary procarcinogen in an activation step leading to carcinogenesis was expounded and elaborated by Wetterhahn *et al.* [20, 25, 27–28, 31–34] and in some of the present our work [11–14, 35–39, 41–46].

ESR of chromium ions of different valencies [21–27]

If chromium is a human carcinogen based on epidemiologic data, which form of chromium Cr^{+6}, Cr^{+5}, Cr^{+4}, or Cr^{+3} is the penultimate or ultimate carcinogen? Cr^{6+} is reduced by ascorbate, glutathione [40] and NADH/NADPH to Cr^{5+} via enzyme complexes such as cytochrome P450, cytochrome b_5, electron transport chains and glutathione reductase. The octahedral Cr^{5+}-NADPH has a vacant site for H_2O_2 to form a long-lived complex, which, in turn, generates ·OH radicals in a Fenton reaction. These ·OH radicals appear to modify either the serum proteins, cell surface lipoproteins or glycoproteins leading to hypersensitivity and cytotoxic reactions. On the other hand, they can also modify specific nucleotides, thus affecting gene expression. Moreover, they can also excise phosphodiester bonds in DNA double helices, thus initiating carcinogenic events in target sites including upper and lower respiratory tract and bladder in humans. NADH (or NADPH) is 2–3 times more effective in forming Cr^{5+} on a molar basis than other potential reducing agents such as ascorbate or glutathione. The experimental evidence underlying this assertion is as follows: hyperfine splitting with 0.095 G spacing was discovered in the 9.6 GHz. X-band EPR showing interactions between protons and oxygen in the diol-Cr^{5+} complex ($g = 1.9796$). This hyperfine splitting is observed when Cr^{5+} is in the presence of NADH/NADPH or GSH reductase; but not with GSH or ascorbate.

Cr^{6+} enters the cell as CrO_4^- via diffusion through the Cl^-, HCO_3^- anionic channel. Cr^{6+} is then reduced intracellularly, sequentially to form Cr^{5+}, Cr^{4+}, Cr^{3+}. Changes in free energy during formation of these paramagnetic chromium radicals are as follows:

> $\Delta E = 3$–4 Kcal/mole for the formation of Cr^{5+} at $g = 1.9950$, and probably 1.9793.

NADH/NADPH and GSH (reductase) are the cellular reductants [22–25]. It is by forming reactive species, especially ·OH ions through 'Fenton-like' reactions, that transcription and replication of DNA occur, leading to cytotoxicity, hypersensitivity and carcinogenesis at the organismic level [26–35, 47–56].

Similar to glutathione, ascorbate generates Cr^{5+} and Cr^{4+} when incubated with Cr^{6+}. In addition, carbon-centered alkyl radicals and formyl radicals are generated simultaneously. Electrophoretic assays show that ascorbate-derived-radicals induce double-strand-breaks in DNA. In the circulatory system, ascorbate is present in high concentration extracelluarly, while glutathione and glutathione reductase [62] are present intracellularly in higher concentrations especially at the plasma membrane which facilitates uptake of amino acids or xenobiotics like CrO_4^{2-}, or $Cr_2O_7^{2-}$ under the study here.

Cr^{4+} generates Cr^{5+}, ·OH and OH – in the presence of H_2O_2. DNA double strands breaks are also induced by the Cr^{4+}, Cr^{5+}-

mediated Fenton-like reaction [36–52]. ·OH radicals from the reaction of H_2O_2 with Cr^{4+}, Cr^{5+} lead to hydroxylation of 2'-dG to form 8-OH-2'dG. The induction of double strand breaks or modification of the heterocyclic structure of nucleotides change the genomic sequence. These alterations can lead to cell death or genetic expression changes. Experimental chromate treatments have also been shown to inhibit cell growth by specific blockage of the progression of cells through S-phase of the cell cycle, that then affect cell proliferation directly via lipid peroxidation. It is likely that these changes will also induce somatic mutations and or somatic recombinations in exposed organisms leading to cancer development [53–56].

While evidence from epidemiologic data strongly suggests chromium is a human carcinogen, it is not known if all chromium ion species are equally potent in human carcinogenesis. Also unknown is whether Cr^{4+} or Cr^{5+} is the penultimate carcinogen. Also it is not yet established whether either or both paramagnetic radicals Cr^{4+}/Cr^{5+} are responsible for the liver cytotoxicity and or carcinogenicity in other target organs. It is hypothesized here that Cr^{5+} is the hepatoxicant, while Cr^{4+} tetravalent compound forms the penultimate carcinogen in the upper and lower respiratory tract and the bladder.

Carcinogenicity/genotoxicity of Cr^{+4}, Cr^{+5}, Cr^{+6}

Using surrogate prokaryote systems in human carcinogenicity [57] testing ignores genomic differences between bacteria and humans. Employing meiosis and chromosomal pairing in rodent genetic studies also has important drawbacks, since somatic mutation and recombination mechanisms associated with carcinogenesis differ from recombination during meiosis or in mutation of gametes. The best genetic experiments reflecting somatic cellular changes in eukaryotes are somatic recombination and mutation studies such as reported here in *D. melanogaster* [19, 51–61].

From epidemiologic studies of chromate production workers, there is evidence for a carcinogenic effect but demonstrated only after a prolonged latency from time of first exposure . This latter finding is likely relevant to information from the ESR data that show Cr^{+6} in the presence of lysosomeal phosphate is converted in cells to form Cr^{+4}, Cr^{+5} and ultimately into relatively insoluble Cr^{+3}. Our interpretative results from cell physiology [62] indicate that Cr^{+6} likely enters cells as $CrO_4^{-2}/Cr_2O_7^{-2}$ via an anion channel, whereas Cr^{+3} enters cells following phagocytosis [63] and is stored in lysosomes [64] in combination with the phosphate ion. In comparing these mechanisms, Cr^{+3} is at least 2–3 orders of magnitude lower in bioavailability than Cr^{+6} [65].

Thus, potent genotoxic effects of $Cr^{4+, 5+}$ observed in *D. melanogaster* somatic wing spot tests suggest that Cr^{+4} is the penultimate (and ·OH the ultimate carcinogen) in Cr^{6+} carcinogenesis. Moreover, this mechanism is an initiation event. Since genetic alterations induced by Cr^{6+} and Cr^{4+} appear to arise from induced mitotic recombination, it is likely the initiation event in chromium carcinogenesis involves the double strand excision of the DNA helix leading to a loss of heterozygosity in various target sites in the nuclear genome.

References

1. Sengstag C: The role of mitotic recombination in carcinogenesis. Crit Rev Toxicol 24: 323–353, 1994
2. Shammas MA, Reis RJS: Recombination and its role in DNA repair, cellular immortalization and cancer. Age 22: 71–88, 1999
3. Cavenee WK, White RL: The genetic basis of cancer. Sci Am 272: 72–79, 1995
4. Tischfield JA: Loss of heterozygosity or: How I learned to stop worrying and love mitotic recombination. Am J Hum Genet 61: 995–999, 1997
5. Shao C, Deng L, Henegariu O, Liang L, Stambrook PJ, Tischfield JA: Chromosomal instability contributes to loss-of-heterozygosity in mice lacking p53. Proc Natl Acad Sci USA 97: 7405–7410, 2000
6. Graf U, Würgler FE, Katz AJ, Frei H, Juon H, Hall CB, Kale PG: Somatic mutation and recombination test in *Drosophila melanogaster.* Environ Mutagen 6: 153–188, 1984
7. Graf U, Frei H, Kagi A, Katz AJ, Würgler FE: Thirty compounds tested in the *Drosophila* wing spot test. Mutat Res 22: 359–373, 1989
8. Frei H, Würgler FE: Statistical methods to decide whether mutagenicity test data from *Drosophila* assays indicate a positive, negative or inconclusive result. Mutat Res 203: 297–308, 1988
9. Tischfield JA: Loss of heterozygosity or: How I learned to stop worrying and love mitotic recombination. Am J Hum Genet 61: 995–999, 1997
10. Gupta PK, Sahota A, Boyadjiev SA, Bye S, Shao C, O'Neill JP, Hunter TC, Albertini RJ, Stambrook PJ, Tischfield JA: High frequency *in vivo* loss of heterozygosity is primarily a consequence of mitotic recombination. Cancer Res 57: 1188–1193, 1997
11. Liu KJ, Jiang JJ, Shi XL, Gabrys H, Wakczaj T, Swartz HM: Low frequency EPR study of Cr (V) formation from Cr (VI) in living plants. Biochem Biophys Res Comm 206: 829–834, 1995
12. Liu KJ, Jiang JJ, Swartz HM, Shi XL: Low frequency EPR detection of Chromium (V) formation by Chromium (VI) reduction in whole live mice. Arch Biochem Biophys 313: 248–255, 1995
13. Liu KJ, Shi XL, Dalal NS: Synthesis of Cr^{4+}-GSH, its identification and its free hydroxyl radical generation: A model compound for Cr^{6+} carcinogenicity. Biochem Biophys Res Comm 235: 54–58, 1997
14. Chiu A, Chiu N, Shi X, Beaubier J, Dalal N: Activation of a procarcinogen by reduction: $Cr^{6+} \rightarrow Cr^{5+} \rightarrow Cr^{4+} \rightarrow Cr^{3+}$: A case study by electron spin resonance spectroscopy (ESR/PMR). J Env Health Part C. Env Carc Ecotox Rev 16: 135–148, 1998
15. Vogel EW, Nivard JM: Performance of 181 chemicals in *Drosophila* assay predominantly monitoring interchromosomal mitotic recombination. Mutagenesis 8: 57–81, 1993
16. Vogel EW, Zijlstra JA: Mechanistic and methodological aspects of chemically-induced somatic mutation and recombination in *D. melanogaster*. Mutat Res 182: 243–264, 1987
17. Vogel EW, Zijlstra JA: Somatic cell mutagenesis in *D. melanogaster* in comparison with genetic damage in early germ-cell stages. Mutat Res 180: 189–200, 1987
18. Würgler FE, Vogel EW: *In vivo* mutagenicity testing using somatic cells of *D. melanogaster*. In: F. de Serres (ed). Chemical Mutagens, Vol. 10. Plenum Press, New York, 1986, pp 1–59

19. Davis PS, Shen MW, Judd BH: Asymmetric pairings of transposons in and proximal to the white locus of *Drosophila* account for four classes of regularly occurring exchange products. Proc Natl Acad Sci USA 84: 174–178, 1987
20. Miller EC: Some current perspectives on chemical carcinogenesis in humans and experimental animals: A presidential address. Cancer Res 38: 1479–1496, 1978
21. Miller EC, Miller JA: Searches for ultimate chemical carcinogens and their reactions with cellular macromolecules. Cancer 47: 2327–2345, 1981
22. Petrilli FL, DeFlora S: Toxicity and mutagenicity of hexavalent chromium in *Salmonella typhimurium*. Appl Environ Microbiol 33: 805–809, 1977
23. Petrilli FL, DeFlora S: Oxidation of inactive trivalent chromium to the mutagenic hexavalent form. Mutat Res 58: 167–178, 1978
24. Petrilli FL, DeFlora S: Metabolic deactivation of hexavalent chromium mutagenicity. Mutat Res 54: 139–147, 1978
25. Liu KJ, Shi X, Jiang JJ, Goda F, Dalal NS, Swartz HM: Chromate-induced chromium (V) formation in live mice and its control by cellular antioxidants: An L-band EPR study. Arch Biochem Biophys 323: 33–39, 1995
26. Bianchi V, Celott L, Lanfranchi G, Majone F, Marin G, Montaldi A, Sponza G, Tamino G, Venier P, Zantedeschi A, Levis, AG: Genetic effects of chromium compounds. Mutat Res 117: 279–300, 1983
27. Cupo DY, Wetterhahn KE: Modification of Cr (VI)-induced DNA damage by glutathione and cytochrome P-450 in chicken embryo hepatocytes. Proc Natl Acad Sci USA 82: 6755–6759, 1985
28. Hamilton JW, Wetterhahn KE: Differential effects of chromium (VI) on constitutive and inducible gene expression in chick embryo liver *in vivo* and correlation with chromium (VI) induced DNA damage. Mol Carcinogen 2: 274–286, 1989
29. Kasprzak KS: The role of oxidative damage in metal carcinogenesis. Chem Res Toxicol 4: 604–615, 1991
30. Klein CB, Frenkel K, Costa M: The role of oxidative processes in metal carcinogenesis. Chem Res Toxicol 4: 592–604, 1991
31. Aiyar J, Berkovits HJ, Floyd RA, Wetterhahn KE: Reaction of chromium (V) with hydrogen peroxide in the presence of glutathione; reactive intermediates and resulting DNA damage. Chem Res Toxicol 3: 595–603, 1990
32. Aiyar J, Borges KM, Floyd RA, Wetterhahn KE: Role of chromium (V) glutathione thiyl radical and hydroxyl radical intermediates in chromium (VI)-induced DNA damage. Toxicol Environ Chem 22: 135–148, 1990
33. Jennette KW: Microsomal reduction of the carcinogen chromate produces chromium (V). J Am Chem Soc 108: 874–875, 1982
34. Tsapakos MJ, Wetterhahn KE: The interaction of chromium with nucleic acid. Chem Biol Interact 46: 265–277, 1983
35. Chen F, Ye JP, Zhang XY, Rojanasakul Y, Shi XL: One-electron reduction of Cr^{6+} by α-lipoic acid and related hydroxyl radical generation, dG hydroxylation and nuclear transcription factor-κB activation. Arch Biochem Biophys 338: 165–172, 1997
36. Craf E, Mahoney JR, Bryant RG, Eaton JW: Iron catalyzed hydroxyl radical formation: Stringent requirement for free iron coordination site. J Biol Chem 259: 3620–3624, 1984
37. Liu KJ, Shi XL, Jiang JJ, Goda F, Dalal N, Swartz HM: Chromate induced Cr (V) formation in live mice and its control by cellular antioxidants; an L-band EPR study. Arch Biochem Biophys 323: 33–39, 1995
38. Liu KJ, Shi XL, Jiang JJ, Goda F, Dalal N, Swartz HM: Low frequency electron paramagnetic resonance investigation on metabolism of Cr^{6+} by whole live mice. Ann Clin Lab Sci 26: 176–184, 1996
39. Furono K; Suetsuga T, Sugihara N: Effects of metal ions on lipid peroxidation in cultured rat hepatocytes loaded with α-linolenic acid. J Toxicol Environ Health 48: 121–129, 1996
40. Meister A, Anderson ME: Glutathione. Ann Rev Biochem 52: 711–760, 1983
41. Shi XL, Dalal NS: On the mechanism of chromate reduction by glutathione: ESR evidence for the glutathionyl radical and an insoluble Cr (V) intermediate. Biochem Biophys Res Comm 156: 137–142, 1988
42. Shi XL, Dalal NS: Chromium (V) and hydroxyl radical formation during the glutathione reductase catalyzed reduction of chromium (VI). Biochem Biophys Res Comm 163: 627–634, 1989
43. Shi XL, Dalal NS: On the hydroxyl radical formation in the reaction between hydrogen peroxide and biologically generated Cr (V) species. Arch Biochem Biophys 277: 342–350; 281-90-5, 1990
44. Shi XL, Dalal NS: Evidence for a Fenton-type mechanism for the generation of ·OH radicals in the reduction of Cr (VI) in cellular media. Arch Biochem Biophys 281: 90–95, 1990
45. Shi XL, Dalal NS: NADPH dependent flavoenzyme catalyzed one electron reduction of metal ions, molecular oxygen and generate hydroxyl radicals. FEBS Lett 276: 189–191, 1990
46. Shi XL, Dalal NS: ESR spin trapping detection of hydroxyl radicals in the reactions of Cr (V) complexes with hydrogen peroxide. Free Radical Res Comm 10: 17–26, 1990
47. Szakmary A, Huang S-M, Chang DT, Beachy PA, Sander M: Overexpression of a Rrp1 transgene reduces the somatic mutation and recombination frequency induced by oxidative DNA damage in *Drosophila melanogaster*. Proc Natl Acad Sci USA 93: 1607–1612, 1996
48. Shi, XL, Dalal N: Generation of hydroxyl radical by chromate in biologically relevant systems: Role of Cr(V) complexes versus Tetra-peroxochromate(V). Environ Health Persp 102: 231–236, 1994
49. Shi XL, Dalal NS, Kasprzak KS: Generation of free radicals from hydrogen peroxide and lipid peroxides in the presence of Cr^{3+}. Arch Biochem Biophys 302: 294–299, 1993
50. Shi XL, Dong ZA, Dalal NS, Gannett PM: Chromate mediated free radical generation from cysteine, penicillamine, hydrogen peroxide, and lipid peroxides. Biochim Biophys Acta 1226: 65–74, 1994
51. Shi XL, Mao Y, Knaptor AD, Ding M, Rojanasakul Y, Gannett PM, Dalal N, Liu KJ: Reaction of Cr^{6+} with ascorbate and hydrogen peroxide generates hydroxyl radicals and causes DNA damage: Role of a Cr^{4+} mediated Fenton-like reaction. Carcinogenesis 15: 2475–2478, 1994
52. Shi XL, Sun XO, Gannett PM, Dalal NS: Deferoxamine inhibition of Cr^{5+}mediated radical generation and deoxyguanine hydroxylation: ESR and HPLC evidence. Arch Biochem Biophys 293: 281–286, 1992
53. Tsuda H, Kato K: Chromosomal aberrations and morphological transformation in hamster embryonic cells treated with potassium dichromate *in vitro*. Mutat Res 46: 87–94, 1977
54. Wedrychowski A, Ward WS, Schmidt WN, Hnilica LS: Chromium-induced cross-linking of nuclear proteins and DNA. J Biol Chem 260: 7150–7155, 1985
55. Xu J, Bubley GJ, Detrick B, Blankenship LJ, Patierno SR: Chromium (VI) treatment of normal human lung cells results in guanine-specific DNA polymerase arrest, DNA–DNA cross-links and S-phase blockade of cell cycle. Carcinogenesis 17: 1511–1517, 1996
56. Ye J P, Zhang XY, Young HA, Mao Y, Shi XL: Cr^{6+} induced nuclear factor-B activation in intact cells via free radical reactions. Carcinogenesis 16: 2401–2405, 1995
57. Ames, BN, McCann J, Yamasaki E: Methods for detecting carcinogens and mutagens with the Salmonella/mammalian-microsome mutagenicity test. Mutat Res 31: 347–364, 1975
58. Guzman-Rinconl J, Graf U: *Drosophila melanogaster* somatic mutation and recombination test as a biomonitor. In: F.M. Butterworth (ed). Biomonitor and Biomarkers as Indicators of Environmental Change. Plenum Press, New York, 1995
59. Spano MA, Rincon JG, Abraham SK, Andrade HH: The wing somatic mutation and recombination test (SMART) in *Drosophila melanogaster*: An efficient tool for the detection of genotoxic activity of pure

compounds or complex mixtures as well as for studies on antigenotoxicity. Afr Newslett Occup Health Safety 6: 9–13, 1995

60. Davis PS, Shen MW, Judd BH: Asymmetric pairings of transposons in and proximal to the white locus of *Drosophila* account for 4 classes of regularly occurring exchange products. Proc Natl Acad Sci USA 84: 174–178, 1987

61. Szakmary A, Huang Shu-Mei, Chang DT, Beachy PA, Sander M: Overexpression of a Rrp1 transgene reduces the somatic mutation and recombination frequency induced by oxidative DNA damage in *Drosophila melanogaster*. Proc Natl Acad Sci USA 93: 1607–1612, 1996

62. Standeven AM, Wetterhahn KE: J Am Coll Toxicol 8: 1275, 1989

62a. Kitagawa S, Seki H, Kametani F, Sakurai H: Uptake of hexavalent chromium by bovine erythrocytes and its interaction with cytoplasmic components: The role of glutathione. Chem Biol Interact 40: 265–274, 1982

63. Buttner B, Beyersmann D: Modification of the erythrocyte anion carrier by chromate. Xenobiotica 15: 735–741, 1985

64. Berry JP: The role of lysosomes in the selective concentration of mineral elements. A microanalytical study. Cell Mol Biol 42: 395–411, 1996

65. Kerger BD, Paustenbach DJ, Corbett GE, Finley BH: Absorption and elimination of trivalent and hexavalent chromium in humans following ingestion of a bolus dose in drinking water. Tox Appl Pharm 141: 145–58, 1996

Molecular and Cellular Biochemistry **222**: 69–76, 2001.

Carcinogenic potential and genomic instability of beryllium sulphate in BALB/c-3T3 cells

Nagalakshmi Keshava, Gu Zhou, Michelle Spruill, Mang Ensell and Tong-man Ong

Health Effects Laboratory Division, National Institute for Occupational Safety and Health, Centers for Disease Control and Prevention, Morgantown, West Virginia, USA

Abstract

Occupational exposure to beryllium (Be) and Be compounds occurs in a wide range of industrial processes. A large number of workers are potentially exposed to this metal during manufacturing and processing, so there is a concern regarding the potential carcinogenic hazard of Be. Studies were performed to determine the carcinogenic potential of beryllium sulfate ($BeSO_4$) in cultured mammalian cells. BALB/c-3T3 cells were treated with varying concentrations of $BeSO_4$ for 72 h and the transformation frequency was determined after 4 weeks of culturing. Concentrations from 50–200 μg $BeSO_4$/ml, caused a concentration-dependent increase (9–41 fold) in transformation frequency. Non-transformed BALB/c-3T3 cells and cells from transformed foci induced by $BeSO_4$ were injected into both axillary regions of nude mice. All ten Be-induced transformed cell lines injected into nude mice produced fibrosarcomas within 50 days after cell injection. No tumors were found in nude mice receiving non-transformed BALB/c-3T3 cells 90 days post-injection. Gene amplification was investigated in K-*ras*, c-*myc*, c-*fos*, c-*jun*, c-*sis*, *erb-B2* and *p53* using differential PCR while random amplified polymorphic DNA fingerprinting was employed to detect genomic instability. Gene amplification was found in K-*ras* and c-*jun*, however no change in gene expression or protein level was observed in any of the genes by Western blotting. Five of the 10 transformed cell lines showed genetic instability using different random primers. In conclusion, these results indicate that $BeSO_4$ is capable of inducing morphological cell transformation in mammalian cells and that transformed cells induced by $BeSO_4$ are potentially tumorigenic. Also, cell transformation induced by $BeSO_4$ may be attributed, in part, to the gene amplification of K-*ras* and c-*jun* and some $BeSO_4$-induced transformed cells possess neoplastic potential resulting from genomic instability. (Mol Cell Biochem **222**: 69–76, 2001)

Key words: cell transformation, tumorigenicity, genomic instability, beryllium sulphate

Introduction

Beryllium (Be) is widely used in industries due to its special properties, such as being lighter than aluminum yet 40% more rigid than steel [1]. Although a relatively small number of workers are potentially exposed to high levels of Be in the refining and machining of the metal and in production of Be-containing products, a growing number of workers are exposed to lower levels of Be in ceramics, production of dental equipment and supplies, golf club manufacturing, soldering and welding as well in the electronics, aircraft, aerospace and nuclear industries [2].

Beryllium is classified as a potential carcinogen in humans and inhaled beryllium causes a high incidence of rat lung tumors. Their mutagenicity and carcinogenicity have been extensively studied in the laboratory and in exposed workers [2–4]. Bacterial cells seem insensitive to Be, and most results from bacterial mutagenicity assay systems have been negative [5–8]. In mammalian cells, however, Be induced sister chromatid exchange (SCE), chromosomal aberrations

Address for offprints: N. Keshava, Toxicology and Molecular Biology Branch, HELD, NIOSH, m/s 3014, 1095 Willowdale Road, Morgantown, WV 26505-2845, USA

(CA), *HPRT* gene mutation and cell transformation [7, 9–13]. The carcinogenicity of Be has been demonstrated in several animal species including rats, mice, rabbits and monkeys [2–5, 14]. A report from the International Agency for Research on Cancer [2] concluded that there is sufficient evidence in humans of the carcinogenicity of Be and Be-compounds, however, little is known regarding the mechanism of Be carcinogenesis.

Morphological cell transformation is a useful short-term test for the *in vitro* detection of potential carcinogens [15]. The athymic nude mouse is an excellent animal model for determining the tumorigenesis of morphologically transformed cells [16]. Cell transformation coupled with the nude mouse/tumorigenesis assay is a useful approach to study the carcinogenic potential of environmentally and occupationally related agents. Moreover, it enables the study of mechanistic and/or sequential processes during carcinogenesis. It is known that genetic alterations including gene amplification, deletion, and point mutation may play an important role in the activation and/or inactivation of cancer-related genes. In this investigation, an effort has been made to sequentially study the morphological transforming activity of $BeSO_4$ in BALB/c-3T3 cells and to determine whether these transformed cells are capable of producing tumors in the nude mouse assay system. Also, the possible mechanism of Be-induced cell transformation related to the amplification and/or expression of cancer-related genes was investigated. In addition, genomic instability in these transformed cells was also studied.

Materials and methods

Chemicals and reagents

Beryllium sulfate tetrahydrate ($BeSO_4{\cdot}4H_2O$, 99.99% pure) was purchased from Aldrich Chemical Company, Inc. (Milwaukee, WI, USA) and was freshly made and diluted into different concentrations with sterile water. Eagle's minimum essential medium (MEM), fetal bovine serum (FBS), L-glutamine, trypsin, and penicillin-streptomycin were purchased from Sigma (St. Louis, MO, USA).

Cell culture

BALB/c-3T3 clone A31-1-13 cells, kindly provided by Dr. M.A. Cifone (Covance Laboratories Inc., Vienna, VA, USA), were used in this study. The cells were maintained in MEM supplemented with 10% FBS, 1% L-glutamine, 100 U/ml penicillin and 100 μg/ml streptomycin and incubated at 37°C with 5% CO_2. The cultures were split before reaching 70% confluence.

Cytotoxicity assay

Exponentially growing cells were seeded into 25 cm^2 flasks at a density of 3×10^4 cells/5 ml of complete medium. After 24 h culturing, the cells were exposed to final concentrations of $BeSO_4$ from 50–200 μg/ml for 72 h. The viable cells were counted under a microscope using trypan blue exclusion assay. The doses for the transformation assay were chosen from a range of 30–90% relative to cell survival.

Transformation assay

BALB/c-3T3 cells (3×10^4) at passage 5–8 were seeded into 25 cm^2 flasks (at least 25 flasks for each group) with 5 ml medium. Exposure was initiated 24 h later with final concentrations of 0, 50, 100, 200 μg $BeSO_4$ /ml for 72 h. After treatment the cells were rinsed 3 times with phosphate buffered saline (PBS) and replenished with fresh maintenance medium (7.5% FBS). The cultures were re-fed twice a week for an expression period of 4 more weeks. At the end of the expression period, cells from foci that were larger than 4 mm in diameter with piling characteristics were isolated and established as transformed cell lines for tumorigenicity testing in nude mice. After isolation, the flasks were washed with PBS, then fixed with methanol and stained with 10% Giemsa. Only Type 3 foci [17] in both control and treatment groups were scored as transformed foci.

Tumorigenicity in athymic nude mice

Athymic female nude mice (nu/nu, 3–4 weeks old) and heterozygotic female mice (+/nu) were purchased from Harlan Sprague Dawley (Indianapolis, IN, USA). Thirty-six nude mice were divided randomly into eleven groups (three mice/group, six mice in the BALB/c-3T3 control group). The transformed cell lines were passed 4–6 times prior to injection. Ten transformed cell lines and a non-transformed BALB/c-3T3 cell line were separately harvested and concentrated to densities of 10^7 cells/ml of medium. Both sites of the axillary region of the nude mice were subcutaneously injected with 0.1 ml of cell suspensions (1×10^6 cells /each site). The nude mice were maintained under sterile conditions according to protocol from the American Association for the Accreditation of Laboratory Animal Care. Autoclaved cages, drinking water and food were changed once a week. Four heterozygotic mice were used as room and bedding controls for serosurveillance of rodent pathogens (Complete Health Monitoring Profile, Charles River Laboratories, Wilmington, MA, USA).

The mice were screened twice a week for the appearance and size of tumors. When tumor size at one site reached 1.5–

2.0 cm in length or width, a mouse was euthanized using sodium pentabarbitol. The tumors were surgically removed under sterile conditions and weighed, then divided into three parts. One portion was used for cell culturing, one piece was fixed in 10% neutral buffered formalin (NBF) for histopathological studies, and the third portion was frozen in liquid nitrogen for future analysis. If no visible tumor was found three months after injection, the skin around the injection site was removed and then examined histopathologically. The lungs were also collected for histopathological monitoring of tumor metastases. The tumor, skin or lung tissue, was fixed with 10% NBF and sectioned onto slides. The slides were then stained with hemotoxylin and eosin. Blind assessment was used for the histopathological analysis.

Differential PCR

DNA was isolated from 10 different Be-transformed cell lines and a non-transformed BALB/c-3T3 cells according to standard protocol [18]. Differential PCR was carried out independently using primers for K-*ras*, c-*fos*, c-*jun*, c-*myc*, c-*sis*, *erb-B2* and *p53* (Table 1). Briefly, template DNA (200 ng), a target gene primer pair (30 pmol each), reference gene (β-actin, 10–30 pmol), dNTPs (0.2 mM) and *Taq* polymerase (1.25 U, Promega, Madison, WI, USA) were added in PCR buffer and run for 30 cycles (denaturation at 94°C for 1 min; annealing at 50–59°C for 1 min; and extension at 72°C for 2 min). Following the PCR reaction, products were electrophoresed on a 2% agarose gel and the images were analyzed using the Eagle Eye II (Stratagene, La Jolla, CA, USA) gel documentation system. NIH Image for Macintosh (NIH, Bethesda, MD, USA) was used to determine the intensities of the resulting bands. The relative ratio of the intensity of the target gene to that of the reference gene was calculated allowing the gene copy number to be estimated.

RAPD analysis

Six arbitrary primers were used for randomly amplified polymorphic DNA (RAPD) analysis. The sequences of these primers are shown in Table 2. DNAs (0.5 μg) from each cell line (non-transformed and transformed BALB/c-3T3 cells) were amplified with each arbitrary primer (50 pmol), *Taq* DNA polymerase (1.25 U; Promega, Madison, WI, USA), dNTPs (100 μM; Promega) and $MgCl_2$ (2.5 mM) in 50 μl of reaction buffer for 40 cycles. PCR conditions were: denaturing at 94°C for 30 sec, annealing at 40°C for 1 min and extension at 72°C for 1 min. PCR products (10 μl each) mixed with loading buffer were loaded on 2% agarose gels and electrophoresed with 100 V for 1 h. The gels were stained with ethidium bromide (0.5 μg/ml), visualized using an Eagle Eye II documentation system and photographed. The banding profiles of non-transformed and transformed cell lines were analyzed based on changes in the intensity of a band, band shifts, missing or new bands.

Table 1. Primer sequences for proto-oncogenes and tumor suppressor gene used in differential PCR studies

Gene	Primer sequence
c-*jun*	5′-AACTCGGACCTTCTCACGTCG-3′ 3′-TGCTGAGGTTGGCGTAGACC-5′
K-*ras*	5′-TCTGTTCGTGCAAACTGTCAGC-3′ 3′-TGTCTGCCTACGCCTCTGAAAG-5′
c-*fos*	5′-CGTTGCAGACTGAGATTGCC-3′ 3′-ACCGGACAGGTCCACATCTG-5′
c-*myc*	5′-TCCATTCCGAGGCCACAGCAAG-3′ 3′-TCAGCTCGTTCCTCCTCTGACG-5′
c-*sis*	5′-GGTTGAGTGTCCCAGCATG-3′ 3′-TGGCCACTAGGATGTGCTAC-5′
erb-B2	5′-ATGCAAAGTGCAATGAAGACC-3′ 3′-TGTTCCTCCCTTTGTACTGTCC-5′
p53	5′-CATCACCTCACTGCATGGACG-3′ 3′-CAGACTTGGCTGTCCCAGACTG-5′
β-actin	5′-AGGCATTGTGATGGACTCCG-3′ 3′-AGTGATGACCTGGCCGTCAG-5′

Western blot analysis

Protein was obtained from 10 Be-transformed and one non-transformed BALB/c3T3 cell lines by standard methods [18]. Five micrograms of each protein sample was mixed with an equal volume of 2X SDS sample loading buffer containing 5% 2-mercaptoethanol and heated to 95°C for 3 min before being subjected to Western blot analysis.

Western blotting was carried out according to an established method [19]. Briefly, samples were loaded on a 5–17%

Table 2. Primers used for RAPD analysis to detect DNA alterations in non-transformed, $BeSO_4$-transformed cell lines

Serial No.	Primer No.	Sequence (5′ – 3′)
1	448	TGGGCATCTG
2	453	AGCTGCCGGG
3	457	AAGGCTAGCG
4	460	AGGCATTCCC
5	485	CGGCCCCTGT
6	488	CAGGCCCTTC

polyacrylamide gradient gel, electrophoretically separated, and transferred to Immobilon PVDF transfer membranes. Primary antibodies were used to detect proteins produced by six different proto-oncogenes: *ras* (pan-*ras* Ab-3, Oncogene Research, Cambridge, MA, USA), K-*ras*, *sis*, *myc*, *erb*-B2 (Quality Biotech, Camden, NJ, USA), *fos* (Ab-1, Oncogene Research) and *jun* (Ab-1, Oncogene Research), and one tumor suppressor gene: *p53* (Ab-3, Oncogene Research). The antibody complexes were visualized using an avidin-alkaline phosphatase detection system. Resulting bands were analyzed using the Whole Band Analyzer (BioImage, Ann Arbor, MI, USA).

Statistical analysis

A *t*-test was used to compare the mean number of transformed foci/flask as well as tumor size or weight. Mean and standard deviations were calculated for all the data recorded. Transformation frequencies were also compared using the χ^2 test.

Results

As shown in Table 3, clear dose-dependent cytotoxicity was demonstrated, indicating that $BeSO_4$, at the concentrations tested, was toxic to BALB/c-3T3 cells. Compared with the control, the relative cell surviving rates decreased from 100 to 17.4% with increasing $BeSO_4$ concentrations from 0–200 µg/ml. Table 3 also shows a dose-response relationship for morphological transformation induced by $BeSO_4$. Transformation frequencies (TF) were 9, 14 and 41 fold greater than the spontaneous TF at doses of 50, 100 and 200 µg/ml, respectively. The mean numbers of transformed foci/flask were 8, 9 and 10 fold greater than that of the non-exposed control cells. The differences between control and test groups were statistically significant ($p < 0.001$).

In the tumorigenicity study, an animal was euthanized when tumor size reached 1.5–2.0 cm in length or width at one site. The earliest sacrifice of animals occurred only 12 days after injection with transformed cells (Fig. 1). The number of sacrificed animals increased rapidly from 10–87% between 12 and 34 days after inoculation with transformed cells. The remaining nude mice inoculated with transformed cells reached the tumor size limit within 50 days. In contrast, the six nude mice injected with non-transformed BALB/c-3T3 cells were kept for 90 days and no tumors were found.

The tumorigenic potential of transformed cells in nude mice is summarized in Table 4. All mice inoculated with transformed cells developed tumors, and tumors were seen at both sites except in one animal. The histopathological results showed that all tumors were fibrosarcomas, and no metastases were found in the lungs. Compared with non-transformed BALB/c-3T3 cells, tumor frequencies increased significantly and latency periods decreased. The tumor frequency, average tumor weight, average tumor size and the average latency period were not statistically different between transformed cell lines.

To further study the mechanism of $BeSO_4$-induced cell transformation, gene amplification, protein expression and genomic instability assays were carried out. Gene amplification was studied in proto-oncogenes (K-*ras*, c-*fos*, c-*jun*, c-*myc*, c-*sis*, *erb-B2*) and a tumor suppressor gene (*p53*) (Fig. 2). Increased copy number was found for c-*jun* and K-*ras*. None of the other genes tested showed amplification. In this study, analysis of gene expression was also conducted in all the genes as in amplification. Also, mutations in K-*ras*, codons 12, 13 and 61 were evaluated using the PCR-RFLP method. No increase in the expression in any of the genes were observed (data not shown) and no mutations were found in K-*ras*, codons 12, 13 or 61 in 10 $BeSO_4$-induced transformed cell lines (data not shown). In addition, Western blot analysis showed no difference in the levels of any of the proteins examined (K-RAS, MYC, FOS, JUN, SIS, ERB-B2 and P53) in Be-transformed cells.

Genomic changes in $BeSO_4$-induced transformed cell lines were detected by RAPD analysis (Table 5). Representative banding patterns of 10 different cell lines of $BeSO_4$-transformed cell DNA using primer #453 are shown in Fig. 3. Six random primers were used to randomly span the whole genome. Non-transformed BALB/c-3T3 cells and 10 transformed cell lines obtained from all three different concen-

Table 3. Cytotoxicity and morphological transformation induced by $BeSO_4$ in BALB/c-3T3 cells

$BeSO_4$ (µg/ml)	Flasks/ group	Type III foci	Mean no. of transformed foci flask[a] (× ± S.D.)	Relative cell surviving[b] (%, × ± S.D.)	Transformation frequency[c] (10^{-5} × ± S.D.)
0.0	54	18	0.33 ± 0.476	100.00	0.70 ± 0.57
50.0	60	156	2.60 ± 1.532*	68.20 ± 16.69	6.05 ± 0.78*
100.0	63	184	2.92 ± 1.619*	43.95 ± 8.13	10.08 ± 0.03*
200.0	72	245	3.40 ± 2.053*	17.40 ± 0.14	28.59 ± 4.82*

*Significantly different from control, $p < 0.001$. [a]Mean number of transformed foci/flask = total number of type III foci/number of flasks; [b]relative cell surviving = (number of live cells/number of live cells in control) × 100%; [c]transformation frequency = total number of type III foci/(total number of plated cells x relative cell surviving).

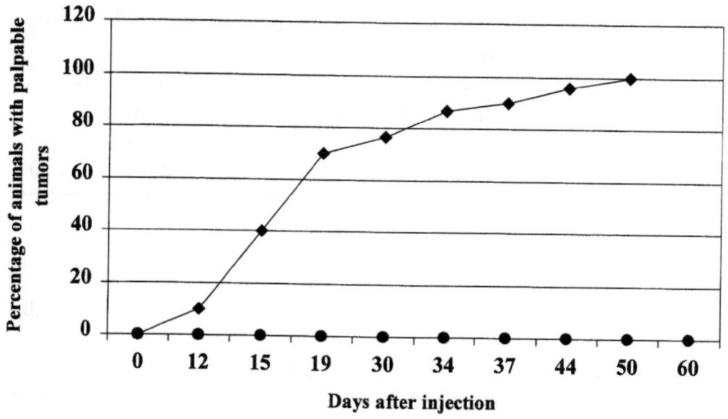

Fig. 1. Tumorigenic potential of Be-transformed cells. Mice injected with transformed cells (◆); mice injected with non-transformed BALB/c-3T3 cells (●).

trations were studied. Band shifts and changes in intensity were found in two transformed cell lines using primer 448. Similar changes including missing bands and the appearance of new bands were seen when primers 485 and 488 were used. Transformed cell lines 2, 3, 6, 8 and 10 also had genomic changes (Table 5).

Discussion

Beryllium sulfate is inactive in most bacterial mutagenesis assays. It induces mutations in neither *Salmonella typhimurium* (strains TA 98, 100, 1530, 1535, 1536, 1537, 1538 and LT_2), in the presence or absence of metabolizing systems, nor in *Escherichia coli* WP2 *uvr*A [5, 6, 20, 21]. Moreover, $BeSO_4$ was not cytotoxic in bacteria, and no growth inhibition was seen even with exposures up to 5 mg/plate with or without S9 metabolic activation in the Ames assay [5]. $BeSO_4$ was, however, cytotoxic to BALB/c-3T3 cells as shown in this study. It has been suggested that unlike mammalian cells, bacterial cells fail to uptake beryllium salts [2]. This seems to be a plausible explanation for the negative results from bacterial systems.

Exogenous metabolic activation was not used in this study, but a dose-response relationship in cell transformation induced by $BeSO_4$ was obtained, and the differences between the control and experimental groups were statistically significant ($p < 0.001$). These results indicate that additional metabolic activation was not necessary for $BeSO_4$ to induce morphological cell transformation in BALB/c-3T3 cells, but these results could not answer whether Be directly induced cell transformation without any metabolism, since BALB/c-3T3 cells still had metabolic capability, although it was much lower than in primary cells [17].

The mean numbers of transformed foci/plate and the transformation frequencies were used to evaluate the cell transformation assay [10, 17, 22–24]. For the former, because the cytotoxicity was not considered as a factor when simply counting the number of foci per plate, the transforming ability may not be accurately estimated. For example, in the current study, if the mean numbers of foci/flask were used to evaluate the transforming ability of $BeSO_4$, it would be only 8, 9 or 10 fold greater than that of the control. But the transformation frequencies were 9, 14 or 41 fold greater than

Table 4. Tumorigenic potential of transformed cells induced by $BeSO_4$ in nude mice

Transformed cells	No. of mice	Sites of inoculation	Characteristics of tumor					
			frequency	size (cm³)	weight (g)	type	latency (days)	metastases
50.0 μg/ml								
C1	3	6	6	2.79 ± 0.46	1.49 ± 0.08	FS	37.3	0
C2	3	6	6	4.20 ± 0.62	2.09 ± 0.15	FS	15.0	0
C3	3	6	6	2.83 ± 0.49	1.53 ± 0.23	FS	19.0	0
Subtotal	9	18	18/18*	3.27 ± 0.83	1.72 ± 0.33		23.78 ± 10.71*	0
100 μg/ml								
C1	3	6	5	2.89 ± 1.22	1.45 ± 0.28	FS	43.7	0
C2	3	6	6	5.02 ± 1.09	1.83 ± 0.59	FS	12.0	0
C3	3	6	6	5.53 ± 2.92	3.34 ± 1.49	FS	19.0	0
C4	3	6	6	3.97 ± 0.87	2.24 ± 0.83	FS	17.7	0
Subtotal	12	24	23/24*	4.35 ± 1.82	2.22 ± 1.07		23.08 ± 13.05*	0
200 μg/ml								
C1	3	6	6	2.59 ± 1.05	1.34 ± 0.37	FS	16.3	0
C2	3	6	6	4.73 ± 2.35	2.28 ± 0.62	FS	31.3	0
C3	3	6	6	4.35 ± 1.36	1.70 ± 0.59	FS	15.0	0
Subtotal	9	18	18/18*	3.89 ± 1.76	1.77 ± 0.59		20.89 ± 8.02*	0
BALB/c 3T3 cells	6	12	0/12	0	0	–	> 90.00	0

FS – Fibrosarcoma. *Significantly different from control; C1, C2, C3, C4 are different transformed cell isolates from the same treatment group.

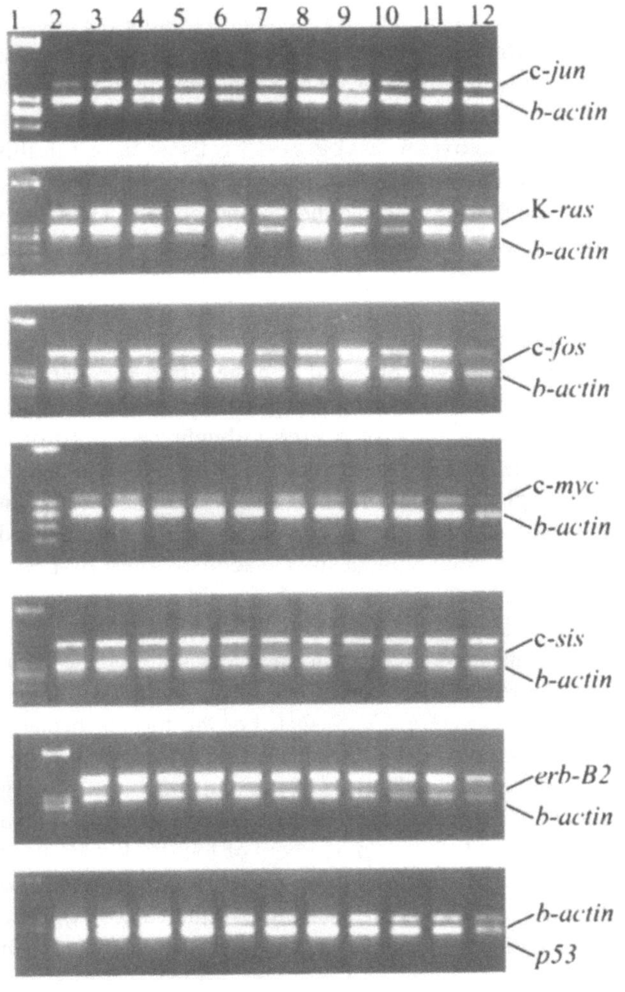

Fig. 2. Gene amplification for c-*jun*, K-*ras*, c-*fos*, c-*myc*, c-*sis*, *erb-B2* and *p53*. Lane 1: Φ174/*Hae III* molecular weight marker; lane 2: non-transformed BALB/c-3T3 cells; lanes 3–12: $BeSO_4$-transformed cells.

that of the control (Table 3). When cytotoxicity was included in the calculation, a much stronger transformation response was obtained at higher concentrations. Therefore, the mean number of foci/plate may not be a good indicator for evaluation of the transforming activity, particularly at toxic doses.

Carcinogenesis is thought to be a multistep process of loss of growth control arising in a single cell, including initiation, promotion and progression [25, 26]. Cell transformation *in vitro* also goes through a complex series of events with patterns of growth control changes similar to carcinogenesis *in vivo* [15, 27]. This relationship allows morphological cell transformation to be considered useful as an *in vitro* model for the investigation of chemical carcinogenesis [15, 28], however, its measured endpoint has an important drawback. Because morphologically transformed foci are not tumors, it cannot be clearly answered whether the morphologically altered foci have the potential of tumorigenesis *in vivo*. Therefore, it is very useful in understanding the carcinogenesis of

Table 5. Genetic instability detected in Be-transformed cell lines when compared to non- transformed BALB/c-3T3 cells

Primers	Genetic alterations in transformed cells lines									
	1	2	3	4	5	6	7	8	9	10
448	–	–	–	–	–	a,b	–	a	–	–
453	–	–	–	–	–	a,b,c	–	b,c	–	–
457	–	b,c	c	–	–	b	–	–	–	–
460	–	–	–	–	–	a,b	–	–	–	a,b
485	–	–	–	–	–	a,c	–	b	–	–
488	–	–	–	–	–	d,b	–	b	–	d,a

[a]band shift; [b]change in intensity; [c]missing band; [d]new band appearance.

chemicals when the cell transformation assay is coupled with a nude mouse study. The combined system has the advantages of an *in vitro* short-term test, while enabling the study of mechanistic and/or sequential processes that occur during carcinogenesis. Although Dunkel *et al.* [10] described that $BeSO_4$-induced BALB/c-3T3 cell transformation, the tumorigenic potential of morphologically transformed cells *in vivo* has not been reported. The results, as shown in our study, indicate that transformed cells induced by $BeSO_4$ have a strong potential for tumorigenesis *in vivo* because transformed cells induced tumors rapidly in nude mice after inoculation.

It has been concluded by IARC that Be is potentially carcinogenic to humans and causes high incidence of tumors in animals, but the mechanism of carcinogenesis is not clearly understood. In mammalian cell systems, the major Be-induced damages reported in the literature are SCEs, CAs and cell transformation [9, 11, 12]. Activation of the K-*ras* proto-oncogene and inactivation of the *p53* tumor suppressor gene are events common to many types of human cancers. No K-*ras* gene mutations or less than 10% K-*ras* codon 12 GGT-GTT transversion, however, were detected in rat lung tumors induced by Be [29–31]. In our study, no mutations in K-*ras*

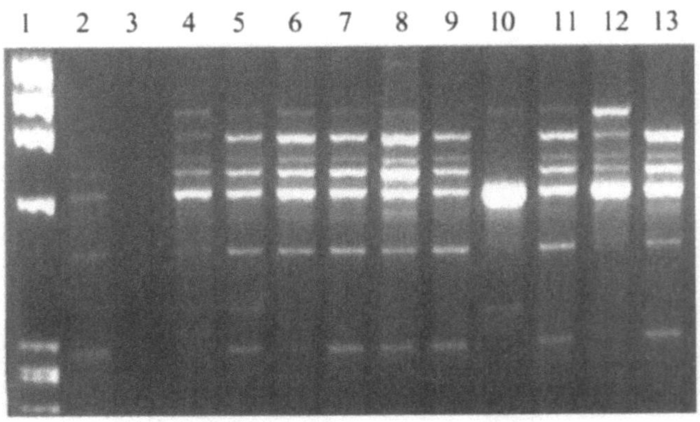

Fig. 3. Random amplified polymorphic DNA analysis of non-transformed and $BeSO_4$-transformed cell lines. Lane 1: Φ174/*Hae III* molecular weight marker; lane 2: non-transformed BALB/c-3T3 cells; lane 3: a blank; lanes 4–13: $BeSO_4$-transformed cells.

codon 12, 13 or 61 were found in any of the transformed cell lines. Therefore, these results suggest that K-*ras* activation is a rare event in the Be carcinogenic process. No *p53* gene mutations were observed in previous studies of animal lung tumors [28, 29]. It seems, therefore, that point mutations of these genes might not play a critical role in Be-induced animal lung tumors. Further studies were performed in our laboratory to determine whether alterations in certain proto-oncogenes and tumor suppressor genes were involved in Be-induced cell transformation and/or tumor development in nude mice. Both K-*ras* and c-*jun* were amplified in most transformed cell lines. We have previously observed c-*jun* overexpression by Western blotting in silica and glass fiber-transformed BALB/c-3T3 cells [32], however this result was not seen in the Be-transformed cells. The level of gene amplification of both K-*ras* and c-*jun* in the present study was relatively low (only 2 times the normal copy number). Therefore, it is possible that the level of increased expression was outside the lower limits of detection by Western blotting, yet still significant enough to influence cell transformation. Also, genomic alterations as observed by RAPD showed that some transformed cell lines were altered. Therefore, it is possible that cell transformation induced by $BeSO_4$ may be attributed, in part, to the gene amplification of K-*ras* and c-*jun* and that some $BeSO_4$-induced transformed cells possess neoplastic potential resulting from genomic instability.

Acknowledgements

We are grateful to Dr. A.F. Hubbs and Ms. L. Battelli for their expert technical assistance in pathology.

References

1. WHO: Beryllium (Environmental Health Criteria 106). Geneva, 1990
2. IARC Monographs on the Evaluation of the Carcinogenic Risk to Humans: Beryllium, cadmium, mercury, and exposures in the glass manufacturing industry. Lyon 58: 41–117, 1993
3. IARC Monographs on the Evaluation of the Carcinogenic Risk of Chemicals to Humans: Some metals and metallic compounds. Lyon 23: 143–204, 1980
4. Flamm WG: Beryllium: Laboratory evidence. IARC Sci Publ 65: 199–201, 1985
5. Ashby J, Ishidate M Jr, Stoner GD, Morgan MA, Ratpan F, Callander RD: Studies on the genotoxicity of beryllium sulphate *in vitro* and *in vivo*. Mutat Res 240: 217–225, 1990
6. Dunkel VC, Zeiger E, Brusick D, McCoy E, McGregor D, Mortelmans K, Rosenkranz HS, Simmon VF: Reproducibility of microbial mutagenicity assays: I. Tests with *Salmonella typhimurium* and *Escherichia coli* using a standardized protocol. Environ Mutagen 6: 1–251, 1984
7. Kuroda K, Endo G, Okamoto A, Yo YS, Horiguchi S-I: Genotoxicity of beryllium, gallium and antimony in short-term assays. Mutat Res 264: 163–170, 1991
8. Arlauskas A, Baker RS, Bonin AM, Tandon RK, Crisp PT, Ellis J: Mutagenicity of metal ions in bacteria. Environ Res 36: 379–388, 1985
9. Larramendy ML, Popescu NC, DiPaolo JA: Induction by inorganic metal salts of sister chromatid exchanges and chromosome aberration in human and Syrian hamster cell strains. Environ Mutagen 3: 597–606, 1981
10. Miyaki M, Akamatsu N, Ono T, Koyama H: Mutagenicity of metal cations in cultured cells from Chinese hamster. Mutat Res 68: 259–263, 1979
11. Dunkel VC, Pienta RJ, Sivak A, Traul KA: Comparative neoplastic transformation responses of Balb/3T3 cells, Syrian hamster embryo cells, and Rauscher murine leukemia virus-infected Fischer 344 rat embryo cells to chemical carcinogens. J Natl Cancer Inst 67: 1303–1315, 1981
12. DiPaolo JA, Casto BC: Quantitative studies of *in vitro* morphological transformation of Syrian hamster cells by inorganic metal salts. Cancer Res 39: 1008–1013, 1979
13. Steel VE, Wilkinson BP, Arnold JT, Kutzman RS: Study of beryllium oxide genotoxicity in cultured respiratory epithelial cells. Inhal Toxicol 1: 95–110, 1989
14. Finch GL, Hoover MD, Hahn FF, Nikula KJ, Belinsky SA, Haley PJ, Griffith WC: Animal models of beryllium-induced lung disease. Environ Health Perspect 104: 973–979, 1996
15. LeBoeuf RA, Kerchaert KA, Aardema MJ, Isfort RJ: The use of short- and medium-term tests for carcinogens and data on genetic effects. In: D.B. McGregor, J.M. Rice, S. Venitt (eds). Carcinogenic Hazard Evaluation. IARC Sci Publ 146: 409–425, 1999
16. Stutman O: In: J. Fogh, B. Giovanella (eds). The Nude Mouse in Experimental and Clinical Research. Academic Press, New York, 1978, pp 411–435
17. Dunkel VC, Rogers C, Swierenga SHH, Brillinger RL, Gilman JPW, Nestmann ER: Recommended protocols based on a survey of current practice in genotoxicity testing laboratories: III. Cell transformation in C3H/10T1/2 mouse embryo cell, BALB/c 3T3 mouse fibroblast and Syrian hamster embryo cell cultures. Mutat Res 246: 285–300, 1991
18. Sambrook J, Fritsch EF, Maniatis T: In: Molecular Cloning: A Laboratory Manual. Cold Spring Harbor Laboratory Press, Cold Spring Harbor, NY, 1980
19. Niman HL, Thompson AMH, Yu A, Markham M, Willems JJ, Herwig KR, Habib NA, Wood CB: Antipeptide antibodies to detect oncogene-related proteins in urine. Proc Natl Acad Sci USA 82: 7924–7928, 1985
20. Simon VF: *In vitro* mutagenicity assays of chemical carcinogens and related compounds with *Salmonella typhimurium*. J Natl Cancer Inst 62: 911–918, 1979
21. Tso WW, Fung WP: Mutagenicity of metallic cations. Toxicol Lett 8: 195–200, 1981
22. Tu AS, Murray TA, Hatch KM, Sivak A, Milman HA: *In vitro* transformation of BALB/c-3T3 cells by chlorinated ethanes and ethylenes. Cancer Lett 28: 85–92, 1985
23. Fitzgerald DJ, Piccoli C, Yamasaki H: Detection of non-genotoxic carcinogens in the BALB/c 3T3 cell transformation/mutation assay system. Mutagenesis 4: 286–291, 1989
24. Colacci A, Vaccari M, Perocco P, Via CD, Silingardi P, Manzin E, Horn W, Grilli S: Enhancement of BALB/c 3T3 cells transformation by 1,2-dibromoethane promoting effect. Carcinogenesis 17: 225–231, 1996
25. MacDonald F, Ford CJ: In: Molecular Biology of Cancer. Bios Scientific Publisher, Oxford, 1997
26. Cooper G, Okenquist S, Silverman L: Transforming activity of DNA of chemically transformed and normal cells. Nature 284: 418–421, 1980
27. Berwald Y, Sachs L: *In vitro* transformation with chemical carcinogens. Nature 200: 1182–1184, 1963
28. Isford RJ, LeBoeuf RA: The Syrian hamster embryo (SHE) cell transformation system: A biologically relevant *in vitro* model – with car-

cinogen predicting capabilities – of *in vivo* multistage neoplastic transformation. Crit Rev Oncol 6: 251–260, 1995
29. Belinsky SA, Swafford DS, Finch GL, Mitchell CE, Kelly G, Hahn FF, Anderson MW, Nikula KJ: Alterations in the K-*ras* and *p53* genes in rat lung tumors. Environ Health Perspect 105: 901–906, 1997
30. Nickell-Brady C, Hahn FF, Finch GL, Belinsky SA: Analysis of K-*ras*, *p53* and *c-raf-1* mutations in beryllium-induced rat lung tumors. Carcinogenesis 15: 257–262, 1994
31. Finch GL, Hoover MD, Hahn FF, Nikula KJ, Belinsky SA, Haley PJ, Griffith WC: Animal models of beryllium-induced lung disease. Environ Health Perspect 104: 973–979, 1996
32. Gao H-G, Brick J, Ong S, Miller M, Whong W-Z, Ong T: Selective hyperexpression of *c-jun* oncoprotein by glass fiber- and silica-transformed BALB/c3T3 cells. Cancer Lett 112: 65–69, 1997

Molecular and Cellular Biochemistry **222**: 77–83, 2001.

Cr (VI) induces cell growth arrest through hydrogen peroxide-mediated reactions

Zhuo Zhang,[1,2] Stephen S. Leonard,[1,2] Suwei Wang,[1,2] Val Vallyathan,[1,2] Vince Castranova[1,2] and Xianglin Shi[1,2]

[1]*Department of Basic Pharmaceutical Sciences, West Virginia University, Morgantown, WV;* [2]*Pathology and Physiology Research Branch, Health Effects Laboratory Division, National Institute for Occupational Safety and Health, Morgantown, WV, USA*

Abstract

Cr (VI) compounds are widely used in industries and are recognized human carcinogens. The mechanism of carcinogenesis associated with these compounds is not well understood. The present study focused on Cr (VI)-induced cell growth arrest in human lung epithelial A549 cells, using flow cytometric analysis of DNA content. Treatment of the cells with Cr (VI) at 1 μM caused a growth arrest at G_2/M phase. An increase in Cr (VI) concentration enhanced the growth arrest. At a concentration of 25 μM, Cr (VI)-induced apoptosis became apparent. Superoxide dismutase (SOD) or sodium formate did not alter the Cr (VI)-induced cell growth arrest. While catalase inhibited growth, indicating H_2O_2 is an important mediator in Cr (VI)-induced G_2/M phase arrest. Electron spin resonance (ESR) spin trapping measurements showed that incubation of cells with Cr (VI) generated hydroxyl radical (˙OH). Catalase inhibited the ˙OH radical generation, indicating that H_2O_2 was generated from cells stimulated by Cr (VI), and that H_2O_2 functioned as a precursor for ˙OH radical generation. The formation of H_2O_2 from Cr (VI)-stimulated cells was also measured by the change in fluorescence of scopoletin in the presence of horseradish peroxidase. The mechanism of reactive oxygen species generation involved the reduction of molecular oxygen as shown by oxygen consumption assay. These results support the following conclusions: (a) Reactive oxygen species are generated in Cr (VI)-stimulated A549 cells through reduction of molecular oxygen, (b) Among the reactive oxygen species generated, H_2O_2 played a major role in causing G_2/M phase arrest in human lung epithelial cells. (Mol Cell Biochem **222**: 77–83, 2001)

Key words: chromium, cell cycle, apoptosis, reactive oxygen species

Introduction

Chromate (Cr (VI)) compounds, widely used in industry, have been shown to have serious toxic and carcinogenic effects on humans [1, 2]. Epidemiological studies in workers occupationally exposed to Cr (VI) compounds provided evidence on the high incidence of respiratory tract cancers [3–5]. Cr (VI) has been demonstrated to induce a variety of DNA lesions, such as single-strand breaks, alkali-labile sites, and DNA protein cross-links. In contrast, most Cr (III) compounds, the final product in the reduction of Cr (VI), are relatively nontoxic, noncarcinogenic, and nonmutagenic [6–8]. Since Cr (VI) does not react with isolated DNA, the reduction of Cr (VI) by cellular reductants to lower oxidation states has been considered to be an important step in the mechanism of Cr (VI)-induced carcinogenesis [7]. Earlier studies have shown that a relatively long-lived Cr (V) species is formed from the reduction of Cr (VI) by various cellular reductants [9]. Since Cr (V) complexes are generally characterized as being labile and reactive, whereas Cr (III) com-

Address for offprints: X. Shi, Pathology and Physiology Research Branch, National Institute for Occupational Safety and Health, 1095 Willowdale Road, Morgantown, WV 26505, USA plexes are relatively inert, the Cr (V) intermediates have been suggested to be the likely candidates as the carcinogenic form of chromium compounds [10].

While several studies have shown that the Cr (VI)-induced DNA damage is strongly dependent on the formation of Cr (V) intermediates, free radicals generated by Cr (V) intermediates may also play an important role [11–18]. We have shown earlier that chromium is able to generate ·OH radical through a Cr (VI)-mediated Fenton and Haber-Weiss cycle [17, 18]. Treatment of Chinese hamster V-79 cells with FAD and Cr (VI) resulted in an increase in DNA strand breaks over that observed upon treatment of cells with Cr (VI) alone [12]. This increase in DNA strand breakage was attributed to enhanced Cr (VI)-related hydroxyl (·OH) radical formation in the presence of FAD. In contrast, incubation of Chinese hamster V-79 cells with an antioxidant, vitamin E, prior to treatment with Cr (VI) led to a decrease in Cr (VI)-induced DNA strand breaks [12, 14]. Participation of free radicals in the generation of Cr (VI)-mediated DNA damage in both noncellular systems and cultured human cells was suggested by the observation that this damage was inhibited by antioxidants [15, 16, 19].

Under normal circumstances the cell cycle proceeds without interruptions. However, when damage particular to DNA occurs, most normal cells have the capacity to arrest proliferation in G_1/S, or G_2/M and then resume proliferation after the damage is repaired [20]. The cell cycle controls the onset of DNA replication and mitosis in order to ensure the integrity of the genome [21, 22]. Lack of fidelity in DNA replication and maintenance can result in deleterious mutations, leading to cell death or, in multicellular organisms, cancer [20]. It has been reported that Cr (VI) not only decreased DNA synthesis in LL 24 cells, but also induced S phase arrest [23]. One to two μM $K_2Cr_2O_7$ produced a prolongation of the G_2 phase of the cell cycle in NHIK 3025 cells [24].

The goal of the present study was to answer the following questions. (a) Can Cr (VI) cause cell growth arrest in A549 cells? (b) If it can, does ROS play a role in Cr (VI)-induced cell growth arrest? (c) Among ROS, which species play a major role?

Materials and methods

Reagents

Potassium dichromate (Cr (VI)) and 5,5-diethylenetriamine-pentaacetic acid (DMPO) were purchased from Aldrich (Milwaukee, WI, USA). RNase A, sodium formate, catalase, and superoxide dismutase (SOD) were purchased from Sigma (St. Louis, MO, USA). Propidium iodide (PI) was from Molecular probes (Eugene, OR, USA). Both F12K nutrient mixture medium and fetal bovine serum (FBS) were purchased from Gibco BRL (Life Technologies, Gaithersburg, MD, USA). The spin trap, DMPO, was purified by charcoal decoloration and vacuum distillation and was free of ESR detectable impurities.

Cell culture

The human lung epithelial cell line, A549, was cultured in F12k nutrient mixture medium containing 10% FBS, 2 mM L-glutamine and 25 mg/ml gentamicin in an incubator at 5% CO_2 and 37°C.

Measurement of cell cycle/DNA content

DNA content in G_1/S and G_2/M phase were analyzed using flow cytometry [25, 26]. Cells were first fixed with 70% ice-cold ethanol for more than 2 h, followed by incubation with freshly prepared staining buffer (0.1% Triton X-100 in PBS, 200 mg/ml RNase A, and 20 mg/ml PI) for 15 min at 37°C. Cell cycle analysis was performed by flow cytometry with at least 10,000 cells for each sample. The percentage of cells in G_1/S and G_2/M phase were then analyzed using ModFit LT software.

Electron spin resonance (ESR) measurements

ESR spin trapping was used to examine hydroxyl radical generation. The spin adduct, a relatively long-lived free radical product formed by the reaction of a short-lived radical with a diamagnetic compound, can be measured by conventional ESR [27]. The intensity of the spin adduct signal corresponds to the amount of short-lived radicals trapped. All samples were measured using a Varian E9 ESR spectrometer and a flat cell assembly as described previously [9]. A549 cells (1×10^6) were mixed with DMPO (200 mM) to a total final volume of 0.5 ml of PBS. The reaction mixture was transferred to a flat cell for measurement for 5 min.

Cellular hydrogen peroxide (H_2O_2)

H_2O_2 was monitored by measuring the change in fluorescence of scopoletin in the presence of horseradish peroxidase [16]. Fluorescence was measured at 37°C using 1×10^6 A549 cells suspended in 1 ml of PBS at an excitation wavelength of 350 nm and an emission wavelength of 460 nm using a PerSeptive Biosystems Cytofluor multiwell plate reader series 4000 (PerSeptive Biosystems Inc., Framingham, MA, USA).

Oxygen consumption assay

The reaction mixtures contained A549 cells (1×10^6 cells/ml) suspended in 1.5 ml of PBS [28]. Oxygen consumption was determined at 37°C using a Gilson oxygraph equipped with a Clark eletrode. The oxygraph was calibrated with medium equilibrated with oxygen of known concentrations.

Statistical analysis

All data were based on at least 3 independent experiments. H_2O_2 formation and oxygen consumption data were presented as means ± S.D. and analyzed using one-way ANOVA with the Scheffe's test. A p value less than 0.05 was considered statistically significant.

Results

Effects of Cr (VI) on the cell cycle

To investigate Cr (VI)-induced cell growth arrest, DNA content was measured by flow cytometry. Figure 1a shows the histogram of the A549 cells without Cr (VI) treatment as a control. It can be noted from this figure that 6.8% of the cell population was in G_2/M phase. Treatment of the cells with Cr (VI) at a concentration of 1 μM for 24 h increased the cell population at G_2/M phase to 10.02% (Fig. 1b). An increase in the Cr (VI) concentrations further increased the percentage of cells in G_2/M phase to 18.10% at 5 μM, and 26.16% at 10 μM (Figs 1c and 1d). At the concentration of 25 μM, cells underwent apparent apoptosis as a sub-G_1 peak appeared (Fig. 1e).

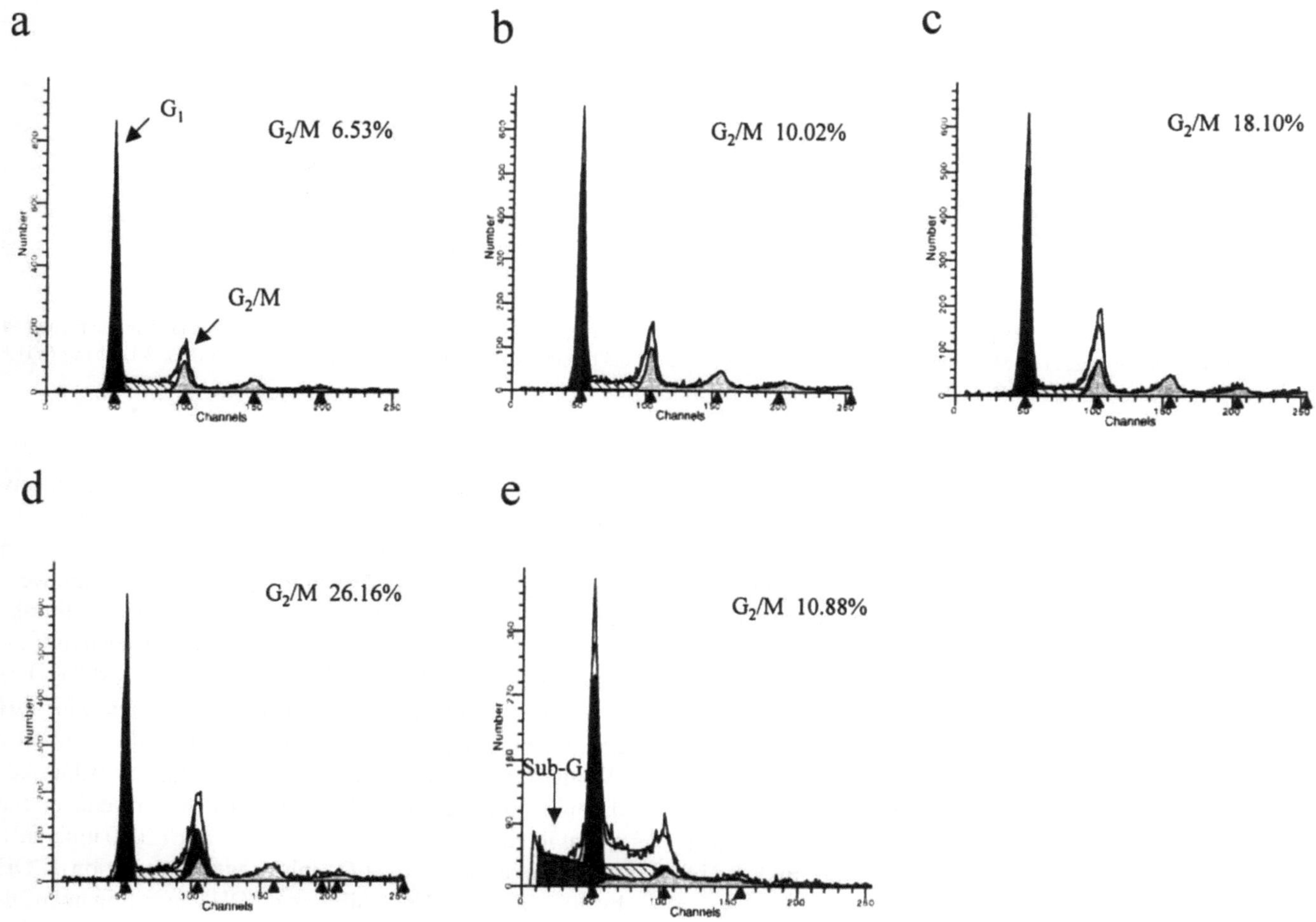

Fig. 1. Cr (VI)-induced cell growth arrest in human lung epithelial A549 cells. DNA content was measured by flow cytometry. A549 cells were suspended in 10% fetal bovine serum (FBS) F12 K nutrient mixture medium in a 100 mm dish. After 80–90% confluence, cells were washed with PBS for three times, and treated with various concentration of Cr (VI) at 37°C for 24 h: (a) cells only; (b) treatment with 1.0 μM Cr (VI); (c) treatment with 5.0 μM Cr (VI); (d) treatment with 10 μM Cr (VI); and (e) treatment with 25 μM Cr (VI). It was noted that at 25 μM Cr (VI), apoptotic cells became apparent as indicated by the sub-G_1 signal.

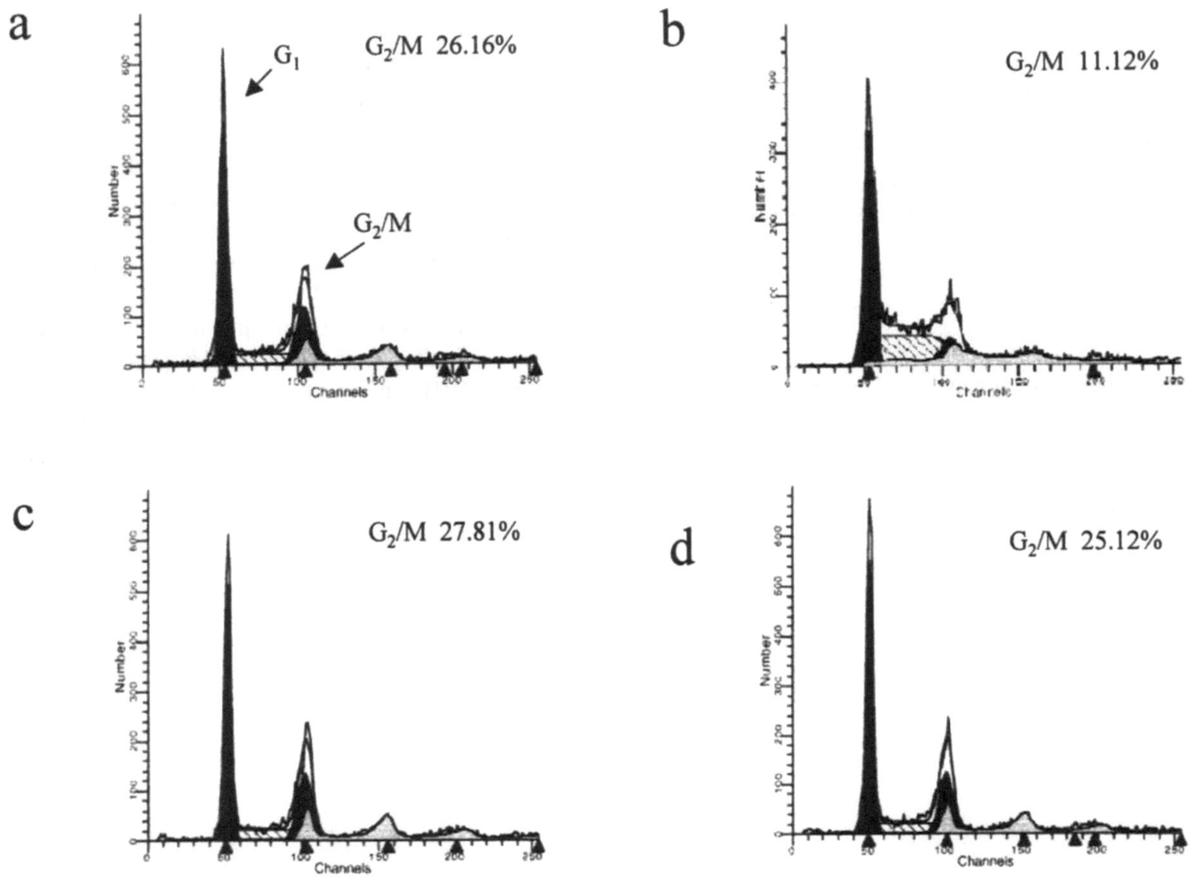

Fig. 2. Effect of antioxidants on Cr (VI)-induced cell growth arrest in human lung epithelial A549 cells. Cells were incubated in a 100 mm dish and pretreated with 50,000 units/ml catalase, 300 μM sodium formate, or 500 units/ml SOD for 0.5 h before Cr (VI) treatment (10 μM) for 24 h. (a) Cr (VI); (b) Cr (VI) + catalase; (c) Cr (VI) + sodium formate; (d) Cr (VI) + SOD.

Effects of antioxidants on Cr (VI)-induced cell growth arrest

Figure 2 shows the effects of antioxidants on Cr (VI)-induced G_2/M phase arrest. SOD, a specific $O_2^{\bullet-}$ radical scavenger, or formate, a specific ·OH radical scavenger, did not exhibit significant effect, i.e. the percentage of G_2/M phase was 26.16, 27.81 and 25.12% for Cr (VI), Cr (VI) plus formate and Cr (VI) plus SOD, respectively (Figs 2a, 2c and 2d), indicating that neither $O_2^{\bullet-}$ nor ·OH was directly involved in the mechanism of Cr (VI)-induced cell growth arrest. Catalase, a scavenger of H_2O_2, decreased Cr (VI)-induced G_2/M phase arrest from 26.16 to 11.12%, indicating that H_2O_2 was involved in Cr (VI)-induced cell growth arrest (Fig. 2b).

Hydroxyl radical formation from Cr (VI)-stimulated cells

ESR study was used to detect the ·OH formation from Cr (VI)-stimulated cells. A549 cells alone did not produce any detectable amount of free radicals (Fig. 3a), whereas addition of 2 mM Cr (VI) generated a 1:2:2:1 quartet ESR spin adduct signal (Fig. 3b). The splittings of this spectrum were $a_H = a_N = 14.9$ G, where a_H and a_N denote hyperfine splittings of the α-hydrogen and the nitroxyl nitrogen, respectively, indicating the DMPO/·OH adduct (Shi and Dalal, 1992a). The detection of this DMPO/·OH spin adduct is evidence for ·OH generation. The peak at the right side was assigned to a Cr (V) intermediate based on the lineshape and the g value. Addition of formate (50 mM), an ·OH scavenger, reduced the signal intensity by 50% (Fig. 3d). Catalase (2000 units/ml), a specific scavenger of H_2O_2, inhibited the generation of ·OH by 90% (Fig. 3c). The inhibition of ·OH generation upon addition of catalase indicates that H_2O_2 was generated, and that it was a precursor for ·OH generation. SOD (5 μg/ml), a scavenger of $O_2^{\bullet-}$ radical, did not show any significant inhibition, showing that the 1:2:2:1 quartet signal is not due to the $O_2^{\bullet-}$ radical trapping.

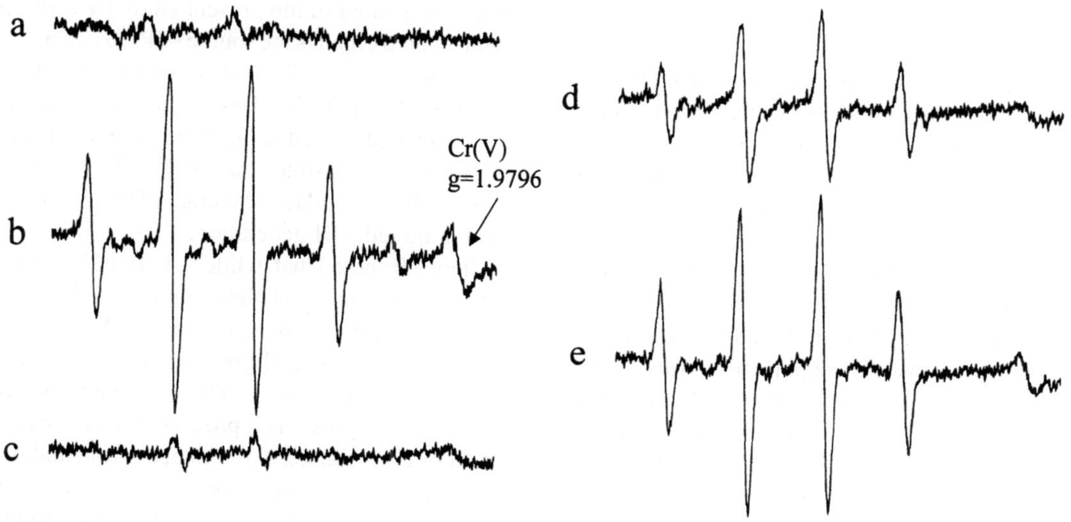

Fig. 3. Cr (VI)-induced free radical formation and the effects of antioxidants. 1 × 10^6 A549 cells were suspended in 0.5 ml of PBS and mixed with 200 mM DMPO and Cr (VI) with or without antioxidants. ESR spectra were recorded for 5 min at operational conditions. (a) A549 cells only; (b) cells + Cr (VI); (c) cells + Cr (VI) + catalase; (d) cells + Cr (VI) + sodium formate; (e) cells + Cr (VI) + SOD. The final concentration were: Cr (VI), 2 mM; catalase, 2,000 units/ml; sodium formate, 50 mM; SOD, 5 μg/ml.

Formation of hydrogen peroxide

As discussed in the previous section, catalase inhibition of ·OH generation from A549 cells stimulated by Cr (VI) implied that H_2O_2 was a precursor of ·OH production. In this section, the H_2O_2 generation was measured directly as the change in fluorescence of scopoletin in the present of horseradish peroxidase. Figure 4 shows that stimulation of A549 cells with 2 mM Cr (VI) increased H_2O_2 production by 3-fold above the sum of the fluorescence for cells alone and Cr (VI) alone.

Oxygen consumption by the Cr (VI)-stimulated cells

Since both H_2O_2 and ·OH were generated in the Cr (VI)-stimulated cells, it would be expected that H_2O_2 was generated from the reduction of molecular oxygen via $O_2^{\cdot-}$ radical as an intermediate. The O_2 consumption from cells was measured using an oxygraph. O_2 consumption was 176 nmol/5 × 10^5 cells in control cells, whereas it was 235 nmol/5 × 10^5 cells after Cr (VI) treatment, i.e. an increase in O_2 consumption by 33% (Fig. 5).

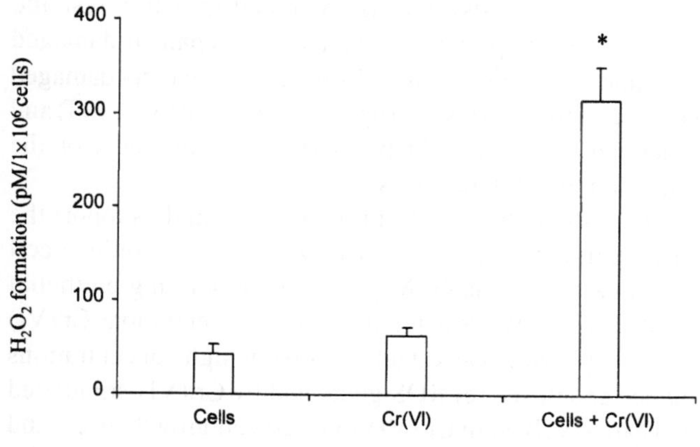

Fig. 4. Cr (VI)-induced H_2O_2 formation from human lung epithelial A549 cells. H_2O_2 was monitored by measuring the change in fluorescence of scopoletin in the presence of horseradish peroxidase. The samples contained 1.0 × 10^6 cells suspended in 1 ml of PBS with and without 2 mM Cr (VI). Fluorescence was monitored at an excitation wavelength of 350 nm and an emission wavelength of 460 nm. *$p < 0.05$ compared to control (one-way ANOVA with Scheffe's test).

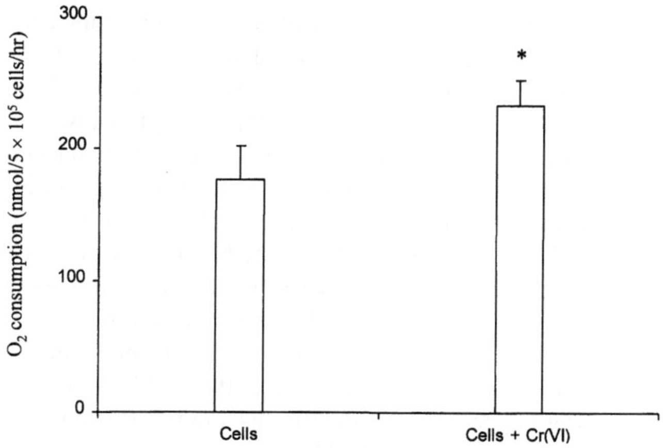

Fig. 5. Cr (VI)-stimulated oxygen consumption from human lung epithelial A549 cells. 1 × 10^6 cells were suspended in 1.5 ml of PBS and of oxygen consumption measured at 37°C using an oxygraph. The final concentration of Cr (VI) was 2 mM. *$p < 0.05$ compared to control (one-way ANOVA with Scheffe's test).

Discussion

ROS-mediated reactions are believed to be involved in various pathological processes. Initiation of carcinogenesis can be associated with an imbalance caused by the excessive generation of ROS. ROS can cause DNA strand breaks, base modification, lipid peroxidation and protein modification, resulting in oxidative stress. Various signal pathways are involved in the mechanism of ROS-induced oxidative stress, including activation of nuclear transcription factors, NF-κB [29], AP-1 [30], and p53 [31]. Under oxidative stress, dividing cells will exhibit a cell cycle checkpoint response [32, 33]. The division of cells must be carefully regulated and coordinated with both cell growth and DNA replication in order to ensure the formation of progeny cells containing intact genomes [20]. In higher eukaryotes, the cell machinery is itself regulated by the growth factors that control cell proliferation, allowing the division of individual cells to be coordinated with the needs of the organism as a whole. Defects in cell cycle regulation are a common cause of the abnormal proliferation of cancer cells. The cell cycle regulation and carcinogenesis are believed to be closely interconnected.

Cell cycle checkpoints monitor movement through the cell cycle, survey for cell damage, and induce a pause in cell cycle progression when necessary. Damage to growing cells causes a temporary pause in G_1/S, or G_2/M phase until the damage is repaired. When damage is severe, cells may either undergo apoptosis or enter a dormant G_0 state. Regulation of cell cycle progression is achieved by events including cyclin accumulation and degradation; phosphorylation of Cdks, cyclins, and other proteins; regulation of cyclin/Cdk dimerization; and the binding of a number of Cdk inhibitory proteins [34–36]. Movement of cells from G_2 to M is regulated by cyclin A and cyclin B/Cdc2. Cylin B/Cdc2 kinase activity peaks in late G_2 and remains high until its degradation [37]. Previous study has shown that Cr (VI) treatment of normal human lung cells results in guanine-specific DNA polymerase arrest, DNA-DNA cross-links and S-phase blockade of the cell cycle [23]. The present study shows that Cr (VI) is able to cause cell growth arrest at G_2/M phase in the human lung epithelial cell line, A549 cells. The percentage of cells in G_2/M phase increased in a dose-dependent manner following treatment with Cr (VI). ROS generated in the reduction of Cr (VI) by the cells are involved in the Cr (VI)-induced cell growth arrest. Among ROS generated by Cr (VI) stimulation of A549 cells, H_2O_2 appears to be the species responsible for Cr (VI)-induced cell growth arrest at the G_2/M phase. The following experimental observations support this conclusion. (a) Catalase, a specific H_2O_2 scavenger, decreased Cr (VI)-induced growth arrest, (b) Neither SOD, a $O_2^{\cdot-}$ scavenger, nor sodium formate, an $^{\cdot}OH$ radical scavenger, exhibited any effect.

The generation of ROS by Cr (VI) stimulated A549 cells is demonstrated in the present study by ESR spin trapping. Molecular oxygen is the source of ROS generation in Cr (VI)-stimulated A549 cells as demonstrated by the oxygen consumption assay. Molecular oxygen was consumed to generate $O_2^{\cdot-}$, which produced H_2O_2 upon dismutation. H_2O_2 produced $^{\cdot}OH$ via a Fenton-like reaction (Cr (V) + $H_2O_2 \rightarrow$ Cr (VI) + $^{\cdot}OH + OH^-$). Catalase scavenged H_2O_2, a precursor of $^{\cdot}OH$, and inhibited $^{\cdot}OH$ generation.

It may be noted that, while at 1–10 μM, Cr (VI) caused cell growth arrest at G_2/M phase, Cr (VI) induced-apoptosis became apparent at a concentration of 25 μM. It appears that at this concentration cell growth arrest was not sufficient for cells to repair the damage. Thus the apoptosis mechanism was initiated. Apoptosis is a process in which cell death occurs in an orderly manner through activation and/or synthesis of gene products necessary for cell destruction. It is a response to physiologic and pathologic stress that disrupts the balance between the rate of cell division and elimination. In diseases such as cancer, there is imbalance between the rate of cell proliferation and cell death. Agents that promote or suppress apoptosis can alter the rates of cell division and death, influencing the anomalous accumulation of neoplastic cells. Using morphological and DNA fragment analyses, our laboratory has shown that Cr (VI) is able to cause apoptosis through both p53-dependent and p53-independent mechanisms [38]. The results obtained from the present study using flow cytometry show that at a relative low concentration, Cr (VI) caused cell growth arrest while at relatively high concentration apoptosis was initiated.

It may also be noted that many other metal ions, mineral particles and chemical carcinogens, such as cobalt, nickel, vanadium, asbestos and silica are known to be cable of generating ROS. It is possible that these agents may have the same function as Cr (VI); i.e. they may cause cell growth arrest and apoptosis. Because both cell growth arrest and apoptosis are important mechanisms in repair of damaged cells and elimination of the cells which are severely damaged cells, cell growth arrest and apoptosis induced by Cr (VI) and other carcinogens could have an important influence on the induction of carcinogenesis.

The results obtained from the present study support the following conclusions: (a) Cr (VI) is able to induce cell growth arrest at the G_2/M phase in human lung epithelial A549 cells, (b) While at relatively low concentrations Cr (VI) causes cell growth arrest and relatively high concentrations induce apoptosis, (c) ROS generated by Cr (VI)-stimulated cells are involved in Cr (VI)-induced cell growth arrest and among ROS, H_2O_2 plays a key role, (d) H_2O_2 is generated by the reduction of molecular oxygen by cells. From the results it can be speculated that other metal carcinogens, such as vanadium, cobalt and nickel, which are ROS-promoting agents, may cause cell growth arrest and apoptosis by a mechanism similar to that of Cr (VI).

Acknowledgements

Research funded under Interagency Agreement Number 98-18-00m2 between the Occupational Safety and Health Administration and the National Institute for Occupational Safety and Health. The views expressed in the paper are those of the authors and do not necessarily reflect the official positions of OSHA. The authors would like to thank Dr. Murali Rao for his critical review of the manuscript.

References

1. De Flora S, Serra D, Basso C, Zanacchi P: Mechanistic aspects of chromium carcinogenicity. Arch Toxicol 13(suppl): 28–39, 1989
2. Freeman N, Stern AH, Lioy PJ: Exposure to chromium dust from homes in a chromium surveillance project. Arch Environ Health 52: 213–219, 1997
3. World Health Organization. IARC monographs on the evaluation of carcinogenic risks to humans. Lyon, France, 1990
4. Enterline PE: Respiratory cancer among chromate workers. J Occup Med 16: 523–526, 1974
5. Norseth T: The carcinogenicity of chromium. Environ Health Perspect 40: 121–130, 1981
6. De Flora S, Wetterhahn KE: Mechanisms of chromium metabolism and genotoxicity. Life Chem Rep 7: 169–244, 1989
7. Connett PH, Wetterhahn KE: Metabolism of the carcinogen chromate by cellular constituents. Struct Bonding 54: 93–124, 1983
8. Venitt S, Levy L: Mutagenicity of chromates in bacteria and its relevance to chromate carcinogenicity. Nature 250: 493–495, 1974
9. Shi X, Dalal NS: Role of superoxide radical in chromium (VI) generated hydroxyl radical: The Cr (VI) Haber-Weiss cycle. Arch Biochem Biophys 292: 323–327, 1992
10. Jennette KE: Microsomal reduction of the carcinogen chromate reduces chromium (V). J Am Chem Soc 104: 874–875, 1982
11. Aiyar J, Berkovits HJ, Floyd RA, Wetterhahn KE: Reaction of chromium (VI) with hydrogen peroxide in the presence of glutathione: Reactive intermediates and resulting DNA damage. Chem Res Toxicol 3: 595–603, 1990
12. Sugiyama M: Effects of vitamin E and vitamin B2 on chromate-induced DNA lesions. Biol Trace Element Res 21: 399–404, 1989
13. Kortenkamp A, Oeken G, Beyersmann D: The DNA cleavage induced by a chromium (V) complex and by chromate and glutathione is mediated by activated oxygen species. Mutat Res 232: 155–161, 1990
14. Sugiyama M, Ando A, Ogura R: Effect of vitamin E on survival, glutathione reductase and formation of chromium (V) in Chinese Hamster V-79 cell treated with sodium chromate (VI). Carcinogenesis 10: 737–741, 1989
15. Shi X, Chiu A, Chen CT, Halliwell B, Castranova V, Vallyathan V: Reduction of chromium (VI) and its relationship to carcinogenesis. J Toxicol Environ Health 2: 87–104, 1999
16. Leonard SS, Wang S, Zang L, Castranova V, Vallyathan V, Shi X: Role of molecular oxygen in the generation of hydroxyl and superoxide anion radicals during enzymatic Cr (VI) reduction and its implication to Cr (VI)-induced carcinogenesis. J Environ Pathol Toxicol Oncol 19: 49–60, 2000
17. Shi X, Dalal NS: Hydroxyl radical generation in the NADH/microsomal reduction of vanadate. Free Rad Res Commun 17: 369–376, 1992
18. Shi X, Dalal NS: Chromium (V) and hydroxyl radical formation during the glutathione reductase-catalyzed reduction of chromium (VI). Biochem Biophys Res Commun 163: 627–634, 1989
19. Snyder RD: Role of active oxygen species in metal-induced DNA strand breakage in human diploid fibroblasts. Mutat Res 193: 237–246, 1988
20. Shackelford RE, Kaufmann WK, Paules RS: Cell cycle control, checkpoint mechanisms, and genotoxic stress. Environ Health Perspect 107: 5–24, 1999
21. Hartwell LH, Kastan MB: Cell cycle control and cancer. Science 266: 1821–1828, 1994
22. Mercer WE: Checking on the cell cycle. J Cell Biochem 30/31(suppl): 50–54, 1998
23. Xu J, Bubley GJ, Detrick B, Blankenship LJ, Patierno SR: Chromium (VI) treatment of normal human lung cells results in guanine-specific DNA polymerase arrest, DNA-DNA cross-links and S-phase blockade of cell cycle. Carcinogenesis 17: 1511–1517, 1996
24. Bakke O, Jakobsen K, Eik-Nes KB: Concentration-dependent effects of potassium dichromate on the cell cycle. Cytometry 5: 482–486, 1984
25. Nicoletti I, Migliorati G, Pagliacci MC, Grignani F, Riccardi C: A rapid and simple method for measuring thymocyte apoptosis by propidium iodide staining and flow cytometry. J Immunol Meth 139: 271–279, 1991
26. Sgonic R, Wick G: Methods for the detection of apoptosis. Int Arch Allergy Immunol 105: 327–332, 1994
27. Rosen GM, Finkelstein E: Use of spin traps in biological systems. Adv Free Radical Biol Med 1: 345–375, 1985
28. Leonard SS, Gannett PM, Rojanasakul Y, Schwegler-Berry D, Castranova V, Vallyathan V, Shi X: Cobalt-mediated generation of reactive oxygen species and its possible mechanism. J Inorg Biochem 70: 239–244, 1998
29. Ye J, Zhang X, Young HA, Mao Y, Shi X: Chromium (VI)-induced nuclear factor-κB activation in intact cells via free radical reactions. Carcinogenesis 16: 2401–2405, 1995
30. Chen F, Ding M, Lu Y, Leonard SS, Vallyathan V, Castranova V, Shi X: Participation of MAP kinase p38 and IκB kinase in chromium (VI)-induced NF-κB and AP-1 activation. J Environ Pathol Toxicol Oncol 19: 231–238, 2000
31. Wang S, Leonard SS, Ye J, Ding M, Shi X: The role of hydroxyl radical as a messenger in Cr (VI)-induced p53 activation. Am J Physiol 279: C868–C875, 2000
32. Russo T, Zambrano N, Esposito F, Ammendola R, Cimino F, Fiscella M, Jackman J, O'Connor PM, Anderson CW, Appella E: A p53-independent pathway for activation of WAF_1/CIP_1 expression following oxidative stress. J Biol Chem 270: 29386–29391, 1995
33. Chen QM, Bartholomew JC, Campisi J, Acosta M, Reagan JD, Ames BN: Molecular analysis of H_2O_2-induced senescent-like growth arrest in normal human fibrosis: p53 and Rb control G_1 arrest but not cell replication. Biochem J 332: 43–50, 1998
34. Morgan DO: Principles of CDK regulation. Nature 374: 131–134, 1995
35. Draetta G: Cell cycle control in eukaryotes: Molecular mechanisms of Cdc2 activation. Trends Biochem Sci 15: 378–383, 1990
36. Norbury C, Nurse P: Animal cell cycles and their control. Annu Rev Biochem 61: 441–470, 1993
37. Pines J, Hunter T: Isolation of a human cyclin cDNA: Evidence for cyclin mRNA and protein regulation in the cell cycle and for interaction with $p34^{cdc2}$. Cell 58: 833–846, 1989
38. Ye J, Wang S, Leonard SS, Sun Y, Butterworth L, Antonini J, Ding M, Rojanasakul Y, Vallyathan V, Castranova V, Shi X: Role of reactive oxygen species and p53 in chromium (VI)-induced apoptosis. J Biol Chem 274: 34974–34980, 1999

Molecular and Cellular Biochemistry **222**: 85–95, 2001.

Comparison of roles of three mitogen-activated protein kinases induced by chromium(VI) and cadmium in non-small-cell lung carcinoma cells

Show-Mei Chuang and Jia-Ling Yang

Molecular Carcinogenesis Laboratory, Department of Life Sciences, National Tsing Hua University, Hsinchu, Taiwan, Republic of China

Abstract

Chromium(VI) [Cr(VI)] and cadmium (Cd) compounds are ubiquitous environmental carcinogens that have been associated with lung tumors and can induce apoptosis in various cell types. Three major mitogen-activation protein kinases (MAPKs), extracellular signal-regulated kinase (ERK), c-JUN N-terminal kinase (JNK) and p38, have been shown to regulate apoptosis. In this study we explore the abilities of Cr(VI) and Cd to activate JNK, p38 and ERK, including their roles in metal-mediated growth inhibition and apoptosis in a human non-small-cell lung carcinoma cell line, CL3. Exposure to $K_2Cr_2O_7$ markedly activated JNK and p38 and moderately activated ERK in a dose- and time-dependent manner. The activated p38 decreased markedly and rapidly and the activated JNK decreased gradually when Cr(VI) was removed from media. At low cytotoxic doses, $CdCl_2$ decreased ERK activity with concurrently transient activation of JNK, whereas at high cytotoxic doses it persistently activated all three MAPKs. The strength and duration of JNK and p38 activated by Cd were higher and longer than Cr(VI) did when compared at similar cytotoxic doses. In comparable experiment conditions Cd is a much stronger apoptotic inducer than Cr(VI) in CL3 cells. Cross-talk of MAPKs was observed in cells exposed to Cr(VI) but not Cd. Both metals could increase JNK activity through MKK7 but not MKK4. The Cd-activated JNK is involved in apoptosis, but the Cr-activated JNK is not. PD98059, an inhibitor of the ERK upstream activators MKK1/2, greatly enhanced the cytotoxicity and apoptosis of cells treated with low Cd doses. SB202190, an inhibitor of p38, decreased the cytotoxicity and apoptosis induced by high Cd doses. Conversely, neither SB202190 nor PD98059 altered Cr(VI)-induced cytotoxicity. The results suggest that JNK and p38 signals cooperatively participate in apoptosis induced by Cd and that the decreased ERK signal by low Cd doses contributes to growth inhibition or apoptosis. Oppositely, activation of ERK, JNK and p38 by Cr(VI) does not affect cytotoxicity. (Mol Cell Biochem **222**: 85–95, 2001)

Key words: carcinogenic metal, JNK, p38, ERK, MKK, apoptosis

Introduction

Extracellular signal-regulated kinases (ERKs), c-JUN N-terminal kinases (JNKs) and p38 kinases are three major groups of mammalian mitogen-activated protein kinases (MAPKs) that are stimulated by DNA damage agents, cytokines and growth factors [1–5]. MAPKs are activated through a protein kinase cascade that involves dual phosphorylation by distinct MAPK kinases (MKKs) on Thr and Tyr within the motif Thr-Glu-Tyr (ERK), Thr-Pro-Tyr (JNK) and Thr-Gly-Tyr (p38) in the catalytic domain. ERK is activated by MKK1 and MKK2, while JNK by MKK4 and MKK7, and p38 by MKK3, MKK4 and MKK6 [3, 4]. Once activated, MAPKs regulate gene expression through phosphorylation of downstream transcriptional factors. Activation of MAPKs plays a pivotal role in controlling cell growth, survival, differentia-

Address for offprints: J.-L. Yang, Molecular Carcinogenesis Laboratory, Department of Life Sciences, National Tsing Hua University, Hsinchu 300, Taiwan, Republic of China tion and apoptosis [1–5]. Usually, activation of the JNK and p38 pathways is involved in growth arrest and apoptosis [2, 3], whereas the ERK cascade is an important pathway for mitogenesis and differentiation [1, 6]. However, the strength and duration of activated-MAPKs and cell types can determine diverse gene expression that result in different physiological consequences [1, 5].

Cr(VI) and Cd compounds are ubiquitous environmental toxicants that have been evaluated as human carcinogens [7, 8]. Occupational exposure to Cr(VI) has been strongly associated with increased incidence of human lung cancer [7]. The main sources of human exposure to Cd are tobacco smoke, food and industrial pollution [8]. For cigarette smoke, up to 60% of Cd deposited in the lung is absorbed [9]. Cr(VI) genotoxicity and carcinogenicity are highly associated with the generation of reactive species including Cr(V), Cr(IV), Cr(III) and reactive oxygen species (ROS) during its intracellular reduction [10, 11]. Similarly, ROS has also been implicated in Cd genotoxicity [12–14]. Both metals promote apoptosis in a variety of cells [15–19].

Cr(VI) has been shown to activate ERK in rat H4 hepatoma cells [20]. At a very high dose (500 μM), Cr(VI) can activate three MAPKs in human bronchial epithelial cells [21]. Cd has also been reported to activate ERK [22, 23], p38 [18, 23] or JNK [24] in certain cultured cells. However, the roles of MAPKs in growth arrest and apoptosis of Cr(VI)- or Cd-treated cells remains unclear. Recently, we have attempted to explore the ability of Cr(VI) and Cd to activate JNK, p38 and ERK and their roles in regulating growth arrest and apoptosis of cells exposed to these two metal carcinogens. Both Cr(VI) and Cd activate JNK, p38 and ERK in a human lung adenocarcinoma line CL3 [25, 26], however, distinct kinetics and functions of their activation are observed between these metal carcinogens. Here we compare and discuss the finding of different roles of MAPKs in controlling growth and death upon Cr(VI) or Cd exposures.

Materials and methods

Cell culture

The CL3 cell line established from a non-small-cell lung carcinoma tumor of a 60-year-old male patient in Taiwan [27] was provided by Dr. Pan-Chyr Yang at the Department of Internal Medicine and Clinical Pathology, National Taiwan University Hospital, Taipei. Cells were routinely cultured in RPMI-1640 media (Gibco Life Technologies, Grand Island, NY, USA) supplemented with sodium bicarbonate (2.2% w/v), L-glutamine (0.03% w/v), penicillin (100 units/ml), streptomycin (100 μg/ml) and fetal calf serum (10%). Cells were maintained at 37°C in a humidified incubator containing 5% CO_2 in air.

Transfection

Plasmids containing dominant negative forms of JNK1 (pCMV-JNK1-APF; JNK1DN), JNK2 (pSR-JNK2-APF; JNK2DN), MKK4 (JNKK[K116R]; MKK4KR), or wild-type forms of JNK1 and JNK2 were kindly provided by Dr. M. Karin (University of California, San Diego, CA, USA) [28–30]. Plasmids containing a dominant negative form of MKK7 (MKK7A) and the wild-type form of p38 were gifts from Dr. J. Han (The Scripps Research Institute, La Jolla, CA, USA) [31, 32]. The final amount of transfected DNA for a 60 mm dish was adjusted to equal amounts using empty vector. Cells (4×10^5) were plated in a p60 mm dish 1 day before transfection. Plasmids were transfected into CL3 cells by the calcium-phosphate precipitation. After incubation for 6 h, the cells were washed with phosphate-buffered saline (PBS) and kept cultured in complete media for 24 h followed by sustenance in serum-free media for 16–18 h. The cells were then subjected for Cr(VI) or Cd treatment.

Treatment

Cells in exponential growth were plated at a density of 1×10^6 cells/100 mm dish and cultured in serum-free media for 16–18 h before metal treatment. Cadmium chloride (Merck, Darmstadt, Germany) and potassium dichromate (Merck) were dissolved in MilliQ-purified water (Millipore, Bedford, MA, USA). Cells were treated with Cr(VI) or Cd in serum-free medium. In experiments to determine the effects of PD98059 (Calbiochem, San Diego, CA, USA) and SB202190 (Calbiochem) on MAPKs, cell growth inhibition, cytotoxicity and apoptosis, cells were treated with these kinase inhibitors for 1 h and then co-exposed to Cr(VI) or Cd for 3 h. At the end of treatment, the drug-containing medium was removed and the cells were washed twice with PBS. PD98059 and SB202190 were prepared in dimethylsulfoxide as a stock of 50 and 10 mM, respectively. The concentration of Cr(VI) expressed in this study is equivalent to one-half that of potassium dichromate.

Cell growth

After treatment, the cells were cultured in media containing 10% serum for 0–3 days before trypsinization. Cells were mixed with 0.4% trypan-blue (Gibco, Life Technologies) for 15 min and the number of unstained cells were determined using a hemocytometer.

Viability assay

Cells were seeded in 96-well plates at a density of 5×10^3/well, cultured in serum-free medium for 16–18 h and then exposed to metal for 3 h. After treatment, the cells were washed with PBS twice and cultured in complete media for 24 h. The cells were treated with 0.5 mg/ml 3-(4,5-dimethyl-thiazol-2-yl) 2,5-diphenyl tetrazolium bromide (MTT) (Sigma Chemical Co., St. Louis, MO, USA) and maintained for 3 h in a CO_2 incubator. Cells having functional mitochondria succinate dehydrogenase can convert MTT to formazan that generates a blue color when dissolved in dimethylsulfoxide. The intensity was determined using a reader for enzyme-linked immunosorbent assay and an absorption wavelength of 565 nm.

Cytotoxicity assay

Immediately after treatment cells were trypsinized, diluted and plated at a density of 200–1000 cells/60 mm-Petri dish in triplicate. The cells were cultured for 10–14 days and stained with 1% crystal violet solution. The colony numbers were counted for cytotoxicity determination, calculated as the number of colonies in the treated cells divided by those colonies obtained in the untreated cells.

Apoptosis

The annexin V-fluorescein isothiocyanate (FITC) binding assay was adopted for determination of apoptosis. Briefly, 2×10^5 cells were washed twice with PBS and suspended in binding buffer (10 mM HEPES, pH 7.4, 140 mM NaCl, and 2.5 mM $CaCl_2$). The cells were stained with annexin V-FITC (1 ng/μl) and propidium iodide (5 ng/μl) for 15 min in the dark. The stained cells were analyzed using flow cytometry and the CellQuest program (Becton Dickinson). Apoptotic cells were detected by those stained with annexin V-FITC but not with propidium iodide.

JNK kinase assay

Cells were lysed in a buffer containing 20 mM HEPES at pH 7.6, 75 mM NaCl, 2.5 mM $MgCl_2$, 0.1 mM EDTA, 0.1% Triton X-100, 0.1 mM Na_3VO_4, 50 mM NaF, 0.5 μg/ml leupeptin, 1 μg/ml aprotinin and 100 μg/ml 4-(2-aminoethyl)-benzenesulfonyl fluoride. The cell lysate was rotated at 4°C for 30 min, centrifuged at 10,000 rpm for 10 min and the precipitates discarded. Protein concentrations were determined with the BCA protein assay kit (Pierce, Rockford, IL, USA) using bovine serum albumin as a standard. The JNK activity assay was performed using GST-cJUN(1–79) as substrate according to the procedure described by Hibi *et al.* [28].

Western blot analysis

Appropriate amounts of proteins were loaded onto 10% SDS-polyacrylamide gels and transferred electrophoretically to PVDF membranes. Membranes were probed with primary antibody, followed by a horseradish peroxidase-conjugated second antibody (Bio-Rad, Hercules, CA, USA). Phospho-specific antibodies for ERK and p38 were purchased from New England BioLabs Inc. (Beverly, MA, USA). The rabbit polyclonal antibodies against p38 MAP kinase (C-20) and ERK2 (C-14) were purchased from Santa Cruz Biotechnology Inc. (Santa Cruz, CA, USA). The mouse monoclonal antibody to JNK1/2 was ordered from Pharmingen (San Diego, CA, USA). The immunocomplexes were detected using the enhanced chemiluminescence detection procedure according to the manufacturer's recommendations (Amersham, Pharmacia Biotech).

Results

Activation of JNK, p38 and ERK by Cr(VI) and Cd

Human lung adenocarcinoma CL3 cells in exponential growth were cultured in serum-free media for 18 h before they were exposed to various concentrations of potassium dichromate or cadmium chloride in serum-free media for 3 h. After PBS washing, proteins in the whole cell extracts were prepared to examine activation of JNK, p38 and ERK. The kinase activity of JNK was determined using [γ-^{32}P]ATP and GST-cJUN(1–79). Activation of p38 or ERK was determined using phospho-specific antibodies. As shown in Fig. 1, both Cr(VI) and Cd dose-dependently induced JNK activity. At 10 (~ 75% survival) and 80 μM (~ 3% survival) Cr(VI), JNK was activated 4- and 69-fold the untreated level (Fig. 1A). Similarly, 60 (~ 80% survival) and 160 μM (~ 4% survival) Cd increased JNK activity 10- and 157-fold the untreated level (Fig. 1B). Low cytotoxic doses of Cr(VI) or Cd did not significantly activate p38, while high cytotoxic doses of the two metals could markedly activate it (Fig. 1). Activation of ERK was moderately increased by Cr(VI); 10–30 μM Cr(VI) induced 2-fold and 40–80 μM Cr(VI) induced 4-fold the untreated level (Fig. 1A). In contrast, low cytotoxic doses of Cd (15–80 μM) attenuated ERK activity while high cytotoxic doses of Cd (130–160 μM) increased 4-fold the untreated level (Fig. 1B). Western blot analyses revealed that no significantly difference in the levels of JNK, p38 and ERK proteins between untreated and metal-treated cells (data not shown).

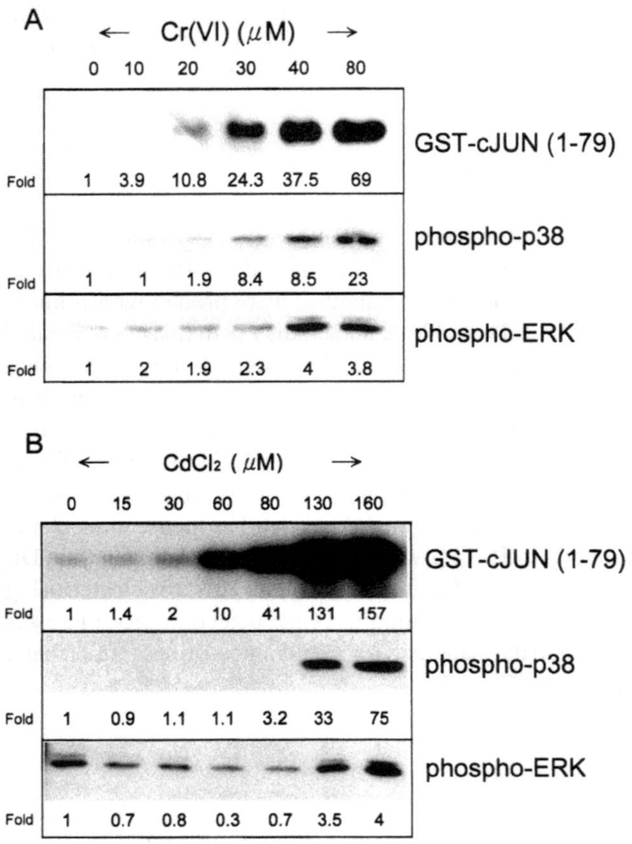

Fig. 1. Dose-dependent activation of JNK, p38 and ERK by Cr(VI) and Cd in CL3 cells. Cells were cultured in serum-free media for 18 h and treated with Cr(VI) (A) or Cd (B) for 3 h. The cells were washed with PBS and the whole cell extracts were isolated for JNK activity assay using GST-cJUN(1–79) as substrate. The same cell extract obtained from metal-treated CL3 cells was examined for the amounts of the dual phosphorylated forms of p38 and ERK using phospho-specific antibodies. The relative radioactivity was quantitated using a computing densitometer equipped with the ImagQuant analysis program. Both p42 and p44 phospho-ERK are quantitated. The relative activities of JNK, p38 and ERK averaged from 3–6 experiments are shown under the autoradiographs.

To determine the persistence of MAPKs activation by these two metals, CL3 cells were exposed to 30 μM Cr(VI) or 80 μM Cd in serum-free media for various times. Figure 2A shows that Cr(VI) markedly activated JNK and p38 kinases in a time-dependent manner. However, the kinetics of their activation was different, i.e. Cr(VI) activated p38 faster than JNK (Fig. 2A). In contrast, the maximum JNK activation by 80 μM Cd was observed after a 3 h exposure, and decreased after exposure for a longer period (6–12 h; Fig. 2B). Cd time-dependently activated p38 kinase that occurred following the maximum JNK activation (Fig. 2B). Furthermore, Cr(VI) moderately activated ERK independent of the exposure time (Fig. 2A), while short-term (0.5–3 h) Cd exposure abated

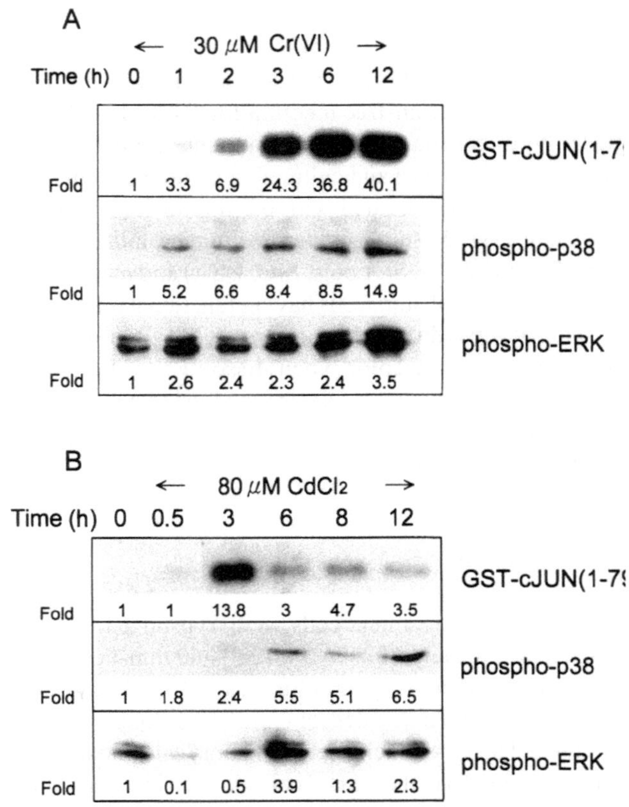

Fig. 2. Time-dependent activation of JNK, p38 and ERK by Cr(VI) and Cd in CL3 cells. Cells were treated with 30 μM Cr(VI) (A) or 80 μM Cd (B) in serum-free media for 0.5–12 h. The whole cell extracts were isolated for determination of the activities of JNK, p38 and ERK as described in Fig. 1. Data averaged from 3 experiments are shown under the autoradiographs.

ERK activity and longer exposure (6–12 h) increased above the control levels (Fig. 2B).

To further investigate the stability of metal-activated MAPKs, cells were treated with 30–50 μM Cr(VI) (25–8% survival) or 80–150 μM Cd (70–6% survival) for 3 h, washed with PBS, cultured in metal-free complete media for another 0, 1 or 8 h and then assayed for activation of JNK and p38. The p38 activated by Cr(VI) decreased markedly and rapidly, while Cr(VI)-activated JNK decreased gradually when the metal was removed from the media (Fig. 3A). On the other hand, the JNK activity induced by low cytotoxic doses of Cd decreased while those induced by high cytotoxic doses of Cd remained at high levels when the metal was removed from the media for 8 h (Fig. 3B). High doses of Cd also persistently activated p38, whereas low Cd doses did not increase it (Fig. 3B). The results suggest that at high cytotoxic doses Cd persistently activates JNK and p38, while Cr(VI) transiently activates these stress-activated kinases.

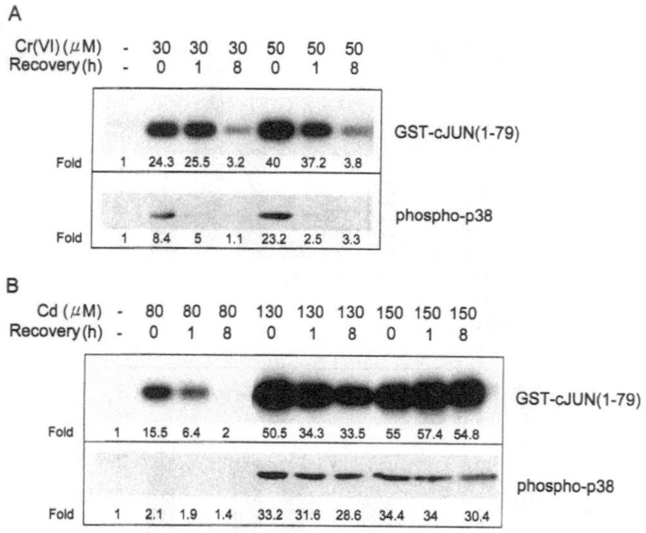

Fig. 3. Duration of activation of JNK and p38 by Cr(VI) and Cd in CL3 cells. Cells were treated with 30–50 μM Cr(VI) (A) or 80–150 μM Cd (B) in serum-free media for 3 h, washed with PBS and cultured continually in RPMI-1640 media containing 10% serum for various times as indicated. The whole cell extracts were isolated for determination of the activities of JNK and p38 as described in Fig. 1. Data averaged from 3 experiments are shown under the autoradiographs.

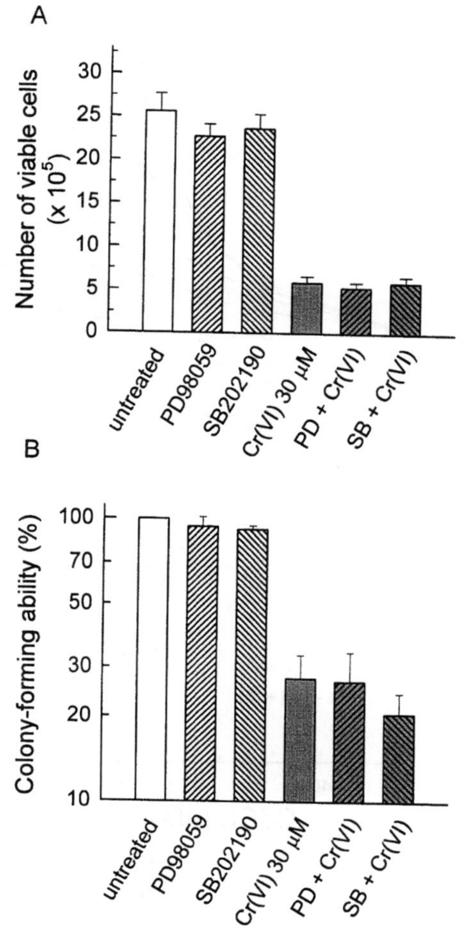

Fig. 4. Effects of the SB202190 and PD98059 on the growth rate and cytotoxicity of Cr(VI)-treated CL3 cells. (A) Cells (2×10^5) were grown in complete media overnight, exposed to PD98059 (50 μM) or SB202190 (10 μM) for 1 h and then treated with Cr(VI) for 3 h. At the end of treatment, the cells were washed with PBS, fed with complete media and cultured for another 3 days. Cells were trypsinized, stained with trypan blue and the numbers of viable cells were determined with a hemocytometer. (B) Cytotoxicity was determined by colony-forming ability assay. The relative survival of cells was calculated from 4 independent experiments and bars represent S.E.M.

Roles of ERK and p38 in cytotoxicity and apoptosis induced by Cr(VI) and Cd

To inspect the roles of ERK and p38 signaling pathways in metal toxicity, CL3 cells were exposed to Cr(VI) or Cd in the presence of kinase inhibitors PD98059 and SB202190 specific for MKK1/2 (ERK upstream kinases) and p38, respectively. Co-administrating SB202190 or PD98059 did not influence the growth rates and cytotoxicity of Cr(VI)-treated cells (Fig. 4). In contrast, PD98059 (50 μM) decreased the viability of cells exposed to 80 μM Cd ($p < 0.01$, day 3), whereas SB202190 (10 μM) potentiated the numbers of viable cells in populations treated with 130 μM Cd ($p < 0.01$, days 2 and 3; Fig. 5A). The effects of these kinase inhibitors on Cd-induced cytotoxicity also showed that PD98059 greatly enhanced cytotoxicity caused by 80–120 μM Cd, while SB202190 decreased the cytotoxicity induced by 160 μM Cd (Fig. 5B). Furthermore, results obtained from the annexin V-FITC binding assay showed that Cr(VI) did not significantly cause apoptosis in CL3 cells, only 2-fold above the background levels was observed at 50 μM (data not shown). Conversely, Cd induced significant numbers of apoptotic cells, ~ 40% of cells undergoing apoptosis 8 h post-exposure to 130 μM Cd for 3 h. Co-administrating SB202190 (10 μM) markedly reduced the frequency of apoptosis in cells exposure to 130 μM Cd, while co-administrating PD98059 (50 μM) enhanced apoptotic cells induced by 80 μM Cd (Fig. 5C).

The activities of JNK, ERK and p38 induced by the two metals in the presence of kinase inhibitors were examined to determine whether cross-talk exists among MAPKs. SB202190 (10 μM) significantly decreased the JNK activity (Fig. 6A) but did not influence the ERK activation by Cr(VI) (Fig. 6B). On the other hand, PD98059 (50 μM) decreased Cr(VI)-activated p38 by ~ 60% (Fig. 6C), but did not significantly alter the Cr(VI)-activated JNK (Fig. 6A). In contrast, co-administration PD98059 with Cd did not affect JNK or p38 activated by Cd; also, SB202190 did not significantly alter JNK or ERK activated by Cd (data not shown). These results indicate that MAPKs activated by Cr(VI) but not Cd in CL3 cells exhibit cross-talk. The above results also indi-

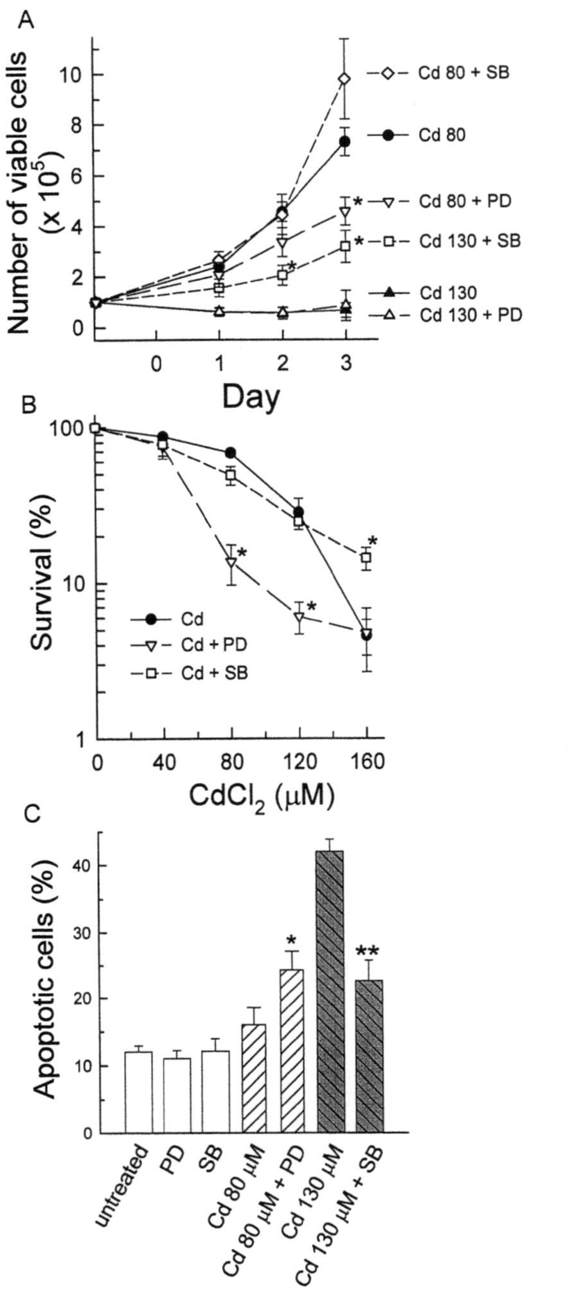

Fig. 5. Effects of SB202190 and PD98059 on the growth rate, viability and apoptosis of $CdCl_2$-treated CL3 cells. Cells at exponential growth were plated and cultured for 24 h in complete medium followed by serum starvation for 18 h. The cells were exposed to PD98059 (50 μM) or SB202190 (10 μM) for 1 h, followed by Cd treatment for 3 h and washed with PBS. (A) The cells were cultured in medium containing 10% serum for 1–3 days and the growth rate was determined by trypan blue exclusion assay. (B) The viability of cells was determined by MTT assay 1 day after exposure to Cd. (C) Cells undergoing apoptosis were examined at 8 h post exposure to Cd using annexin V-FITC binding assay. Results were obtained from 3–6 experiments and bars represent S.E.M. *$p < 0.05$ and **$p < 0.01$ using Student's *t*-test for comparison between Cd-treated cells and those exposed to the same doses of Cd in the presence of PD98059 or SB202190.

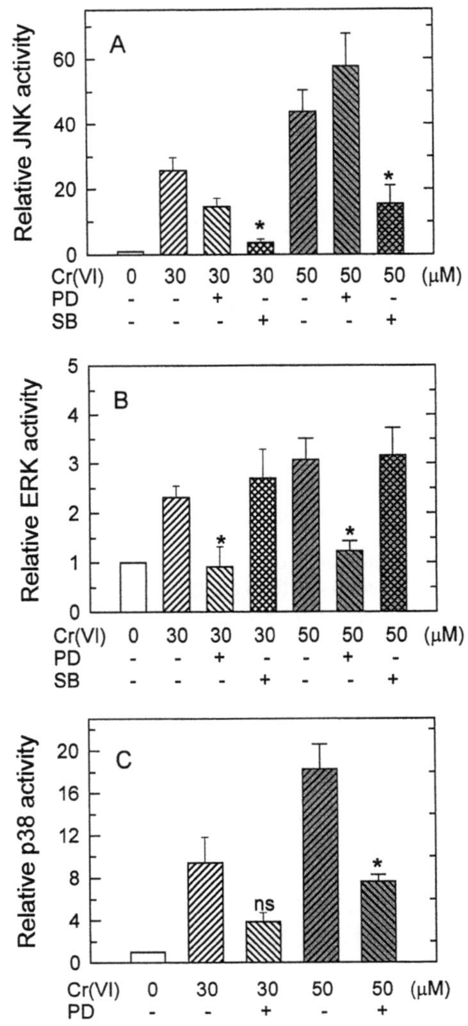

Fig. 6. Effects of SB202190 and PD98059 on the activities of JNK, ERK and p38 induced by Cr(VI) in CL3 cells. PD98059 (50 μM) or SB202190 (10 μM) were added to cells 1 h before Cr(VI) treatment for 3 h. Whole cell extracts were isolated for determination of the activities of JNK, p38 and ERK as described in Fig. 1. The induced levels of ERK, JNK and p38 activities were quantified from at least 3 experiments. *$p < 0.05$ using Student's *t*-test for comparison between Cr(VI)-treated cells and those exposed to the same doses of Cr(VI) in the presence of PD98059 or SB202190. ns, not significant.

cate that ERK and p38 signals play opposite roles in Cd-mediated toxicity, while these stress-activated kinases have limited role in Cr(VI) toxicity that may due partly to the cross-talk effect.

Roles of JNK in cytotoxicity and apoptosis induced by Cr(VI) and Cd

Both MKK4 and MKK7 are the major upstream activators of JNK by various stimuli [3, 4]. To determine the effect of these MKKs on JNK activation by Cr(VI) or Cd, we transfected CL3 cells with a kinase-defective form of MKK4 (MKK4KR) or MKK7 (MKK7A). Two-day after transfection, the cells were exposed to Cr(VI) or Cd for 3 h in serum-free medium and assayed for JNK activation. Figure 7 shows that JNK activation by 30 μM Cr(VI) and 80 μM Cd were 50 and 35% suppressed by expression of MKK7A in CL3 cells, respectively. Conversely, expression of MKK4KR moderately increased JNK activation by Cr(VI) or Cd (Fig. 7). These results are in contrast to the mechanism of JNK activation observed in cells irradiated with UV light (254 nm) (Fig. 7). These results are consistent with JNK activation by various stimuli being mediated differentially through MKK4 and MKK7, e.g. TNFα and interleukin-1 activate MKK7 much more than MKK4, whereas, anisomycin and UV light activate MKK4 more than MKK7 [33, 34]. The results suggest that MKK7, but not MKK4, is an upstream activator of JNK in cells treated with Cr(VI) or Cd. However, transient expression of MKK7A did not significantly alter the cytotoxicity induced by Cr(VI) and Cd (data not shown). Putative JNK upstream activators other than MKK7 and MKK4 may be involved. Such an uncharacterized JNK upstream activator has been observed in cells exposed to hyperosmolarity [34].

To inspect the roles of JNK isoforms in metal toxicity, we transfected CL3 cells with dominant-negative forms of JNK1 and JNK2 (JNK1DN and JNK2DN) or wild-type JNK1 and JNK2. Expression of JNK1DN or JNK2DN blocked Cd-activated JNK in CL3 cells (Fig. 8A), however, they did not alter Cr(VI)-induced JNK activity (data not shown). Additionally, JNK1DN was more efficient than JNK2DN in arresting JNK activation by Cd (Fig. 8A), suggesting that Cd activated JNK1 more than JNK2. Expression of JNK1DN or JNK2DN did not affect the activities of p38 and ERK induced by Cd

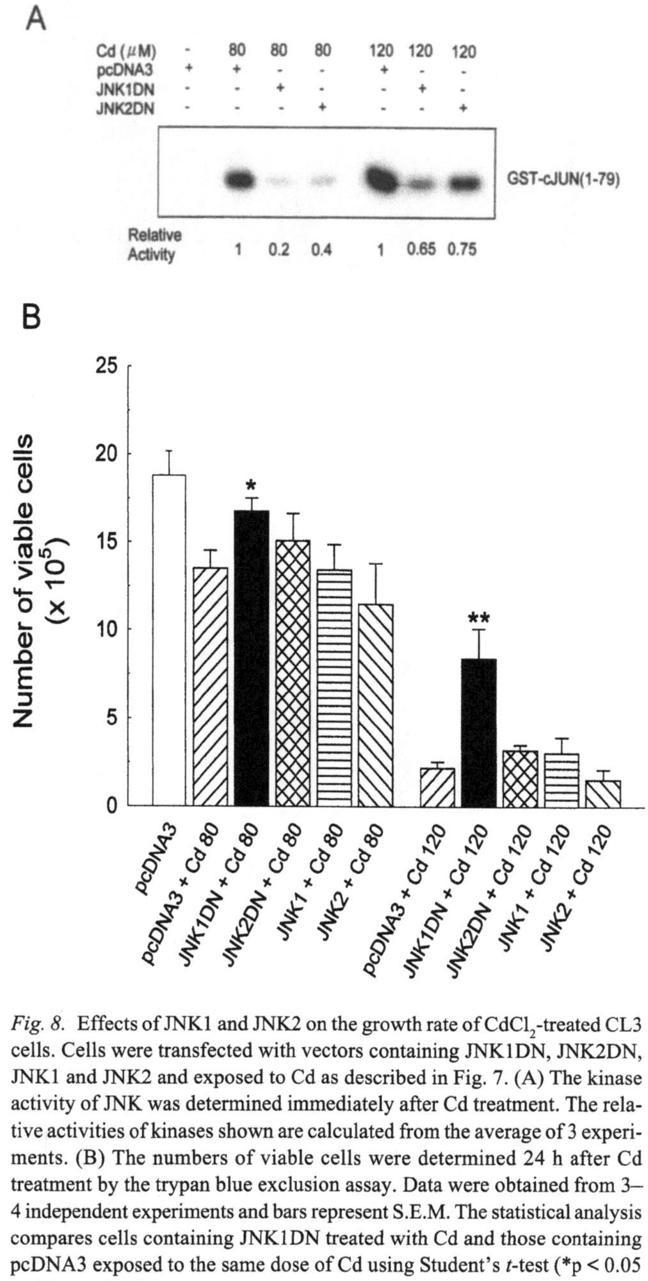

Fig. 8. Effects of JNK1 and JNK2 on the growth rate of $CdCl_2$-treated CL3 cells. Cells were transfected with vectors containing JNK1DN, JNK2DN, JNK1 and JNK2 and exposed to Cd as described in Fig. 7. (A) The kinase activity of JNK was determined immediately after Cd treatment. The relative activities of kinases shown are calculated from the average of 3 experiments. (B) The numbers of viable cells were determined 24 h after Cd treatment by the trypan blue exclusion assay. Data were obtained from 3–4 independent experiments and bars represent S.E.M. The statistical analysis compares cells containing JNK1DN treated with Cd and those containing pcDNA3 exposed to the same dose of Cd using Student's *t*-test (*$p < 0.05$ and **$p < 0.01$).

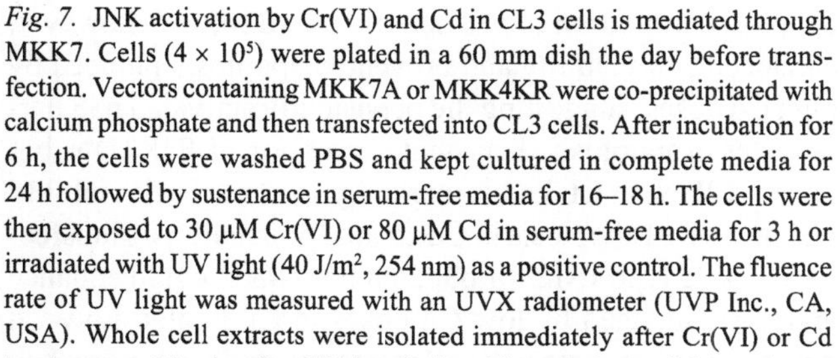

Fig. 7. JNK activation by Cr(VI) and Cd in CL3 cells is mediated through MKK7. Cells (4×10^5) were plated in a 60 mm dish the day before transfection. Vectors containing MKK7A or MKK4KR were co-precipitated with calcium phosphate and then transfected into CL3 cells. After incubation for 6 h, the cells were washed PBS and kept cultured in complete media for 24 h followed by sustenance in serum-free media for 16–18 h. The cells were then exposed to 30 μM Cr(VI) or 80 μM Cd in serum-free media for 3 h or irradiated with UV light (40 J/m^2, 254 nm) as a positive control. The fluence rate of UV light was measured with an UVX radiometer (UVP Inc., CA, USA). Whole cell extracts were isolated immediately after Cr(VI) or Cd treatment or 30 min after UV irradiation. The relative activities of JNK shown are calculated from the average of 3–9 experiments.

(data not shown). The effects of Cd-activated JNK1 and JNK2 on the cell growth inhibition and apoptosis were determined using the trypan-blue exclusion and annexin V-FITC binding assays. The numbers of viable cells in Cd-treated populations were significantly increased by expression of JNK1DN in CL3 cells, whereas JNK2DN had no effect (Fig. 8B). Similarly, JNK1DN significantly abated Cd-induced apoptosis but JNK2DN did not (Fig. 9). Conversely, transfection of wild-type JNK1 into CL3 cells influenced neither growth inhibition nor apoptosis caused by Cd (Figs 8B and 9). In contrast, transfection of wild-type JNK2 or p38 could enhance apopto-

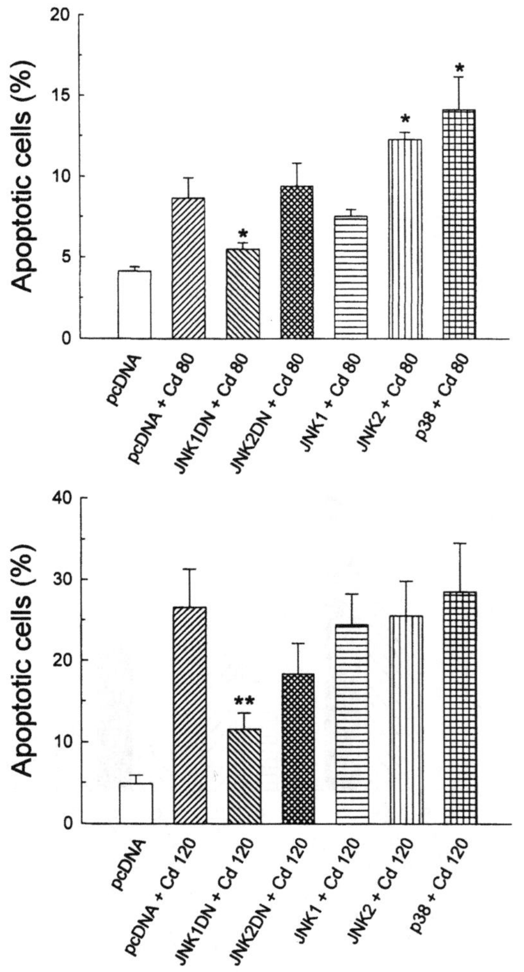

Fig. 9. Effects of JNK1, JNK2 and p38 on apoptosis induced by $CdCl_2$ in CL3 cells. Transient transfection and expression of JNK1DN, JNK2DN, JNK1, JNK2 or p38 in cells followed by Cd treatment were performed as described in Fig. 7. Cells undergoing apoptosis were examined 24 and 8 h post-exposure to 80 and 120 µM of Cd, respectively, using annexin V-FITC binding assay. Results were obtained from 4-6 experiments and bars represent S.E.M. The statistical analyses compare cells containing JNK1DN, JNK2 or p38 treated with Cd and those containing pcDNA3 exposed to the same dose of Cd using Student's *t*-test (*$p < 0.05$ and **$p < 0.01$).

sis induced by low Cd doses in CL3 cells ($p < 0.05$, Fig. 9). These results suggest that JNK1 in combination with JNK2 and p38 transmit a signal for Cd-induced growth inhibition and apoptosis.

Discussion

In this study, we have demonstrated that Cr(VI) at sub-lethal doses can activate JNK, p38 and ERK in human lung adenocarcinoma CL3 cells. JNK and p38 could be persistently activated provided that Cr(VI) was constantly present in the culture medium. However, activated p38 decreased rapidly and activated JNK reduced gradually after Cr(VI) withdrawal. The duration of activation of JNK and p38 may be correlated with the half-life of the reactive species generated through the intracellular reduction of Cr(VI), which may be diminished after Cr(VI) withdrawal. On the other hand, Cd at low cytotoxic doses transiently activated JNK and concomitantly reduced ERK activity, whereas high cytotoxic doses of Cd persistently activated JNK and p38 in CL3 cells. Cd induces stronger activities of JNK and p38 with longer half-life as compared with those activated by Cr(VI) at similar cytotoxic dose ranges. The distinct kinetics of MAPK activation by Cr(VI) and Cd could be attributed in part to that Cr(VI) may activate negative feedback pathways, e.g. dual-specific MAPK phosphatases [35], to down-regulate JNK and p38 activities, while Cd may inhibit such negative regulating systems.

Although, the pyridinyl imidazole inhibitors of p38 have been reported to have no effect on the activity of other MAPK family members [36–38], SB202190 was found to significantly decrease JNK activity induced by Cr(VI). This finding suggests that p38 activity is required for the JNK activation by Cr(VI). The result is also consistent with the observation that p38 activation precedes JNK activation by Cr(VI) in time-dependent experiments. On the other hand, specific inhibition of the ERK pathway by PD98059 could reduce Cr(VI)-activated p38 by ~ 60%, suggesting that ERK activity is upstream of the p38 activation by Cr(VI). Cross-talk between ERK and stress-activated p38 and JNK has been reported. For examples, ERK activation by arsenite (500 µM) is highly dependent on the MKK6/p38 pathway [39]. Inhibition of ERK activation by PD98059 can potentiate JNK1 activation induced by ionizing radiation [40, 41]. However, activation of the three MAPKs in Cd-treated CL3 cells does not exhibit cross-talk. These results indicate the interactions of MAPK signal pathways are variable that may be stimulus- and cell type-dependent.

Co-administrating Cd with specific inhibitors to block ERK or p38 signals showed that these signal pathways have opposing effects on the growth arrest and apoptosis induced by Cd. Inhibition of ERK activity by PD98059 accelerated growth inhibition and apoptosis induced by low Cd doses, suggesting ERK is involved in cell proliferation and anti-apoptosis. Inhibition of p38 activity by SB202190, however, enhanced cell viability and prevented apoptosis induced by high Cd doses, suggesting the persistently activated p38 participates in apoptosis. Transient expression of JNK1DN, but not JNK2DN, in CL3 cells markedly increased cell viability and prevented apoptosis in Cd-treated populations. Transfection of wild type JNK2 or p38, but not JNK1, could enhance apoptosis induced by low Cd doses in CL3 cells. Low Cd doses did not activate p38 and may activate JNK1 more than JNK2. This would explain in part why transfection of wild-type p38 or JNK2 enhanced apoptosis when the cells were exposed to low Cd doses. The results shown here suggest that Cd may sequentially activate JNK1, JNK2 and p38 in a dose-dependent manner to cooperatively induce apoptosis. Decreased ERK activity may combine with increased JNK activity to transmit a signal temporally causing growth arrest by low Cd doses in CL3 cells.

On the other hand, transient activation of these three MAPKs by Cr(VI) does not influence its cytotoxicity. The cytotoxicity of Cr(VI) compounds are associated with their ability to enter cells rapidly and to generate reactive species during Cr(VI) intracellular reduction that subsequently result in various types of DNA damage, including strand breaks, DNA-protein crosslinks, DNA-DNA crosslinks, Cr-DNA adducts and base modifications [10, 11]. It is known that DNA-protein or DNA-DNA crosslinks are poorly repaired in comparison with DNA strand breaks generated in Cr(VI)-treated mammalian cells [42, 43]. Cr(VI)-induced DNA-protein or DNA-DNA crosslinks, which are the major causative lesions of Cr(VI) cytotoxicity, may not contribute to the activation of p38 and JNK signals. Alternatively, transient activation of these kinases by Cr(VI) may not affect the levels of those lethal lesions, thereby does not influence cytotoxicity.

It is well documented that both Cr(VI) and Cd promote apoptosis in a variety of cells [15–19]. However, in our comparable experiment conditions Cd is a much stronger apoptotic inducer than Cr(VI) in CL3 cells at similar cytotoxic doses. The results correlate with that Cr(VI) transiently activates JNK and p38 while Cd persistently induces their activities. This finding is also agreed with that persistent, but not transient, activation of JNK and p38 triggers apoptosis. Apoptosis is a defense system to eliminate damaged cells upon exposure to carcinogens thereby reduces carcinogenesis. The fact that Cd causes more apoptotic cells than Cr(VI) in CL3 lung cancer cells at equal cytotoxic dose ranges suggests Cr(VI) may be more effective than Cd to induce lung cancer. Interestingly, several reports have shown the ability of Cd to suppress tumor growth and progression in rodent lung and liver [44, 45]. The relationship between apoptosis and tumor suppression by Cd is not clear [45]. Our finding that JNK and p38 signal pathways induced by Cd participate in growth arrest and apoptosis of human lung cancer cells suggests a potential role of these MAPKs in Cd-mediated tumor suppression.

JNK and p38 signals could transmit some physiological consequences other than apoptosis in different cell types. For examples, JNK signal has been implicated in protection against alkylating agents in 3T3 cells [46]. Additionally, JNK1 is selective activated by the apoptosis inhibitor protein hILP for protection against ICE-induced apoptosis [47]. Activation of p38 by interleukin-2 or interleukin-7 is required to transmit the mitogenic signal in T cells [48]. The p38 pathway activated by phorbol ester may contribute to a more invasive phenotype in a human squamous cell carcinoma cell line [49]. Moreover, isoforms of MAPKs may possess opposing functions, e.g. while p38α induces apoptosis, p38β inhibits it [50]. All the above evidences have indicated that activation of JNK or p38 can be a protective mechanism to support cell survival and may contribute to carcinogenesis. The epigenetic role of JNK and p38 activation by Cr(VI) in carcinogenesis deserved further studies.

In summary, Cr(VI) induces JNK, p38 and ERK in a dose-dependent manner, yet their activation does not associate with cytotoxicity. This may be attributed to the fact that: (i) the short duration of activation of these kinases by Cr(VI) may not affect the levels of lethal lesions, such as DNA cross-links; (ii) cross-talk between MAPKs may obstruct observation of the effect of each kinase; (iii) Cr(VI) may activate other signal transduction pathways to counteract the effects of MAPKs. In contrast, low-cytotoxic doses of Cd activate JNK, reduce ERK and do not affect p38, producing convergent signals transiently inhibiting cell growth, whereas high-cytotoxic doses of Cd persistently activate JNK and p38 to induce apoptosis. Although both metals are classified as group 1 carcinogens [7, 8], Cd induces higher apoptosis than Cr(VI) in CL3 lung cancer cells. Together, this study provides evidence suggesting that sub-lethal doses of Cr(VI) or Cd induced three MAPK signals play distinct epigenetic roles in metal carcinogenesis.

Acknowledgements

The authors would like to thank Dr. P.-C. Yang for providing the CL3 cells and Dr. M. Karin and Dr. J. Han for the expression vectors. This work was supported by the National Science Council, Republic of China under Contract No. NSC87-2311-B007-032 and NSC88-2311-B007-028.

Reference

1. Marshall CJ: Specificity of receptor tyrosine kinase signaling: transient versus sustained extracellular signal-regulated kinase activation. Cell 80: 179–185, 1995
2. Whitmarsh AJ, Davis RJ: Transcription factor AP-1 regulation by mitogen-activated protein kinase signal transduction pathways. J Mol Med 74: 589–607, 1996
3. Ip YT, Davis RJ: Signal transduction by the c-Jun N-terminal kinase (JNK)-from inflammation to development. Curr Opin Cell Biol 10: 205–219, 1998
4. Garrington TP, Johnson GL: Organization and regulation of mitogen-activated protein kinase signaling pathways. Curr Opin Cell Biol 11: 211–218, 1999
5. Nebreda AR, Porras A: p38 MAP kinases: Beyond the stress response. Trends Biochem Sci 25: 257–260, 2000

6. Hill CS, Treisman R: Transcriptional regulation by extracellular signals: Mechanisms and specificity. Cell 80: 199–211, 1995
7. IARC: Chromium, nickel, and wedding. In: IARC Monographs on the Evaluation of the Carcinogenic Risks of Chemicals to Humans, vol 19. International Agency for Cancer Research, Lyon, 1990
8. IARC: Cadmium and cadmium compounds. In: IARC Monographs on the Evaluation of Carcinogenic Risks to Humans: Cadmium, Mercury, Beryllium and Exposures in the Glass Industry, vol. 58. International Agency for Research on Cancer, Lyon, 1993, pp 119–237
9. Degraeve N: Carcinogenic, teratogenic and mutagenic effects of cadmium. Mutat Res 86: 115–135, 1981
10. Wetterhahn KE, Hamilton JW, Aiyar J, Borges KM, Floyd R: Mechanism of chromium(VI) carcinogenesis: Reactive intermediates and effect on gene expression. Biol Trace Element Res 21: 405–411, 1989
11. Shi X, Chiu A, Chen CT, Halliwell B, Castranova V, Vallyathan V: Reduction of chromium(VI) and its relationship to carcinogenesis. J Toxicol Environ Health B Crit Rev 2: 87–104, 1999
12. Ochi T, Ohsawa M: Participation of active oxygen species in the induction of chromosomal aberrations by cadmium chloride in cultured Chinese hamster cells. Mutat Res 143: 137–142, 1985
13. Yang JL, Chao JI, Lin JG: Reactive oxygen species may participate in the mutagenicity and mutational spectrum of cadmium in Chinese hamster ovary-K1 cells. Chem Res Toxicol 9: 1360–1367, 1996
14. Hwua YS, Yang JL: Effect of 3-aminotriazole on anchorage independence and mutagenicity in cadmium- and lead-treated diploid human fibroblasts. Carcinogenesis 19: 881–888, 1998
15. Ye J, Wang S, Leonard SS, Sun Y, Butterworth L, Antonini J, Ding M, Rojanasakul Y, Vallyathan V, Castranova V, Shi X: Role of reactive oxygen species and p53 in chromium(VI)-induced apoptosis. J Biol Chem 274: 34974–34980, 1999
16. Pritchard DE, Singh J, Carlisle DL, Patierno SR: Cyclosporin A inhibits chromium(VI)-induced apoptosis and mitochondrial cytochrome c release and restores clonogenic survival in CHO cells. Carcinogenesis 21: 2027–2033, 2000
17. Carlisle DL, Pritchard DE, Singh J, Patierno SR: Chromium(VI) induces p53-dependent apoptosis in diploid human lung and mouse dermal fibroblasts. Mol Carcinogen 28: 111–118, 2000
18. Galan A, Garcia-Bermejo ML, Troyano A, Vilaboa NE, de Blas E, Kazanietz MG, Aller P: Stimulation of p38 mitogen-activated protein kinase is an early regulatory event for the cadmium-induced apoptosis in human promonocytic cells. J Biol Chem 275: 11418–11424, 2000
19. Robertson JD, Orrenius S: Molecular mechanisms of apoptosis induced by cytotoxic chemicals. Crit Rev Toxicol 30: 609–627, 2000
20. Kim G, Yurkow EJ: Chromium induces a persistent activation of mitogen-activated protein kinases by a redox-sensitive mechanism in H4 rat hepatoma cells. Cancer Res 56: 2045–2051, 1996
21. Samet JM, Graves LM, Quay J, Dailey LA, Devlin RB, Ghio AJ, Wu W, Bromberg PA, Reed W: Activation of MAPKs in human bronchial epithelial cells exposed to metals. Am J Physiol 275: L551–L558, 1998
22. Wang Z, Templeton DM: Induction of c-*fos* proto-oncogene in mesangial cells by cadmium. J Biol Chem 273: 73–79, 1998
23. Hung JJ, Cheng TJ, Lai YK, Chang MD: Differential activation of p38 mitogen-activated protein kinase and extracellular signal-regulated protein kinases confers cadmium-induced HSP70 expression in 9L rat brain tumor cells. J Biol Chem 273: 31924–31931, 1998
24. Matsuoka M, Igisu H: Activation of c-Jun NH_2-terminal kinase (JNK/SAPK) in LLC-PK1 cells by cadmium. Biochem Biophys Res Commun 251: 527–532, 1998
25. Chuang SM, Liou GY, Yang JL: Activation of JNK, p38 and ERK mitogen-activated protein kinases by chromium(VI) is mediated through oxidative stress but does not affect cytotoxicity. Carcinogenesis 21: 1491–1500, 2000
26. Chuang SM, Wang IC, Yang JL: Roles of JNK, p38 and ERK mitogen-activated protein kinases in the growth inhibition and apoptosis induced by cadmium. Carcinogenesis 21: 1423–1432, 2000
27. Yang PC, Luh KT, Wu R, Wu CW: Characterization of the mucin differentiation in human lung adenocarcinoma cell lines. Am J Respir Cell Mol Biol 7: 161–171, 1992
28. Hibi M, Lin A, Smeal T, Minden A, Karin M: Identification of an oncoprotein- and UV-responsive protein kinase that binds and potentiates the c-Jun activation domain. Genes Dev 7: 2135–2148, 1993
29. Kallunki T, Su B, Tsigelny I, Sluss HK, Dérijard B, Moore G, Davis R, Karin M: JNK2 contains a specificity-determining region responsible for efficient c-Jun binding and phosphorylation. Genes Dev 8: 2996–3007, 1994
30. Lin A, Minden A, Martinetto H, Claret FX, Lange-Carter C, Mercurio F, Johnson GL, Karin M: Identification of a dual specificity kinase that activates the Jun kinases and p38-Mpk2. Science 268: 286–290, 1995
31. Han J, Lee JD, Bibbs L, Ulevitch RJ: A MAP kinase targeted by endotoxin and hyperosmolarity in mammalian cells. Science 265: 808–811, 1994
32. Huang S, Jiang Y, Li Z, Nishida E, Mathias P, Lin S, Ulevitch RJ, Nemerow GR, Han J: Apoptosis signaling pathway in T cells is composed of ICE/Ced-3 family proteases and MAP kinase kinase 6β. Immunity 6: 739–749, 1997
33. Lawler S, Cuenda A, Goedert M, Cohen P: SKK4, a novel activator of stress-activated protein kinase-1 (SAPK1/JNK). FEBS Lett 414: 153–158, 1997
34. Moriguchi T, Toyoshima F, Masuyama N, Hanafusa H, Gotoh Y, Nishida E: A novel SAPK/JNK kinase, MKK7, stimulated by TNFα and cellular stresses. EMBO J 16: 7045–7053, 1997
35. Keyse SM: Protein phosphatases and the regulation of mitogen-activated protein kinase signalling. Curr Opin Cell Biol 12: 186–192, 2000
36. Lee JC, Laydon JT, McDonnell PC, Gallagher TF, Kumar S, Green D, McNulty D, Blumenthal MJ, Heys JR, Landvatter SW, Strickler JE, McLaughlin MM, Siemens IR, Fisher SM, Livi GP, White JR, Adams JL, Young PR: A protein kinase involved in the regulation of inflammatory cytokine biosynthesis. Nature 372: 739–746, 1994
37. Cuenda A, Rouse J, Doza YN, Meier R, Cohen P, Gallagher TF, Young PR, Lee JC: SB 203580 is a specific inhibitor of a MAP kinase homologue which is stimulated by cellular stresses and interleukin-1. FEBS Lett 364: 229–233, 1995
38. Foltz IN, Lee, JC, Young PR, Schrader JW: Hemopoietic growth factors with the exception of interleukin-4 activate the p38 mitogen-activated protein kinase pathway. J Biol Chem 272: 3296–3301, 1997
39. Ludwig S, Hoffmeyer A, Goebeler M, Kilian K, Hafner H, Neufeld B, Han J, Rapp UR: The stress inducer arsenite activates mitogen-activated protein kinases extracellular signal-regulated kinases 1 and 2 via a MAPK kinase 6/p38- dependent pathway. J Biol Chem 273: 1917–1922, 1998
40. Carter S, Auer KL, Reardon DB, Birrer M, Fisher PB, Valerie K, Schmidt-Ullrich R, Mikkelsen R, Dent P: Inhibition of the mitogen activated protein (MAP) kinase cascade potentiates cell killing by low dose ionizing radiation in A431 human squamous carcinoma cells. Oncogene 16: 2787–2796, 1998
41. Reardon DB, Contessa JN, Mikkelsen RB, Valerie K, Amir C, Dent P, Schmidt-Ullrich RK: Dominant negative EGFR-CD533 and inhibition of MAPK modify JNK1 activation and enhance radiation toxicity of human mammary carcinoma cells. Oncogene 18: 4756–4766, 1999
42. Sugiyama M, Wang XW, Costa M: Comparison of DNA lesions and cytotoxicity induced by calcium chromate in human, mouse, and hamster cell lines. Cancer Res 46: 4547–4551, 1986
43. Xu J, Bubley GJ, Detrick B, Blankenship LJ, Patierno SR: Chromium (VI) treatment of normal human lung cells results in guanine-specific

DNA polymerase arrest, DNA-DNA crosslinks and S-phase blockade of cell cycle. Carcinogenesis 17: 1511–1517, 1996
44. Waalkes MP, Diwan BA, Weghorst CM, Ward JM, Rice JM, Cherian MG, Goyer RA: Further evidence of the tumor-suppressive effects of cadmium in the B6C3F1 mouse liver and lung: Late stage vulnerability of tumors to cadmium and the role of metallothionein. J Pharmacol Exp Ther 266: 1656–1663, 1993
45. Waalkes MP, Diwan BA: Cadmium-induced inhibition of the growth and metastasis of human lung carcinoma xenografts: Role of apoptosis. Carcinogenesis 20: 65–70, 1999
46. Liu ZG, Baskaran R, Lea-Chou ET, Wood LD, Chen Y, Karin M, Wang JYJ: Three distinct signalling responses by murine fibroblasts to genotoxic stress. Nature (Lond) 384: 273–276, 1996
47. Sanna MG, Duckett CS, Richter BWM, Thompson CB, Ulevitch RJ: Selective activation of JNK1 is necessary for the anti-apoptotic activity of hILP. Proc Natl Acad Sci USA 95: 6015–6020, 1998
48. Crawley JB, Rawlinson L, Lali FV, Page TH, Saklatvala J, Foxwell MJ: T cell proliferation in response to interleukins 2 and 7 requires p38 MAP kinase activation. J Biol Chem 272: 15023–15027, 1997
49. Simon C, Goepfert H, Boyd D: Inhibition of the p38 mitogen-activated protein kinase by SB 203580 blocks PMA-induced M_r 92,000 type IV collagenase secretion and *in vitro* invasion. Cancer Res 58: 1135–1139, 1998
50. Nemoto S, Xiang J, Huang S, Lin A: Induction of apoptosis by SB202190 through inhibition of p38β mitogen-activated protein kinase. J Biol Chem 273: 16415–16420, 1998

Molecular and Cellular Biochemistry **222**: 97–106, 2001.

Differential zinc and DNA binding by partial peptides of human protamine HP2

Wojciech Bal,[1,2] Marcin Dyba,[1] Zbigniew Szewczuk,[1] Małgorzata Jeżowska-Bojczuk,[1] Jan Lukszo,[3] Gayatri Ramakrishna[2] and Kazimierz S. Kasprzak[2]

[1]*Faculty of Chemistry, University of Wroclaw, Wroclaw, Poland;* [2]*Laboratory of Comparative Carcinogenesis, National Cancer Institute at Frederick, Frederick, MD;* [3]*Peptide Analysis and Synthesis Unit, Research Technology Branch, National Institute of Allergy and Infectious Diseases, Rockville, MD, USA*

Abstract

The Zn(II) binding by partial peptides of human protamine HP2: $HP2_{1-15}$; $HP2_{1-25}$, $HP2_{26-40}$, $HP2_{37-47}$, and $HP2_{43-57}$ was studied by circular dichroism (CD). Precipitation of a 20-mer DNA by these partial peptides and the effects of Zn(II) thereon were investigated using polyacrylamide gel electrophoresis (GE). The results of this study suggest that reduced HP2 (thiol groups intact) can bind Zn(II) at various parts of the molecule. In the absence of DNA, the primary Zn(II) binding site in reduced HP2 is located in the 37–47 sequence (involving Cys-37, His-39, His-43, and Cys-47), while in the presence of DNA, the strongest Zn(II) binding is provided by sequences 12–22 (by His-12, Cys-13, His-19, and His-22) and 43–57 (His-43, Cys-47, Cys-53, and His-57). In its oxidized form, HP2 can bind zinc through His residues of the 7–22 sequence. Zn(II) markedly enhances DNA binding by all partial peptides. These findings suggest that Zn(II) ions may be a regulatory factor for sperm chromatin condensation processes. (Mol Cell Biochem **222**: 97–106, 2001)

Key words: protamine P2, protamine peptides, DNA, zinc, protamine-zinc binding

Introduction

DNA in vertebrate sperm is stabilized through the binding to protamines, small basic proteins rich in arginine, which provide positive charges required for the neutralization of DNA phosphates [1, 2]. In mammals, protamines also contain cysteine residues, which form intermolecular disulfide bridges, increasing the stability of the DNA-protein complex. Consequently, protamine-bound DNA in sperm head is compacted severalfold compared to somatic nuclei [3]. This function is exerted by protamine P1, expressed in all mammalian species. A few species, including mouse and man, express an additional protamine, P2, also containing histidine, [1, 2]. Strikingly, its presence at 50–70% of total protamine is required for male fertility in humans [4]. The sequence of human P2 (HP2) is as follows [1, 2]:

RTHGQ-SHYRR-RHCSR-RRLHR-IHRRQ-HRSCR-RRKRR-SCRHR-RRHRR-GCRTR-KRTCR-RH

Zn(II) binding in sperm chromatin is necessary for maintaining sperm viability [5]. HP2 is believed to provide these zinc binding site(s). A recent paper indicated that Zn(II) content of mammalian sperm cells (including human) correlates with protamine 2 abundance and it is sufficient in the majority of cases to form 1:1 complex with P2 [6]. The identity and localization of this site(s) within HP2 is, however, uncertain. One proposal, based on the effects exerted by Zn(II) on HP2 CD spectrum and structure prediction algorithms, suggests that two Zn(II) ions are coordinated to four separate histidine nitrogens each, producing multiple loops in the HP2 molecule [7]. This suggestion, neglecting the participation of Cys sulfur donors in the binding, is contrary to known coordination

Address for offprints: W. Bal, Faculty of Chemistry, University of Wroclaw, F. Joliot-Curie 14, 50-383 Wroclaw, Poland preferences of Zn(II). The other proposal, based on Co(II) substitution and UV-vis spectra, suggests a single Cys2/His2 binding site, more or less analogous to zinc-finger domains [8]. Such a coordination set is chemically viable, but the criteria of selection of the particular coordinating residues out of many available (two out of nine His and two out of five Cys side chains) are not clear. Our preliminary study, using CD spectroscopy, indicated that HP2 may bind two Zn(II) ions *in vitro* with a relatively high affinity [9].

Human protamine HP2 possesses another metal-binding domain. It is the N-terminal tripeptide Arg-Thr-His, a specific Cu(II)- and Ni(II)-binding site, analogous to the physiological Cu(II) carrier in human serum albumin, and also involved in nickel toxicity [10]. In our recent studies of Cu(II) and Ni(II) interactions with HP2, we used its N-terminal pentadecapeptide ($HP2_{1-15}$) as a functional model. We have shown that the binding of either of these metal ions is very strong and specific [11]. Also, the Ni(II) complex can activate oxygen and hydrogen peroxide, leading to the oxidative damage of DNA and the peptide itself [12, 13], while the activity of the Cu(II) complex is limited to peptide destruction. In a subsequent work, we used NMR to elucidate the structure of the Ni(II) complex of $HP2_{1-15}$ [14]. We found that the Ni(II) binding results in long-range peptide structuring, which may be responsible for the enhancement of DNA binding and precipitation upon the N-terminal metal ion coordination, seen previously [12, 13].

Previous studies suggested that the Zn(II) binding has no effect on HP2-DNA interactions [15]. Taking into account the essentiality of Zn(II) for sperm viability on the one hand, and clear enhancement of DNA-$HP2_{1-15}$ binding by Cu(II)/Ni(II) on the other hand, we have decided to reanalyze this issue. We undertook an attempt to localize the Zn(II) binding site(s) within the HP2 molecule in the absence and presence of DNA. The multiplicity of potential Zn(II) donor groups, combined with an absence of a secondary structure, makes the identification of the actual binding site(s) a very difficult task. We assumed, based on previous CD studies [7, 9] that the highly charged HP2 molecule (+29 per 57 residues, from arginine and lysine side chains) would maintain an extended conformation, as the one minimizing electrostatic repulsion. Therefore, the most likely binding sites would be formed locally, involving four consecutive His and/or Cys donor groups. His/Cys metal binding domains are most specific to Zn(II) when the binding is provided by two Cys and two His residues, and such domains are most frequently encountered in DNA binding [16, 17]. Therefore, we decided to dissect the HP2 sequence into five partial peptides containing such residues, and to study their properties comparatively.

Cys residues are present in HP2 in positions 13, 29, 37, 47 and 54. We assumed that Cys-13 is too distant from Cys-29 to be combined in the same peptide. On the other hand, it is surrounded by many His residues; thus the selection of peptides A ($HP2_{1-15}$) and B ($HP2_{1-25}$). The other four Cys residues could be combined in three sequential pairs, allowing for the design of three partially overlapping peptides: C ($HP2_{26-40}$), D ($HP2_{37-47}$), and E ($HP2_{43-57}$). Each of them contains just two His residues. In this way each of our partial peptides provided just one potential Zn(II) binding site with a Cys/His3 or Cys2/His2 donor set. We synthesized these peptides and studied their Zn(II) binding properties in the absence and presence of a short double-stranded oligonucleotide to create a dataset that should allow to propose the localization of one or more Zn(II) binding sites in the whole HP2. This paper presents the results of our studies.

Materials and methods

Materials

$ZnCl_2$ (99.99% purity) was purchased from Aldrich (Milwaukee, WI, USA). Other simple chemicals were obtained from Sigma Chemical Co. (St. Louis, MO, USA) or E. Merck (Darmstadt, Germany).

Peptide synthesis

Peptides $HP2_{1-15}$ (A), $HP2_{1-25}$ (B), and $HP2_{26-40}$ (C) were synthesized by Fmoc strategy [18] on a solid support, using an Applied Biosystems Inc. (ABI, Foster City, CA, USA) model 430A automated peptide synthesizer. The substrates, N-Fmoc protected amino acids, Fmoc-amide resin, and all other peptide synthesis reagents were obtained from ABI. Acetylation of the N-terminal was performed using N-acetylimidazole (Sigma) in dimethylformamide. Peptides $HP2_{37-47}$ (D) and $HP2_{43-57}$ (E) were also synthesized by Fmoc strategy on a solid support, but using a manual setup. Materials for theses syntheses were obtained from Calbiochem-Novabiochem AG, Läufelfingen, Switzerland (N-Fmoc protected amino acids, amide MBHA resin, Fmoc-Cys Wang resin and benzotriazol-1-yl-oxy-tripyrrolidinophosphonium hexafluorophosphate (PyBOP)), and Sigma or Riedel-deHaën GmbH, Seeize, Germany) (other standard peptide synthesis reagents). Cleavage was effected on a 0.25 mmol scale, using a mixture of trifluoroacetic acid (TFA), thioanisole and ethanedithiol (v/v/v = 92/5/3), over a period of 2 h for automated synthesis or on a 0.05 mmol scale, using a mixture of trifluoroacetic acid (TFA), phenol, water, thioanisole and triisopropylsilane (v/v/v/v/v = 80/5/5/5/5), over a period of 9 h for manual synthesis. The crude peptides were purified by preparative HPLC on the Waters Delta Pak (Waters, Milford, MA, USA) C_{18} column, 300 Å pore size, 15 μm particle size, 19 × 300 mm, eluting with 0.1% TFA/water (solvent A) and 0.1% TFA/80% acetonitrile/water (solvent B), using a gradient of 5–25% B/

40 min at a flow rate of 12.5 mL/min with detection at 215 nm. Fractions collected across the main peak were assessed by HPLC analysis on a Rainin Dynamax (Ridgefield, NJ, USA) C_{18}, 300 Å, 5 µm, 4.6 × 250 mm column, using gradient of 0%–35% B over 20 min, flow rate of 1 mL/min and detection at 215 nm. Correct fractions were pooled and lyophilized to yield white solids of purity exceeding 95% as assessed by HPLC analysis of the final materials. Identity of all peptides was confirmed by mass spectrometry, utilizing a Voyager RP MALDI-TOF instrument from PerSeptive Biosystems, Inc. (Framingham, MA, USA).

While the four shorter sequences were found to be homogenous, the longer, very basic peptide B was still contaminated by the truncated sequences missing one or more arginine residues. That peptide was further purified by a semi-preparative HPLC on an ion exchange column (PolyCAT A; 300 Å, 5 µm, 4.6 × 200 mm) from The Nest Group, Inc. (Southborough, MA, USA), eluting with 10 mM ammonium acetate/10% acetonitrile/water (solvent A) and 10 mM ammonium acetate/1 M sodium perchlorate/10% acetonitrile/water (solvent B), using a gradient of 35–100% B/35 min at a flow rate of 4 mL/min with detection at 220 nm. Fractions collected across the main peak were assessed by HPLC analysis on the same column and solvent system, at the flow rate of 1 mL/min. Correct fractions were pooled and desalted by preparative HPLC on the Waters Delta Pak C_{18} column, same as above, eluted with 0.1% TFA/water (solvent A) and 0.1% TFA/80% acetonitrile/water (solvent B), using a gradient of 5–35% B/40 min at a flow rate of 12.5 mL/min with detection at 215 nm. Fractions collected across the main peak were pooled and lyophilized to yield a white solid of purity exceeding 95% as assessed by HPLC and MALDI-TOF MS analysis.

Oligonucleotide synthesis

The complementary 20-mer oligodeoxynucleotides, 5′-GGAGGCGTCCGGTTCCAATA-3′ (SS1) and 5′-TATTGGAACCGGACGCCTCC-3′ (SS2) were obtained purified from SAIC-Frederick and annealed to form a double-stranded 20-mer nucleotide (DS). The annealing occurred with a 100% efficacy. However, the final preparation contained an 18% surplus of SS1 and was used as such. This composition appeared to be advantageous by allowing us to make comparisons of peptide binding to the two forms of DNA in single experiments.

Potentiometry

The protonation constants of the peptides in the presence of 0.1 M KNO_3 were determined at 25ºC using pH-metric titrations over the pH range 3–11 (Molspin automatic titrator, Molspin Ltd., Newcastle-upon-Tyne, UK) with NaOH as titrant. Changes in pH were monitored with a combined glass-calomel electrode calibrated daily in hydrogen concentrations by HNO_3 titrations [19]. Sample volumes of 1.5 mL and peptide concentrations of 0.3–0.5 mM were used. All measurements were performed under argon. The data were analyzed using the SUPERQUAD program [20].

CD spectra

The spectra were recorded on a Jasco J-720 spectropolarimeter (Japan Spectroscopic Company, Hiroshima, Japan) in the spectral range of 190–350 nm, for 2.5–10 µM peptides dissolved in 10 mM sodium phosphate buffer, containing 20 µM DTT. Without the latter, the spectra could not be measured due to autooxidation and precipitation of the peptides under ambient atmosphere. Duplicate runs were averaged to reduce measurement noise at short wavelengths.

Gel electrophoresis

The samples dissolved in the modified TB running buffer (90 mM Tris-borate, pH 7.4, no EDTA) containing oligonucleotides, 1 µM DS and 0.18 µM SS1 (high concentration) or 0.4 µM DS and 0.072 µM SS1 (low concentration) and combinations of various concentrations of HP2 partial peptides and Zn(II) in a total volume of 12 µL were incubated for 15 min, mixed with 2 µL of loading buffer (30% Ficoll and 0.05% bromophenol blue and xylene cyanol in 10 × running buffer) and loaded on Novex Pre-Cast 12% polyacrylamide gels (Invitrogen, Carlsbad, CA, USA). The electrophoresis was run for 2 h at 120 V, using the running buffer. In some experiments, the samples also contained 0.5 mM DTT. The gels were stained with ethidium bromide (Bio-Rad, Hercules, CA, USA) and photographed and processed with an IS-1000 Digital Imaging System (Alpha Innotech Corp., San Leandro, CA, USA). DNA in individual lanes was quantified using reference lanes of untreated oligonucleotides, present in each gel.

Results

Potentiometry

Protonation constants of peptides A, C, D and E are presented in Table 1. Peptide B was not measured because we were not able to obtain it in amounts required by potentiometry. Average values of protonation constants of His residues are also presented in Table 1.

Table 1. Protonation constants of HP2 partial peptides in 0.1 M KNO_3 at 25°C (pK values)[a]

Peptide	His_1	His_2	His_3	Cys_1	Cys_2	Average His	Other
A	5.48	5.90	6.47		8.33	5.95	7.13[b], 9.83[c]
C	5.64	6.56		7.67	8.82	6.10	9.97[d]
D	5.45	5.94		7.56	8.15	5.70	6.54[b]
E	5.38	5.97		6.83	8.67	5.68	11.06[d]

[a]S.D. based on statistical error distributions yielded by SUPERQUAD are within 0.01–0.02 log units; [b]terminal amine; [c]Tyr phenol; [d]Lys side-chain amine.

Circular dichroism

In order to measure the Zn(II) binding to protamine partial peptides directly, CD spectra of peptides A–E, dissolved in 10 mM sodium phosphate buffer containing 20 μM DTT were recorded in the presence of increasing amounts of Zn(II), until the saturation of spectral changes. DTT, an efficient peptide thiol reductant [21], was used to preserve peptide thiols in their reduced state and thus assure peptide solubility. Two saturable effects were seen in the spectra (Fig. 1): a decrease and a slight redshift of the short wavelength spectral minimum, and the appearance of a new, positive band at ca. 220 nm. These two effects allowed for the construction of two independent titration curves for each peptide, from which log K_c values (conditional constants for pH 7.4, which ignore peptide protonation) for ZnP complexes (P = A, B, C, D, E), were calculated. These constants take into account the competition for Zn(II) binding from DTT [22] and phosphate [23]. Formal stability constants (β) were obtained from K_c values using potentiometry-derived protonation constants for the peptides. Both are presented in Table 2.

The log K_c values may be somewhat overestimated due of uncertainties in actual concentrations of reduced DTT, e.g. a 2-fold decrease of DTT concentration results in a decrease of the calculated log K_c value by 0.2–0.4 units. The curves for peptides B, D and E did not exhibit deviations from the 1:1 binding model. On the contrary, an apparent increase of log K_c values was found for initial titration points for peptides A and C. This may be interpreted as formation of 2:1 peptide-to-zinc complexes at high peptide surplus over free Zn(II) ions. These points were not used in the calculations of K_c values.

Gel electrophoresis

The binding of HP2 partial peptides to DS and SS1 oligonucleotides was studied by polyacrylamide gel electrophoresis. Two oligonucleotide concentrations were used: 1 μM DS + 0.18 μM SS1 (high concentration) or 0.4 μM DS + 0.072 μM SS1 (low concentration). In order to correlate the

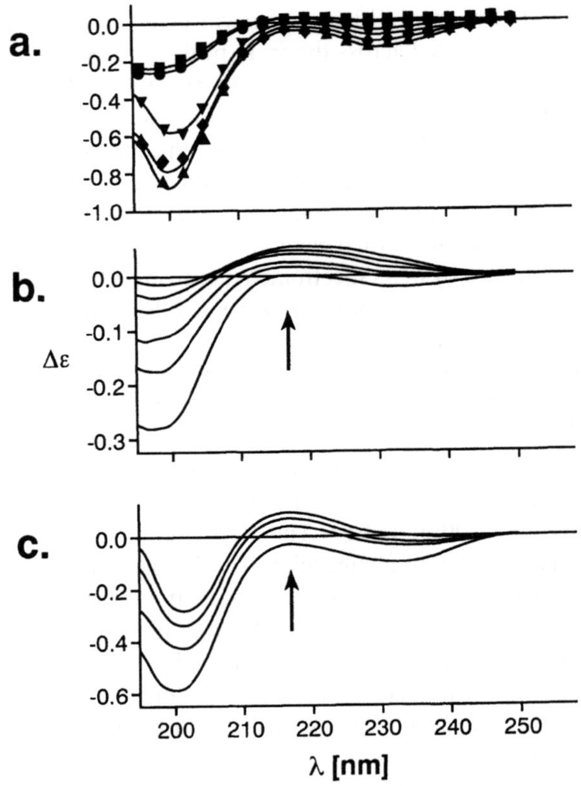

Fig. 1. CD of HP2 partial peptides (2.5–10 μM) in 10 mM Na phosphate buffer, pH 7.4. Arrows indicate the directions of spectral changes upon Zn(II) additions. (a) Spectra in the absence of Zn(II): ■, peptide A; ●, peptide B; ▲, peptide C; ▼, peptide D; ◆, peptide E. (b) Zn(II) titration of peptide B (0–4 mol equivalents of Zn(II), 0.8 mol equivalent steps). (c) Zn(II) titration of peptide D (0–3 mol equivalents of Zn(II), 1.0 mol equivalent steps).

redox status of the thiol groups of the peptides with their DNA binding abilities, the experiments were done in the presence and absence of DTT. The nucleotide mixtures were incubated with varying amounts of individual peptides for 15 min, then loaded onto gels and analyzed. In all cases only the free (soluble) oligonucleotides, migrating in the gel, and not their non-migrating peptide complexes (insoluble), were measured. Examples of the gels are presented in Fig. 2, and examples of the titration curves obtained from them by densitometry are presented in Fig. 3. It was uniformly found that the titration curves for SS1 followed those for DS, indicating the same affinities of both forms of DNA to the individual reduced peptides. The sigmoidal curves obtained were very similar to those seen by Bianchi *et al.* in their study of HP2 binding to DNA fragments of various lengths [15]. These authors demonstrated that the binding of HP2 to DNA caused precipitation of the resulting complex. Their results, obtained using DNA fragments much longer than ours, indicated that the precipitation occurred at the arginine/phosphate ratio of 1.2, independently of the length of DNA. This interpretation does not allow for calculation of binding constants. We ob-

Table 2. Conditional stability constants $(K_c)^a$ at pH 7.4 and overall stability constants $(\beta)^b$ of Zn(II) complexes with HP2 partial peptides, calculated from CD spectra

Peptide	log K_c (minimum)[c]	log K_c (CT)[c]	Average log K_c	Log β_{MHP}	Log β_{MP}
A	6.0(1)[d]	6.0(2)	6.0(2)	17.0	7.2
B	7.0(1)	7.1(3)	7.1(3)	–	–
C	4.9(4)	5.4(2)	5.2(4)	17.1	7.1
D	7.6(1)	7.8(4)	7.7(3)	–	8.9
E	6.0(2)	5.9(2)	6.0(2)	18.5	7.4

[a]K_c = [MP]/[M][P]; where [MP] is the total molar concentration of the metal-peptide complexes; [M] is the molar concentration of uncomplexed metal; [P] is the total concentration of uncomplexed peptide in its all protonation states; [b]$\beta = [MH_nP]/[M][H]^n[P]$; [c]From the minimum and charge transfer (CT) wavelengths, respectively; [d]S.D. of the last digit are given in parentheses.

served DNA precipitation at the excess of peptide charges over DNA phosphates higher than 1.2 (see Fig. 3). High ionic surplus of peptide in the precipitate is impossible for electrostatic reasons. One possible explanation is based on an assumption that the titration curves represent an equilibrium phenomenon that can be described by a binding constant of some kind.

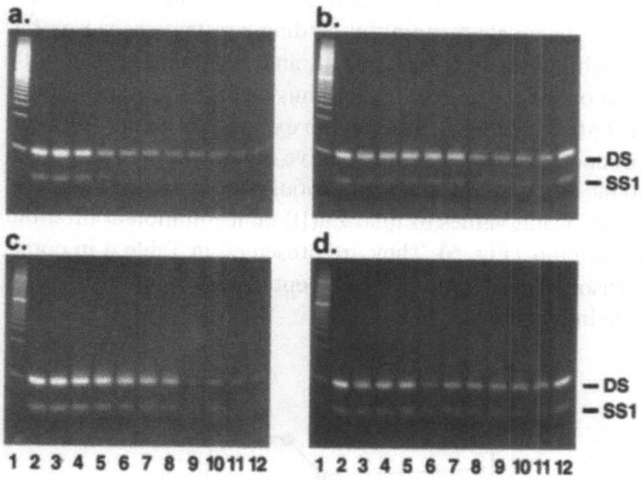

Fig. 2. Examples of polyacrylamide gels of DNA interactions with HP2 partial peptides and Zn(II). (a) DNA titration with peptide D in the absence of DTT. Lane 1: DNA molecular weight markers; lane 2: DNA control (1 μM DS + 0.18 μM SS1); lanes 3–12: DNA as in lane 2, with incrementally increasing amounts of peptide D, from 2.95–29.5 μM. (b) DNA titration with Zn(II) in the presence of peptide D and the absence of DTT. Lane 1: DNA molecular weight markers; lane 2: DNA control (1 μM DS + 0.18 μM SS1); lane 3: DNA as in lane 2, with 14.75 μM peptide D; lanes 4–11: DNA and peptide D as in lane 3, with increasing amounts of Zn(II): 1, 2, 10, 20, 50, 100, 150, 200 × 8.33 μM, respectively; lane 12: DNA as in lane 2, with Zn(II) as in lane 11 (1.67 mM). (c) DNA titration with peptide A in the presence of DTT. Lane 1: DNA molecular weight markers; lane 2: DNA control (1 μM DS + 0.18 μM SS1); lanes 3–12: DNA as in lane 2, with incrementally increasing amounts of peptide A, from 2.9–29 μM. (d) DNA titration with Zn(II) in the presence of peptide A and DTT. Lane 1: DNA molecular weight markers; lane 2: DNA control (1 μM DS + 0.18 μM SS1); lane 3: DNA as in lane 2, with 8.7 μM peptide A; lanes 4–11: DNA and peptide A as in lane 3, with increasing amounts of Zn(II): 1, 2, 10, 20, 50, 100, 150, 200 × 8.33 μM, respectively; lane 12: DNA as in lane 2, with Zn(II) as in lane 11 (1.67 μM).

The attempts to analyze the titration curves in this way presented difficulties. We found just one formula that yielded consistent values of constants for any (and all) of the peptides in both the absence and presence of DTT. This formula represents a modified solubility product (K_{so}), with the peptide concentration in the first power instead of (formally appropriate) stoichiometric coefficient derived from the peptide charge (between 4 and 8 for individual peptides):

$$K_{so} = [DS]_{free} \times [P]_{free} \ (P = A, B, C, D, \text{ or } E)$$

The K_{so} values were consistent between high and low concentration experiments, supporting the validity of this approach. We might also presume that the alternative cause of the observed deviation from the approximate 1:1 charge stoichiometry could be coprecipitation of very high amounts of the buffer anion (borate) with the DNA-peptide complex. However, if this were true then such coprecipitation would systematically decrease the K_{so} values at lower concentrations of peptides and DNA; but this was not observed.

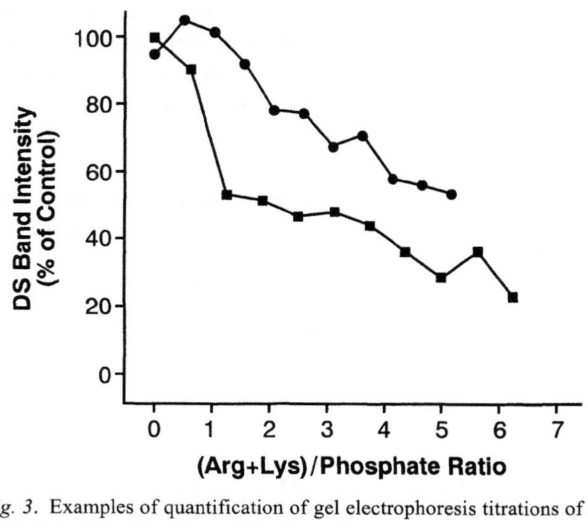

Fig. 3. Examples of quantification of gel electrophoresis titrations of DS DNA (0.4 μM) with HP2 partial peptides in the presence of DTT: ■, peptide B; ●, peptide C.

From the electrostatic neutrality condition, the stoichiometry of the DNA-peptide complex should be based on the balance of positive peptide charges and negatively charged DNA phosphates (40 for DS). Bianchi et al. included only Arg residues as charged at neutral pH [15]. In fact, Lys residues, which are charged at pH 7.4, should also be included. The question was open for His residues, which are relatively acidic in HP2 partial peptides (Table 1), with average pK_a values below 6. This might, however, change in the vicinity of a highly negatively charged oligonucleotide, making His side chains effective DNA binding sites [24]. The calculations were done for all these possibilities. We found that in our experimental conditions of high molar peptide excess over oligonucleotides, the inclusion of His side chains had a negligible impact on calculations of K_{so} values (increase by less than 0.05 log units, well within the S.D.) The presence of SS1 in our DS preparation was taken into account in the calculations by assuming identical K_{so} values (as indicated directly by titration curves) and using the precipitate stoichiometry provided by the electroneutrality principle. The log K_{so} values thus obtained for reducing (with DTT) and oxidizing (without DTT) conditions are presented in Table 3. Figure 4 shows the relationship between the peptide charge (taken as number of Arg and Lys residues) and log K_{so}. Note that in reducing conditions, the dependence of log K_{so} value on peptide charge is perfectly linear (R = 0.996). In the absence of DTT, this linearity is lost.

Another set of gel electrophoresis experiments was aimed at estimating the effect of Zn(II) on the interaction of HP2 partial peptides and oligonucleotides. These were performed as Zn(II) titrations of oligonucleotide-peptide mixtures, using peptide excess that previously precipitated of 10–20% of the added DNA. In all cases titration curves had sharp sigmoidal shapes (see Fig. 5 for examples). The enhancement of DNA precipitation was invariably seen for higher Zn(II) concentrations. In some cases, the concentration of free oligonucleotide fell below the detection limit. This effect was absent in the absence of peptides (lanes 12 of Zn(II) titration gels). The appropriate log K_{so} values at Zn(II) saturation,

Table 3. Log K_{so} values determined for oligonucleotide precipitation by HP2 partial peptides (free) and their Zn(II) complexes in the absence and presence of DTT[a,b]

	DS		DS/DTT	
Peptide	Free	Δ Zn	free	Δ Zn
A	–11.8(2)	1.2(5)	–11.5(3)	0.5(2)
B	–12.2(4)	1.4(3)	–12.1(2)	> 1.8
C	–12.3(3)	> 1.6	–12.0(2)	> 2.3
D	–11.9(2)	0.4(1)	–11.6(2)	1.2(6)
E	–12.5(2)	2.1(4)	–11.9(2)	> 2.0

[a]Δ Zn = log K_{so} (free) – log K_{so} (Zn); [b]S.D. on the last digit are given in parentheses.

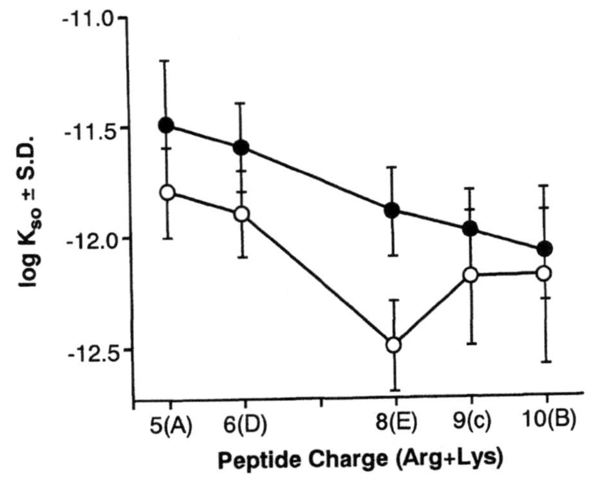

Fig. 4. The dependence of log K_{so} values of peptide-DS DNA complexes on peptide charge: ■, with DTT; O, without DTT.

compared to Zn(II)-free situation (or their lower limits when DNA was precipitated out) are presented in Table 3.

The results of gel electrophoresis experiments did not allow to calculate meaningful binding constants because of a possibility for free Zn(II), DTT, and their complexes to diffuse out of the gel wells (and thus change the equilibrium) just after loading. However, the experiments were fully reproducible. Therefore, the relative orders of Zn(II) affinities to the DNA complexes with various peptides could be established using values of total Zn(II) concentration at titration mid-points (Fig. 5). They are presented in Table 4 in comparison with affinity orders for peptide-DNA and Zn(II)-peptide interactions.

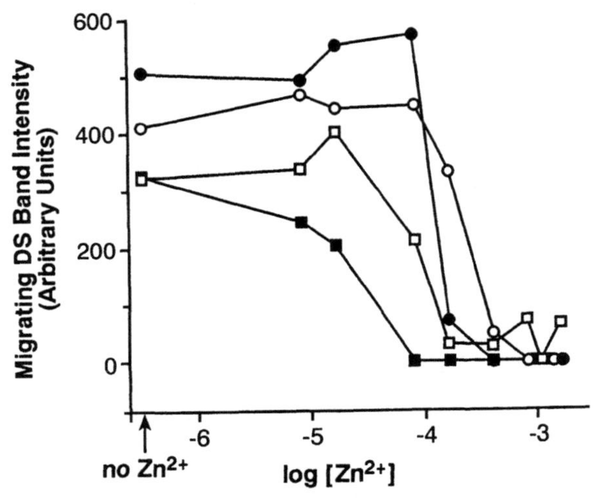

Fig. 5. Examples of the effects of Zn(II) on DS precipitation by HP2 partial peptides (expressed as the amount of DS migrating in the gel); ■, peptide B with DTT; ●, peptide C with DTT; □, peptide B without DTT; O, peptide C without DTT.

Discussion

Zn(II) binding to partial peptides in the absence of DNA

CD titrations were evaluated using two mutually independent, direct features of Zn(II) coordination to the peptides. The minimum at 200 nm, which decreased in the presence of increasing amounts of Zn(II), originates in coupled peptide carbonyl chromophores and reflects the overall conformation of the peptide's main chain [25]. Analogous decreases of such bands were seen previously in Zn(II) titrations of the whole HP2 [7, 9], as well as zinc finger proteins [26, 27]. The maximum at 220 nm, which appeared upon Zn(II) addition to each of the peptides studied in this work, is a charge transfer (CT) band of the Zn(II)-S bond [27, 28]. Therefore, these spectral changes reflected the peptide folding due to multidentate Zn(II) binding, as well as specific contributions to the binding from thiolates. As shown in Table 2, both effects independently yielded the same values of K_c, thus assuring that the formation of zinc binding domains was really measured.. Peptide D provided the strongest Zn(II) binding, with log K_c of 7.7; peptide B came second with log K_c of 7.0; and the remaining HP2 partial peptides bound Zn(II) much more weakly. Interestingly, there was no preference for 2Cys/2His (C, D and E) over 3His/Cys sites (A and B). The presence of the CT band at 220 nm assures Cys sulfur usage by peptide B for Zn(II) binding and excludes an alternative His4 coordination. The values found for HP2 partial peptides are much below the lower limit of log K_c values measured for regular zinc fingers, 9–12 [16, 17, 26, 29-31]. This is likely due to the electrostatic repulsion from many charged Arg residues, counteracting the peptide folding required for tetradentate Zn(II) coordination as opposed to cooperative protein folding present in typical zinc fingers [17]. Our results correspond well with the affinity constants obtained for HP2, equal to 7.1 (log K_c for the primary site) and 5.6 (log K_c for the secondary site) [9]. The two highest log K_c values for partial peptides are 7.7 and 7.0. Noting that these values may be somewhat overestimated, and that the electrostatic repulsion should be stronger in the whole peptide than in its fragments, the localization of the primary Zn(II) site of HP2 within peptide D is more than likely. Our results also suggest that the secondary Zn(II) binding site is localized in peptide B. It can *a priori* contain any three of five His residues (the involvement of Cys-13 residue was confirmed directly). The 1 log unit enhancement of binding compared to peptide A suggests that His-19 and/or His-22 belong to the binding site. Assuming its sequentiality, it may be composed of either His-12, Cys-13, His-19, and His-22, or His-7, His-12, Cys-13, and His-19. The Zn(II) complex of peptide B may be a mixture of both. The +1 charge of the Zn(II) site, resulting from the 3His/Cys coordination mode, less favorable in the highly charged whole HP2, may be responsible for the stability difference between it and peptide B.

According to a previous proposal [8], Zn(II) should bind to two Cys donors from our peptide E (Cys-47 and Cys-54) along with two His donors from peptide B (His-19 and His-22). Our results indicate that peptide D offers a much higher zinc-binding affinity than peptide E, and thus do not support that proposal.

Table 4. Relative affinity orders for HP2 partial peptides interactions with DNA and Zn(II) in the absence and presence of DTT[a]

Components	DS	DS/DTT
Zn + P + DNA; Zn binding	B ≥ A >> C ≥ D > E	B = E > D > A >> C
Zn + P + DNA; Zn enhancement of DNA precipitation	E ~ C > B > A > D	E ~ C ~ B > D > A
Zn + P	–	D > B >> A = E > C
P + DNA	E > B = C > D ≥ A	B ≥ C ≥ E > D ≥ A

[a] ≥, difference within S.D.; ~, probably of similar magnitude; >, difference within 1 log unit; >>, difference higher than 1 log unit.

DNA interaction with partial peptides of HP2

DNA binding is the fundamental physiological function of HP2. We aimed at gaining insight into the details of this interaction by studying the binding of its partial peptides to a short oligonucleotide. We hoped that its small size would prevent precipitation seen in previous studies of the whole HP2 [15] or $HP2_{1-15}$ [12, 13] binding to DNA of various sizes. This failed, as DNA precipitation was the only effect we could measure. Nevertheless, valuable information was obtained. Table 3 presents the apparent solubility products calculated from gel electrophoresis experiments performed at reducing (DTT present) and oxidizing (DTT absent; ambient atmosphere) conditions. Note that the K_{so} values obtained by us are valid only for our particular experimental settings, and are used here only for comparative purposes.

Under the reducing conditions, the precipitating capacity of the peptides is a linear function of the number of their positively charged arginine and lysine residues (Fig. 4). Arginines and lysines are distributed differently in the sequences of particular partial peptides, and so the adoption of any non-extended conformation would result in deviations from the linearity observed. This finding is in agreement with models of DNA binding by various protamines in the absence of Zn(II) [32-34]. The linearity is lost in the absence of DTT, but the binding becomes somewhat stronger, by ca. 0.3 log units on average, apparently due to the formation of intra- and/or intermolecular disulfide bridges. Such enhancement has been relatively highest for peptide E, which correlates with a low pK value for one of its cysteines (Table 1), bearing in mind that thiols are oxidized to disulfides as anions [35]. This effect corresponds to the process of sperm chromatin condensation by the formation of disulfide bridges.

Zn(II) interactions with peptide-DNA complexes

The titrations of the oligonucleotide/peptide mixtures with Zn(II) showed that, at sufficient excess, Zn(II) is able to enhance DNA precipitation by all the investigated peptides. This enhancement typically exceeded 1 log unit of apparent K_{so} in both the presence and absence of DTT (Table 3). The Zn(II) titration curves exhibited very sharp sigmoidal shapes (Fig. 5). Such effect is characteristic for two-state systems and thus indicates the formation of specific peptide-Zn(II)-DNA complexes, which are less soluble (i.e. stronger) than the parent peptide-DNA aggregates. Zn(II) added to peptide-free DNA at the highest concentration used in titrations did not cause DNA precipitation, nor affected its mobility (Fig. 2, lane 12 in all Zn(II) titration gels), indicating that Zn(II) binds to the peptide rather than the DNA part of peptide-DNA complexes. Zn(II) affinities to peptide-DNA complexes could be ordered using total Zn(II) concentrations at transition midpoints, thus providing insight into Zn(II) binding preferences in the DNA-bound peptides (Table 4). Peptides B and E provide strongest Zn(II) binding for reducing conditions. Peptide B remained dominant for oxidizing conditions, while E lost its binding capabilities due to sulfur oxidation. The total Zn(II) concentrations at transition midpoints were lower in the presence of DTT even though a large proportion of Zn(II) was sequestered by the latter. Therefore, Zn(II) binding involving Cys residues was much stronger than with His residues only. Peptide specific Zn(II) effects on DNA binding suggest that peptide-Zn(II) complexes retain specific, partially folded, conformations in the presence of DNA. Interestingly, these complexes precipitate DNA more efficiently than free peptides, with stoichiometries approaching the 1:1.2 value found previously for the whole HP2 [15].

We have obtained relative Zn(II) affinities for individual peptides covering the whole sequence of HP2 for reduced, DNA-free conditions, as well as for the reduced and oxidized DNA-bound states. Our results suggest that Zn(II) may bind at various sites in HP2, depending on external conditions. These possibilities are presented graphically in Fig. 6.

Kvist *et al.* demonstrated the importance of Zn(II) bound to thiols in sperm chromatin for maintaining sperm viability [5, 36–39]. In particular, the Zn(II) level in sperm head increases upon ejaculation as a result of Zn(II) transfer from prostatic fluid [36]. Balhorn and coworkers studied the roles of HP1 and HP2 in DNA condensation, including formation of intra- and intermolecular disulfide bridges [40–43] and recently showed that Zn(II) was abundant enough to bind whole HP2 in human sperm at a 1:1 stoichiometry [6]. Our results suggest that Zn(II) binding to HP2 in sperm chromatin may be heterogeneous, depending on the involvement of particular Cys residues in the disulfide formation, and on DNA binding. The enhancement of DNA binding by HP2 partial peptides in the presence of Zn(II) indicates that Zn(II) ions may play a regulatory role in specific structuring of sperm chromatin.

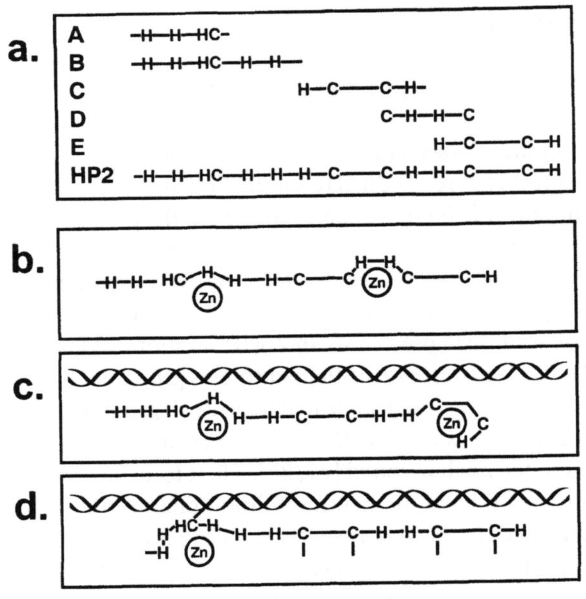

Fig. 6. A proposal of patterns of Zn(II) binding to HP2 at various conditions. Two highest affinity binding sites are shown for reduced HP2, and one for oxidized HP2, where binding constants are much lower. See text for details. (a) Schematic representation of the peptides; (b) Reduced HP2 in the absence of DNA; (c) Reduced HP2 in the presence of DNA; (d) Oxidized HP2 in the presence DNA.

Conclusion

We described Zn(II) binding to a series of peptides representing partial sequences of human protamine HP2. DNA precipitation by these peptides and the effects of Zn(II) thereon were investigated as well. The results enabled us to suggest possible binding sites for Zn(II) in the whole HP2 molecule under various conditions. The major conclusion emerging from these investigations is that HP2 may exhibit heterogeneity of Zn(II) binding *in vivo*. Zn(II) enhanced DNA precipitation by all partial peptides. This effect indicates that Zn(II) may participate in the physiological regulation of sperm chromatin condensation. However, more definitive data regarding the positions and coordination modes of the Zn(II) binding sites and their physiological significance to sperm chromatin structure can be only provided by future studies under conditions securing solubility of the DNA-protamine complexes.

Acknowledgements

This work was supported by the Polish Committee for Scientific Research (KBN) grant no. 6 P04A 024 13 and grant no. SPUB/COST/T-9/DZ 44/99, within the framework of the COST D8 action. A subsistence grant from Fogarty International Center, National Institutes of Health, USA, to W.B. and a Domestic Grant for Young Scholars from the Foundation of Polish Science to M.D. are also gratefully acknowledged.

Reference

1. McKay DJ, Renaux BS, Dixon GH: Human sperm protamines. Amino-acid sequences of two forms of protamine P2. Eur J Biochem 156: 5–8, 1986
2. Ammer H, Henschen A, Lee CH: Isolation and amino-acid sequence analysis of human sperm protamines P1 and P2. Occurrence of two forms of protamine P2. Biol Chem Hoppe-Seyler 367: 515–522, 1986
3. Pogany GC, Corzett M, Weston S, Balhorn R: DNA and protein content of mouse sperm. Exp Cell Res 136: 127–136, 1981
4. Balhorn R, Reed S, Tanphaichitr N: Aberrant protamine 1/protamine 2 ratios in sperm of infertile human males. Experientia 44: 52–55, 1988
5. Bjorndahl L, Kvist U: Influence of seminal vesicular fluid on the zinc content of human sperm chromatin. Int J Androl 13: 232–237, 1990
6. Bench G, Corzett MH, Kramer CE, Grant PG, Balhorn R: Zinc is sufficiently abundant within mammalian sperm nuclei to bind stoichiometrically with protamine 2. Mol Reprod Dev 56: 512–519, 2000
7. Gatewood JM, Schroth GP, Schmid CW, Bradbury EM: Zinc-induced secondary structure transitions in human sperm protamines. J Biol Chem 265: 20667–20672, 1990
8. Bianchi F, Rousseaux-Prevost R, Sautiere P, Rousseaux J: P2 protamines from human sperm are zinc-finger proteins with one Cys2/His2 motif. Biochem Biophys Res Commun 182: 540–547, 1992
9. Quintanilla-Vega B, Hoover DJ, Bal W, Silbergeld E, Waalkes MP, Anderson LD: Lead interaction with human protamine HP2 as a mechanism of male reproductive toxicity. Chem Res Toxicol 13: 594–600, 2000
10. Kozlowski H, Bal W, Dyba M, Kowalik-Jankowska T: Specific structure-stability relations in metallopeptides. Coord Chem Rev 184: 319–346, 1999
11. Bal W, Jeżowska-Bojczuk M, Kasprzak KS: Binding of nickel(II) and copper(II) to the N-terminal sequence of human protamine HP2. Chem Res Toxicol 10: 906–914, 1997
12. Bal W, Lukszo J, Kasprzak KS: Mediation of oxidative DNA damage by nickel(II) and copper(II) complexes with the N-terminal sequence of human protamine HP2. Chem Res Toxicol 10: 915–921, 1997
13. Liang R, Senturker S, Shi X, Bal W, Dizdaroglu M, Kasprzak KS: Effects of Ni(II) and Cu(II) on DNA interaction with the N-terminal sequence of human protamine P2: Enhancement of binding and mediation of oxidative DNA strand scission and base damage. Carcinogenesis 20: 893–898, 1999
14. Bal W, Wójcik J, Maciejczyk M, Grochowski P, Kasprzak KS: Induction of a secondary structure in the N-terminal pentadecapeptide of human protamine HP2 through Ni(II) coordination. An NMR study. Chem Res Toxicol 13: 823–830, 2000
15. Bianchi F, Rousseaux-Prevost R, Bailly C, Rousseaux J: Interaction of human P1 and P2 protamines with DNA. Biochem Biophys Res Commun 201: 1197–1204, 1994
16. Krizek BA, Merkle DL, Berg JM: Ligand variation and metal ion binding specificity in zinc finger peptides. Inorg Chem 32: 937–940, 1993
17. Berg JM, Godwin HA: Lessons from zinc-binding peptides. Annu Rev Biophys Biomol Struct 26: 357–371, 1997
18. Meienhofer J, Waki M, Heimer EP, Lambros TJ, Makofske RC, Chang CD: Solid phase synthesis without repetitive hydrolysis. Preparation of leucylalanyl-glycyl-valine using 9-fluorenylmethyloxy- carbonylamino acids. Int J Peptide Prot Res 13: 35–42, 1979
19. Irving H, Miles MG, Pettit LD: A study of some problems in determining the stoichiometric proton dissociation constants of complexes by potentiometric titrations using a glass electrode. Anal Chim Acta 38: 475–488, 1967
20. Gans P, Sabatini A, Vacca A: SUPERQUAD: An improved general program for computation of formation constants from potentiometric data. J Chem Soc Dalton Trans 1195–1199, 1985
21. Lees WJ, Whitesides GM: Equilibrium constants for thiol-disulfide interchange reactions: A coherent, corrected set. J Org Chem 58: 642–647, 1993
22. Krężel A, Leśniak W, Jeżowska-Bojczuk M, Młynarz P, Brasuń J, Kozłowski H, Bal W: Coordination of heavy metals by dithiothreitol, a commonly used thiol group protectant. J Inorg Biochem 84: 77–88, 2001
23. Banerjea D, Kaden T, Sigel H: Enhanced stability of ternary complexes in solution through the participation of heteroaromatic N bases. Comparison of the coordination tendency of pyridine, imidazole, ammonia, acetate and hydrogen phosphate toward metal ion nitrilotriacetate complexes. Inorg Chem 20: 2586–2590, 1981
24. Vu HM, Minch MJ: α-(Ac)AKRHRKV, a model of the histone H4 amino terminus, uses an unprotonated histidine in phosphate binding. J Pep Res 51: 162–170, 1998
25. Woody RW: Circular dichroism and conformation of unordered peptides. Adv Biophys Chem 2: 37–79, 1992
26. Roehm PC, Berg JM: Sequential metal binding by the RING finger domain of BRCA1. Biochemistry 36: 10240–10245, 1997
27. Basile LA, Coleman JE: Optical activity associated with the sulfur to metal charge transfer bands of Zn and Cd GAL4. Prot Sci 1: 617–624, 1992.
28. Vašák M, Kägi JHR, Hill HAO: Zinc(II), cadmium(II) and mercury(II) thiolate transitions in metallothionein. Biochemistry 20: 2852–2856, 1981
29. Worthington MT, Amann BT, Nathans D, Berg JM: Metal binding properties and secondary structure of the zinc-binding domain of Nup475. Proc Natl Acad Sci USA 93: 13754–13759, 1996
30. Posewitz MC, Wilcox DE: Properties of the Sp1 zinc finger 3 peptide: Coordination chemistry, redox reactions, and metal binding competition with metallothionein. Chem Res Toxicol 8: 1020–1028, 1995
31. Predki PF, Sarkar B: Effect of replacement of 'zinc finger' zinc on estrogen receptor DNA interactions. J Biol Chem 267: 5842–5846, 1992
32. Fita I, Campos JL, Puigjaner LC, Subirana, JA: X-ray diffraction study of DNA complexes with arginine peptides and their relation to nucleoprotamine structure. J Mol Biol 167: 157–177, 1983
33. Balhorn R: A model for the structure of chromatin in mammalian sperm. J Cell Biol 93: 298–305, 1982
34. Hud NV, Milanovich FP, Balhorn R: Evidence of novel secondary structure in DNA-bound protamine is revealed by Raman spectroscopy. Biochemistry 33: 7528–7535, 1994
35. Wallace TJ, Schriesheim A: Solvent effects in the base-catalyzed oxidation of mercaptans with molecular oxygen. J Org Chem 27: 1514–1516, 1962
36. Kvist U: Importance of spermatozoal zinc as temporary inhibitor of sperm nuclear chromatin decondensation ability in man. Acta Physiol Scand 109: 79–84, 1980

37. Bjorndahl L, Kjellberg S, Roomans GM, Kvist U: The human sperm nucleus takes up zinc at ejaculation. Int J Androl 9: 77–80, 1986
38. Kvist U, Bjorndahl L, Kjellberg S: Sperm nuclear zinc, chromatin stability, and male fertility. Scanning Microsc 1: 241–1247, 1987
39. Kvist U, Kjellberg S, Bjorndahl L, Soufir JC, Arver S: Seminal fluid from men with agenesis of the Wolffian ducts: Zinc-binding properties and effects on sperm chromatin stability. Int J Androl 13: 245–252, 1990
40. Balhorn R, Corzett M, Mazrimas JA: Formation of intraprotamine disulfides *in vitro*. Arch Biochem Biophys 296: 384–393, 1992
41. Zalensky AO, Allen MJ, Kobayashi A, Zalenskaya IA, Balhorn R, Bradbury EM: Well-defined genome architecture in the human sperm nucleus. Chromosoma 103: 577–590, 1995
42. Allen MJ, Bradbury EM, Balhorn R: AFM analysis of DNA-protamine complexes bound to mica. Nucleic Acids Res 25: 2221–2226, 1997
43. Brewer, LR, Corzett M, Balhorn R: Protamine-induced condensation and decondensation of the same DNA molecule. Science 286: 120–123, 1999

Molecular and Cellular Biochemistry **222**: 107–118, 2001.

Reduction of Cr (VI) by cysteine: Significance in human lymphocytes and formation of DNA damage in reactions with variable reduction rates

George Quievryn, Margarida Goulart, Joseph Messer and Anatoly Zhitkovich
Department of Pathology and Laboratory Medicine, Brown University, Providence, RI, USA

Abstract

The induction of genotoxicity by Cr (VI) is dependent on its reductive activation inside the cell. Our recent studies have found that reduction of Cr (VI) by cysteine resulted in the formation of mutagenic Cr (III)–DNA adducts in the absence of oxidative DNA damage. In this work, we examined the formation of oxidative and Cr (III)-dependent types of DNA damage under a broader range of Cr (VI) and cysteine concentrations and investigated a potential role of this reducer in intracellular metabolism of Cr (VI). Peripheral lymphocytes from unexposed humans had 7.8-fold excess of glutathione over cysteine, whereas lymphocytes from stainless steel welders contained only 3 times higher amount of glutathione ($p = 0.0009$) which was entirely caused by the decrease in the concentration of glutathione. A strong correlation ($r = 0.72$) between the levels of both thiols was found in lymphocytes from controls. The number of DNA–protein crosslinks in lymphocytes from welders was 4.1 times higher than among controls, indicating the presence of Cr (VI)-dependent DNA damage. The average rate of Cr (VI) reduction by cysteine was approximately 5 times faster than that by glutathione. Higher reduction rate combined with the decrease in the intracellular concentration of glutathione should make cysteine a predominant Cr (VI)-reducing thiol in lymphocytes of welders. Analysis of the initial rates of Cr (VI) reduction by different concentrations of cysteine suggested the presence of one- and two-electron pathways, with one-electron mechanism dominating in the physiological range of concentrations. There was no detectable formation of DNA breaks or abasic sites under a broad range of Cr (VI) and cysteine concentrations, resulting in up to 68-fold differences in the rates of reduction and the production of as many as 3 Cr (III)–DNA adducts per 10 bp. The reactions with slow reduction rates (low concentrations of cysteine) led to the most extensive formation of Cr (III)-DNA adducts. In summary, these results further establish Cr (III)–DNA adducts as the major form of DNA damage resulting from Cr (VI) metabolism by cysteine. The role of cysteine in reduction of Cr (VI) becomes more significant under conditions of occupational exposure to Cr (VI)-containing welding fumes. (Mol Cell Biochem **222**: 107–118, 2001)

Key words: chromium, cysteine, DNA adduct, DNA damage, glutathione, welders

Introduction

Epidemiological studies have strongly linked human exposure to Cr (VI) with elevated incidence of respiratory cancers [1–3]. Genotoxic activity of Cr (VI) has been confirmed in transformation, mutagenesis and clastogenesis assays [4–6]. Cr (VI) is a pro-carcinogen that requires reductive activation to cause DNA damage and binding of Cr to intracellular components [7, 8]. The final oxidation state of Cr in all biological systems is Cr (III). Exposure of cells to Cr (VI) has been found to result in the formation of various forms of DNA damage, including strand breaks [9], alkali-labile sites

Address for offprints: A. Zhitkovich, Department of Pathology and Laboratory Medicine, Brown University, Box G-B511, Providence, RI 02912, USA

[10], DNA–DNA crosslinks [11], DNA–protein crosslinks [12, 13] and ternary Cr (III)–DNA adducts with glutathione (GSH), cysteine (Cys) or histidine [14]. Ternary amino acid–Cr (III)–DNA adducts are formed through a direct coordination of Cr atom to oxygen of the phosphate group [15]. Induction of DNA strand breaks in Cr (VI)-exposed cells can result from either a direct DNA cleavage by intermediate Cr forms [16–18] or through the attack of the sugar-phosphate backbone by hydroxyl radicals [19]. Alkali-labile sites may indicate the presence of abasic sites, but they could also be formed by a subset of Cr (III)–DNA adducts [20]. There was no increase in the amount of 8-oxodG in DNA isolated from tissues of Cr (VI)-exposed animals [21]. Synthetic Cr (III)–DNA adducts [22] and DNA adducts produced during metabolism of Cr (VI) with GSH [23] or Cys [24] have all been found to be mutagenic in human cells. Ternary Cr (III)–DNA adducts were significantly more mutagenic than binary Cr (III)–DNA complexes [22, 24].

Ascorbate, GSH and Cys are considered to be the major intracellular reducers of Cr (VI) [25–27]. The reduction mechanisms and the spectrum of intermediate species formed in the reactions driven by each individual reducer appear to be different [28–30]. This also raises the possibility that the extent and the forms of resulting DNA damage could also be quite different. Reductions of Cr (VI) by GSH and ascorbate have been extensively studied, whereas Cys-dependent reactions are investigated in much less detail. Previous studies have found that the initial rate of Cr (VI) reduction by Cys was higher than that by a more abundant intracellular thiol, GSH [25, 27]. Hence, the contribution of Cys to reduction of Cr (VI) could be significant even if its concentration is lower than that of GSH. The relative roles of Cys and GSH in reduction of Cr (VI) could not be deduced from the comparison of the initial rates of reactions and the intracellular concentrations of these thiols because GSH-driven reactions exhibit non-linear kinetics [25, 26].

Reduction of Cr (VI) by 2 mM Cys led to the formation of Cr (III)–DNA adducts without detectable production of DNA breakage or abasic sites [20]. The absence of oxidative DNA damage during Cys-dependent reduction of Cr (VI) was further confirmed in experiments with replication of shuttle-vector plasmids in human cells [24]. Reductive metabolism of Cr (VI) with Cys led to the induction of mutagenicity and replication blockage in the shuttle-vector plasmids and Cr (III)–DNA adducts were found to be entirely responsible for both types of genotoxicity [24]. In this work, we investigated the formation of oxidative DNA lesions and Cr (III)–DNA adducts under a much broader range of reaction conditions since the formation of reactive intermediates could vary depending on several experimental parameters, such as ligand concentrations and rates of reduction. Additionally, a relative importance of Cys in reduction of Cr (VI) was examined by comparing the reduction activities of Cys and GSH in the physiological range of concentrations and by determining intracellular concentrations of these thiols in lymphocytes isolated from control and Cr (VI)-exposed individuals.

Materials and methods

Chemicals

Glutathione, HPLC-grade methanol, monobromobimane, buffers and all reagents for electrophoresis were from Sigma (St. Louis, MO, USA). Chelex-100 resin and Bio-Gel P-30 columns were purchased from Bio-Rad (Hercules, CA, USA). Stock of 2′,7′-dichlorofluoroscin diacetate (DCFH-DA) was obtained from Molecular Probes (Eugene, OR, USA). Potassium chromate (A.C.S. reagent) and N,N′-dimethylethylenediamine (DED) were from Aldrich (Milwaukee, WI, USA), ^{51}Cr-labeled chromate was from Amersham (Arlington Heights, IL, USA). Supercoiled and relaxed forms of ΦX174 DNA were obtained from New England Biolabs (Beverly, MA, USA). L-cysteine was obtained from Gibco BRL. Solutions of Cys and GSH were used within 30 min after preparation. All components of Cr (VI) reduction reactions were purified from trace amounts of Fe and Cu using Chelex-100 columns as described previously [20].

Human subjects and isolation of lymphocytes

All human subjects were from Portugal living in the vicinity of Lisbon. Cr (VI)-exposed workers (n = 5) were male stainless steel welders who used the acetylene flame method on chromium/nickel alloys in an open environment. Welders did not use any protective masks over nose/mouth. Blood samples from age-matched control male subjects (n = 22) were obtained from a local blood center. Peripheral blood lymphocytes were isolated from heparin-stabilized blood within 2–3 h after the blood draw. Purification of lymphocytes was performed by a standard Ficoll-based technique using Histopaque-1077 from Sigma. A portion of lymphocytes was frozen in a small vol. of Ca, Mg-free PBS while remaining cells were lysed in 1% SDS, 20 mM Tris-HCl (pH 8.0) and then stored at –80°C.

Determination of cysteine and glutathione in lymphocytes

Intracellular concentrations of GSH and Cys were measured by HPLC as described previously [31] with minor modifications. In brief, cells were suspended in a PBS-DPTA solution followed by acidification with 50 mM methanesulfonic acid. The samples were subjected to two cycles of freezing/thawing and low molecular weight thiols were recovered in the supernatants after centrifugation at 12,000 × g, 10 min at room temperature. The standard derivatization reaction contained 50 µl cell extracts, 50 mM HEPES, pH 8.0, 5 mM DPTA, 15 mM NaOH, 2 mM dithiotreitol (DTT) and 5 mM monobromobimane. A set of reactions was also performed in the absence of DTT. After 10 min incubation at room temperature, derivatization reactions were terminated by the addition of 25 mM methanesulfonic acid. HPLC separation of fluorescent derivatives of GSH and Cys was performed using a Shimadzu LC-10ADvp liquid chromatograph, equipped with SIL-10ADvp autoinjector, Ultrasphere ODS column (5 µm, 250 × 4.6 mm), and an RF-10AxL fluorescence detector. Chromatographic conditions were as described previously [31]. Thiol concentrations were normalized per amount of protein in each sample. Protein concentrations were determined using an SDS-compatible kit from Bio-Rad.

Measurements of DNA–protein crosslinks

The number of DNA–protein crosslinks (DPC) was determined by a K-SDS precipitation assay [32], as recently modified [31]. To minimize the batch effect, controls and cases were run in parallel and each set of samples also contained an internal standard. The size of DNA fragments was determined by agarose electrophoresis and the final results were expressed as the number of DNA–protein crosslinks per 10^7 bp [31].

Cr (VI) reduction

Disappearance of Cr (VI) was monitored by following absorbance at 372 nm. Experiments that compared the reduction rates by Cys and GSH were performed in the presence of 50 mM MOPS or phosphate buffer (pH 7.0) and 50 µM Cr (VI). It was important to use high buffer concentrations because GSH solutions are very acidic and the use of 1:5 or higher ratio of GSH to buffer resulted in the significant shifts of the pH. The reduction reaction was initiated by the addition of 0.5 ml of 100 µM Cr (VI) solution to 0.5 ml of prewarmed 2 × buffer/thiol mixture, which ensured rapid mixing of the reagents in the cuvette. The temperature of all reactions was set to 37°C and maintained electronically. Thiol solutions were always freshly prepared. Reduction kinetics by different concentrations of Cys was studied using 25 mM MOPS (pH 7.0) and 100 µM chromate. The use of 25 or 50 mM MOPS buffer typically resulted in similar rates of Cr (VI) reduction by Cys. The presence or absence of DNA in the reaction mixtures did not affect reduction rates. The initial rates of reduction were determined from plots of $\ln(A_t\text{-}A_\infty)$ vs. time. Absorbance measurements were recorded every 12–15 sec in Schimadzu UV 1601 spectrophotometer.

Measurements of DCFH oxidation

Generation of oxidizing species during Cr (VI) reduction was monitored by the formation of fluorescent 2′,7′-dichlorofluoroscein which is a product of DCFH oxidation. Hydrolysis of DCFH-diacetate ester was carried out by incubation with 10 mM NaOH for 30 min followed by neutralization with Chelex-treated phosphate buffer (pH 7.0) [33]. The formation of fluorescent 2′,7′-dichlorofluoroscein was measured in samples that contained 10 µM DCFH, 25 mM buffer (pH 7.0) and various concentrations of chromate and Cys. All incubations were performed at 37°C in 96-well plates. Fluorescence was measured with 485 nm excitation and 535 nm emission filters in a microplate fluorometer (SpectraFluor Plus, Tecan).

Assays for DNA strand breaks and abasic sites

Detection of DNA strand breaks was based on the determination of the relative amounts of supercoiled and nicked forms of ΦX174 plasmids following reduction of Cr (VI). A standard reaction mixture contained 0.3 µg supercoiled DNA, 25 mM MOPS or phosphate buffer (pH 7.0) and various concentrations of Cys and chromate (total vol. = 25 µl). After incubation for 2 h at 37°C, the samples were placed on ice, mixed with 6 × Ficoll buffer (15% Ficoll 400, 0.25% bromphenol blue, 0.25% xylene cyanol) and loaded onto 1% agarose gels. Electrophoresis was carried out in 50 mM TAE buffer (50 mM Tris-acetate, 1 mM EDTA, pH 8.0). DNA bands were stained with ethidium bromide and photographed using Bio-Rad Gel-Doc 2000 gel documentation system. The amounts of DNA in individual bands were determined using the Quantity One software package.

The determination of abasic sites was based on the quantitation of nicked DNA following the conversion of these sites into strand breaks through β-elimination reaction with N,N′-dimethylethylenediamine (DED) [34]. Cr (VI) reduction was performed as described above for strand breakage analysis. DNA was purified on Bio-Gel P-30 columns and then treated with 100 mM DED (pH 7.4) at 37°C for 30 min. Control experiments with acid-depurinated DNA demonstrated that cleavage reaction was complete after 10 min. The amount of DNA breaks in DED-treated DNA was determined by agarose electrophoresis. Preparation of depurinated DNA standards that contained approximately 1.4 abasic sites per plasmid was done according to Pierre and Laval [35].

Measurements of Cr–DNA adducts

A number of Cr–DNA adducts was determined by the inclusion of trace amounts of radiolabeled ^{51}Cr (VI) in the reaction mixtures. Approximately 100,000–200,000 c.p.m. of ^{51}Cr

were added to each sample. Reaction mixtures (total vol. = 50 µl) contained 25 mM MOPS (pH 7.0), 0.6 µg supercoiled ΦX174 DNA, 0.5–5 mM Cys and different concentrations of chromate. Samples were incubated for 120 min at 37°C followed by the removal of unbound components on Bio-Gel P-30 spin columns. DNA was precipitated with 2 vol. of cold ethanol in the presence of 200 mM NaCl and then washed twice with 70% ethanol. This purification protocol removes ionically associated Cr–DNA complexes allowing the determination of stable Cr (III)–DNA adducts [20]. DNA pellets were dissolved in H_2O and the amounts of DNA-associated ^{51}Cr radioactivity were determined using a scintillation counter.

Statistics

The linear relationship between data sets was evaluated by the Pearson's correlation test, whereas the comparison between groups was performed using the non-parametric Mann–Whitney two-sample test (Wilcoxon rank-sum test).

Results

Cysteine and glutathione concentrations in peripheral lymphocytes of control and Cr (VI)-exposed workers

The determination of Cys and GSH was based on HPLC separation of their fluorescent derivatives formed in the reaction with monobromobimane. Cys and GSH peaks were well resolved from non-specific products (Figs 1A and 1B) and the fluorescence-based detection afforded a very high sensitivity of thiol measurements. The Cys peak had a retention time of 6.7 min, whereas GSH was eluted at 10.7 min. One of the major concerns in the determinations of reduced forms of cellular thiols is their potential autooxidation during sample preparations which can be particularly significant when thiols are present at low concentrations. To avoid underestimation of reduced thiols in cell lysates due to autooxidation processes, we added 2 mM DTT to the derivatization reactions which was found to lead to the complete conversion of oxidized glutathione into its reduced form (Figs 1C and 1D). The autooxidation reactions were also minimized by the preparation of cellular extracts at acidic pH and using Fe-chelator DTPA.

Figure 2 shows that the relative amounts of Cys and GSH in control and Cr (VI)-exposed individuals were significantly different. The ratio of GSH to Cys among welders was 3.0 whereas the mean value for controls was 7.8 (2.6-fold difference, $p = 0.0009$). This difference was caused by a 2.3-fold lower concentration of GSH in lymphocytes from welders (98.8 ± 39.4 nmol/mg protein in controls vs. 43.8 ± 11.9 nmol/mg protein in welders, $p = 0.002$). Average Cys concentrations were practically identical between two groups of subjects (14.3 ± 6.9 nmol/mg protein and 14.9 ± 2.7 nmol/mg protein for controls and welders, respectively). There was a strong correlation between intracellular concentrations of both thiols in lymphocytes from unexposed individuals ($r = 0.72$, $p = 0.0001$) (Fig. 3). The number of Cr (VI)-exposed workers was too small to perform a similar type of analysis. After exclusion of one subject with unusually low values for both thiols, the control population exhibited 2.9- and 6.8-fold range of GSH and Cys concentrations, respectively. Because of the strong correlation between concentrations of both thiols, the variation in the ratio of GSH to Cys was only 3.1-fold.

The significant differences in the thiol composition of lymphocytes from welders could potentially be a result of a physiological response of these cells to the presence of toxic Cr (VI). To test whether lymphocytes from welders experienced exposure to Cr (VI), we determined the level of DPC in these cells. Measurements of DPC have been previously found to be a sensitive indicator of human exposure to Cr (VI) [13, 36]. We found that welders had 4.1 times more crosslinks as compared to the control population (Fig. 4). With the exception of one subject, there was no overlap in DPC values between exposed and unexposed groups. The difference in the number of DNA–protein crosslinks between controls and welders was highly significant with $p = 0.01$. The mean number of DPC among welders was 11.4 ± 5.8 per 10^7 bp, whereas DPC in controls averaged 2.8 ± 1.2 per 10^7 bp. The results strongly suggested that lymphocytes from welders had elevated levels of genotoxic damage caused by Cr (VI).

Reduction of Cr (VI) by cysteine and glutathione

To determine a potential toxicological significance of the observed changes in thiol concentrations in lymphocytes from welders, we compared Cr (VI) reduction rates in the presence of Cys and GSH (Fig. 5). Reduction kinetics of 50 µM Cr (VI) in the presence of 1 mM Cys was described by a single exponential function and the reaction was complete after approximately 150 min. The reduction curves in the presence of GSH revealed two components and the overall rate of reduction was substantially slower than that for Cys. Complete reduction of 50 µM Cr (VI) during 150 min incubation time required at least 5 mM GSH. Similar results were also obtained when reduction reactions were studied in 50 mM phosphate buffer, pH 7.0 (not shown). Reactions with the lowest GSH concentrations exhibited the greatest differences in the rates of reduction for the fast and slow components (20–35-fold depending on buffer). Increased GSH concentrations led to approximately 2 times more rapid rise in the reduction rate of the slow component relative to the increase in the initial

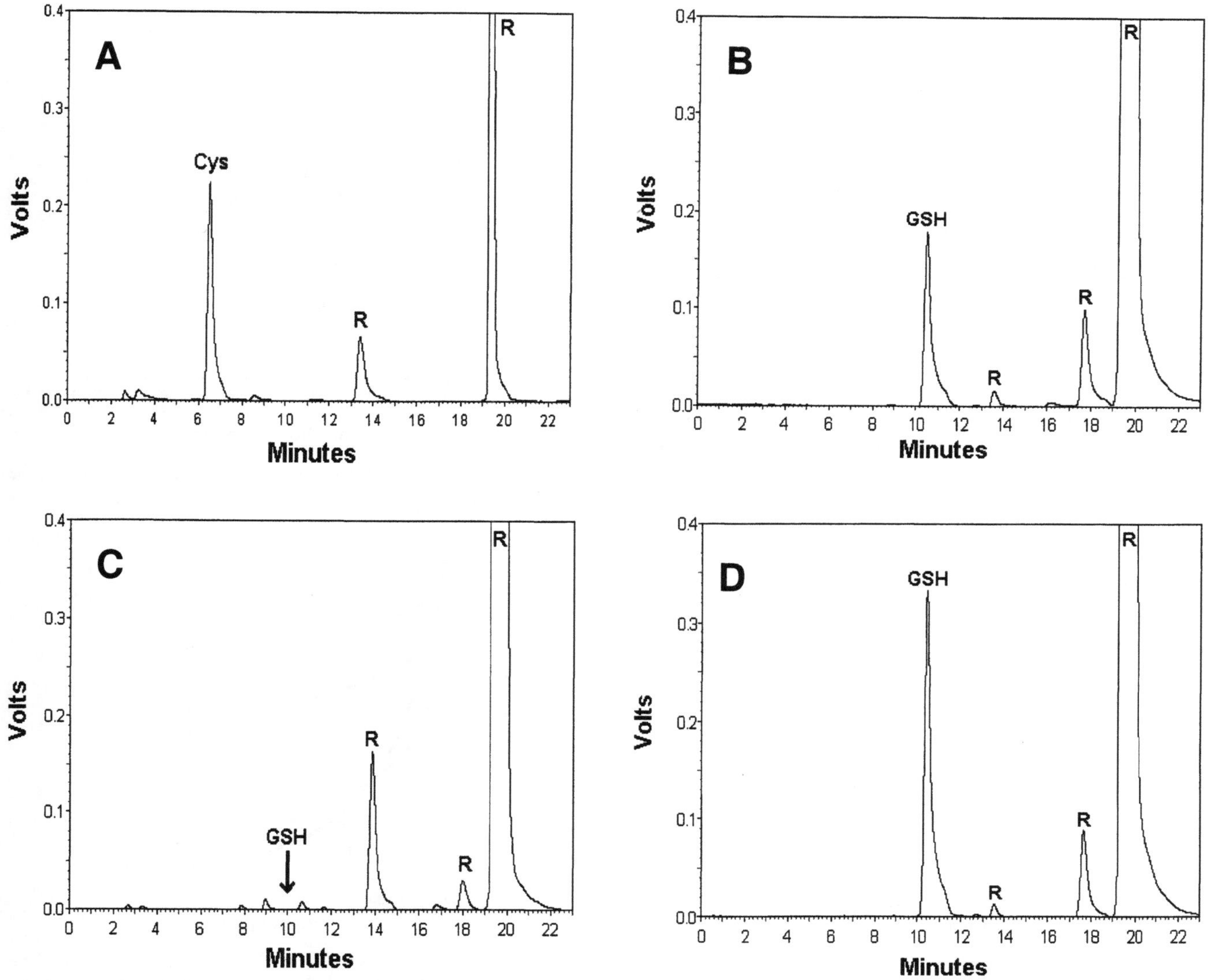

Fig. 1. HPLC analysis of monobromobimane derivatives of cysteine and glutathione. (A) cysteine standard (0.5 nmol); (B) glutathione standard (0.5 nmol); (C) glutathione disulfide (0.5 nmol); (D) glutathione disulfide (0.5 nmol) derivatized in the presence of 2 mM DTT. Reduction of 0.5 nmol GSSG gives 1 nmol GSH. R = non-specific reagent products.

rate (fast component). Assuming that the reduction rates *in vivo* and *in vitro* are similar, the results suggest that both thiols would contribute equally to the metabolism of Cr (VI) when the ratio of GSH to Cys is about 5:1.

Kinetics of Cr (VI) reduction by different concentrations of cysteine

Reductive metabolism of Cr (VI) in the presence of 2 mM Cys has previously been reported to result in the generation of binary and ternary Cr (III)–DNA adducts without detectable formation of oxidative lesions [20, 24]. Since the different rates of Cr (VI) reduction may produce variable amounts of the reactive intermediates, we investigated the formation of oxidative DNA lesions and stable Cr (III)–DNA adducts using a 10-fold range of Cys concentrations. We also employed conditions with the low ratios of Cys to Cr (VI), which was expected to increase the stability of intermediate Cr forms due to the limited amounts of the reducer. Complete reduction of 100 μM Cr (VI) during 120 min incubation period was achieved only with 2 mM and 5 mM Cys whereas samples containing 0.5 mM and 1 mM Cys still had 7.9 and 36.2% Cr (VI) unreduced at the end of the incubation time (Fig. 6). Analysis of Cr (VI) disappearance during the first 3 min of the reaction showed that 2-fold increase in the Cys concentrations on average resulted in 2.1 times faster initial rates of reduction. The reduction rates at later time intervals exhib-

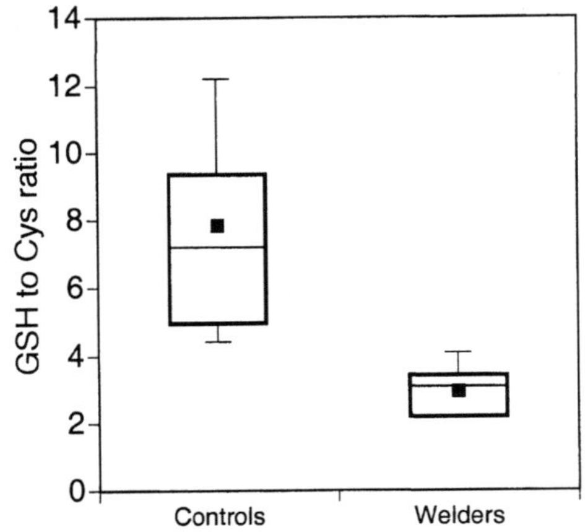

Fig. 2. Relative amounts of glutathione and cysteine in peripheral lymphocytes of controls and stainless steel welders. Intracellular concentrations of both thiols were determined by HPLC.

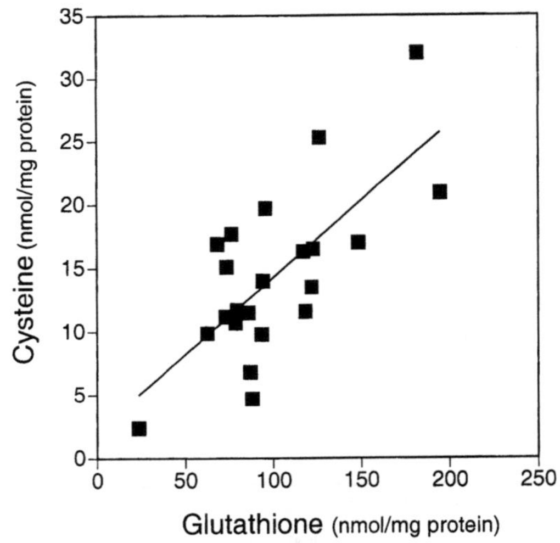

Fig. 3. Linear relationship between intracellular concentrations of cysteine and glutathione in lymphocytes of Cr-unexposed subjects (r = 0.72).

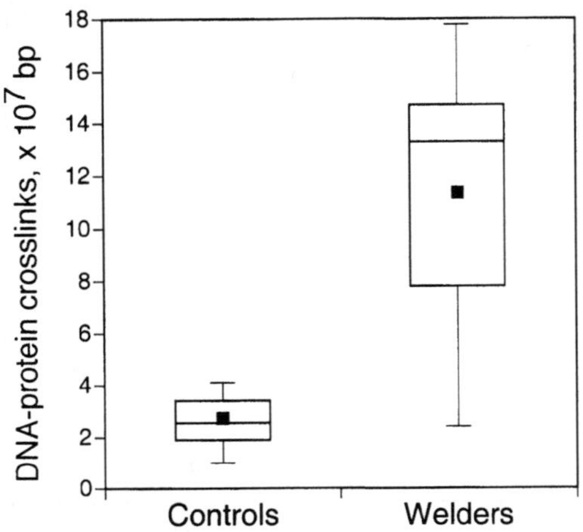

Fig. 4. DNA–protein crosslinks in peripheral lymphocytes of controls and stainless steel welders. The number of crosslinks was determined by the K-SDS precipitation assay.

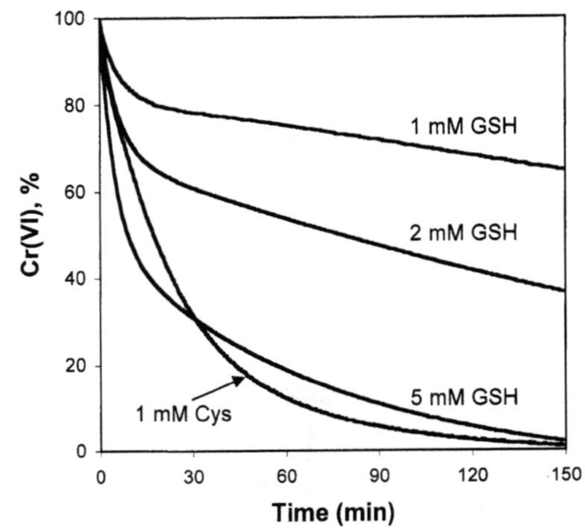

Fig. 5. Time-course of Cr (VI) reduction by cysteine and glutathione. Samples contained 50 mM MOPS (pH 7.0), 50 μM chromate and indicated concentrations of the reducers.

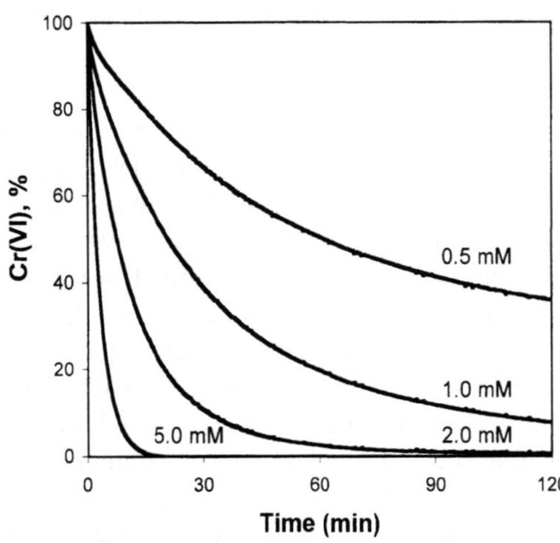

Fig. 6. Kinetics of Cr (VI) reduction by different concentrations of cysteine. Samples contained 25 mM MOPS (pH 7.0), 100 μM chromate and indicated concentrations of Cys.

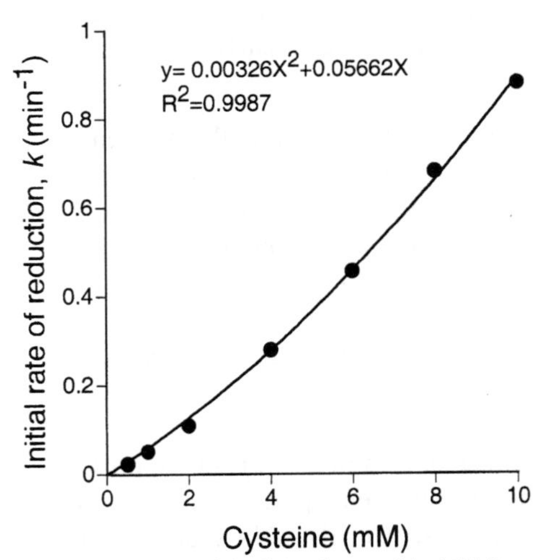

Fig. 7. The least-squares fit of the dependence of the initial rate of reduction on the concentration of cysteine. Each data point represents a mean of 2–3 independent experiments.

ited a different dependence on Cys concentration which was caused by a significant drop in the amount of reduced Cys in samples with its low initial concentrations. Plots of $\ln(A_t\text{-}A_\infty)$ vs. time were linear for samples containing 2 and 5 mM Cys but they were curved for 0.5 and 1 mM Cys. The initial rate of reduction was 13.1 times higher in the presence of 5 mM Cys compared to that for 0.5 mM Cys, whereas the differences in the reduction rates during the last quintile of the reaction were as large as 68-fold. The least-squares fit of the initial rates vs. Cys concentrations (Fig. 7) gave a non-linear function ($R^2 = 0.9987$):

$$k = 0.00326[\text{Cys}]^2 + 0.05662[\text{Cys}] \qquad \text{(Eq. 1)}$$

where k is the initial rate of reduction (min^{-1}) and [Cys] is Cys concentration in mM.

The presence of the linear and quadratic terms in Eq. 1 suggests that reduction of Cr (VI) is a mixed-order reaction with respect to cysteine. The first order reaction describes one-electron reduction of Cr (VI) by cysteine, whereas the second order function indicates the involvement of two-electron reduction mechanism.

Oxidation of DCFH during reduction of Cr (VI) by cysteine

Reduction of Cr (VI) by all concentrations of Cys resulted in the formation of oxidizing species as evidenced by strong increases in fluorescence of oxidized DCFH (Figs 8A and 8B). The generation of DCFH-oxidizing species and reduction of Cr (VI) appeared to follow a very similar kinetics. The total yield of oxidized DCFH was progressively higher with decreasing concentrations of Cys at all Cr (VI) doses (Fig. 8B). Reactions containing 0.5 mM Cys generated on average 2.6 times higher levels of DCFH oxidation than those with 5 mM Cys. The greatest differences in the oxidation of DCFH were found at high Cr (VI) concentrations (3.1-fold). The adjustment for the amount of unreduced Cr (VI) gave approximately 5 times higher yield of oxidized DCFH in samples with 0.5 mM Cys in comparison with 5 mM Cys-based reactions. In general, the differential yield of oxidized DCFH at low and high Cys concentrations was consistent with the expected increase in the stability of intermediate Cr species.

DNA damage resulting from reduction of Cr (VI) by cysteine

To test whether the formation of DCFH-oxidizing species also leads to oxidative DNA damage, we measured the amount of strand breaks and abasic sites in control and Cr (VI)-treated DNA. Figure 9 shows that the relative amounts of intact supercoiled plasmids and nicked forms were very similar in all samples, indicating the absence of DNA breakage. At the highest Cr (VI) concentrations, the intensity of both supercoiled and relaxed DNA bands in MOPS-based reactions was slightly lower relative to control (Fig. 9A), which probably reflects a diminished binding of ethidium bromide to Cr (III)-modified DNA [24]. Reduction of Cr (VI) in MOPS, but not in phosphate buffer, also resulted in the reduced mobility of supercoiled DNA molecules which was particularly noticeable in samples containing 1 and 2 mM Cys. This change in the electrophoretic mobility is caused by unwinding of supercoiled plasmids following binding of Cr (III) [24]. Reduction of Cr (VI) in MOPS buffer by either 0.5 or 5 mM Cys resulted only in small changes in the electrophoretic mobility of supercoiled DNA. The absence of any changes in the

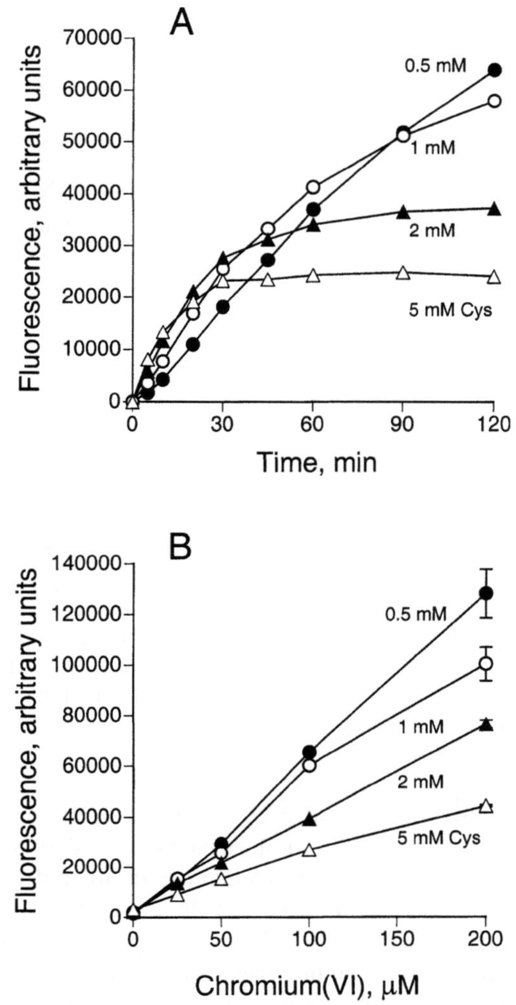

Fig. 8. Oxidation of DCFH in the course of Cr (VI) reduction by cysteine. Samples contained 25 mM MOPS (pH 7.0), 20 μM DCFH and indicated concentrations of chromate and Cys. All incubations were performed at 37°C. Shown are means ± SD of three independent measurements. (A) time-course of DCFH oxidation by 100 μM Cr (VI); (B) dose-dependence of DCFH oxidation in the presence of different concentrations of Cys and Cr (VI).

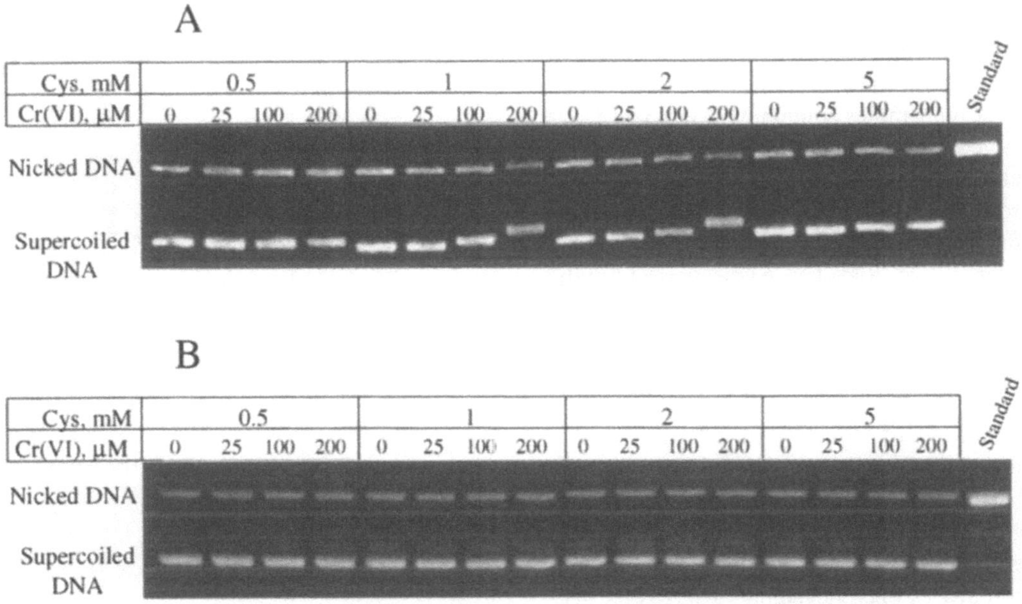

Fig. 9. Agarose gels of ΦX174 DNA incubated in the presence of Cr (VI) and cysteine. DNA strand breakage would be detected as an increase in the amount of nicked DNA with the accomponied decrease in the intensity of supercoiled bands. Samples contained 0.3 μg DNA and were incubated for 2 h at 37°C. DNA was separated in 1% agarose gels and stained with ethidium bromide. (A) Reactions were performed in 25 mM MOPS (pH 7.0); (B) Reactions were performed in 25 mM Na-phosphate (pH 7.0).

electrophoretic behavior of DNA samples treated with Cr (VI)/Cys in the presence of phosphate is caused by the lack of formation of Cr (III)-DNA adducts in this buffer [20]. Reductive metabolism of Cr (VI) by 0.5–5 mM Cys also failed to cause a loss of DNA bases (Figs 10A and 10B). This conclusion is based on the practically identical amounts of the nicked DNA form following a treatment of control and Cr (VI)-exposed DNA samples with DED. The efficient conversion of abasic sites into DNA breaks in the presence of DED was confirmed using an acid-depurinated DNA standard.

The formation of total Cr (III)–DNA adducts was dependent on both Cr (VI) and Cys concentrations (Fig. 11). The results showed a strong inhibition of Cr-DNA binding in the presence of increased concentrations of Cys. A low yield of Cr (III)–DNA adducts in the presence of 5 mM Cys is consistent with the relatively small changes in the mobility of supercoiled plasmids in comparison to the samples containing 1 mM and 2 mM Cys (Figs 9A and 9B). However, a high level of Cr–DNA binding in 0.5 mM Cys-based reactions did not result in the extensive unwinding of supercoiled plasmids. In fact, despite approximately 2 times higher number of Cr (IIII)–DNA adducts, the mobility of supercoiled plasmids from samples containing 0.5 mM Cys was much less affected in comparison to the reactions with 2 mM Cys. Thus, the total number of Cr (III)–DNA adducts does not always result in a high degree of structural changes in supercoiled DNA molecules.

Discussion

Relative roles of cysteine and glutathione in Cr (VI) reduction

HPLC measurements of GSH and Cys concentrations in freshly isolated peripheral lymphocytes from unexposed human subjects found that GSH was the major thiol in these cells with almost 8-fold molar excess over Cys. However, the ratio of GSH to Cys averaged 3.0 in cells obtained from Cr (VI)-exposed welders. Previous studies have shown that both GSH and Cys were capable of reducing of Cr (VI), but the initial rate of reduction by Cys has been consistently found to be significantly higher [28, 30]. The presence of the negative charge on GSH has been proposed to lead to the decreased formation of thioesters with chromate anion [30]. The kinetics of Cr (VI) reduction by Cys has been described by a single component [20, 37], whereas reduction reactions driven by GSH were characterized by the presence of the fast initial reaction and the slow late component [28, 29]. One approach to compare a relative activity of two reducers with the different mechanisms of the reaction was to determine their concentrations required to complete Cr (VI) reduction over a specified period of time. The comparison of the reaction rates at biologically relevant concentrations indicates that the contribution of both thiols to reduction of Cr (VI) would be equal when the ratio of GSH to Cys is approximately 5:1.

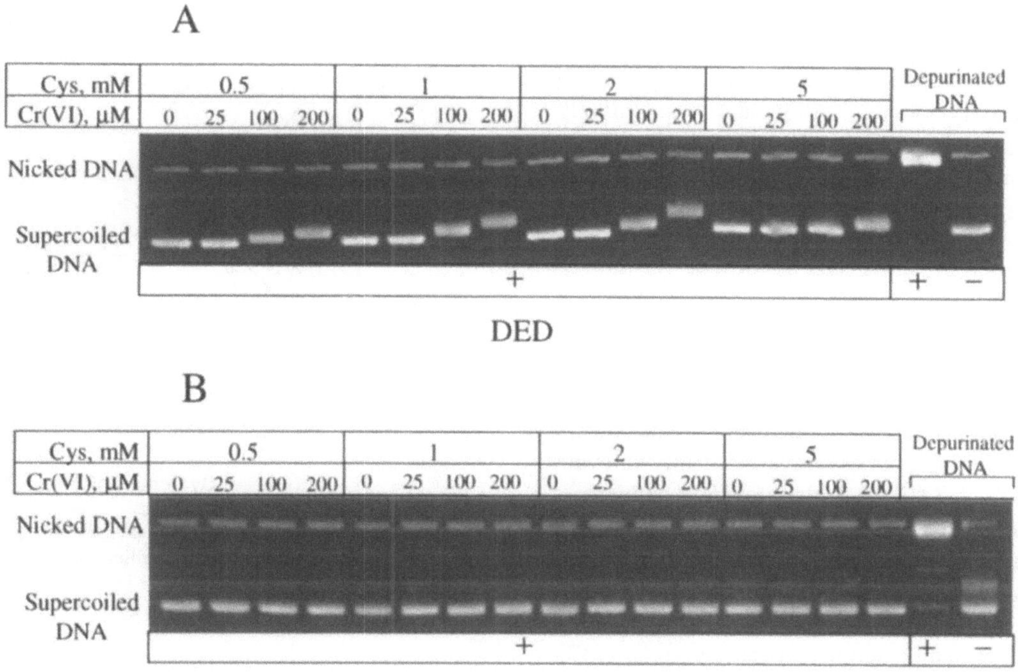

Fig. 10. Detection of abasic sites in DNA. DNA was incubated in the presence of Cr (VI)/Cys for 2 h at 37°C followed by Bio-Gel P-30 chromatography and treatment with DED. DED reacts with abasic sites converting them into strand breaks. Nicked and intact supercoiled forms of DNA were separated in 1% agarose and then stained with ethidium bromide. (A) Reactions were performed in 25 mM MOPS (pH 7.0); (B) Reactions were performed in 25 mM Na-phosphate (pH 7.0).

Because of the depressed levels of GSH causing the ratio of GSH:Cys to fall to 3.0, Cys is expected to be a dominant thiol involved in Cr (VI) reduction in lymphocytes of stainless steel

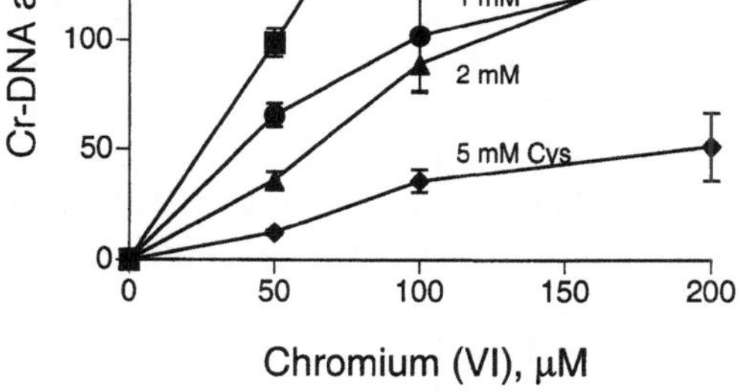

Fig. 11. Formation of Cr-DNA adducts during reduction of Cr (VI) by different concentrations of cysteine. Unreacted and ionically bound Cr were removed by Bio-Gel P-30 chromatography and ethanol precipitation. The amount of DNA-bound Cr was determined using [^{51}Cr]chromate. Shown are means ± S.D. of 3–6 independent measurements.

welders. This result suggests that cells with the thiol ratio similar to that of welders could be the most appropriate experimental model for the identification of Cr (VI)-induced DNA lesions that can be used as potential biomarkers of Cr (VI) exposure. Given that all controls had the ratio of GSH to Cys significantly higher than 3:1, it may be somewhat difficult to find a human cell line with the required composition of the intracellular thiols. Cultured human diploid fibroblasts have been reported to have 7.5-fold excess of GSH over Cys [38], which is essentially identical to our results for control lymphocytes.

There are two possible reasons for low concentrations of GSH in lymphocytes of welders: (1) welders represent a subpopulation of individuals with inherently low concentrations of GSH in lymphocytes or (2) the decrease in the level of GSH was caused by occupational exposure. The first possibility seems unlikely considering that there were no control subjects who had the ratio of GSH to Cys in the range found among welders. A second possibility implies that the decrease in the level of GSH was a result of exposure to Cr (VI) and/or other metals present in welding fumes. Cr and Ni are known to bind to GSH producing stable coordinate complexes that could be excreted from cells [39]. The presence of elevated concentrations of Cr, Ni and Fe can also result in oxidative stress, which would result in the depletion of GSH due to its role in radical scavenging. A combination of the stress-mediated depletion and the participation of GSH in binding and excretion of metals could potentially be responsible for the observed decrease in the amount of free intracellular GSH in lymphocytes from welders. The presence of the elevated levels of DPC in lymphocytes from welders was unlikely caused by the observed decrease in the intracellular concentrations of GSH since there was no correlation between these two parameters among unexposed controls (r = 0.07).

Reduction of Cr (VI) by cysteine and resulting DNA damage

Analysis of the initial rates of reduction in the presence of different concentrations of Cys revealed the presence of two reaction mechanisms (Eq. 1). At physiologically relevant concentrations of Cys (≤ 2 mM), the prevalent pathway is one-electron reduction (≥ 90%), whereas two-electron reduction dominates at much higher concentrations (Fig. 12). Approximately equal contribution of both pathways would occur at 17.5 mM Cys. Two-electron reduction of Cr (VI) has been proposed as the dominant reaction mechanism for Cys [37]. This suggestion was based on a low yield of Cr (V) [28] and the inability to fit kinetic data into a linear function. However, measurements of Cr (V) were performed in the presence of 30 mM Cys [37], a concentration associated with the predominantly two-electron reduction according to our model (Fig. 12). ESR measurements are not sufficiently sensitive to perform quantitation of Cr (V) at biologically relevant Cys concentrations that are predicted to generate a much more significant amount of Cr (V). The yield of reactive intermediates detected by oxidation of DCFH was much higher when reduction of Cr (VI) proceeded at a slow rate, which could be related either to a greater stability of reactive Cr intermediates in the presence of low Cys concentrations or higher production of the oxidants. Although the exact nature of species responsible for DCFH oxidation during Cr (VI) reduction is unknown, Cr (V) is one of the likely candidates. Cr (V) has been detected during reduction of Cr (VI) by Cys [28] and model Cr (V) complexes were capable of direct oxidation of DCFH [33]. The predominance of the single-electron reduction mechanism at low Cys concentrations would result in a higher yield of Cr (V) which lends a support for the possibility that Cr (V) is one of the reactive species responsible for oxidation of DCFH during reduction of Cr (VI).

Electrophoretic analysis of supercoiled plasmids did not detect even minor increases in the amount of single strand breaks or abasic sites following incubation of DNA in Cr (VI)–Cys reactions. These results stand in the clear contrast with a very extensive oxidation of DCFH, demonstrating that DNA is much less susceptible to oxidation than the commonly used oxidant probe, DCFH. The lack of strand breakage and abasic sites formation over a wide range of Cr (VI) and reducer concentrations strongly suggests that reduction

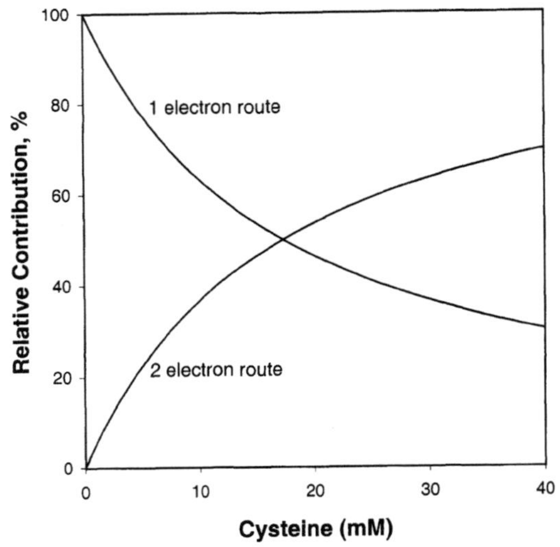

Fig. 12. Relative contribution of one- and two-electron pathways to reduction of Cr (VI) at different concentrations of cysteine. Plots are drawn using eq. 1.

of Cr (VI) by Cys does not produce oxidants capable of damaging DNA. The plasmid relaxation assay employed here for the detection of breakage and base loss can reliably detect 1 break or 1 abasic site per 50,000 bp (10% increase in the amount of nicked DNA). Although the presence of oxidized bases has not been analyzed, it is very unlikely that they were formed. This conclusion is based on the fact that strand breaks and abasic sites are the most common products of DNA damage by oxidants and it would be difficult to propose a reaction mechanism leading exclusively to base damage. Oxidation of DNA by model Cr (V) complexes leads primarily to strand breakage and abasic sites [16–18], whereas hydroxyl radicals would cause all forms of damage, including extensive damage to the sugar-phosphate backbone [19]. The absence of mutagenicity and replication blockage in shuttle-vector plasmids treated under conditions blocking the formation of Cr (III)–DNA adducts confirms the absence of base oxidation during reduction of Cr (VI) with 2 mM Cys [24]. In the absence of direct measurements or biological testing, however, the formation of oxidized DNA bases at Cys concentrations other than 2 mM cannot be completely ruled out.

In contrast to oxidative lesions, reduction of Cr (VI) by all concentrations of Cys led to a very extensive formation of Cr (III)–DNA adducts. The yield of Cr (III)–DNA adducts in the presence of 200 μM Cr (VI) was as high as 3 Cr per 10 bp, a number of lesions more than 10,000 times higher than potentially undetectable levels of DNA breaks or apurinic sites. This number of adducts is clearly incompatible with survival [24] and, therefore, the absence of oxidative lesions even under these conditions illustrates the insignificant role of oxidative mechanisms in the induction of genotoxic damage during Cys-dependent reduction of Cr (VI). The highest yield of Cr (III)–DNA adducts was found in samples containing the lowest concentration of Cys, potentially reflecting a less efficient formation of Cr (III)–$(Cys)_2$ complexes that exhibit weak reactivity toward DNA [40]. In addition to a variable yield of adducts, the use of different Cys concentrations probably resulted in the different spectra of Cr (III)–DNA adducts since structural changes in supercoiled DNA were largely unrelated to the amount of total adducts. Reduction of Cr (VI) by 2 mM Cys has been found to produce binary Cr (III)–DNA and ternary Cys–Cr (III)–DNA adducts, as well as interstrand DNA–DNA crosslinks [20]. At present, it is not known what forms of Cr–DNA adducts cause unwinding of supercoiled DNA molecules.

Acknowledgements

This work was supported by grant ES08786 from the National Institute of Environmental Health Sciences (Anatoly Zhitkovich) and a graduate fellowship from the Portuguese Foundation for Science and Technology (Margarida Goulart).

References

1. Sorahan T, Burges DC, Hamilton L, Harrington JM: Lung cancer mortality in nickel/chromium platers, 1946–1995. Occup Environ Med 55: 236–242, 1998
2. Mancuso TF: Chromium as an industrial carcinogen: Part I. Am J Ind Med 31: 129–139, 1997
3. Langardt S: One hundred years of chromium and cancer: Review of epidemiological evidence and selected case reports. Am J Ind Med 17: 189–215, 1990
4. Sen P, Conway K, Costa M: Comparison of the localization of chromosome damage induced by calcium chromate and nickel compounds. Cancer Res 47: 2142–2147, 1987
5. Biedermann KA, Landolph JR: Role of valence state and solubility of chromium compounds on induction of cytotoxicity, mutagenesis and anchorage independence in diploid human fibroblasts. Cancer Res 50: 7835–7842, 1990
6. Itoh S, Shimada H: Clastogenicity and mutagenicity of hexavalent chromium in lacZ transgenic mice. Toxicol Lett 91: 229–233, 1997
7. Landolph JR: Role of free radicals in metal-induced carcinogenesis. In: A. Sigel, H. Sigel (eds). Metal Ions in Biological Systems. Marcel Dekker, New York, 1999, pp 445–483
8. Snow E: Metal carcinogenesis: Mechanistic implications. Pharmacol Ther 53: 31–65, 1992
9. Sugiyama M, Wang X-W, Costa M: Comparison of DNA lesions and cytotoxicity induced by calcium chromate in human, mouse, and hamster cell lines. Cancer Res 46: 4547–4551, 1986
10. Cantoni O, Costa M: Analysis of the induction of alkali sensitive sites in the DNA by chromate and other agents that induce single strand breaks. Carcinogenesis 5: 1207–1209, 1984
11. Xu J, Bubley GJ, Detrick B, Blankenship LJ, Patierno SR: Chromium (VI) treatment of normal human lung cells results in guanine-specific DNA polymerase arrest, DNA–DNA crosslinks and S-phase blockage of cell cycle. Carcinogenesis 17: 1511–1517, 1996
12 Costa, M: Analysis of DNA-protein complexes induced by chemical carcinogens. J Cell Biochem 92: 127–135, 1990
13. Zhitkovich A, Voitkun V, Kluz T, Costa M: Utilization of DNA–protein crosslinks as a biomarker of chromium exposure. Environ Health Perspect 106 (suppl 4): 969–974, 1998
14. Zhitkovich A, Voitkun V, Costa M: Glutathione and free amino acids form stable complexes with DNA following exposure of intact mammalian cells to chromate. Carcinogenesis 16: 907–913, 1995
15. Zhitkovich A, Voitkun V, Costa M: Formation of the amino acid–DNA complexes by hexavalent and trivalent chromium *in vitro*: Importance of trivalent chromium and the phosphate group. Biochemistry 35: 7275–7282, 1996
16. Farrel RP: Chromium (V)-induced cleavage of DNA: Are chromium (V) complexes the active carcinogens in chromium (VI)-induced cancers? Chem Res Toxicol 2: 227–229, 1989
17. Bose RN, Moghaddas S, Mazzer PA, Dudones LP, Joudah L: Oxidative damage of DNA by chromium (V) complexes: Relative importance of base versus sugar oxidation. Nucleic Acids Res 27: 2219– 2226, 1999
18. Sugden K: Formation of modified cleavage termini from the reaction of chromium (V) with DNA. J Inorg Biochem 77: 177–183, 1999
19. Shi X, Ding M, Ye J, Wang S, Leonard SS, Zang L, Castranova V, Vallyathan V, Chiu A, Dalal N, Liu K: Cr (IV) causes activation of nuclear transcription factor-kappa B, DNA strand breaks and dG hydroxylation via free radical reactions. J Inorg Biochem 75: 37–44, 1999
20. Zhitkovich A, Messer J, Shrager S: Reductive metabolism of Cr (VI) by cysteine leads to the formation of binary and ternary Cr–DNA adducts in the absence of oxidative DNA damage. Chem Res Toxicol 13: 1114–1124, 2000
21. Yuann JM, Liu KJ, Hamilton JW, Wetterhahn KE: *In vivo* effects of ascorbate and glutathione on the uptake of chromium, formation of chromium (V), chromium-DNA binding and 8-hydroxy-2'-deoxyguanosine in liver and kidney of Osteogenic Disorder Shionogi rats following treatment with chromium (VI). Carcinogenesis 20: 1267–1275, 1999
22. Voitkun V, Zhitkovich A, Costa M: Cr (III)-mediated crosslinks of glutathione or amino acids to the DNA phosphate backbone are mutagenic in human cells. Nucl Acids Res 26: 2024–2030, 1998
23. Liu S, Dixon K: Induction of mutagenic DNA damage by chromium (VI) and glutathione. Environ Mol Mutagen 28: 71–79, 1996
24. Zhitkovich A, Song Y, Quievryn G, Voitkun V: Non-oxidative mechanisms are responsible for the induction of mutagenesis by reduction of Cr (VI) with cysteine: The role of ternary DNA adducts in Cr (III)-dependent mutagenesis. Biochemistry 40: 549–560 (2001)
25. Connett P, Wetterhahn, KE: Metabolism of the carcinogen chromate by cellular constituents. Struct Bond 54: 93–124, 1983
26. Suzuki Y, Fukuda K: Reduction of hexavalent chromium by ascorbic acid and glutathione with special reference to the rat lung. Arch Toxicol 64: 169–176, 1990
27. O'Brien P, Wang G, Wyatt PB: Studies on the kinetics of the reduction of chromate by glutathione and related thiols. Polyhedron 11: 3211–3216, 1992
28. Kitagawa S, Seki H, Kametani F, Sakurai H: EPR study on the interaction of hexavalent chromium with glutathione or cysteine: Production of pentavalent chromium and its stability. Inorg Chim Acta 152: 251–255, 1988
29. Stearns DM, Wetterhahn KE: Reaction of Cr (VI) with ascorbate produces chromium (V), chromium (IV), and carbon-based radicals. Chem Res Toxicol 7: 219–230, 1994
30. Shi X, Dong Z, Dalal NS, Gannett PM: Chromate-mediated free radical generation from cysteine, penicillamine, hydrogen peroxide, and lipid hydroperoxides. Biochim Biophys Acta 1226: 65–72, 1994

31. Quievryn G, Zhitkovich A: Loss of DNA–protein crosslinks from formaldehyde-exposed cells occurs through spontaneous hydrolysis and an active repair process linked to proteosome function. Carcinogenesis 21: 1573–1580, 2000
32. Zhitkovich A, Costa M: A simple, sensitive assay to detect DNA–protein crosslinks in intact cells and *in vivo*. Carcinogenesis 13: 1485–1489, 1992
33. Martin BD, Schoenfeld JA, Sugden KD: Hypervalent chromium mimics reactive oxygen species as measured by the oxidant-sensitive dyes 2′,7′-dichlorofluorescin and dihydrorhodamine. Chem Res Toxicol 11: 1402–1410, 1998
34. McHugh PJ, Knowland J: Novel reagents for chemical cleavage at abasic sites and UV photoproducts in DNA. Nucleic Acids Res 23: 1664–1670, 1995
35. Pierre J, Laval J: Specific nicking of DNA at apurinic sites by peptides containing aromatic residues. J Biol Chem 256: 10217–10220, 1981
36. Werfel U, Langen V, Eickhoff I, Schoonbrood J, Vahrenholtz C, Brauksiepe A, Popp W, Norpoth K: Elevated DNA single-strand breakage frequencies in lymphocytes of welders exposed to chromium and nickel. Carcinogenesis 19: 413–418, 1998
37. Kwong DWJ, Pennington DE: Stoichiometry, kinetics, and mechanisms of the chromium (VI) oxidation of L-cysteine. Inorg Chem 23: 2538–2532, 1984
38. Dypbukt JM, Atzori L, Edman CC, Grafstrom RC: Thiol status and cytopathological effects of acrolein in normal and xeroderma pigmentosum skin fibroblasts. Carcinogenesis 14: 975–980, 1993
39. Rabenstein DL: Metal complexes of glutathione and their biological significance. In: D. Dolphin, R. Poulson, O. Avramovic (eds). Glutathione: Chemical, Biochemical and Medical Aspects, Vol. A. Wiley and Sons, New York, 1989
40. Hneihen A, Standeven AM, Wetterhahn KE: Differential binding of chromium (VI) and chromium (III) complexes to salmon sperm nuclei and nuclear DNA and isolated calf thymus DNA. Carcinogenesis 14: 1795–1803, 1993

Molecular and Cellular Biochemistry **222**: 119–125, 2001.

Transactivation of RARE and GRE in the cellular response to arsenic

Chuanshu Huang,[1] Jingxia Li,[1] Min Ding,[2] Max Costa,[1] Vincent Castranova,[2] Val Vallyathan,[2] Gong Ju[3] and Xianglin Shi[2]

[1]*Nelson Institute of Environmental Medicine, New York University School of Medicine, NY;* [2]*Health Effects Laboratory Division, National Institute for Occupational Safety and Health, Morgantown, WV, USA;* [3]*The Institute of Neuroscience, The Fourth Military Medical University, Xi'an, P.R. China*

Abstract

Arsenic compounds are a somewhat unique class of metals, which have been considered as both carcinogens and chemotherapeutic agents for cancers. Tumor promotion effects of arsenic are believed to be associated with its transactivational activities on transcription factors, such as AP-1 and NFκB, while the induction of cell apoptosis and differentiation by arsenic is considered to be a mechanism for the chemotherapeutic effects of arsenic. Here, we found that exposure of cells to arsenite and arsenate leads to transactivation of retinoic acid response elements (RARE) and glucocorticoid response elements (GRE) in mouse epidermal JB6 cells. These inductions occur in a time-dependent manner. Furthermore, induction of RARE activity by arsenic was synergistically enhanced by co-treatment of cells with retinoic acid, while GRE activation by arsenic was not affected by combined treatment of cells with fluocinolone acetonide (FA). In consideration of the important role of RARE and GRE in induction of cell differentiation, we speculate that transactivation of RARE and GRE by arsenic may be involved in its induction of cell differentiation and anti-cancer activities in addition to its induction of apoptosis. (Mol Cell Biochem **222**: 119–125, 2001)

Key words: arsenic, RARE, GRE, signal transduction

Abbreviations: RARE – retinoic acid response elements; RAR – retinoic acid receptor; RA – trans-retinoic acid; GRE – glucocorticoid response elements; FA – fluocinolone acetonide; NFκB – nuclear factor-κB; AP-1 – activator protein-1; Erks – extracellular signal-regulated protein kinases; MAPK – mitogen-activated protein kinase; JNKs – c-Jun N-terminal kinases; FBS – fetal bovine serum; MEM – minimal essential medium.

Introduction

A growing amount of evidence demonstrates that arsenic is a human carcinogen [1–5]. The International Agency for Research on Cancer (IARC) concluded in 1980 and again in 1987 that arsenic is a human carcinogen according to the available epidemiological findings and experimental data [6–8]. One of the most sensitive targets of arsenic toxicity is the skin. Both inorganic and organic forms are absorbed by human and animal skin [9, 10]. Animal experiments show that following chronic exposure, arsenic accumulates in the skin and hair [9, 10]. Although arsenic can be ingested and absorbed from the diet, the skin is a major target organ. Previous studies suggested that the accumulation of arsenic in the skin increases dermal sensitivity to ultraviolet (UV) light which may lead to enhanced carcinogenic effects [11, 12]. Many cases of skin cancer have been reported in people exposed to arsenic through medicinal use [9, 10]. Epidemiological evidence has also shown an excess of skin cancers following arsenic exposure. Interestingly, arsenic-containing

Address for offprints: C. Huang, Nelson Institute of Environmental Medicine, New York University School of Medicine, NY 10016, USA

compounds have been used for treatment of cancer and skin disease for hundreds of years in both western and traditional Chinese medicine [13–18]. Arsenite was routinely used to control elevated leukocyte counts in chronic myelogenous leukemia in the early 1990s [14, 15]. For more than 1000 years, the Chinese medications pi shuang and xiong huang (Realgar), which contain As_2O_3 and As_2S_2, respectively, have been used to treat cancers and other skin diseases [19]. Recently, As_2O_3 was reported to induce complete remission in a high proportion of patients with refractory acute promyelocytic leukemia [17]. Therefore, arsenic compounds are somewhat unique metalloids, which have been considered as both carcinogenic and cancer chemotherapeutic agents.

The carcinogenic effects of arsenic are believed to act at the level of tumor promotion by activating signal transduction pathways leading to activation of transcription factors, which in turn modulate gene expression [20–25]. It has been reported that arsenite is a potent stimulator for induction of c-fos and c-jun gene expression and AP-1 transactivational activity and JNK activity in Hela cells [22]. Ludwig *et al.* [23] reported that arsenite induces Erks activation through MAPK kinase 6/p38-dependent pathways. Because the skin is one of the major targets of arsenic, we recently addressed effects of arsenic on signal transduction pathways leading to activation of transcription factors and kinase in the mouse epidermal cell line, JB6 cells, and mouse skin [21, 26]. We showed that low concentrations of arsenite induced Erks activation and that Erks activation was required for arsenite-induced cell transformation [21]. However, high concentration of arsenite caused activation of JNKs, which mediated cell apoptosis in mouse epidermal JB6 cells [27]. It has been proposed that apoptosis and differentiation are two major mechanisms for the therapeutic effects of arsenic [17, 18, 27, 29, 30]. The arsenic-induced apoptosis was mediated by a JNK-dependent pathway [27], targeting tubulins [28] and down-regulation of Bcl-2 [18]. However, the molecular mechanism for induction of cell differentiation has not been investigated yet. In this report, we found that treatment of cells with arsenic resulted in significant activation of retinoic acid response elements (RARE) and glucocorticoid response elements (GRE).

Materials and methods

Cell culture and reagents

Mouse epidermal JB6, Cl41 cells, and their RARE-luciferase reporter as well as GRE-luciferase stably transfectants were cultured in monolayers in Eagle's minimal essential medium containing 5% fetal bovine serum (FBS), 2 mM L-glutamine, and 25 µg of gentamicin/ml [31]. Fetal bovine serum was from Life Technologies, Inc. (Rockville, MD, USA); Eagle's minimal essential medium (MEM) was from Calbiochem (San Diego, CA, USA); luciferase assay substrate was from Promega (Madison, WI, USA); both arsenite and arsenate were from Sigma (St. Louis, MO, USA).

Generation of stable transfectants with RARE-luciferase reporter

Cl41 cells were cultured in a 6-well plate until they reached 85–90% confluence. One µg of CMV-neo vector with 12 µg of RARE-luciferase reporter plasmid DNA and 15 µl of LipofectAMINE reagent, were used to transfect each well in the absence of serum. After 10–12 h, the medium was replaced with 5% FBS MEM. Approximately 30–36 h after the beginning of the transfection, the cells were freed from the plate with 0.033% trypsin, and cell suspensions were plated onto 75 ml culture flasks and cultured for 24–28 days with G418 selection (800 µg/ml). The stable transfectants were identified by measuring the basal level of luciferase activity. The stable transfectant, Cl41 RARE mass, was established and cultured in G418-free MEM for at least two passages before each experiment.

Generation of stable transfectants with GRE-luciferase reporter

Cl41 cells were cultured in a 6-well plate until they reached 85–90% confluence. One µg of CMV-neo vector with 12 µg of GRE-luciferase reporter plasmid DNA and 15 µl of LipofectAMINE reagent, were used to transfect each well in the absence of serum. After 10–12 h, the medium was replaced with 5% FBS MEM. Approximately 30–36 h after the beginning of the transfection, the cells were freed from the plate with 0.033% trypsin, and cell suspensions were plated onto 75 ml culture flasks and cultured for 24–28 days with G418 selection (800 µg/ml). The stable transfectants were identified by measuring the basal level of luciferase activity. The stable transfectant, Cl41 GRE mass, was established and cultured in G418-free MEM for at least two passages before each experiment.

Assay for activation of RARE by arsenic in JB6 Cl41 cells

Confluent monolayers of JB6 Cl41 RARE mass cells were trypsinized, and 8×10^3 viable cells suspended in 100 µl of 5% FBS MEM were added to each well of a 96-well plate. Plates were incubated at 37°C in a humidified atmosphere of 5% CO_2. Twelve to 24 h later, the cells were exposed to arsenic (diluted in 0.1% of FBS MEM at the concentration indicated; storage concentration of arsenic is 50 mM dissolved in PBS) for RARE induction and maintained in culture. After different time periods, cultures were extracted with lysis buffer, and luciferase activity was measured using a luminometer (Monolight 2010). The results are expressed as relative RARE activity [32].

Determination of GRE activation in JB6 Cl41 cells by arsenic

Confluent monolayers of JB6 Cl41 GRE mass cells were trypsinized, and 8×10^3 viable cells suspended in 100 µl of 5% FBS MEM were added to each well of a 96-well plate. Plates were incubated at 37°C in a humidified atmosphere of 5% CO_2. Twelve to 24 h later, the cells were exposed to arsenic (diluted in 0.1% of FBS MEM at the concentration indicated; storage concentration of arsenic is 50 mM dissolved in PBS) for GRE induction and maintained in culture. After different culture periods, the cells were extracted with lysis buffer, and luciferase activity was measured using a luminometer (Monolight 2010). The results are expressed as relative GRE activity [32].

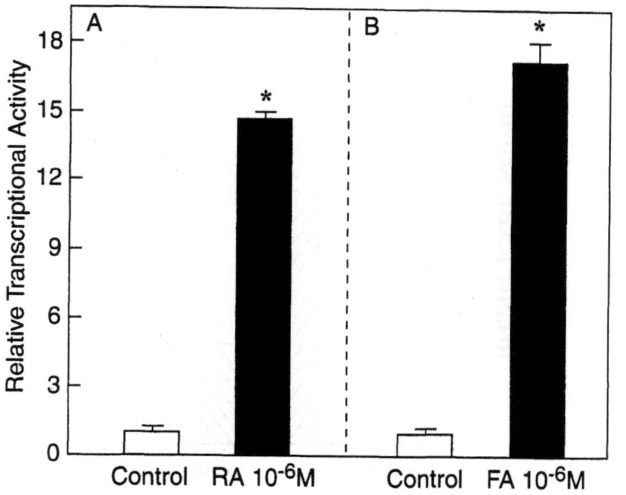

Fig. 1. Identification of RARE- and GRE-luciferase reporter stable transfectants. 8×10^3 Cl41 RARE mass1 (A) or Cl41 RARE mass1 (B) cells were seeded into each well of 96-well plates. After being cultured at 37°C overnight, the cells were starved for 12 h by replacing medium with 0.1% FBS MEM. Then the cells were treated with either (A) RA (1 µM) or (B) FA (1 µM). After being cultured for 24 h, the luciferase activity was measured. The results are presented as relative RARE or GRE activity. Each bar indicates the mean and S.D. of 4 identically treated assay wells. *Indicates a significant increase from control ($p < 0.05$).

Results

Establishment of stable transfectants with RARE-luciferase and GRE-luciferase reporter in mouse epidermal JB6 cells

In order to study the activation of RARE and GRE in JB6 cell culture models, we used a LipofectAMINE kit to transfect CMV-neo vector and RARE-luciferase reporter plasmid DNA, and CMV-neo vector and GRE-luciferase reporter plasmid into JB6 cells, respectively. After G418 selection for 24–28 days, the stable transfectants were identified by measuring basal levels of luciferase activity. Stable transfectants, Cl41 RARE mass1 and Cl41 GRE mass1, were established. Both cells showed good responses to RA and FA, respectively (Fig. 1).

Induction of RARE in mouse epidermal JB6 cells by arsenic

Both retinoic acid and arsenic were reported to induce cell differentiation [33]. It is well accepted that retinoic acid (RA) induces leukemia cell differentiation through retinoic acid receptors resulting in RA-mediated RARE activation [34–36]. To investigate the molecular mechanism of arsenic-induced cell differentiation, we tested the effects of arsenic on RARE activation in JB6 RARE-luciferase stable transfectants, Cl41 RARE mass. We incubated Cl41 RARE mass1 cells, with either arsenite or arsenate. The results showed that treatment of Cl41 cells with either arsenite or arsenate, led to significant RARE induction (Fig. 2). Interestingly, the induction of RARE by arsenate (55 fold) was much higher than that induced by RA (22 fold) or arsenite (8 fold). A time course study suggested that treatment of cells with arsenic resulted in the maximum induction of RARE activity at 12–24 h post-arsenic treatment of cells (Fig. 3). After which the RARE activity returned to near basal levels (Fig. 3). In contrast, RA-induced RARE activity reached the peak between 12 and 48 h after RA treatment, then returned toward the basal level (Fig. 3). These results demonstrate that both arsenite and arsenate induce very strong activation of RARE activity in mouse epidermal cells.

Activation of GRE in the cellular response to arsenic

Previous studies indicated that arsenic compounds have effects on the function of the glucocorticoid receptor (GR) [37, 38]. However, the results from different groups were not consistent [37, 38]. Therefore, we investigated the GRE activation after JB6 cells were exposed to arsenic. As shown in Fig. 4, treatment of cells with arsenic results in marked GRE transactivation in mouse JB6 cells, with arsenate being more effective than arsenite. This activation appears to be time-dependent, with a maximal response at 24 h of exposure (Fig. 5). These data are not only consistent with previous findings

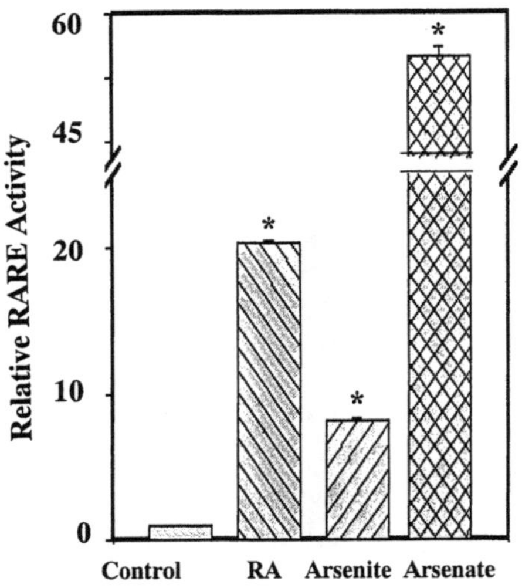

Fig. 2. Induction of RARE activity by arsenic in mouse epidermal JB6 cells. 8 × 10^3 Cl41 RARE mass1 cells were seeded into each well of 96-well plates. After being cultured at 37°C overnight, the cells were starved for 12 h by replacing medium with 0.1% FBS MEM. Then the cells were treated with RA (1 μM), arsenite (100 μM) or arsenate (100 μM). After being cultured for 24 h, the luciferase activity was measured. The results are presented as relative RARE activity. Each value indicates the mean and S.D. of 4 identically treated assay wells. *Indicates a significant increase from control ($p < 0.05$).

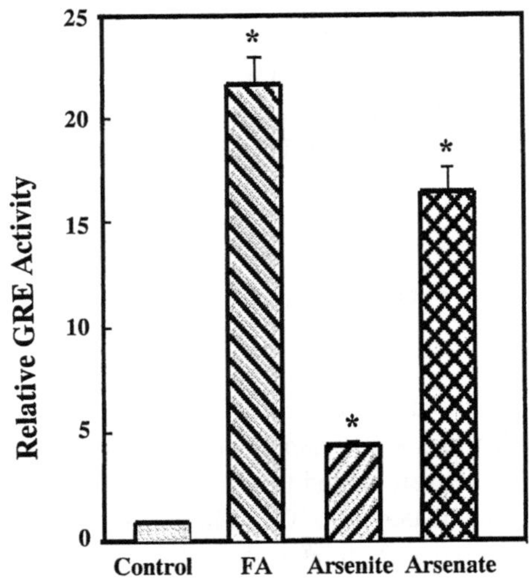

Fig. 4. Induction of GRE activity by arsenic in mouse epidermal JB6 cells. 8 × 10^3 Cl41 GRE mass1 cells were seeded into each well of 96-well plates. After being cultured at 37°C overnight, the cells were starved for 12 h by replacing medium with 0.1% FBS MEM. Then the cells were treated with FA (1 μM), arsenite (100 μM) or arsenate (100 μM). After being cultured for 24 h, the luciferase activity was measured. The results are presented as relative GRE activity. Each bar indicates the mean and S.D. of 4 identically treated assay wells. *Indicates a significant increase from control ($p < 0.05$).

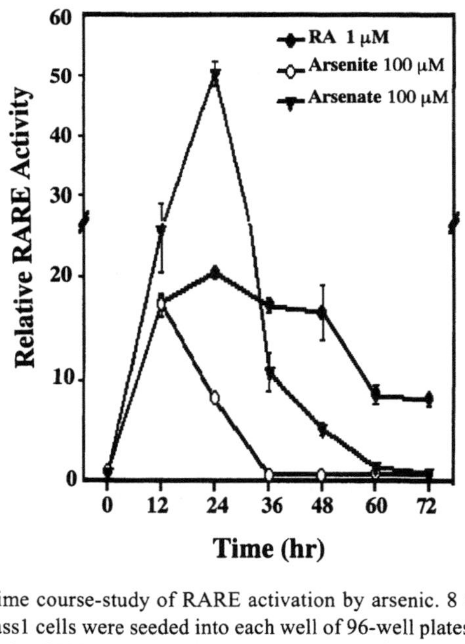

Fig. 3. Time course-study of RARE activation by arsenic. 8 × 10^3 Cl41 RARE mass1 cells were seeded into each well of 96-well plates. After being cultured at 37°C overnight, the cells were starved for 12 h by replacing medium with 0.1% FBS MEM. Then, the cells were treated with RA (1 μM), arsenite (100 μM) or arsenate (100 μM) for various times as indicated. The luciferase activity was measured as described in Materials and methods. The results are presented as relative RARE activity. Each value indicates the mean and S.D. of 4 identically treated assay wells.

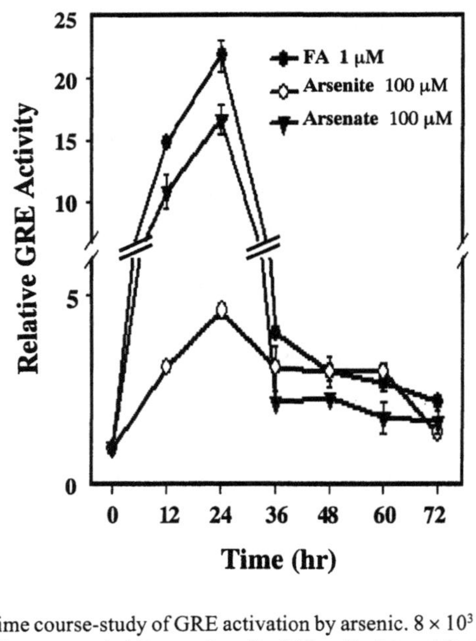

Fig. 5. Time course-study of GRE activation by arsenic. 8 × 10^3 Cl41 GRE mass1 cells were seeded into each well of 96-well plates. After being cultured at 37°C overnight, the cells were starved for 12 h by replacing medium with 0.1% FBS MEM. Then the cells were treated with FA (1 μM), arsenite (100 μM) or arsenate (100 μM) for various times as indicated and the luciferase activity was measured. The results are presented as relative GRE activity. Each value indicates the mean and S.D. of 4 identically treated assay wells.

that treatment of cells with arsenic caused a translocation of GR to the nucleus in an hormone independent manner [37, 38], but also demonstrate that arsenic induced the nuclear function of GR as a transcription factor.

Arsenic has synergistic effects on RA-induced RARE activation, but not on FA-induced GRE activation

It has been demonstrated that RA and arsenic have a synergistic effect on induction of NB4 cell differentiation [33]. To determine whether arsenic has synergistic effects on RA-induced RARE activation and FA-induced GRE activation, we exposed the cells to a combination of arsenic and RA or arsenic and FA. Results show that arsenic had a significant synergistic effect on RA-induced RARE activation (Fig. 6), while it did not show any effects on FA-induced GRE activation (data not shown). This result provides insight into the molecular mechanism involved in combined RA and arsenic therapy.

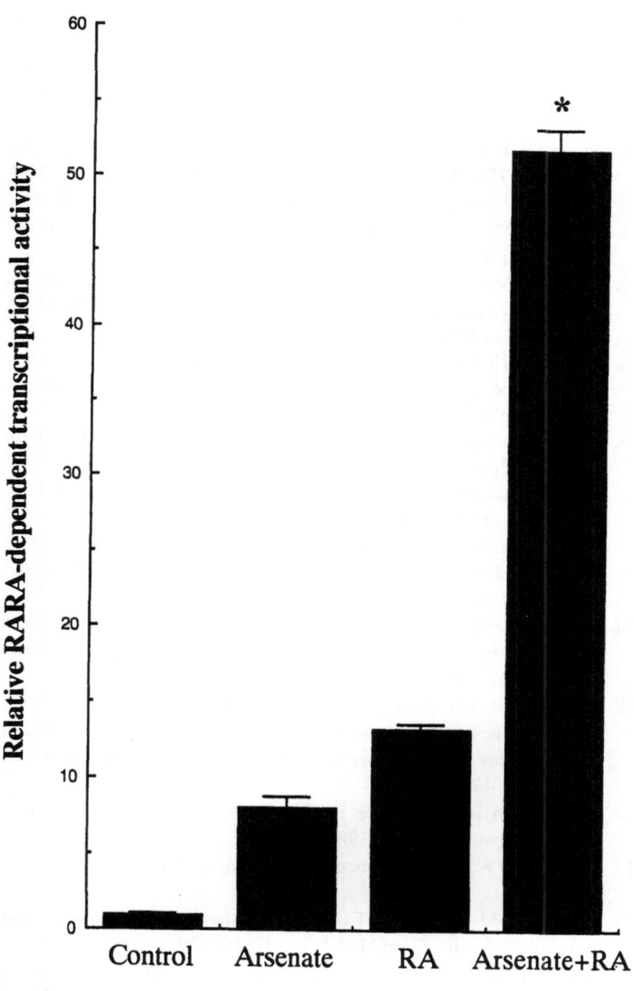

Fig. 6. Synergistic effects of arsenate on RA-induced RARE activation. 8 × 10^3 C141 RARE mass1 cells were seeded into each well of 96-well plates. After being cultured at 37°C overnight, the cells were starved for 12 h by replacing medium with 0.1% FBS MEM. Then the cells were treated with RA (1 μM) and arsenate (50 μM). After being cultured for 24 h, the luciferase activity was measured. The results are presented as relative RARE activity. Each bar indicates the mean and S.D. of 4 identically treated assay wells. *Indicates a significant increase from control ($p < 0.05$).

Discussion

In the present study, we investigated the effects of arsenic on RARE transactivation and GRE activation in mouse epidermal JB6 cell lines. The results show that treatment of mouse epidermal Cl41 cells with either arsenite or arsenate leads to significant transactivation of RARE and GRE. This activation appears to be time-dependent. The induction levels of activation of RARE and GRE by arsenate appears to be much higher than those by arsenite. The induction of RARE by arsenic could be synergistically enhanced with co-treatment of cells with RA, while arsenic did not show any effect on FA-induced GRE activation. These results demonstrate that both arsenite and arsenate could induce transactivation of RARE and GRE in mouse epidermal cells, and this effect may be responsible for induction of cell differentiation by arsenic.

Retinoids are a class of chemical compounds structurally related to vitamin A, which includes retinoic acid and its natural and synthetic analogs [34–36, 40]. Vitamin A has long been recognized as an indispensable nutritional factor necessary for the promotion of general growth, maintenance of visual function, regulation of differentiation of epithelial tissues, and embryonic development [34–36, 40]. The effect of vitamin A on epithelial tissue has attracted much attention for decades because vitamin A deficiency in animals and humans leads to hyperkeratosis of the skin and to hyperplastic and metaplastic changes in the epithelia of mucous membranes, which are observed in many skin diseases [34–37]. Previous studies have demonstrated that retinoids have a wide array of biological functions, such as inhibition of cell proliferation, induction of cell differentiation and apoptosis, immunomodulation, inhibition of angiogenesis, and influence on the expression of oncogenes and tumor suppressor genes [34–36, 40]. Retinoids can inhibit tumor cell growth and induce the differentiation and reversal of certain malignant cells to normal phenotype [34–36, 40]. Retinoic acid has been shown to be an effective treatment in inhibiting papilloma formation in the mouse model and tumor promoter-induced cell transformation in mouse epidermal JB6 cells [40, 41]. Thus, retinoids have been used for the treatment of many skin diseases, and as promising drugs for treatment of several cancers [34–41]. All of these biological activities of RA are believed to be mediated by transcriptional activation of RARE and inhibition of AP-1 activity, acting through distinct nuclear receptors, namely the retinoic acid receptors (RARs) and the retinoid X receptors (RXRs) [36, 40, 41]. The binding of RA to RARs and/or RXRs leads to formation and activation of a RAR-RXR heterodimer. These activated RAR-RXR heterodimer complexes bind to RARE and regulate its targeted gene expression in a RA-dependent manner [34, 36, 42]. In the present study, we found that exposure of cells to arsenic led to marked transactivation of RARE in a time-dependent manner. This result indicates that arsenic compounds are stimulators for RARE activation. Considering the role of RARE activation in cell differentiation and treatment of many skin diseases by RA, we speculate that the arsenic-induced RARE activation may play some role in arsenic-induced cell differentiation and the efficiency of arsenic in treatment of skin disease and cancers. Our study also demonstrates that arsenic had a significant synergistic effect on RA-induced RARE activation. This may explain the synergism of cell differentiation induced by combined arsenic and RA treatment, and may provide a mechanistic basis for combined RA and arsenic therapies.

The glucocorticoid hormones are known to serve a variety of important functions in cells and tissues [43, 44]. They can promote cell differentiation and inhibit cell proliferation [43, 44]. Previous studies have also shown that glucocorticoid hormones prevent TPA-induced inflammation, skin hyperplasia and skin tumor formation [40, 43, 44]. The cellular response to glucocorticoid hormones is mediated through highly specific cytoplasmic glucocorticoid receptors (GR). Upon activation by the binding of glucocorticoid hormones, GR form a homodimer and migrates to the nucleus of the cell, where the GR homodimer binds to a glucocorticoid response element (GRE) and regulates gene expression [44, 45]. During this study, we found that exposure of mouse epidermal cells to arsenic resulted in marked activation of GRE. Unlike RARE, arsenic-induced GRE activation was not enhanced by the synthetic glucocorticoid fluocinolone acetonide (FA). Because GRE response is involved in FA-mediated anti-inflammatory and anti-cancer effects, we suggest that arsenic-induced GRE activation may be associated with the molecular mechanism for arsenic treatment of skin diseases and prevention of cancers, in addition to induction of apoptosis and RARE.

Arsenic, used as a drug in traditional Chinese medicine, has attracted a great deal of attention in recent years because of its ability to induce complete remission of disease in patients with acute promyelocytic leukemia [17, 18, 33]. It is well known that arsenic is a widely used traditional Chinese medicine for treatment of various cancers and skin diseases [13–18]. It is proposed that the anti-carcinogenic activity of arsenic is mainly through the induction of cell apoptosis and differentiation [17, 18, 27, 29, 30]. Previous studies have reported that several molecular mechanisms may be responsible for the apoptosis induction by arsenic, such as a JNK-dependent pathway, targeting tubulins and down-regulation of Bcl-2 [18, 27, 28]. However, the molecular mechanisms for induction of cell differentiation by arsenic remain unclear. In this study, we demonstrate that treatment of mouse epidermal JB6 cells with arsenic caused marked activation of RARE and GRE, which are considered to play important roles in induction of cell differentiation by RA and FA, respectively [34–36, 44, 45]. Therefore, we suggest that arsenic-induced activation of RARE and GRE may play an important role in its induction of cell differentiation and may be involved in the therapeutic effects of arsenic on cancers and skin diseases. We will test this hypothesis in our future studies.

References

1. Chen C-J, Wang C-J: Ecological correlation between arsenic levels in well water and age-adjusted mortality from malignant neoplasms. Cancer Res 50: 5470–5474, 1990
2. Gilman JPW, Swierenga SHH: Inorganic carcinogenesis. In: C.E. Searle (ed). Chemical Carcinogenesis, 2nd edn. American Cancer Society Monograph, Washington, DC, 1985, pp 577–630
3. Smith AH, Hopenhayn-Rich C, Bates MN, Goeden HM, Hertz-Picciotto P, Duggan HM, Wood R, Kornett MJ, Smith MT: Cancer risks from arsenic in drinking water. Environ Health Perspect 97: 259–267, 1992
4. Snow ET: Metal carcinogenesis: Mechanistic implications. Pharmacol Ther 53: 31–65, 1992
5. Huang C, Ma W-Y, Li J, Goranson A, Dong Z: Requirement of Erks, but not JNKs, for arsenite-induced cell transformation. J Biol Chem 274: 14595–14601, 1999
6. Chan PC, Huff J: Arsenic carcinogenesis in animals and in humans: Mechanistic, experimental, and epidemiological evidence. Environ Carcinogen Ecotox Rev C15: 83–122, 1997
7. International Agency for Research on Cancer: IARC Monograph on Evaluation of Carcinogenic Risk to Humans 23: 39–324, 1980
8. International Agency for Research on Cancer: IARC Monograph on Evaluation of Carcinogenic Risks to Humans 7(suppl): 100–134, 1987
9. National Institute for Occupational Safety and Health: National Occupational Exposure Survey (1980–1983). Department of Health and Human Services, Cincinnati, OH, 1984
10. Lansdown ABG: Physiological and toxicological changes in the skin resulting from the action and interaction of metal ions. Crit Rev Toxicol 25: 397–462, 1995
11. Luchtrath H: The consequences of chronic arsenic poisoning among Moselle wine growers. Pathoanatomical investigations of post-mortem examinations performed between 1960 and 1977. J Cancer Res Clin Oncol 105: 173–182, 1983
12. Philipp R: Arsenic exposure: Health effects and the risk of cancer. Rev Environ Health 5: 27–57, 1985
13. Bainbridge WS: In: The Cancer Problem. Macmillan, New York, 1914, p 271
14. Forkner C, McNair-Scott TF: Arsenic as a therapeutic agent in chronic myeloid leukemia. J Am Med Assoc 97: 305, 1931
15. Kandel EV, Leroy GV: Chronic arsenic poisoning during the treatment of chronic myeloid leukemia. Arch Intern Med 60: 846–866, 1937
16. Konig A, Wrazel L, Warrell RP Jr, Rivi R, Pandolfi PP, Jakubowski A, Gabrilove JL: Comparative activity of melarsoprol and arsenic trioxide in chronic B-cell leukemia lines. Blood 90: 562–570, 1997
17. Chen GQ, Shi XG, Tang W, Xiong SM, Zhu J, Cai X, Han ZG, Ni JH, Shi GY, Jia PM, Liu MM, He KL, Niu C, Ma J, Zhang P, Zhang TD, Paul P, Naoe T, Kitamura K, Miller W, Waxman S, Wang ZY, de The H, Chen SJ, Chen Z: Use of arsenic trioxide (As2O3) in the treatment

of acute promyelocytic leukemia (APL): I. As2O3 exerts dose-dependent dual effects on APL cells. Blood 89: 3345–3353, 1997
18. Chen GQ, Zhu J, Shi XG, Ni JH, Zhong HJ, Si GY, Jin XL, Tang W, Li XS, Xong SM, Shen ZX, Sun GL, Ma J, Zhang P, Zhang TD, Gazin C, Naoe T, Chen SJ, Wang ZY, Chen Z: *In vitro* studies on cellular and molecular mechanisms of arsenic trioxide (As2O3) in the treatment of acute promyelocytic leukemia: As2O3 induces NB4 cell apoptosis with downregulation of Bcl-2 expression and modulation of PML-RAR alpha/PML proteins. Blood 88: 1052–1061, 1996
19. Jiang Su New Medical College: In: Encyclopedia of Chinese Medicine. Shanghai Scientific Publishing House, Shanghai, China, 1986, pp 1620–1622
20. Snow ET: Metal carcinogenesis: Mechanistic implications. Pharma Ther 53: 31–65, 1992
21. Chen NY, Ma WY, Huang C, Ding M, Dong D. Activation of PKC is required for arsenite-induced signal transduction. J Environ Pathol Toxicol Oncol 19: 297–305, 2000
22. Cavigelli M, Li WW, Lin A, Su B, Yoshioka K, Karin M: The tumor promoter arsenite stimulates AP-1 activity by inhibiting a JNK phosphatase. EMBO J 15: 6269–6279, 1996
23. Ludwig S, Hoffmeyer A, Goebeler M, Kilian K, Häfner H, Neufeld B, Han J, Rapp UR: The stress inducer arsenite activates mitogen-activated protein kinases extracellular signal-regulated kinases 1 and 2 via a MAPK kinase 6/p38-dependent pathway. J Biol Chem 273: 1917–1922, 1998
24. Simeonova PP, Luster MI: Mechanisms of arsenic carcinogenicity: Genetic or epigenetic mechanisms? J Environ Pathol Toxicol Oncol 19: 281–286, 2000
25. Simeonova PP, Wang S, Toriuma W, Kommineni V, Matheson J, Unimye N, Kayama F, Harki D, Ding M, Vallyathan V, Luster MI: Arsenic mediates cell proliferation and gene expression in the bladder epithelium: Association with activating protein-1 transactivation. Cancer Res 60: 3445–3453, 2000
26. Huang C, Li J, Ma W-Y, Chen N-Y, Dong Z: Activation of AP-1 by arsenate and arsenate in both cell and AP-1 luciferase-transgenic mice. FASEB J (in press)
27. Huang C, Ma WY, Li J, Dong Z: Arsenic induces apoptosis through a c-Jun NH2-terminal kinase-dependent, p53-independent pathway. Cancer Res 59: 3053–3058, 1999
28. Li YM, Broome JD: Arsenic targets tubulins to induce apoptosis in myeloid leukemia cells. Cancer Res 59: 776–780, 1999
29. Rousselot P, Labaume S, Marolleau JP, Larghero J, Noguera MH, Brouet JC, Fermand JP: Arsenic trioxide and melarsoprol induce apoptosis in plasma cell lines and in plasma cells from myeloma patients. Cancer Res 59: 1041–1048, 1999
30. Zhu XH, Shen YL, Jing YK, Cai X, Jia PM, Huang Y, Tang W, Shi GY, Sun YP, Dai J, Wang ZY, Chen SJ, Zhang TD, Waxman S, Chen Z, Chen GQ: Apoptosis and growth inhibition in malignant lymphocytes after treatment with arsenic trioxide at clinically achievable concentrations. J Natl Cancer Inst 91: 772–778, 1999
31. Huang C, Ma W-Y, Bowden GT, Dong Z: Ultraviolet B-induced activated protein-1 activation does not require epidermal growth factor receptor but is blocked by a dominant negative PKCλ/ι. J Biol Chem 49: 31262–31268, 1996
32. Huang C, Ma W-Y, Young MR, Colburn N, Dong Z: Shortage of mitogen-activated protein kinase is responsible for resistance to AP-1 transactivation and transformation in mouse JB6 cells. Proc Natl Acad Sci USA 95: 156–161, 1998
33. Gianni M, Koken MH, Chelbi-Alix MK, Benoit G, Lanotte M, Chen Z, de The H: Combined arsenic and retinoic acid treatment enhances differentiation and apoptosis in arsenic-resistant NB4 cells. Blood 91: 4300–4310, 1998
34. Chambon P: A decade of molecular biology of retinoic acid receptors. FASEB J 10: 940–954, 1996
35. Glass CK, Rosenfeld MG, Rose DW, Kurokawa R, Kamei Y, Xu L, Torchia J, Ogliastro MH, Westin S: Mechanisms of transcriptional activation by retinoic acid receptors. Biochem Soc Trans 25: 602–605, 1997
36. Lehmann JM, Jong L, Fanjul A, Cameron JF, Lu XP, Haefner P, Dawson MI, Pfahl M: Retinoids selective for retinoid X receptor response pathways. Science 258: 1944–1946, 1992
37. Simons SS Jr, Chakraborti PK, Cavanaugh AH: Arsenite and cadmium (II) as probes of glucocorticoid receptor structure and function. J Biol Chem 265: 1938–1945, 1990
38. Sanchez ER: Heat shock induces translocation to the nucleus of the unliganded glucocorticoid receptor. J Biol Chem 267: 17–20, 1992
39. Hamilton JW, Kaltreider RC, Davis AM, Green RA: Arsenite (II) and chromium (VI) alter glucocorticoid receptor (GR) function and GR-dependent gene regulation. Metal Ions Biol Med 6: 66–68, 2000
40. Huang C, Ma W-Y, Dawson MI, Rincon M, Flavell RA, Dong Z: Blocking activator protein-1 activity, but not activating retinoic acid response element, is required for the antitumor promotion effect of retinoic acid. Proc Natl Acad Sci USA 94: 5826–5830, 1997
41. Li JJ, Dong Z, Dawson MI, Colburn NH: Inhibition of tumor promoter-induced transformation by retinoids that transrepress AP-1 without transactivating retinoic acid response element. Cancer Res 56: 483–489, 1996
42. Minucci S, Leid M, Toyama R, Saint-Jeannet JP, Peterson VJ, Horn V, Ishmael JE, Bhattacharyya N, Dey A, Dawid IB, Ozato K: Retinoid X receptor (RXR) within the RXR-retinoic acid receptor heterodimer binds its ligand and enhances retinoid-dependent gene expression. Mol Cell Biol 17: 644–655, 1997
43. Budunova IV, Carbajal S, Kang H-I, Viaje A, Slaga TJ: Altered glucocorticoid receptor expression and function during mouse skin carcinogenesis. Mol Carcinogen 18: 177–185, 1997
44. Adcock IM: Molecular mechanisms of glucocorticosteroid actions. Pulm Pharmacol Ther 13: 115–126, 2000
45. Weigel NL: Steroid hormone receptors and their regulation by phosphorylation. Biochem J 319: 657–667, 1996

Molecular and Cellular Biochemistry **222**: 127–136, 2001.

Orthopaedic implant related metal toxicity in terms of human lymphocyte reactivity to metal-protein complexes produced from cobalt-base and titanium-base implant alloy degradation

Nadim J. Hallab, Katalin Mikecz, Csaba Vermes, Anastasia Skipor and Joshua J. Jacobs

Department of Orthopedic Surgery, Rush-Presbyterian-St. Lukes Medical Center, Chicago, IL, USA

Abstract

Metal toxicity from sources such as orthopaedic implants was investigated in terms of immune system hyper-reactivity to metal implant alloy degradation products. Lymphocyte response to serum protein complexed with metal from implant alloy degradation was investigated in this *in vitro* study using primary human lymphocytes from healthy volunteers (n = 10). Cobalt chromium molybdenum alloy (Co-Cr-Mo, ASTM F-75) and titanium alloy (Ti-6Al-4V, ASTM F-136) beads (70 µm) were incubated in agitated human serum at 37 degrees Celsius to simulate naturally occurring metal implant alloy degradation processes. Particulate free serum samples, which were incubated with metal, were then separated into molecular weight based fractions. The amounts of soluble Cr and Ti within each serum fraction were measured and correlated with lymphocyte proliferation response to the individual serum fractions. Lymphocytes from each subject were cultured with 11 autologous molecular weight based serum fractions either with or without added metal. Two molecular weight ranges of human serum proteins were associated with the binding of Cr and Ti from Co-Cr-Mo and Ti implant alloy degradation (at < 30 and 180–330 kDa). High molecular weight serum proteins (≈ 180 kDa) demonstrated greater lymphocyte reactivity when complexed with metal released from Co-Cr-Mo alloy and Ti alloy than with low (5–30 kDa) and midrange (30–77 kDa) serum proteins. When the amount of lymphocyte stimulation was normalized to both the moles of metal and the moles of protein within each fraction (Metal-Protein Complex Reactivity Index, MPCRI), Cr from Co-Cr-Mo alloy degradation demonstrated approximately 10 fold greater reactivity than Ti in the higher molecular weight serum proteins (≈ 180–250 kDa). This *in vitro* study demonstrated a lymphocyte proliferative response to both Co-Cr-Mo and Ti alloy metalloprotein degradation products. This response was greatest when the metals were complexed with high molecular weight proteins, and with metal-protein complexes formed from Co-Cr-Mo alloy degradation. (Mol Cell Biochem **222**: 127–136, 2001)

Key words: bioreactivity, immune response, implants, biocompatibility, metal protein complex, lymphocyte

Introduction

Metal toxicity is not a phenomena typically associated with orthopedic implants. This may be due in large part to the enormous success of total joint replacements in restoring mobility and quality of life to over a million people worldwide every year. However over the long term (> 7 years), total joint replacements have been associated with adverse local and remote tissue responses. Moreover, while these responses are largely ascribed to particulate-based inflammatory reactions of macrophages the degree to which these reactions are mediated by classical metal toxicity responses, is not known.

Address for offprints: N.J. Hallab, Department of Orthopedic Surgery, Rush Medical College, 1653 W. Congress Parkway, Chicago, IL 60612, USA

The degradation products of prosthetic biomaterials mediate these adverse effects [1, 2]. A significant amount of biomaterial degradation takes place over time resulting in locally and systemically elevated levels of metal [3–6]. Normal human serum levels of prominent implant metals are shown in Table 1 [3–9]. Homogenates of remote organs and tissue obtained postmortem from subjects with cobalt and titanium base alloy total joint replacement components have indicated that significant increases in Co, Cr, Ti and Al concentrations can occur in the heart, liver, kidney, spleen, and lymphatic tissue (see Table 2) [10–12]. This metallic debris may exist as micrometer to nanometer size particles, colloidal and ionic forms of metal (specifically or non-specifically bound by protein) [13], inorganic metal salts/oxides, and/or in an organic storage form such as hemosiderin. Elevated concentrations of circulating metal derived from orthopedic implant degradation may have direct and accumulatory toxicologic effects over the long term. However, the significance of these circulating metal implant degradation products remains unknown in terms of their short and long term toxicologic/pathophysiologic effects and bioavailability *in vivo*. One effect associated with metal implant components, which until now has remained uncharacterized as a form of metal toxicity, is the unpredictable inducement of overaggressive immunologic responses (specifically of type IV, delayed type hypersensitivity) [14–19] leading to untoward clinical effects such as periprosthetic bone loss. Metals reported as sensitizers have included, Ni, Co, Cr and Ti, all constituents of orthopaedic implants [20]. The immunogenic potential of metal degradation products, which purportedly form hapten or hapten-like complexes with serum proteins, has been previously demonstrated [21, 25].

The specific T cell subpopulations, the cellular mechanism of recognition/activation and the antigenic metal-protein determinants created by these metals remain incompletely characterized. The subsets of participating lymphocytes of nickel sensitive individuals were found to be primarily CD4+ and CD45RO+ cells, whereas CD8+ and CD8+CD11b+ lymphocytes were shown to be underrepresented [26]. Sensitive T-cells have been shown to recognize metals such as nickel in the context of major histocompatibility complex (MHC) Class II molecules [26, 27]. The Langerhans cells of the dermis are well characterized as the primary antigen presenting cells (APC) associated with dermal hypersensitivity. The dominant APC (if any) responsible for mediating an implant related hypersensitivity response remains unknown. Candidate APC's in the periprosthetic region include macrophages, endothelial cells, lymphocytes, langerhans, dendritic cells and to lesser extent parenchymal tissue cells. While there is general consensus implicating the T cell receptor (TCR) in metal induced activation, there are conflicting reports of which region/receptor specificity is responsible for dominating metal reactivity [26–30]. Some investigators report no preferential receptor selection [26], while others have shown the CDR3B region of the VB17+ T cell receptor to be critical in the sense that without this region metal reactivity is abrogated [28, 29]. Metals have also been shown to act as facilitating agents in the crosslinking of receptors (e.g. VB17 of CDR1

Table 1. Approximate average concentrations of metal in human body fluids with and without total joint replacement [3–9]

		Ti	Al	V	Co	Cr	Mo	Ni
					(ng/ml or ppb)			
Serum	Normal	2.7	2.2	<0.8	0.18	0.05	*	0.4–3.6
	THA	4.4	2.4	1.7	0.2–0.6	0.3	*	<9.1
	THA-F	8.1	2.2	1.3	*	0.2	*	*
	TKA	3.2	1.9	<0.8	*	*	*	*
	TKA-F	135.6	3.7	0.9	*	*	*	*
Urine	Normal	<1.9	6.4	0.5	*	0.06	*	*
	TJA	3.55	6.53	<0.4	*	0.45	*	*
Synovial fluid	Normal	13	109	5	5	3	21	5
	TJA	556	654	62	588	385	58	32
Joint capsule	Normal	723	951	122	25	133	17	3996
	TJA	1540	2053	288	1203	651	109	2317
	TJA-F	19173	1277	1514	821	3329	447	5789
Whole blood	Normal	17	13	6	0.1–0.1.2	2.0–4.0	0.5–1.8	2.9–7.0
	TJA	67	218	23	20	110	10	29

Normal – subjects without any metallic prosthesis (not including dental); THA – subjects with well functioning total hip arthroplasty; THA-F – subjects with a poorly functioning total hip arthroplasty (needing surgical revision); TKA – subjects with well functioning total knee arthroplasty; TKA-F – subjects with a poorly functioning total knee arthroplasty (needing surgical revision); TJA – subjects with well functioning total joint arthroplasty; TJA-F – subjects with a poorly functioning total joint arthroplasty (needing surgical revision). *Not tested.

Table 2. Concentrations of metal in body tissue of humans with and without total joint [10–12]

		Cr	Co	Ti	Al	V
				replacements (μg/g)		
Skeletal muscle	Normal	<12	<12	*	*	*
	TJA	570	160	*	*	*
Liver	Normal	<14	120	100	890	14
	TJA	1130	15200	560	680	22
Lung	Normal	*	*	710	9830	26
	TJA	*	*	980	8740	23
Spleen	Normal	10	30	70	800	<9
	TJA	180	1600	1280	1070	12
Psuedocapsule	Normal	150	50	<65	120	<9
	TJA	3820	5490	39400	460	121
Kidney	Normal	<40	30	*	*	*
	TJA	<40	60	*	*	*
Lymphatic tissue	Normal	690	10	*	*	*
	TJA	690	390	*	*	*
Heart	Normal	30	30	*	*	*
	TJA	90	280	*	*	*

TJA – Subjects with a well functioning total joint arthroplasty. *Not tested.

T-cell receptor) to create superantigen-like enhancement of T cell receptor-protein contact [28, 29], whereby metalloproteins or metal-peptide complexes that would not otherwise be antigenic, are able to provoke a response. Furthermore, others have shown that entirely independent of an altered 'self' protein antigen, metal has been reported to crosslink thiols of cell surface proteins of murine thymocytes (i.e. CD3, CD4 and CD45) which have been reported to result in the activation of a tyrosine kinase (p56lck), involved with the activation of T cells through the T cell receptor (TCR) [14, 19, 31, 32]. However, despite reports of non-hapten related mechanisms of metal induced lymphocyte activation, clonal lymphocyte specificity associated with type IV delayed type hypersensitivity remains the dominant mechanism associated with implant related hypersensitivity responses [14, 19, 32]. It remains unknown which metal-protein complexes associated with orthopedic biomaterials are responsible for mediating untoward immunologic effects.

In the current investigation, we hypothesize that metal degradation products are immunologically reactive when complexed with proteins and that this reactivity is a function of the composition of the metal and the binding protein(s). To test this hypothesis, isolated primary healthy human lymphocyte/monocyte populations were treated with 11 molecular weight ranges of fractionated serum preconditioned with Co-Cr-Mo alloy (ASTM F-75) and titanium alloy (Ti-6Al-4V, F-136) degradation products. Lymphocyte bioreactivity (as measured by lymphocyte proliferation) to serum protein fractions with and without metal degradation products was then determined as was the total protein within serum fractions and the concentration of Cr and Ti within these serum fractions.

Materials and methods

Serum and lymphocytes were obtained from 10 individuals (5 female, 5 male, average age 34, range 23–59) by peripheral venipuncture after obtaining (Internal Review Board approved) informed consent. All subjects were healthy, with no history of metal allergy and no history of receiving a metal implant (other than dental). Serum samples were allowed to complex with metal released from Co-Cr-Mo and Ti alloy degradation, then separated based on molecular size into 11 fractions. These fractions were tested for Cr or Ti metal content, concentrated, and used in cell cultures of lymphocytes isolated from the same subject. Lymphocyte responses were then measured using a proliferation (or lymphocyte transformation) assay.

Blood collection and cell isolation

Human primary lymphocytes and monocytes were isolated from 30 ml of peripheral blood obtained by venous puncture using a 30 ml heparinized syringe (10 units/ml blood). Density gradient separation (Ficoll-isopaque, Pharmacia, Piscataway, NJ, USA) was used to obtain mononuclear leukocyte cell fractions from heparinized whole blood, yielding between approximately 15–30 × 10^6 mononuclear leukocytes per 30 milliliters of heparinized blood collected per subject.

Metal-protein complex formation

Metal-protein complexes were produced by incubating 0.5 ml of spherical metal particles (70 μm diameter, Starmet Corp., Concord, MA, USA) in 4 ml of serum under constant rocking motion (LabQuake, Sigma Co.) at 37°C for 1 week. For each subject, 4 ml of serum was used to incubate with titanium alloy (Ti6Al4V, ASTM F-136) and Co-Cr-Mo alloy (ASTM F-75) beads (70 μmm diameter). Residual colloidally suspended metal particles were removed from the serum samples using centrifugation at 8,000 rpm for 30 min (Sorvall, DuPont) and sterile filtration through a 0.22 μm filter. Metal-treated and control untreated serum samples were fractionated using FPLC and analyzed for metal content.

Serum fractionation

Serum protein samples were separated into 11 fractions based on molecular size, using fast-protein-liquid-chromatography (FPLC). FPLC, using an all polymer environment, was selected for serum protein separation to minimize metal contamination by processing. The FPLC system was verified to be contamination free through preliminary testing of the eluent. A P-500 Pump (Pharmacia, Piscataway, NJ, USA) was used to pump 0.5 ml samples of serum through two 24 ml Superdex™ 200 columns (Pharmacia, Piscataway, NJ, USA) mounted in series at a flow rate of 5 ml/h, using double deionized water (> 18 MOhms) from a Milli-Q water system (Millipore, Bedford, MA, USA). Other Ultrex grade eluent buffers such as KPO_4 and NaOH were found to contain metal within the parts per million range, precluding their use in these studies. Detection of protein exiting the columns was carried out using ultra-violet absorbence monitoring (UV-1 cell, Pharmacia, Piscataway, NJ, USA). High molecular weight standard protein calibrants (Pharmacia, Piscataway, NJ, USA) were used both with FPLC and in conjunction with polyacrylamide gel electrophoresis to determine the approximate molecular weight of prominent proteins within each FPLC fraction. Metal content within FPLC serum fractions was analyzed using a graphite furnace Zeeman atomic absorption spectrophotometer (GFZ-AAS) (Perkin-Elmer, Norwalk, CT, USA). Serum fractions of 2.5 ml were concentrated using 5 kDa concentration tubes (Centricon, Millipore) at 2000 rpm for 30 min, for a final volume of 50 μl. The method detection limit after FPLC was 0.6 ng/ml for Cr and 40 ng/ml for Ti (detection limits in undiluted serum are 0.03 ng/ml for Cr and 2.0 ng/ml for Ti) [3–6]. All collection containers and apparati were triple acid-washed with Ultrex-grade chemicals (Baker, Chicago, IL, USA) or verified to be contamination free by AAS.

The analysis of metal content within fractionated and unfractionated serum was limited to Cr and Ti as representative metals of Co-Cr-Mo and Ti alloy degradation, respectively. These metals were chosen as representatives because they were the two most prominently elevated metals of each alloy detected in the serum of patients with normally functioning and failed total joint replacements (see Table 1). Other metals such as Co, although more prominent in the composition of Co-Cr-Mo alloy (see Table 3), were not as elevated as Cr in the serum of TJR patients, presumably due to high levels of clearance from the serum.

Proliferation assay (lymphocyte transformation tests)

All of the 15–30 × 10^6 mononuclear leukocytes obtained from each patient were used for proliferation assay analysis. This limited biologic reactivity analysis to this single assay. Proliferation of cells was measured by [^{3}H]-thymidine (Amersham International, Arlington Heights, IL, USA) incorporation into DNA in a 96-well microplate system. The amount of proliferation resulting from each treatment was quantified using the average of three radioactivity incorporation values. This average for each treatment was normalized to that of the negative control (no treatment) producing a ratio, generally termed either a proliferation factor, proliferation index, proliferation ratio or stimulation index, SI. In this paper this ratio will be referred to as the stimulation index (SI). The SI was used to compare lymphocyte reactivity to the two different

Table 3. Weight percent of different metals within the three most prominent implant alloys

Alloy	% Ni	% Ta	% Co	% Cr	% Ti	% Mo	% Al	% Fe	% Mn	% Cu	% V
Stainless steel (ASTM F138)	13–15.5	*	*	17–19	*	2–4	*	Balance	*	0.5	*
Co Alloy (ASTM F75)	1	*	Balance	27–30	*	5–7	*	*	1	*	*
Ti Alloy (ASTM F136)	*	*	*	*	Balance	*	5.5–6.5	*	*	*	3.5–4.5

*indicates less than 0.1.

metals as well as the metal-protein complexes. The lower limit of this stimulation index is zero indicating all cells stopped dividing before addition of [^{3}H]-thymidine, after 5½ days.

Proliferation assays were performed using Ficol separated mononuclear/lymphocyte cell fractions collected from 30 milliliters of peripheral blood per subject. These lymphocytes were cultured in 96-well cell-culture plates (Sigma), at a density of 0.1–0.3 × 10^6 cells/well for a period of 6 days in 150 μL of DMEM/well, 10% autologous serum at 37°C and 0.5% CO_2, with either metal treatments (i.e. 15 μl/well of fractionated and concentrated metal treated serum), a positive control (0.01 mg/ml PHA) and a negative control (untreated). Each treatment was conducted in triplicate (3 wells/treatment). Approximately 72 wells of a 96-well plate were used per subject (11 treatments in triplicate for Co- and Ti-alloy treated serum, 3 positive and negative controls). [^{3}H]-thymidine was added during the last 12 h of incubation after 5½ days of treatment. At day six [^{3}H]-thymidine uptake (1 μCi/culture well) was measured using liquid scintillation. The SI was calculated using measured radiation counts per minute (cpm):

$$\text{Simulation Index} = (\text{mean cpm with treatment})/(\text{mean cpm without treatment}) \quad (1)$$

Six days of incubation were chosen to reproduce, *in vitro*, the time lag associated with *in vivo* lymphocyte proliferation in a DTH response. Radiolabeled lymphocytes were collected onto membranes using a cell harvester (Tomtech Mach 2, Orange, CT, USA) and the amount of differential radiation incorporation was measured using liquid scintillation (Wallac 1205 Betaplate, Gaithersburg, MD, USA).

Total Protein Assay

Assessment of total protein was conducted on serum post fractionation using Pierce Protein reagents (Rockford, IL, USA). Total protein assays were conducted using 10 μl of sample with 200 μl of substrate in 96-well plates, with serial dilutions of bovine serum albumin (Sigma) used as standard calibrants on every plate. The method detection limit was approximately 5 mg/ml total protein.

Statistical analysis

Metal content analysis

By convention, to calculate group means, Cr or Ti concentrations below the detection limit were expressed as one-half the method detection limit. Intergroup comparisons, independent of these means, were made using Kruskall-Wallis non-parametric analysis of variance. The Wilcoxon-Mann-Whitney test was then used if the Kruskall-Wallis test revealed significant differences at $p < 0.05$.

Lymphocyte reactivity

Measured data were subjected to statistical analysis using Student's *t*-tests. Student's *t*-tests for independent samples with unequal or equal variances were used to test equality of the mean values at a 95% confidence interval ($p < 0.05$). Comparisons between group reactivities are limited to individual comparison of reactivity at each specific molecular weight ranges. All treatment specific reactivity measurements were determined to be normally distributed.

Results

Serum separation

Eleven fractions of sequential molecular weight ranges were collected from each subject's serum, after treatment with Co-Cr-Mo alloy beads, Ti alloy beads or no metal. The average amount of total protein within each fraction (as measured in pretreated fractionated serum of all 10 subjects) is shown in Fig. 1. Note that over 80% of all serum protein was contained within fractions from 32–77 kDa while 15% was within the 180–250 kDa fraction. The greatest amount of total protein was detected in the 60–68 kDa fraction, which contained over 50% of the serum protein.

Chromium levels within serum

Cr levels were determined within the Co-Cr-Mo alloy treated and non-treated serum fractions of all subjects. The levels of Cr were used as an indicator of which protein fraction(s) preferentially bind soluble metal released from Co-Cr-Mo alloy degradation and to what degree. The average Cr level within unfractionated serum was 680 ng/ml (ppb) after treated with metal beads. The Cr levels within treated fractionated serum were measured and averaged for all subjects and compared to those from non-treated fractionated serum from the same individual (Fig. 2). Chromium released from Co-Cr-Mo particles exhibited a bimodal binding pattern, binding to both low molecular weight protein(s) (< 32 kDa), and to higher molecular weight protein(s) in the 180–330 kDa range (Fig. 2). Cr values within the nontreated serum fractions were two orders of magnitude lower than the treated serum concentrations and demonstrated less distinct peaks in protein binding. Levels of Cr within the serum treated fractions were statistically elevated above that of the nontreated serum fractions (as indicated in Fig. 2). Serum proteins of approximately 77–180 kDa did not appear to bind Cr degradation products.

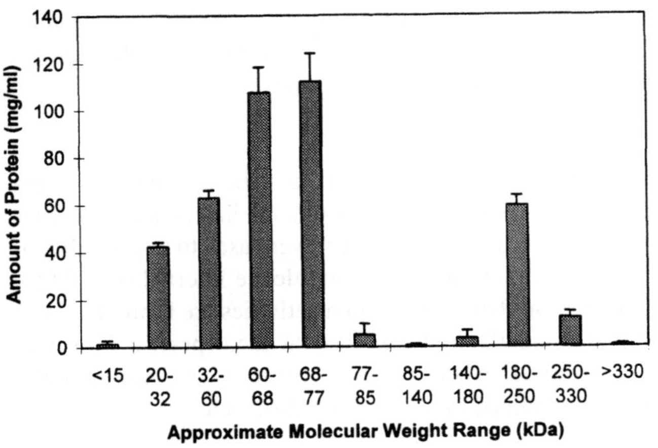

Fig. 1. Mean of total protein content within each of the molecular weight-based serum fractions of all subjects after fast protein liquid chhromatography. Note: The method detection limit is approximately 5 mg/ml. Note: error bars represent S.D.

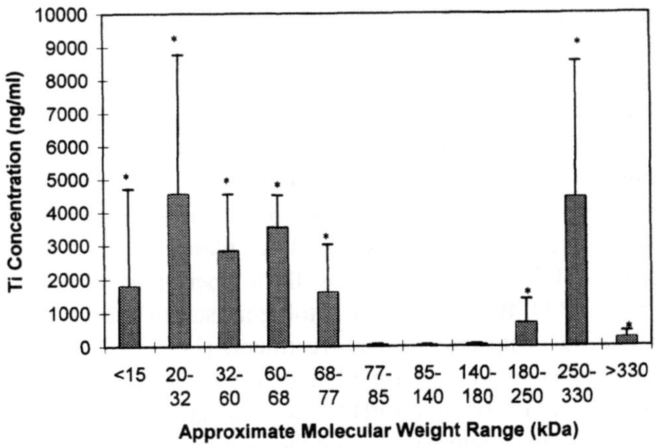

Fig. 3. Average Ti content in serum fractions after Ti-alloy bead treatment. There is a bimodal pattern of protein binding. The Ti concentration within each fraction of non-treated serum was below the method detection limit (40 ng/ml Ti). Note: *indicates significant increase above nontreated fractions (below the detection limit, not shown) at $p < 0.05$. Error bars represent S.D.

Titanium levels within serum

Titanium levels within each of the serum fractions after treatment with Ti alloy beads demonstrated similar distribution to that of Cr with a bimodal pattern of Ti binding (Fig. 3). The average Ti level within unfractionated serum was 1,700 ng/ml (ppb) after treated with Ti alloy metal beads. Ti levels within all nontreated samples, both pre- and post-fractionation, were below the method detection limit of the atomic adsorption spectrometer (< 40 ng/ml Ti post FPLC). Levels of Ti within the 15–77 and 180–330 kDa fractions were statistically elevated above that of the nontreated serum fractions as indicated in Fig. 3. The greatest levels of Ti were concentrated within the low to mid-molecular weight ranges (15–77 kDa) containing approximately 4800 ng/ml. These concentrations of Ti were over three orders of magnitude above that of nontreated serum. Serum proteins of approximately 77–180 kDa did not bind Ti, in a similar pattern as that found with Cr released from Co-Cr-Mo implant degradation.

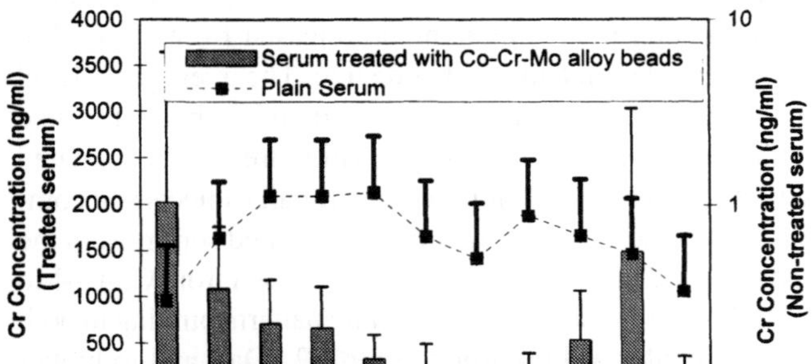

Fig. 2. Comparison of the average Cr content of serum fractions both with and without Co-Cr-Mo alloy bead. A bimodal pattern of binding can be seen for Cr in the Co-Cr-Mo alloy treated serum fractions. Note: The method detection limit is 0.6 ng/ml Cr. All treated fractions had significantly greater Cr than non-treated fractions at a $p < 0.03$. Error bars represent S.D.

Lymphocyte reactivity

The stimulation index for both Co-Cr-Mo and Ti alloy treated fractions were averaged for all subjects and the resulting pattern of activation is shown in Fig. 4. A stimulation index greater than 1.0 indicates lymphocyte activation (i.e. proliferation greater than that of non-treated control serum fractions), while a stimulation index less than 1.0 indicates lymphocyte suppression. Generally, lymphocytes demonstrated greater reactivity to: (1) serum fractions with metal compared to those without, (2) high molecular weight serum fractions (140–330 kDa), and (3) serum fractions with Co-Cr-Mo degradation products as compared with those with Ti alloy degradation products (Fig. 4). Lymphocytes were most greatly stimulated by serum fractions with Cr at 140–330 kDa. All subjects lymphocytes demonstrated increased reactivity to 0.01 mM PHA, exhibiting a range of 10–100 fold increase in thymidine incorporation (cpm) compared to nontreated lymphocytes from the same individual. PHA induced an average SI of 20 (Fig. 4).

Activated lymphocytes demonstrated the greatest proliferation (SI) response to higher molecular weight serum fractions within the 140–330 kDa range, Fig. 4. However, the stimulation index used to measure this proliferation does not take into account the amount of metal or protein within each serum fraction. Normalizing for the amount of metal and pro-

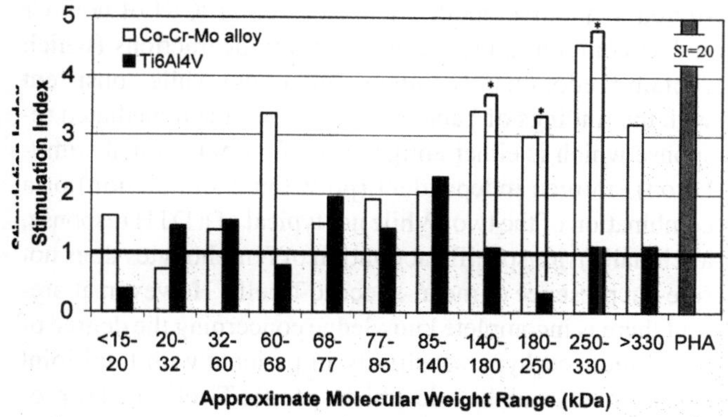

Fig. 4. The mean stimulation index of lymphocytes to autologous serum fractions which were treated with Co-Cr-Mo and Ti alloy beads and compared to nontreated serum fraction. An SI greater than 1.0 indicates a proliferation response. Serum treated with Co-Cr-Mo alloy beads demonstrated significantly greater reactivity within the ranges of 140–330 kDa, *indicates $p < 0.05$, Student's *t*-test. Note: Phytohemaglutin (PHA) was used as a positive control and resulted in an SI = 20, as indicated.

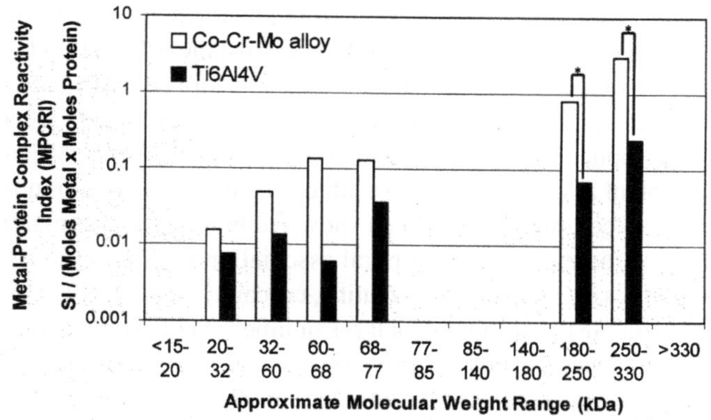

Fig. 5. A comparison of the Metal-Protein Complex Reactivity Index, (MPCRI) for all subjects (n = 10) in reaction to autologous serum fractions which were (1) non-treated, (2) Co-Cr-Mo alloy treated and (3) Ti alloy treated. Note: This is an approximate measure of reactivity based on an average molecular weight determined for each serum fraction and an average of total protein content within that fraction. The MPCRI could not be calculated where concentrations of metal or protein were below method detection limits (i.e. serum fractions < 15, 77, 85, 140, and > 330 kDa). Note: *indicates $p < 0.05$, Student's *t*-test.

tein within each serum fraction facilitates comparative analysis between metal-protein complexes comprised of either different serum proteins or different metals. Therefore, a measure of stimulation/activation was calculated to adjust for both the number of moles of metal and the corresponding moles of protein within the treated fractions. This measure, the Metal-Protein Complex Reactivity Index, (MPCRI), was calculated as shown below.

$$\text{MPCRI} = (\text{Stimulation index})/(\underset{\text{moles/ml}}{\text{Molar concentration of metal}} \times \underset{\text{moles/ml}}{\text{Molar concentration of protein}}) \quad (1)$$

The MCPRI values for Cr and Ti treated lymphocytes are shown in Fig. 5. The MPCRI (Fig. 5) differs from that of the SI (Fig. 4) in the relative levels of activation of Cr and Ti and for both low and high molecular weight proteins. The MPCRI demonstrates less reactivity for midrange proteins (such as albumin and transferrin) due to the large amount of total protein in fractions 3, 4, and 5, (32–77 kD). Cr containing higher molecular weight proteins, dominated by immunoglobulins, are associated with the greatest reactivity.

Discussion

This study demonstrates that metal-protein-complexes produced from the degradation of metal implant alloys can be immunologically reactive *in vitro* in that they are capable of inducing lymphocyte activation through proliferation responses. This hyper-reactivity as a severe or subtle chronic immune response has the potential to promote local and systemic toxicity. Moreover, different serum protein fractions and different metals demonstrate differential patterns of bioreactivity. The specific mechanisms by which metal degradation products activate lymphocytes remains unclear. The three current non-mutually exclusive mechanisms by which metal-protein complexes can activate lymphocytes are: (1) antigen-independent, (2) antigen-dependent and (3) superantigen-like, i.e. a synergistic combination of the first two mechanisms [14, 19]. Experimental support of all three mechanisms has been previously described [14, 19].

Independent of antigen

Metal has been reported to cross-link thiols of cell surface molecules of murine thymocytes (i.e. CD3, CD4 and CD45) which resulted in the activation of a tyrosine kinase (p56lck), involved with the activation of T cells through the T cell receptor (TCR). In addition to these reports of polyclonal activation, other reports implicate anti-chemotactic non-specific immune suppression [14, 19, 32–40].

Antigen dependent metal activation

Metal specific T-cell reactivity remains the primary mechanism purportedly associated with metal-lymphocyte activation [14, 19, 32, 38, 39, 41]. Implant related hypersensitivity reactions have long been considered to be a cell mediated delayed type IV hypersensitivity [25, 50–54], characterized by antigen activation of sensitized T_{DTH} lymphocytes releasing various cytokines which result in the accumulation and activation of macrophages. T_{DTH} lymphocytes are subset populations of T helper (T_H,) lymphocytes purported to be of the CD4+ T_{H-1} subtype (and in rare instances CD8+, cytotoxic T cells, T_C). In this scheme a metal combines with a self-protein and becomes antigenic after being processed by a macrophage or other antigen presenting cell, (APC). Metal implant alloy degradation products as moieties in immunoactive complexes have been shown to be antigenic in many case studies through temporal association with specific responses such as severe dermatitis, urticaria, vasculitis [23, 35, 36, 42–48]. Additionally a large number of cohort investigations have indicated a correlation between metal implants and metal hypersensitivity reactions [25, 49–60].

Combination of antigen dependent and independent mechanisms

Metals may act with serum proteins to crosslink lymphocyte receptors (e.g. BV17 of CDR1 T-cell receptor) without the presence of an antigen presenting cell, i.e. in a superantigen enhancement of T cell receptor-protein contact [28, 29]. In this circumstance proteins or peptides that would not otherwise be antigenic are able to provoke a response. While this hypothesis has not been extensively supported, the lymphocyte reactions in the current investigation are consistent with such nonspecific mitogenic or 'polyclonal' activation mechanisms (i.e. nonspecific to a single or small population of clonally 'sensitized' or differentiated lymphocytes).

Metal-protein complex induced activation of primary human lymphocytes

The MPCRI is an approximate measure of metal-protein bioreactivity based on an average molecular weight determined for each serum fraction, the moles of metal measured in each fraction, and moles of protein within that fraction. For this analysis it was necessary to assume that all metal was bound to serum proteins and that all binding was between serum proteins and metal ions, not nanometer size particles. This measure of metal-protein reactivity could not be calculated where concentrations of metal or protein were below the method detection limit. Figure 5 demonstrates the utility of this index highlighting the higher reactivity associated with metal-protein complexes containing Cr and the greater reactivity of higher molecular weight metal-protein complexes.

Earlier investigations have indicated that metal degradation products may elicit an immune response and combine with immunoglobulins. Yang *et al.* [21] have shown binding of humoral antibodies (i.e. IgG, IgE, IgM and IgA) to metal complexed with glutathione (GSH) suggesting that humoral hypersensitivity reactions may be induced by elevated levels of circulating metal. However, this has not been demonstrated in a human model. The elevated MPCRI of both Cr and Ti containing high molecular weight fractions (which contain predominantly immunoglobulins) while consistent with the findings of Yang *et al.*, suggests a cell mediated response which is either antigen specific (monoclonal stimulation), antigen independent (polyclonal stimulation) or a combination of the two. While not typical of a DTH response, a (clonally) nonspecific activation of lymphocytes does not rule out presence of metal-specific T-cells. However, at present there is incomplete knowledge concerning the degree of metal induced hypersensitivity in patients with total joint replacements and the role which specific T-cells, metal-protein binding and humoral immunity plays in the host response and clinical outcome. The results of this study demonstrate that the most bioreactive metal-protein complexes are those within the higher molecular weight immunoglobulin range.

The majority of released implant metal within and around the joint space and remote from the implant may take the form of particles. Particulate debris is difficult to differentiate from other forms of metal (i.e. soluble metal-protein complexes) when using analytical techniques such as atomic absorption spectroscopy alone [10–12]. Nevertheless, the overlap in the concentration range at which metal ions provoked lymphocyte activation in the current investigation and that reported locally around implants are consistent with the hypothesis that adverse local and remote tissue responses associated with released implant metal particles may be due in part to immunogenic metal ions released from implants.

Conclusions

In this investigation, we have identified two molecular weight ranges of human serum proteins associated with the binding of Cr and Ti from Co-Cr-Mo and Ti implant alloy degradation as potentially toxic. High molecular weight serum proteins (≈ 180 kDa) demonstrated greater lymphocyte bioreactivity when complexed with Cr and Ti than low to midrange (30–77 kDa) serum proteins. This *in vitro* model demonstrates immunogenic potential of both Co-Cr-Mo and Ti alloy degradation products complexed with high molecular weight proteins. A quantifiable measure of metal-protein complex reactivity (MPCRI), demonstrated Co-Cr-Mo alloy degradation products were associated with approximately an order of magnitude greater lymphocyte reactivity than Ti alloy products in the higher molecular weight ranges which are dominated by immunoglobulins (≈ 180–330 kDa). Metal sensitivity may exist as an extreme complication in only a few highly susceptible patients (i.e. less than 1% of joint replacement recipients), or it may be a more common subtle contributor to implant failure. In addition to direct immunogenic responses, metal degradation products may mediate indirect immunologic effects due to immune cell toxicity. Mechanisms by which *in vivo* metal sensitivity occurs have not been well characterized [61, 62]. Thus, we put forward the view of metal implant degradation products as potentially toxic, albeit not by classical interference with cell functionality but through indirect and undesired activation of lymphocytes which then leads to untoward peri-implant and systemic effects. The importance of this line of investigation is growing, as the use of metallic implants is increasing and as expectations of implant durability and performance increase [63, 64].

References

1. Jacobs JJ, Gilbert JL, Urban RM: Corrosion of metallic implants. In: M. Stauffer (ed). Total Revision Hip Arthroplasty. R.N. Mosby, St. Louis, 1994, p 279
2. Black J: Systemic effects of biomaterials. Biomaterials 5: 12, 1984
3. Jacobs JJ, Skipor AK, Black J, Hastings MC, Schavocky J, Urban RM, Galante JO: Metal release and excretion from cementless titanium total knee replacements. Trans Orthoped Res Soc New Orleans, LA: 558, 1991
4. Jacobs JJ, Skipor AK, Black J, Manion L, Urban RM, Galante JO: Metal release in patients with loose titanium alloy total hip replacements. Trans Fourth World Biomaterials Conference Berlin: 266, 1992
5. Jacobs JJ, Skipor AK, Black J, Urban RM, Galante GO: Release and excretion of metal in patients who have a total hip-replacement component made of titanium-base alloy. Am J Bone Joint Surg 73-A: 1475, 1991
6. Jacobs JJ, Skipor AK, Black J, Urban RM, Galante JO: Metal release and excretion in patients with titanium base alloy total hip replacement components. Am J Bone Joint Surg 73A:1475, 1991
7. Stulberg BN, Merritt K, Bauer T: Metallic wear debris in metal-backed patellar failure. J Biomed Mat Res Appl Biomat 5: 9, 1994
8. Stulberg BN, Watson JT, Shull SL, Merritt K, Crowe TD, Stulberg SD, Guidry CH: Metal ion release after porous total knee arthroplasty: A prospective controlled assessment. Trans Orthoped Res Soc 12: 314, 1987
9. Dorr LD, Bloebaum R, Emmanual J, Meldrum R: Histologic, biochemical and ion analysis of tissue and fluids retrieved during total hip arthroplasty. Clin Orthoped Rel Res 261: 82, 1990
10. Jacobs JJ, Skipor AK, Urban RM, Black J, Manion LM, Starr A, Talbert LF, Galante JO: Systemic distribution of metal degradation products from titanium alloy total hip replacements: An autopsy study. Trans Orthoped Res Soc New Orleans: 838, 1994
11. Michel R, Hoffman J, Loer F, Zilkens J: Trace element burdening of human tissue due to corrosion of hip-joint prostheses made of cobalt-chromium alloys. Arch Orthoped Trauma Surg 103: 85, 1984
12. Michel R, Nolte M, Reich M, Loer F: Systemic effects of implanted prostheses made of cobalt-chromium alloys. Arch Orthoped Trauma Surg 110: 61, 1991
13. Hallab NJ, Jacobs JJ, Skipor A, Black J, Mikecz K, Galante JO: Systemic metal-protein binding associated with total joint replacement arthroplasty. J Biomed Mat Res 49: 353, 2000
14. Griem P, Gleichmann E: Metal ion induced autoimmunity. Curr Opin Immunol 7: 831, 1995
15. Silvennoinen-Kassinen S, Ikaheimo I, Karvonen J, Kauppinen M, Kallioinen M: Mononuclear cell subsets in the nickel-allergic reaction *in vitro* and *in vivo*. J Allergy Clin Immunol 89: 794, 1992
16. Silvennoinen-Kassinen S, Poikonen K, Ikaheimo I: Characterization of nickel-specific T cell clones. Scand J Immunol 33: 429, 1991
17. Warner GL, Lawrence DA: The effect of metals on IL-2-related lymphocyte proliferation. Int J Immunopharmacol 10: 629, 1988
18. Warner GL, Lawrence DA: Cell surface and cell cycle analysis of metal-induced murine T cell proliferation. Eur J Immunol 16: 1337, 1986
19. Griem P, von Vultee C, Panthel K, Best SL, Sadler PJ, Shaw CF: T cell cross-reactivity to heavy metals: Identical cryptic peptides may be presented from protein exposed to different metals. Eur J Immunol 28: 1941, 1998
20. Jacobs J, Goodman S, Sumner DR, Hallab N: Biologic response to orthopedic implants. In: AnonymousOrthopedic Basic Science. American Academy of Orthopedic Surgeons, Chicago, 1999, p 402
21. Yang J, Merritt K: Detection of antibodies against corrosion products in patients after Co-Cr total joint replacements. J Biomed Mat Res 28: 1249, 1994
22. Yang J, Merritt K: Production of monoclonal antibodies to study corrosion of Co-Cr biomaterials. J Biomed Mat Res 31: 71, 1996
23. Wang JY, Wickland BH, Gustilo RB, Tsukayama DT: Prosthetic metals impair immune response and cytokine release *in vivo* and *in vitro*. J Orthoped Res 15: 688, 1997
24. Merritt K, Rodrigo JJ: Immune response to synthetic materials. Sensitization of patients receiving orthopaedic implants. Clin Orthoped Rel Res 326: 71, 1996
25. Merritt K, Brown S: Metal sensitivity reactions to orthopedic implants. Int J Dermatol 20: 89, 1981
26. Silvennoinen-Kassinen S, Ikaheimo I, Karvonen J, Kauppinen M, Kallioinen M: Mononuclear cell subsets in the nickel-allergic reaction *in vitro* and *in vivo*. J Allergy Clin Immunol 89: 794, 1992
27. Moulon C, Vollmer J, Weltzien HU: Characterization of processing requirements and metal cross- reactivities in T cell clones from patients with allergic contact dermatitis to nickel. Eur J Immunol 25: 3308, 1995
28. Vollmer J, Fritz M, Dormoy A, Weltzien HU, Moulon C: Dominance of the BV17 element in nickel-specific human T cell receptors relates to severity of contact sensitivity. Eur J Immunol 27: 1865, 1997
29. Vollmer J, Weltzien HU, Moulon C: TCR reactivity in human nickel allergy indicates contacts with complementarity-determining region 3 but excludes superantigen-like recognition. J Immunol 163: 2723, 1999
30. Saito K: Analysis of a genetic factor of metal allergy – polymorphism of HLA-DR, -DQ gene. Kokubyo Gakkai Zasshi 63: 53, 1996
31. Nakashima I, Pu MY, Nishizaki A, Rosila I, Ma L, Katano Y, Ohkusu K, Rahman SM, Isobe K, Hamaguchi M: Redox mechanism as alternative to ligand binding for receptor activation delivering disregulated cellular signals. J Immunol 152: 1064, 1994
32. Kubicka-Muranyi M, Griem P, Lubben B, Rottmann N, Luhrmann R, Gleichmann E: Mercuric-chloride-induced autoimmunity in mice involves up-regulated presentation by spleen cells of altered and unaltered nucleolar self antigen. Int Arch Allergy Immunol 108: 1, 1995
33. Bravo I, Carvalho GS, Barbosa MA, deSousa M: Differential effects of eight metal ions on lymphocyte differentiation antigens *in vitro*. J Biomed Mat Res 24: 1059, 1990
34. Gillespie WJ, Frampton CM, Henderson RJ, Ryan PM: The incidence of cancer following total hip replacement (published erratum appears in Br J Bone Joint Surg 78: 680, 1996 July). Br J Bone Joint Surg 70: 539, 1988
35. Merritt K, Brown SA: Biological Effects of Corrosion Products from Metal. American Society for Testing and Materials, Philadelphia, 1985
36. Poss R, Thornhill TS, Ewald FC, Thomas WH, Batte NJ, Sledge CB: Factors influencing the incidence and outcome of infection following total joint arthoplasty. Clin Orthoped 182: 1984
37. Wang JY, Wicklund BH, Gustilo RB, Tsukayama DT: Prosthetic metals interfere with the functions of human osteoblast cells *in vitro*. Clin Orthoped 216: 1997

38. Kubicka-Muranyi M, Goebels R, Goebel C, Uetrecht J, Gleichmann E: T lymphocytes ignore procainamide, but respond to its reactive metabolites in peritoneal cells: Demonstration by the adoptive transfer popliteal lymph node assay. Toxicol Appl Pharmacol 122: 88, 1993
39. Kubicka-Muranyi M, Kremer J, Rottmann N, Lubben B, Albers R, Bloksma N, Luhrmann R, Gleichmann E: Murine systemic autoimmune disease induced by mercuric chloride: T helper cells reacting to self proteins. Int Arch Allergy Immunol 109: 11, 1996
40. Saoudi A, Castedo M, Nochy D, Mandet C, Pasquier R, Druet P, Pelletier L: Self-reactive anti-class II T helper type 2 cell lines derived from gold salt-injected rats trigger B cell polycolonal activation and transfer autoimmunity in CD8-depleted normal syngeneic recipients. Eur J Immunol 25: 1972, 1995
41. Kubicka-Muranyi M, Behmer O, Uhrberg M, Klonowski H, Bister J, Gleichmann E: Murine systemic autoimmune disease induced by mercuric chloride (HgCl2): Hg-specific helper T-cells react to antigen stored in macrophages. Int J Immunopharmacol 15: 151, 1993
42. Merle C, Vigan M, Devred D, Girardin P, Adessi B, Laurent R: Generalized eczema from vitallium osteosynthesis material. Contact Dermatitis 27: 257, 1992
43. King J, Fransway A, Adkins RB: Chronic urticaria due to surgical clips. New Engl J Med 329: 1583, 1993
44. Barranco VP, Solloman H: Eczematous dermatitis from nickel. J Am Med Assoc 220: 1244, 1972
45. Thomas RH, Rademaker M, Goddard NJ, Munro DD: Severe eczema of the hands due to an orthopaedic plate made of Vitallium. Br Med J 294: 106, 1987
46. Abdallah HI, Balsara RK, O'Riordan AC: Pacemaker contact sensitivity: Clinical recognition and management. Annals Thoracic Surg 57: 1017, 1994
47. Halpin DS: An unusual reaction in muscle in association with a vitallium plate: A report of possible metal hypersensitivity. J Bone Joint Surg 57-B: 451, 1975
48. Gillespie WJ, Frampton CM, Henderson RJ, Ryan PM: The incidence of cancer following total hip replacement. Am J Bone Joint Surg 70-B: 539, 1988
49. Benson MK, Goodwin PG, Brostoff J: Metal sensitivity in patients with joint replacement arthroplasties. Br Med J 4: 374, 1975
50. Brown GC, Lockshin MD, Salvati EA, Bullough PG: Sensitivity to metal as a possible cause of sterile loosening after cobalt-chromium total hip-replacement arthroplasty. J Bone Joint Surg 59-A: 164, 1977
51. Deutman R, Mulder TH, Brian R, Nater JP: Metal sensitivity before and after total hip arthroplasty. Am J Bone Joint Surg 59-A: 862, 1977
52. Kubba R, Taylor JS, Marks KE: Cutaneous complications of orthopedic implants. A two-year prospective study. Arch Dermatol 117: 554, 1981
53. Mayor MB, Merritt K, Brown SA: Metal allergy and the surgical patient. Am J Surg 139: 477, 1980
54. Merritt K, Brown S: Tissue reaction and metal sensitivity. Acta Orthoped Scand 51: 403, 1980
55. Merritt K: Role of medical materials, both in implant and surface applications, in immune response and in resistance to infection. Biomaterials 5: 53, 1984
56. Pinkston JA, Finch SC: A method for the differentiation of T and B lymphocytes and monocytes migrating under agarose. Stain Technol 54: 233, 1979
57. Rooker GD, Wilkinson JD: Metal sensitivity in patients undergoing hip replacement. A prospective study. J Bone Joint Surg 62-B: 502, 1980
58. Carlsson AS, Macnusson B, Moller H: Metal sensitivity in patients with metal-to-plastic total hip arthroplasties. Acta Orthoped Scand 51: 57, 1980
59. Cramers M, Lucht U: Metal sensitivity in patients treated for tibial fractures with plates of stainless steel. Acta Orthoped Scand 48: 245, 1977
60. Fischer T, Rystedt I, Safwenberg J, Egle I: HLA-A, -B, -C and -DR antigens in individuals with sensitivity to cobalt. Acta Derm Venereol (Stockh) 64: 121, 1984
61. Hensten-Pettersen A: In: Allergy and Hypersensitivity. Raven Press, New York, 1993
62. Boyan BD: In: Discussion of Toxicity and Allergy. Raven Press, New York, 1993
63. Black J: In: Prosthetic Materials. VCH Publishers, Inc., New York, 1996
64. Jacobs JJ, Skipor AK, Doorn PF, Campbell P, Schmalzried TP, Black J, Amstutz HC: Cobalt and chromium concentrations in patients with metal on metal total hip replacements. Clin Orthoped Rel Res S329: S256, 1996

Molecular and Cellular Biochemistry **222:** 137–140, 2001.

Arsenic contamination of groundwater and prevalence of arsenical dermatosis in the Hetao plain area, Inner Mongolia, China

Xiaojuan Guo,[1,2] Yoshihisa Fujino,[2] Satoshi Kaneko,[2] Kegong Wu,[1] Yajuan Xia[1] and Takesumi Yoshimura[2]

[1]*Institute for Control and Treatment of Endemic Diseases in Inner Mongolia, China;* [2]*Department of Clinical Epidemiology, Institute of Industrial Ecological Sciences, University of Occupational and Environmental Health, Japan*

Abstract

An investigation was carried out on arsenic contamination of groundwater and prevalence of arsenical dermatosis in the Hetao plain of Inner Mongolia Autonomous Region, China. Based on the screening of water samples from 96 randomly selected wells in this Region, two areas (Wuyuan and Alashan) were chosen as highly contaminated areas because arsenic in the water samples was higher than 50 µg/l. Arsenic was measured using an arsenic silver diethyl dithiocarbamate method for 326 water samples from all the wells in these areas. The results show arsenic contaminated groundwater from tubule-type wells of depths about 15–30 m was serious compared with open-type wells where depth is about 3–5 m. In the Wuyuan area, 96.2% of water samples from tubule-type wells contained arsenic above 50 µg/l and 69.3% in Alashan area; the highest value was 1354 µg/l and 1088 µg/l, respectively. In these two areas, a health survey was carried out for arsenical dermatosis. The results show the prevalence of arsenical dermatosis in the Wuyuan area was 44.8%, higher than 37.1% prevalence of arsenical dermatosis in the Alashan area. The prevalence of arsenical dermatosis was highest in the over 40-year-old age group. There was no sex difference in the prevalence. Further investigation is needed to clarify the actual situation of arsenic pollution of groundwater in Inner Mongolia, China in order to reduce the adverse health effect among residents exposed to arsenic. (Mol Cell Biochem **222**: 137–140, 2001)

Key words: groundwater, arsenic, contamination, health effects, arsenical dermatosis

Introduction

In Inner Mongolia, China, arsenic poisoning from groundwater has been reported since 1990 [1]. It has been found that the arsenic concentration in groundwater came up to 50 µg/l or higher; the Chinese drinking water quality standard value, in 627 villages of Inner Mongolia [2]. There are two regions with arsenic contaminated groundwater. One is a consequence of poisonous sand mining pollution in a small water basin of Kesiketeng County, and the other is located on the Hetao plain (Fig. 1). However, health problems related to arsenic pollution in Inner Mongolia have not been well investigated to date. In the present study, we aimed to investigate the actual situation of arsenic contamination by drinking water and the prevalence of arsenical dermatosis on the Hetao plain of Inner Mongolia.

Materials and methods

Investigation of arsenic contamination by groundwater

Based on the screening of water samples from 96 randomly selected wells in the Hetao Region in 1996, two areas (Wuyuan

Address for offprints: X. Guo, Department of Clinical Epidemiology, Institute of Industrial Ecological Sciences, University of Occupational and Environmental Health, Japan, 1-1 Iseigaoka Yahatanishi-ku, Kitakyushu, Japan 807-8555

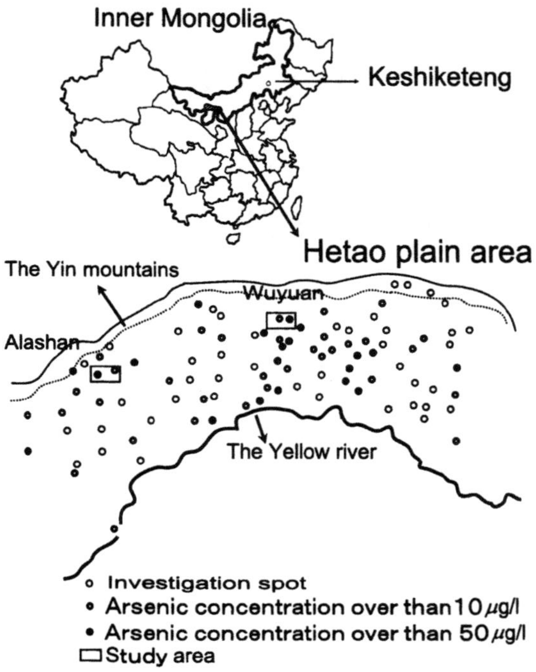

Fig. 1. Geographical location of the wells for screening of groundwater on the Hetao plain area, Inner Mongolia in 1996. Rectangles indicate the areas, Wuyuan and Alashan areas, chosen for this study.

and Alashan) were chosen as highly contaminated areas with arsenic because arsenic in the water samples was higher than 50 μg/l (Fig. 1). In the two chosen areas, an investigation for arsenic contamination of water from all the wells in the areas was conducted. We collected water samples from 4 open-wells where depth is about 3–5 m, and 322 tubule-type wells, where depth is about 15–30 m, and we conducted water quality analyses applying an arsenic silver diethyl dithiocarbamate method (FI-HG-AAS) [3] for arsenic measurement.

Investigation of arsenical dermatosis

Among 1802 residents in the two areas, 1609 people (826 male, 783 female) were included for a general and a dermatological detailed investigation. We conducted interviews and medical health surveys according to household, and recorded the results onto an investigation sheet. The interviews included a questionnaire regarding name, age, sex, date of birth, place of birth, resident history, past illnesses or disease, family history of disease, lifestyles (daily activities, and dietary habit), smoking history, occupation, physical information, reproductive history for females, and information related to drinking water availability and usage. An investigation for arsenical dermatosis was carried out by an experienced dermatologist and the authors (X.G and Z.L) between 1996 and 2000. The investigation comprised dermatological detailed physical check-ups that included the whole body skin, hair, mouth and genitalia mucosa.

Patients suffering from arsenical dermatosis were identified by referring to the clinical diagnostic standard for endemic arsenic poisoning in Inner Mongolia shown in Table 1. Stat View statistical package (version 5.0) was used for data management and analysis of the data.

Results

Table 2 shows the result of arsenic concentration analyses of water samples in the study areas of Alashan and Wuyuan, where concentrations came up to or over the standard value (50 μg/l) of arsenic contamination in China for drinking water and WHO guideline (10 μg/l) [4]. Among the 189 investigated tubule-type wells in the Alashan area, the number of wells with more than 10 μg/l arsenic concentration was 144 (76.2%), and those with more than 50 μg/l was 131(69.3%). The highest arsenic concentration recorded was 1088 μg/l. Among the 133 tubule-type wells in the Wuyuan area, the number of wells with more than 10 μg/l and 50 μg/l were 132 (99.2%) and 128 (96.2%), respectively. The highest arsenic concentration was recorded as 1354 μg/l. Moreover, the arsenic concentrations of drinking water in tubule-type wells (1354 μg/l) were higher than those in open-wells (30 μg/l).

Table 3 shows the prevalence of arsenical dermatosis in the two areas. The prevalence of arsenical dermatosis in the Wuyuan area was 44.8%, and that in the Alashan area was 37.1%. In males, prevalence of arsenical dermatosis was higher (36.4%) in the Wuyuan area than in the Alashan area (52.8%).

Table 4 shows the prevalence of arsenical dermatosis for age groups and sex. The age of patients ranged from 7–79 years old. A progressive increase in the prevalence of arsenical dermatosis was found from the youngest age group (0–9 years) to the 40–49 year age group. The prevalence of arsenical dermatosis was high in the group aged 40+, both for males and females. There was no statistical difference between the prevalence of dermatosis by sex ($p = 0.1177$).

Table 1. Diagnostic criteria for the chronic arsenicism from drinking water

1	A history of having residence in an arsenic concentration area or a history of drinking arsenic concentrated water
2	Keratosis on the palms of the hands and/or soles of the feet
3	Pigmentation and leukomelanosis on the body

Cases were defined when diagnostic criteria 1 was fulfilled along with criteria 2 and/or 3.

Table 2. Investigation of arsenic concentrations in drinking water in two areas

Areas	Well type	Number of wells	0–9 μg/l	%	10–49 μg/l	%	≥ 50 μg/l	%	Highest value (μg/l)
Alashan	Open	0							
	Tubule	189	45	23.8	13	6.9	131	69.3	1088
Wuyuan	Open	4	3	75.0	1	25.0	0	0.0	30
	Tubule	133	1	0.8	4	3.0	128	96.2	1354
Total	Wells	326	49	15.0	18	5.5	259	79.4	1354

Table 3. Prevalence of arsenical dermatosis in two areas

	Alashan			Wuyuan		
	No. of subjects	No. of patients	Prevalence (%)	No. of subjects	No. of patients	Prevalence (%)
Male	610	222	36.4	216	114	52.8
Female	566	214	37.8	217	80	36.9
Total	1176	436	37.1	433	194	44.8

Discussion

According to this study, 79.4% of the wells contained arsenic above China drinking water standard 50 μg/l. Water pollution might be affected by geographic environment and human activities. As a result of screening for the 96 wells on the Hetao plain, we identified that arsenic concentrations of groundwater were higher in the low-lying land of the central area of the plain compared with the area close to the mountains. Furthermore, it was also shown that water of tubule-type wells was more likely to be contaminated with arsenic than that of open-wells. Considering that the strata around mines in the Yin Mountains have been reported to contain arsenic with about 10–60 ppm [5], and arsenic rich aquifers are situated at 10–30 m under the ground, this phenomenon could be explained by collection of groundwater wider contaminated by arsenic from the Yin Mountains. Additionally, 5 billion m^3/year of irrigate water from the Yellow River which are located the higher topography to the low-lying area in the Hetao plain might be modifying the arsenic pollution of the Hetao plain because arsenic pollution seems to be relatively severe in irrigated areas from river not only in Inner Mongolia, but also in Xinjiang, China [6].

The prevalence of arsenical dermatosis was also higher in the Wuyuan area (44.8%) than that in the Alashan area (37.1%) and no dermatosis cases were found in areas where groundwater arsenic concentrations were <50 μg/l [7, 8]. Although a dose-response relationship between dermatosis and arsenic concentration of drinking water is not clear to date, this might be explained by the difference in arsenic concentrations of drinking water between the two areas, because the situation of arsenic contaminated groundwater (> 50 μg/l) was more severe in the Wuyuan area (96.2%) than in the Alashan area (69.3%).

Long-term exposure to high concentrations of arsenic-rich drinking water causes several adverse health effects [9]. In the two areas, the most prominent changes are alterations in keratoses, which are localized primarily on the palms of the

Table 4. Prevalence of arsenical dermatosis by sex and age group

	Male			Female		
Age group (year)	No. of subjects	No. of patients	Prevalence (%)	No. of subjects	No. of patients	Prevalence (%)
0–9	156	6	3.8	140	8	5.7
10–19	236	48	20.3	196	27	13.8
20–29	91	43	47.3	103	32	31.1
30–39	120	67	55.8	128	67	52.3
40–49	104	81	77.9	117	91	77.8
50–59	59	45	76.3	53	41	77.4
60 and over	60	46	76.7	46	28	60.9
Total	826	336	40.7	783	294	37.5

hands and on the soles of the feet, and pigmentation and leucoderma on the torso. In more severe cases with keratoses, cancer may develop on the torso and the limbs and the interval organs [9]. Among residents exposed to arsenic for 15–20 years since the 1980s, only one patient with skin cancer was identified in Wuyuan area. An extended follow-up of the residents would detect more cases with skin cancer or other cancers.

In conclusion, the arsenic contamination of drinking water in Inner Mongolia is serious, and it has been caused by many specific factors especially the geographic environment and human activities; the mines, and irrigation by the yellow River in the study area. Most of the contaminated water was from tubule-type wells, available since the 1980s, with which deeper groundwater can be pumped up. The prevalence of arsenic skin lesions was as high as 40%, and it was related to the percentage of arsenic contaminated wells in the study areas. The prevalence of arsenical dermatosis was found to be highest in persons 40 years of age or older, and there were no sex differences. A further detailed investigation is needed to clarify the actual situation of arsenic pollution of groundwater in Inner Mongolia, China in order to reduce the adverse health effect among residents exposed to arsenic.

References

1. Gao HX: The first discovered endemic arsenic poisoning in Inner Mongolia. Inner Mongolia J Endem Dis Cont Treat 15: 1, 1990 (in Chinese)
2. Sun TZ, Wu KG, Xing CM, Ma HZ, Yu GJ, Li YH, Xia YJ, Guo XJ, Zheng Z: Epidemiological survey of endemic arsenism in Inner Mongolia. Chinese J Endemiol 9: 1–4, 1995 (in Chinese)
3. Wang LF: In: Endemic Arsenic Poisoning and Blackfoot Diseases. Xinjiang Science Hygiene Publishing, Wulumuqi, 1997, pp 193–195 (in Chinese)
4. Buchet J, Lison D: Clues and uncertainties in the risk assessment of arsenic in drinking water. Food Chem Toxicol 38: 81–85, 2000
5. Li SF, Li HJ: Study of geological characteristic and source of the arsenic pollution in Hetao area of Inner Mongol. Inner Mongolia J Endem Dis Cont Treat 19: 1–4, 1994, (in Chinese)
6. Wang LF, Shun XZ, Ai HT Xu XF, Lin FF, Liu HD, Feng ZR: Epidemiology investigation endemic arsenic poisoning in Zhunger area of Xinjiang. Inner Mongolia J Endem Dis Cont Treat 19: 37–40, 1994 (in Chinese)
7. He M, Jing L, Lu SY, Jin T, Niu PJ, Xu GS: Epidemiology investigation endemic arsenic poisoning in Langshan area of Linhe city. Inner Mongolia J Endem Dis Cont Treat 19: 34–36, 1994, (in Chinese)
8. Guo XJ, Ma HZ, Lu Y, Li ZZ, Li ZY, Wu KG, Zheng Z, Xia YJ, Li YH, Zhou HJ, Qin YX, Yu GJ, Dang RH: Study of skin lesion with chronic arsenic polluted. Inner Mongolia J Endem Dis Cont Treat 19: 88–90, 1994, (in Chinese)
9. National Research Council. In: Arsenic in Drinking Water. National Academy Press, Washington, DC, 1999, pp 131

Molecular and Cellular Biochemistry **222**: 141–147, 2001.

Involvement of Erks activation in cadmium-induced AP-1 transactivation *in vitro* and *in vivo*

Chuanshu Huang,[1] Qunwei Zhang,[1] Jingxia Li,[1] Xianglin Shi,[2] Vincent Castranova,[2] Gong Ju,[3] Max Costa[1] and Zigang Dong[4]

[1]*Nelson Institute of Environmental Medicine, New York University School of Medicine, NY;* [2]*Health Effects Laboratory Division, National Institute for Occupational Safety and Health, Morgantown, WV;* [3]*The Hormel Institute, University of Minnesota, Austin, MN, USA;* [4]*The Institute of Neuroscience, The Fourth Military Medical University, Xi'an, P.R. China*

Abstract

Cadmium is a potent and effective carcinogen in rodents and has recently been accepted by IARC (International Agency for Research on Cancer) as a category 1 carcinogen. Cadmium-induced up-regulation of intracellular signaling pathways leading to increased mitogenesis is thought to be a major mechanism for the carcinogenic activity following chronic cadmium exposure. In the present study, we found that exposure of cells to cadmium induced significant activation of AP-1 and all three members of the MAP kinase family in mouse epidermal JB6 cells. The induction of AP-1 activity by cadmium appears to involve activation of Erks, since the induction of AP-1 activity by cadmium was blocked by pretreatment of cells with PD98058. Interestingly, the induction of AP-1 by cadmium was greatly enhanced by the chemical tumor promoter, TPA and the growth factor EGF, but not by ultraviolet C radiation. *In vivo* studies demonstrated that cadmium could also induce transactivation of AP-1 in AP-1-luciferase report transgenic mice. Considering the role of AP-1 activation in tumor promotion, the results presented in this study provide a possible molecular mechanism for cadmium-induced carcinogenesis. (Mol Cell Biochem **222**: 141–147, 2001)

Key words: cadmium, AP-1, transgenic and MAP kinase

Abbreviations: AP-1 – activator protein-1; Erks – extracellular signal-regulated protein kinases; MAPKs – mitogen-activated protein kinases; JNKs – c-Jun N-terminal kinases; TPA – 12-O-tetradecanoylphorbol-13-acetate; EGF – epidermal growth factor; FBS – fetal bovine serum; MEM – minimal essential medium

Introduction

Cadmium is a toxic transition metal that is widely used in industry. Cadmium is a potent and effective carcinogen in rodents, and has recently been accepted by the International Agency for Research on Cancer as a Category 1 (human) carcinogen (IARC, 1994) [1]. The main sources of human exposure to cadmium are tobacco smoke, food and industrial pollution. Cadmium is absorbed by inhalation and ingestion and has a biological half-life > 25 years [2]. For cigarette smoke, up to 60% of the cadmium deposited in the lung is absorbed. Epidemiological studies have identified lung, prostate, and, to a lesser extent, kidney and stomach as primary targets for cadmium-induced tumorigenesis [1].

Carcinogenesis is a lengthy, multistep process that may be arbitrarily divided into initiation, promotion, and progression of the tumor. DNA mutations and alterations in cell signaling are presumed to play a major role in the entire process [3, 4]. Cadmium is not a strong initiator because it is only a weak mutagen [5]. The effect of enhancement of endogenous mutations through inhibition of DNA repair may be a major mechanism for tumor initiation by cadmium [4]. With regard to the promotion and progression stages, the induction of immediate early genes by cadmium points to a probable

Address for offprints: C. Huang, Nelson Institute of Environmental Medicine, New York University School of Medicine, NY 10016, USA

mechanism of stimulation of proliferation of mutated cells and progression to malignancy [4, 5]. It has been reported that several protooncogenes, such as c-jun and c-myc, have been shown to be activated by cadmium in cell culture models [6–13]. The induction of these protooncogens is suggested to be mechanistically important for cadmium-induced carcinogenicity [4, 14, 15]. It was also reported that cadmium could induce the activation of protein kinase C (PKC) and mitogen-activated protein kinases (MAP kinases) [3, 16]. In the present study, we investigated the involvement of transcription factor, AP-1, in cadmium-induced signaling both *in vitro* and *in vivo*.

Materials and methods

Cell culture and agents

Mouse epidermal JB6 Cl41 and its AP-1-luciferase reporter stable transfectant, $P^{+}1$-1, were cultured in monolayers in Eagle's minimal essential medium containing 5% fetal bovine serum (FBS), 2 mM L-glutamine, and 25 μg of gentamicin/ml [17, 18]. Fetal bovine serum was from Life Technologies, Inc. (Rockville, MD, USA); Eagle's minimal essential medium (MEM) was from Calbiochem (San Diego, CA, USA); luciferase assay substrate was from Promega (Madison, WI, USA); cadmium chloride was purchased from Sigma (St. Louis, MO, USA); the phospho-specific antibodies against phosphorylated forms or nonphosphorylated forms of Erks, p38 kinase and JNKs were purchased from New England Biolabs (Beverly, MA, USA).

Animals

2X TRE-luciferase reporter transgenic mice were originally established by R.A. Flavell *et al.* [19]. A C57BL/6 male mouse carrying the 2X TRE-luciferase transgene was crossed with DBA/2 (SASCO, Omaha, NE, USA) females [19, 20]. The offspring were screened by testing both the basal level and TPA-induced level of luciferase activity for the presence of the AP-1-luciferase reporter gene. Males and females were housed separately in solid bottom polycarbonate cages on ventilated animals racks (~ 4–5 mice/cage, individualized by incisions in the ears) under temperature-, humidity-, and light-controlled conditions. Food and water were available *ad libitum* and the dorsal skin of the mice was shaved every week during the experiment period.

Assay for AP-1 activity in vitro

Confluent monolayers of JB6 $P^{+}1$-1 cells were trypsinized, and 8×10^3 viable cells suspended in 100 l of 5% FBS MEM were added to each well of a 96-well plate. Plates were incubated at 37°C in a humidified atmosphere of 5% CO_2. Twelve to 24 h later, cells were starved by culturing them in 0.1% FBS MEM for 24 h. Then, the cells were exposed to cadmium for AP-1 induction and maintained in culture. After different periods of time in culture, the cells were extracted with lysis buffer, and luciferase activity was measured using a luminometer (Monolight 2010). The results are expressed as relative AP-1 activity [21].

Phosphorylation analysis for Erks, P38 kinase, and JNKs

The cells were seeded in 6-well plates and cultured until 80–90% confluent. Then the cells were starved for 48 h in 0.1% FBS MEM and treated with arsenic at the concentration indicated. The cells were then extracted with SDS sample buffer. Immunoblots for detection of phosphorylated proteins were carried out using phospho-specific MAPK antibodies against phosphorylated sites of Erks, P38 Kinase, or JNKs as described previously [20, 21]. Antibodies were purchased from New England Biolabs and used according to manufacturer's recommendations. Antibody proteins were detected by chemiluminescence (ECL, New England Biolabs, Beverly, MA, USA).

Assay of AP-1 activity in vivo

All the mice were characterized by testing both the basal level and UVB-induced level of luciferase activity. The AP-1-luciferase reporter-bearing male and female mice (6–9 weeks old) were randomly divided into two groups (19–21 mice/group), including a negative control group (control) and cadmium-treated group. Two weeks after grouping, the mice were exposed to cadmium by topical application. The dorsal skin of the mice was punch biopsied at different time points to determine the induction of AP-1 activity in the epidermis. The luciferase activity of punch biopsied epidermis was measured as described previously [19, 20].

Statistical analysis

The significance of the difference in the AP-1 activity was determined with the Student's *t*-test.

Results

Induction of AP-1 in AP-1 luciferase transgenic mice by cadmium. To determine activation of AP-1 *in vivo* by cadmium, each AP-1 luciferase reporter transgenic mouse was treated topically with 500 μg of cadmium chloride, and the dorsal skin of the mice was punch biopsied to determine the luciferase activity at 0, 16, 30 and 48 h after treatment with cadmium. The results show that exposure of mice to cadmium induced significant increases of AP-1 activity in the skin of mice (Fig. 1). Maximal induction of AP-1 activity in skin of mice was observed from 16–48 h after cadmium exposure (Fig. 1). These results provide the first evidence that cadmium could induce AP-1 transactivation *in vivo*.

Cadmium induces AP-1 activation in mouse epidermal JB6 cells

In order to investigate the signal transduction pathways involved in cadmium-induced AP-1 transactivation, the mouse epidermal JB6 cell was employed as cell culture model. As shown in Fig. 2, cadmium treatment of P+1-1 cells, a stable JB6 Cl41 transfectant with the AP-1-luciferase reporter, markedly increased of AP-1 activity in dose-dependent manner (Fig. 2). The dose for maximal induction of AP-1 activity ranged between 0.8 and 3.2 μM (Fig. 2). A time course study suggested that treatment of cells with cadmium resulted in increased AP-1 activity within 12 h after treatment (Fig. 3). Maximal induction of AP-1 activity was observed between 24 and 48 h post-cadmium treatment. AP-1 activity gradually returned toward basal levels after 3 days (Fig. 3). These results are consistent with those from *in vivo* studies, suggesting that cadmium is a strong stimulator for AP-1 transactivation both in a cell culture model and in the mouse skin *in vivo* model.

Activation of Erks, JNKs, and P38 kinase by cadmium

AP-1 is a known target of the MAP kinase family, including Erks, JNKs, and p38 kinases. A previous study reported that

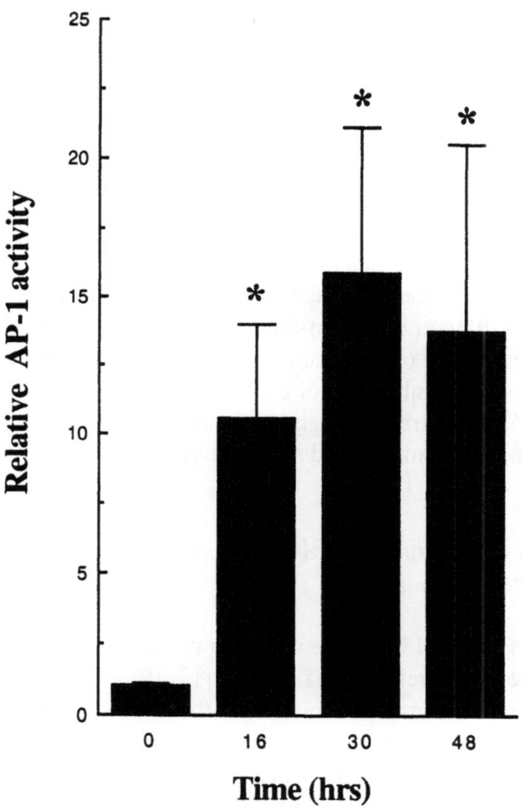

Fig. 1. Induction of AP-1 by cadmium in AP-1-luciferase transgenic mice. The AP-1-luciferase transgenic mice were described under 'Materials and methods.' The mice were randomly divided into two groups, including negative controls and cadmium-treated groups. Two weeks after grouping, the mice were exposed topically to cadmium chloride (500 μg/mouse). The dorsal skin of the mice was punch biopsied to determine the induction of AP-1 activity in the epidermis at different time points. The luciferase activity of punch biopsied epidermis was measured as described previously [25, 26]. The results were presented as relative AP-1 activity (means ± S.E. of 4 independent experiments. *indicates a significant increase from control ($p < 0.05$).

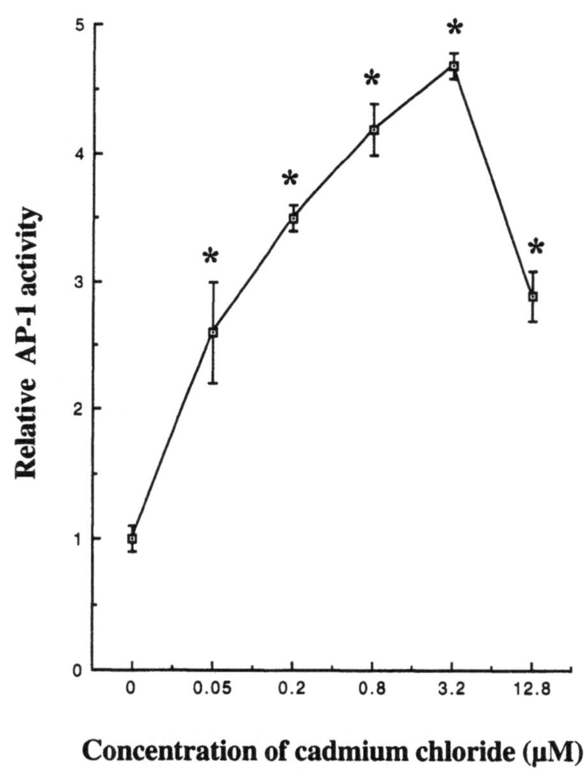

Fig. 2. Transactivation of AP-1 by cadmium in mouse epidermal JB6 cells. 8×10^3 P+1-1 cells were seeded into each well of 96-well plates. After being cultured at 37°C overnight, the cells were starved for 12 h by replacing medium with 0.1% FBS MEM. Then the cells were treated with different concentrations of cadmium chloride. After being cultured for 24 h, the luciferase activity was measured, and the results presented as relative AP-1 activity. Each value indicates the mean and S.E. of 4 identically treated assay wells. *indicates a significant increase from control ($p < 0.05$).

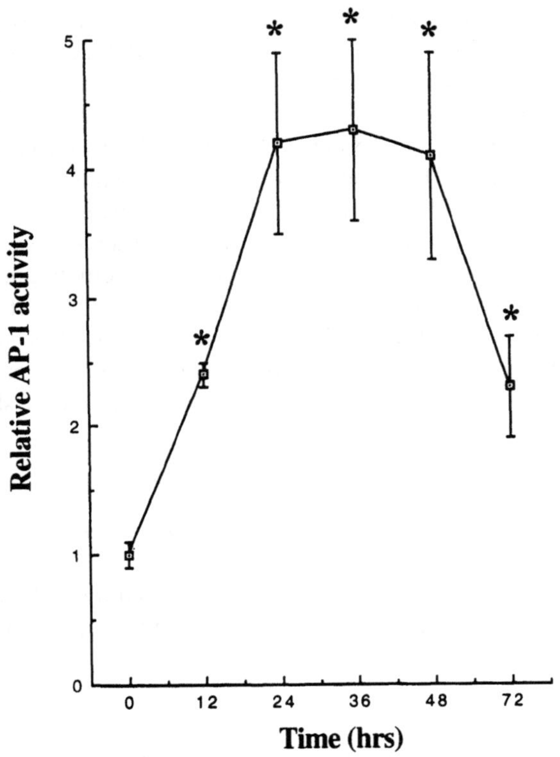

Fig. 3. Time-course studies on cadmium-induced AP-1 activity. 8×10^3 P+1-1 were seeded, cultured and starved as described for Fig. 2. The cells were then treated with cadmium chloride (1 μM) and harvested at the time points indicated for luciferase activity assay. The results are presented as relative AP-1 activity. Each value indicates the mean and S.E. of 4 identically treated assay wells. *indicates a significant increase from control ($p < 0.05$).

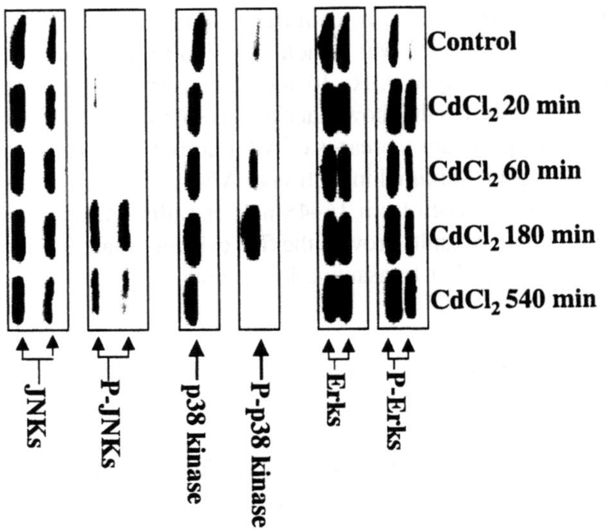

Fig. 4. Activation of Erks, JNKs, and p38 kinase by cadmium. JB6 P+1-1 cells were seeded into each well of 6-well plates. After being cultured until 80–90% confluent, the cells were starved for 48 h in 0.1% FBS MEM. Then the cells were treated with of cadmium chloride (1 μM) and incubated for different times as indicated. The cells were extracted with SDS sample buffer. Immunoblots for detection of nonphosphorylated and phosphorylated proteins were carried out using antibodies against nonphosphorylated Erks, JNKs or p38 kinases and phosphorylated sites of Erks, JNKs, and p38 kinase, respectively.

treatment of non-small cell lung carcinoma cell line CL3 with high doses (> 80 μM) of cadmium induced activation of Erks, JNKs and p38 kinases, while a low dose of cadmium caused inhibition of Erk and had no effect on p38 activity [14]. To determine the effects of cadmium on MAP kinase activation in mouse epidermal cells, we incubated the JB6 cells with 1 μM of cadmium chloride. The activation of Erks, JNKs and p38 kinases was determined using specific antibodies recognizing the dual phosphorylation sites of the specific kinase. The results showed that exposure of cells to 1 μM cadmium chloride caused significant activation of Erks, JNKs, as well as p38 kinases (Fig. 4). The Erks activation appeared to be sustained in all time points observed, while phosphorylation of JNKs and p38 kinases was delayed and transient (Fig. 4). The peak activation of JNKs and p38 kinase was observed at 180 min post-cadmium treatment (Fig. 4).

Blockade of cadmium-induced AP-1 activation by PD98059

PD98059 is a specific inhibitor of MEK1, a kinase immediately upstream of Erks. To investigate the role of Erks in cadmium-induced AP-1 activation, we determined the effects of PD98059 on cadmium-induced AP-1 activation. The results show that pretreatment of cells with PD98059 resulted in significant inhibition of cadmium-induced AP-1 activity (Fig. 5). These results suggest that activation of Erks plays a role in cadmium-induced AP-1 activation.

Effect of cadmium on TPA-, EGF- or UV-induced AP-1 activation

TPA, EGF, and UVC are well known tumor promoters and induce high levels of AP-1 activity [17, 19–21]. To assess the combined effects of cadmium with these tumor promoters on AP-1 activation, P+1-1 cells were treated with combination of cadmium and TPA, EGF or UVC, respectively. The results show that cadmium had an additive effect on either TPA- or EGF-induced AP-1 activity, while it did not enhance UVC-induced the AP-1 activation (Fig. 6). These data suggest that cadmium may enhance the carcinogenic effect of TPA and EGF.

Discussion

Cadmium has been classified as a human carcinogen by IARC in 1993 [1]. Several mechanisms to explain cadmium

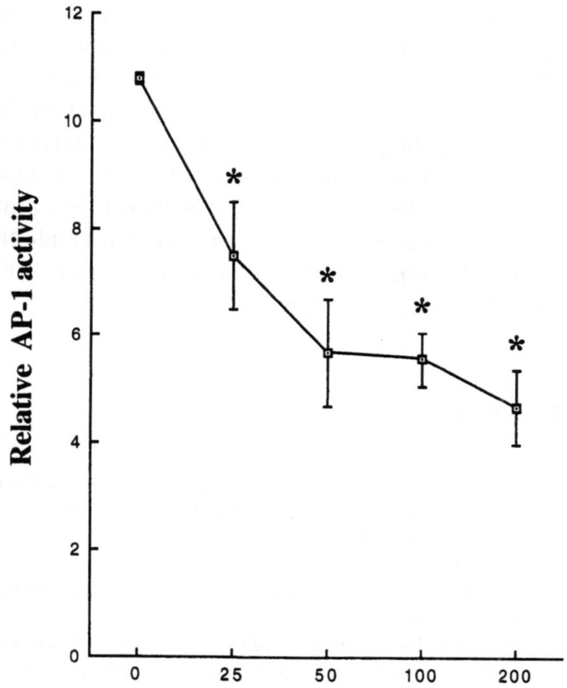

Fig. 5. Inhibition of cadmium-induced AP-1 activity by PD98059. JB6 P^+1-1 cells were seeded, cultured and starved as described for Fig. 2. The cells were first treated with different concentrations of PD98059 as indicated for 45 min. Then the cells were exposed to 2 μM of cadmium chloride for 24 h, and the luciferase activity was measured. The results are presented as relative AP-1 activity (means ± S.E. of 4 independent experiments). *indicates a significant increase from control ($p < 0.05$).

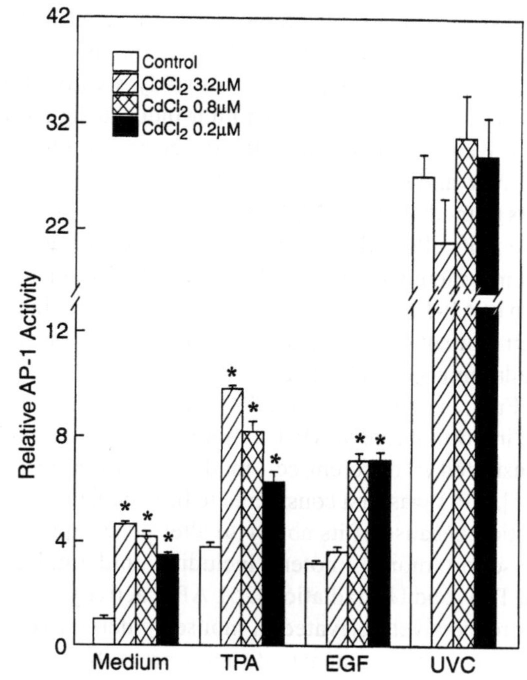

Fig. 6. Effects of cadmium on TPA-, EGF- or UVC-induced AP-1 activation. JB6 P^+1-1 cells were seeded and cultured as described for Fig. 2. After being cultured overnight, the cells were starved for 12 h. Then the cells were exposed to cadmium chloride (concentrations as indicated) plus TPA (10 ng/ml), EGF (10 ng/ml) or UVC (60 J/m^2). The cells were then incubated for 24 h. The luciferase activity was measured and results are presented as relative AP-1 activity (means ± S.E. of 4 independent experiments). *indicates a significant increase from the respective control ($p < 0.05$).

carcinogenesis have been proposed. It is likely that multiple mechanisms are involved in cadmium-induced tumor development [3, 6–18]. The present study focused on the effects of cadmium on activation of AP-1, a transcription factor that is required for tumor promotion in many cases. The results show that cadmium exposure caused transactivation of AP-1 in both a cell culture model and a transgenic mouse model. Exposure of cells to cadmium also induced activation of Erks, p38 and JNKs. Pretreatment of cells with PD98059 significantly inhibited cadmium-induced AP-1 activity. Furthermore, cadmium had an additive effect on TPA- and EGF-induced AP-1 activation. The data from this study demonstrate that cadmium is a strong inducer of AP-1 activation, which may be one of the mechanisms involved in cadmium-induced carcinogenesis.

Cadmium is widely used in industry in plating, batteries, plastics and semiconductors [22]. It is not only highly toxic to human, but also is a human carcinogen [1, 22]. Exposure of cadmium is associated with cancer of lung and prostate in human [23, 24]. In rats, cadmium exposure by subcutaneous injection gives rise to tumors at the site of injection, also in the kidney and in the testes [22, 24]. Inhalation and other methods of cadmium exposure give rise to tumors of the lung and prostate [22, 24]. Cadmium exposure can promote altered gene expression and modify intra- or inter-cellular communication [22]. These and other alterations in signal transduction are generally considered to be related to tumor promotion [22]. It was also found that the enhancement of endogenous mutations through inhibition of DNA repair is a major mechanism reinforcing gene mutation [4]. Thus, cadmium is not considered a strong tumor initiator [5], and it is believed that the carcinogenic activity of cadmium is mainly through tumor promotion rather than initiation [22]. Previous investigations showed that cadmium could induce the activation of protein kinase C (PKC) and MAP kinase [3, 16]. Recently, Chuang *et al.* [14] reported that cadmium induced persistent activation of JNKs or p38 kinases and moderate activation of Erks at high cytotoxic doses, while activities of Erks were decreased by treatment of cells with low doses of cadmium. The results from the same report suggested that JNKs and p38 kinase cooperatively participate in apoptosis induction by cadmium, while decreased Erk activity induced by low doses of Cd contributed to growth arrest and apoptosis [14]. MAP kinases, including Erks, JNKs and p38 kinases, are direct upstream kinases that regulate the transcription factor, AP-1, in cell response to extracellular stimulations, such as EGF, TPA, and UV radiation . In this study, we investigated the involvement of AP-1 in cadmium-induced signal transduction. We found that cadmium induced activation of AP-1 both *in vitro* and *in vivo*. The activation of AP-1 by cadmium appears to mediated by MAP kinases, at least Erks, because treatment of cells with cadmium not only induced activation of all three members of the MAP kinase family, but also induction of AP-1 by cadmium could be dramatically inhibited by pretreatment of cells with PD98059. These data suggest that cadmium-induced the activation of MAP kinases, especially Erks, is involved in its AP-1 induction.

The important role of AP-1 for tumor promotion has been demonstrated in different cell models by different groups [25–35]. AP-1 was first considered to be a mediator of tumor promotion because of its ability to alter gene expression in response to tumor promoters, including epidermal growth factor, TPA, and UV radiation [25]. AP-1 activity was found to be progressively elevated in mouse epidermal JB6 cells representing various stages of tumor promotion [27]. The activation of AP-1 appears to be required for the preneoplastic-to-neoplastic progression of JB6 cells. Overexpression of c-Jun in P^+ cells caused neoplastic transformation and the introduction of a dominant negative mutant of c-Jun blocked tumor promoter-induced cell transformation [27]. Also, blocking AP-1 activity by either pharmacological inhibitors or molecular biological inhibitors could impair tumor development. These inhibitors include PD98059, SB202190, Wortmannin, Ly294002, Insp6, fluocinolane acetonide, retinoic acid, ascorbic acid, dominant negative Erks, dominant negative JNKs and dominant negative P1-3 kinase [19, 21, 28–31]. Moreover, acquisition of a tumor promotion-resistant phenotype is associated with the loss of responsiveness to tumor promoter-induced AP-1 activation [32]. Introducing wild-type Erk_2 converts tumor promotion-resistant phenotype to tumor promotion-sensitive phenotype [21]. In contrast, overexpression of dominant negative mutant Erk_2 blocks TPA- or EGF-induced AP-1 activation and cell transformation [33]. The requirement of AP-1 in tumor promotion was further supported by findings that the retinoid SR11302, an AP-1 inhibition-specific retinoid, markedly inhibited TPA-induced papilloma formation and AP-1 activation *in vivo*, whereas repeated application of another retinoid, SR11235, a transcriptional activator of the retinoic acid response element, did not inhibit papilloma formation and AP-1 activation [16]. The promoter region of most TPA-response genes includes several protooncogenes, such as c-fos, c-Jun, and matrix degrading metalloproteinases contain the AP-1 binding sequence or TRE [25, 34, 35]. Downstream genes with a regulatory element AP-1 may be involved in the process of tumor promotion. Wilson *et al.* [36] reported that an absence of metalloproteinase matrilysin resulted in a reduction in mean tumor multiplicity of ~ 60% and metalloproteinase matrilysin contributed to the very early stage of tumor development [36]. In light of the important role of AP-1 activation in tumor development, we investigated the signal transduction pathways leading to AP-1 activation by cadmium *in vitro* and *in vivo*. The data obtained from this study may provide a molecular mechanism to explain the induction of some proto-oncogenes, such as c-jun, in previous reports. In addition cadmium-induction of AP-1 activation may play a role in cadmium-induced carcinogenesis.

References

1. International Agency for Research on Cancer: Beryllium, cadmium, mercury and exposures in the glass manufacturing industry. IARC Monographs on the evaluation of carcinogenesis. Risks to Humans 58: 41–117, 1993
2. Degraeve N: Carcinogenic, tertogenic, and mutagenic effects of cadmium. Mutat Res 86: 115–135, 1981
3. Beyersmann D, Hechtenberg S: Cadmium, gene regulation and cellular signaling in mammalian cells. Toxicol Appl Pharmacol 144: 247–261, 1997
4. Buzard GS, Kasprzak KS: Possible roles of nitric oxide and redox cell signaling in metal-induced toxicity and carcinogenesis: A review. J Environ Pathol Toxicol Oncol 19: 179–199, 2000
5. Rossman TG, Roy NK, Lin W-C: Is cadmium genotoxic? In: G.F. Nordberg, R.F.M. Herber, L. Alessio (eds). Cadmium in the Human Environment: Toxicity and Carcinogenicity. International Agency for Research on Cancer, Lyon 22, pp 367–375
6. Meplan C, Mann K, Hainaut P: Cadmium induced conformational modifications of wide-type p53 and suppresses p53 response to DNA damage in culture cells. J Biol Chem 274: 31663–31670, 1999
7. Ishikawa T, Igarashi T, Hata K, Fujita T: C-fos induction by heat, arsenite, and cadmium is mediated by a heat shock element in its promoter. Biochem Biophys Res Commun 254: 566–571, 1999
8. Waalkees MP, Rehm S: Cadmium and prostate cancer. J Toxicol Environ Health 43: 251–269, 1994
9. Achanzar WE, Achanzar KB, Lewis JG, Webber MM, Waalkees MP: Cadmium induced c-myc, p53, and c-jun expression in normal human prostate epithelial cells as a prelude to apoptosis. Toxicol Appl Pharmacol 164: 291–300, 2000
10. Abshire MK, Buzard GS, Shiraishi N, Waalkes MP: Induction c-myc and c-jun proto-oncogene expression in rat L6 myoblasts by cadmium in inhibited by zinc preinduction of the metallothionein gene. J Toxicol Environ Health 48: 359–377, 1996
11. Zheng H, Liu J, Choo KHA, Michalska AE, Klaassen CD: Metallothionein-I and -II knock-out mice are sensitive to cadmium-induced liver mRNA expression of c-jun and p53. Toxicol Appl Pharmacol 136: 229–235, 1996
12. Ding W, Templeton DM: Stress-activated protein kinase-dependent induction of c-fos by Cd^{2+} is mediated by MKK7. Biochem Biophys Res Commun 273: 718–722, 2000
13. Coogan TP, Shiraishi N, Wallkes MP: Minimal basal activity and lack of metal-induced activation of the metallothionein gene correlates with lobe-specific sensitivity to the carcinogenic effects of cadmium in the rat prostate. Toxicol Appl Pharmacol 132: 164–173, 1995
14. Chuang S-M, Wang I-C, Yang J-L: Role of JNK, p38 and ERK mitogen-activated protein kinases in the growth inhibition and apoptosis induced by cadmium. Carcinogenesis 7: 1423–1432, 2000

15. Elbirt KK, Whitmarch AJ, Davis RJ, Bonkovsky HL: Mechanism of sodium arsenite-mediated induction of heme oxygenase-1 in hepatoma cells: Role of mitogen-activated protein kinases. J Biol Chem 273: 8922–8931, 2000
16. Matsuoka M, Igisu H: Activation of c-jun NH_2-terminal kinase (JNK/SAPK) in LLC-PK1cells by cadmium. Biochem Biophys Res Commun 251: 527–532, 1998
17. Huang C, Ma W-Y, Boden GT, Dong Z: Ultraviolet B-induced activated AP-1 activation does not require epidermal growth factor receptor but is blocked by a dominant negative PKC. J Biol Chem 49: 31262–31268, 1996
18. Rincon M, Flavell RA: AP-1 transcriptional activity requires both T-cell receptor-mediated and co-stimulatory signals in primary T lymphocytes. EMBO J 13: 4370–4381, 1994
19. Huang C, Ma W-Y, Dawson MI, Rincon M, Flavell RA, Dong Z: Blocking activator protein-1 activity, but not activating retinoic acid response element, is required for the antitumor promotion effect of retinoic acid. Proc Natl Acad Sci USA 94: 5826–5830, 1997
20. Huang C, Ma W-Y, Hanenberger D, Cleary MP, Bowden GT, Dong Z: Inhibition of ultraviolet B-induced activator-protein-1 (AP-1) activity by aspirin in AP-1-luciferase transgenic mice. J Biol Chem 272: 26325–26331, 1997
21. Huang C, Ma W-Y, Young MR, Colburn N, Dong Z: Shortage of mitogen-activation protein kinase in responsible for resistance to AP-1 transactivation and transformation in mouse JB6 cells. Proc Natl Acad Sci USA 95: 156–161, 1998
22. Snow ET: Metal carcinogenesis: mechanistic implications. Pharmacol Therapeut 53: 31–65, 1992
23. Piscator M: Role of cadmium in carcinogenesis with special reference to cancer of the prostate. Environ Health Perspect 40: 107–120, 1981
24. Waalkes MP, Oberdorster G: Cadmium carcinogenesis. In: E.C. Foulkes (ed). Biological Effects of Heavy Metals, vol. 2. CRC Press, Boca Raton, FL, pp 129–158
25. Angel P, Karin M: The role of Jun, Fos, and the AP-1 complex in cell-proliferation and transformation. Biochim Biophy Acta 1072: 129–157, 1991
26. Dong Z, Watts RG, Sun Y, Colburn NH: Progressive elevation of AP-1 activity during preneoplastic-to-neoplastic progression as modeled in mouse JB6 cell variants. Int J Oncol 7: 359–364, 1995
27. Dong, Z, Birrer MJ, Watts RG, Matrisian LM, Colburn NH: Blocking tumor promoter induced AP-1 activity inhibits induced transformation in JB6 cells. Proc Natl Acad Sci USA 91: 609–613, 1994
28. Huang C, Ma W-Y, Dong Z: Inhibitory effects of ascorbic acid on AP-1 activity and transformation in JB6 cells. Int J Oncol 8: 389–393, 1996
29. Huang C, Ma W-Y, Hecht SS, Dong Z: Inositol hexaphosphate inhibits cell transformation and AP-1 activation by targeting phosphatidylinositol-3 kinase. Cancer Res 57: 2873–2878
30. Huang, C, Ma W-Y, Dong Z: Requirement of phosphatidylinositol-3 kinase in epidermal growth factor-induced AP-1 transactivation and transformation in JB6 P^+ cells. Mol Cell Biol 16: 6427–6435, 1996
31. Huang C, Li J, Ma W-Y, Dong Z: JNKs activation is required for JB6 cell transformation induced by TNF-α but not by TPA. J Biol Chem 274: 29672–29676, 1999
32. Bernstein LR, Colburn NH: AP-1/jun function is differentially induced in promotion-sensitive and resistant JB6 cells. Science 244: 566–569, 1989
33. Watts RG, Huang C, Young MR, Li JJ, Dong Z, Pennie WD, Colburn NH: Expression of dominant negative Erk2 inhibits AP-1 transactivation and neoplastic transformation. Oncogene 17: 3493–3498, 1998
34. Angel P, Imagawa M, Chiu R, Stein B, Imbra RJ, Rahmsdorf HJ, Jonat C, Herrlich P, Karin M: Phorbol ester-inducible genes contain a common cis element recognized by a TPA-modulated trans-acting factor. Cell 49: 729–739, 1987
35. Cohen DR, Curran T: Fra-1: A serum-inducible, cellular immediate-early gene that encodes a fos-related antigen. Mol Cell Biol 8: 2063–2069, 1988
36. Wilson CL, Heppner KJ, Labosky PA, Hogan BL, Matrisian LM: Intestinal tumorigenesis is suppressed in mice lacking the metalloproteinase matrilysin. Proc Natl Acad Sci USA 94: 1402–1407, 1997

Molecular and Cellular Biochemistry **222:** 149–158, 2001.

Chromium (VI)-induced oxidative stress, apoptotic cell death and modulation of p53 tumor suppressor gene

Debasis Bagchi, Manashi Bagchi and Sidney J. Stohs
Creighton University School of Pharmacy and Allied Health Professions, Omaha, NE, USA

Abstract

Chromium (VI) is a widely used industrial chemical, extensively used in paints, metal finishes, steel including stainless steel manufacturing, alloy cast irons, chrome, and wood treatment. On the contrary, chromium (III) salts such as chromium polynicotinate, chromium chloride and chromium picolinate, are used as micronutrients and nutritional supplements, and have been demonstrated to exhibit a significant number of health benefits in rodents and humans. However, the cause for the hexavalent chromium to induce cytotoxicity is not entirely understood. A series of *in vitro* and *in vivo* studies have demonstrated that chromium (VI) induces an oxidative stress through enhanced production of reactive oxygen species (ROS) leading to genomic DNA damage and oxidative deterioration of lipids and proteins. A cascade of cellular events occur following chromium (VI)-induced oxidative stress including enhanced production of superoxide anion and hydroxyl radicals, increased lipid peroxidation and genomic DNA fragmentation, modulation of intracellular oxidized states, activation of protein kinase C, apoptotic cell death and altered gene expression. In this paper, we have demonstrated concentration- and time-dependent effects of sodium dichromate (chromium (VI) or Cr (VI)) on enhanced production of superoxide anion and hydroxyl radicals, changes in intracellular oxidized states as determined by laser scanning confocal microscopy, DNA fragmentation and apoptotic cell death (by flow cytometry) in human peripheral blood mononuclear cells. These results were compared with the concentration-dependent effects of chromium (VI) on chronic myelogenous leukemic K562 cells and J774A.1 murine macrophage cells. Chromium (VI)-induced enhanced production of ROS, as well as oxidative tissue and DNA damage were observed in these cells. More pronounced effect was observed on chronic myelogenous leukemic K562 cells and J774A.1 murine macrophage cells. Furthermore, we have assessed the effect of a single oral LD_{50} dose of chromium (VI) on female C57BL/6Ntac and p53-deficient C57BL/6TSG p53 mice on enhanced production of superoxide anion, lipid peroxidation and DNA fragmentation in the hepatic and brain tissues. Chromium (VI)-induced more pronounced oxidative damage in p53 deficient mice. This *in vivo* study highlighted that apoptotic regulatory protein p53 may play a major role in chromium (VI)-induced oxidative stress and toxicity. Taken together, oxidative stress and oxidative tissue damage, and a cascade of cellular events including modulation of apoptotic regulatory gene p53 are involved in chromium (VI)-induced toxicity and carcinogenesis. (Mol Cell Biochem **222**: 149–158, 2001)

Key words: chromium (VI), oxidative stress, *in vitro*, *in vivo*, superoxide anion, hydroxyl radicals, MTT assay, DNA damage, cell cycle analysis, apoptosis, p53 gene

Introduction

Chromium occurs in the workplace predominantly in the oxidation states 3 (III) and 6 (VI). Chromium (VI) salts find extensive application in diverse industries, while trivalent chromium salts such as chromium polynicotinate and chromium picolinate, are used as micronutrients and dietary supplements. It has also been reported that trivalent chromium

Address for offprints: D. Bagchi, Creighton University School of Pharmacy and Allied Health Professions, 2500 California Plaza, Omaha, NE 68178, USA salts are poorly absorbed through the gastrointestinal tract. Chromium (VI) is widely known to cause allergic dermatitis, as well as toxic and carcinogenic effects in humans and animals [1, 2]. Von Burg and Liu [2], and Barceloux [3] have demonstrated the acute and chronic toxicity, neurotoxicity, dermatotoxicity, genotoxicity, carcinogenicity, immunotoxicity, and general environmental toxicity of hexavalent chromium.

Soluble and insoluble hexavalent chromium salts have been demonstrated to induce morphological and neoplastic transformation and mutagenicity in murine and human cells [4]. Soluble hexavalent chromium compound calcium chromate ($CaCrO_4$) has been shown to induce dose-dependent cytotoxicity in C3H/10T1/2 mouse embryo cells and dose-dependent cytotoxicity and mutagenesis in Chinese hamster ovary cells, while insoluble salt lead chromate ($PbCrO_4$) induced cytototoxicity as well as morphological and neoplastic transformation in C3H/10T1/2 mouse embryo cells [4].

Hexavalent chromium compounds induce dose-dependent cytotoxicity and anchorage independence in cultured human diploid fibroblasts [5]. Furthermore, hexavalent chromium compounds have been demonstrated to be approximately 1,000-fold more cytotoxic and mutagenic than trivalent chromium compounds in cultured diploid human fibroblasts, but both hexavalent and trivalent chromium compounds induce dose-dependent anchorage independence in human diploid fibroblasts [6].

Shi and Dalal [7, 8] have used ESR to demonstrate the formation of long-lived Cr (V) intermediates in the reduction of Cr (VI) by glutathione reductase in the presence of NADPH, and the generation of hydroxyl radical. Hydrogen peroxide suppresses Cr (V) and enhances the formation of hydroxyl radicals through a Cr (V) catalyzed Fenton-like reaction. Subsequent investigations with superoxide dismutase showed no significant participation of superoxide anion in the production of hydroxyl radicals. Related studies by Jones *et al.* [9] have provided evidence that suggests that hydroxyl radicals are generated from a Cr (V) intermediate that is responsible for causing DNA single strand breaks. Kawanishi *et al.* [10] have demonstrated that chromium (VI) produces noxious ROS including superoxide anion, singlet oxygen and hydroxyl radicals, through the formation of Cr (V) intermediates. These investigators examined the mechanism of DNA cleavage induced by Cr (VI) in the presence of hydrogen peroxide. ROS were produced by the decomposition of $Cr(V)(O_2)_4^{-3}$ ion, resulting in DNA damage. The generation of hydroxyl radical was also detected by ESR. These results indicate that Cr (V) complexes produced in the reduction of Cr (VI) by cellular reductants react with hydrogen peroxide to generate hydroxyl radicals, which may be the initiators of the primary events in chromium (VI) cytotoxicity. The chromate ion $(CrO_4)^{-2}$, the dominant form of chromium (VI) in neutral aqueous solutions, can readily cross cellular membranes via non-specific anion carriers [11], while chromium (III) is poorly transported across membranes. The differences in membrane transport may explain the differences in the abilities of these two oxidation states of chromium to induce the formation of ROS and produce oxidative tissue damage.

The present work highlights the results of a series of *in vitro* and *in vivo* studies which demonstrate that oxidative stress, apoptosis and the apoptotic regulatory gene p53 are involved in the toxicity and carcinogenicity of chromium (VI) ions.

Materials and methods

Chemicals

Sodium dichromate (chromium (VI) or Cr (VI)), 2,3-dihydroxybenzoic acid, 2,5-dihydroxybenzoic acid, salicylic acid, and HPLC-grade water were purchased from Aldrich Chemical Co. (Milwaukee, WI, USA). Sodium acetate trihydrate and citric acid used for HPLC analysis were purchased from Fluka (Buchs, Switzerland). 2,7-dichlorofluorescein acetate (DCFD) was purchased from Molecular probes (Eugene, OR, USA). All other chemicals were obtained from Sigma Chemical Co. (St. Louis, MO, USA), and were of analytical grade or higher and used without purification.

Cell cultures and treatment

Human chronic myelogenous leukemic K562 cells and murine macrophage J774A.1 cells were obtained from the American Type Culture Collection (Rockville, MD, USA). The human K562 and normal human donor peripheral blood mononuclear cells (HPBM) were maintained and grown in RPMI 1640 medium with 10% fetal calf serum, while J774A.1 cell line was maintained and grown in Dulbecco's modified Eagle's medium (DMEM) containing 10% fetal bovine serum. The cultured K562 and HPBM cells (1×10^6 cells/35 mm petri dish) were incubated with 0, 12.5 and 25 μM concentrations of chromium (VI) for 0, 24, or 48 h. Cultured J774A.1 cells were plated at 25×10^4 cells per 35 mm petri dish (Corning Inc., Corning, NY, USA), and incubated with either 0, 0.20, 0.40 and 0.60 μM concentrations of chromium (VI) at 37°C for 0, 24 and 48 h. Cell viability was determined by trypan blue exclusion technique.

Animals and treatment

Female C57BL/6Ntac and p53-deficient C57BL/6TSG p53 mice (4 weeks old) were obtained from Taconic, Germantown, NY, USA). The animals were housed in a controlled environment at 25°C with a 12 h light/dark cycle, and acclimated for at least 3–5 days before use. The animals were treated with a single oral 0.50 LD_{50} dose (95 mg/kg body wt) of sodium dichromate (chromium (VI) or Cr (VI)) and sacrificed after 24 h. Hepatic and brain tissues were collected and preserved properly for biochemical assays.

Cytochrome c *reduction*

Superoxide anion production was determined on the basis of cytochrome c reduction following the method of Babior *et al.* [12] and as previously described by us [13]. Absorbance values were converted to nmoles of cytochrome *c* reduced/15 min/mg protein or nmoles cytochrome *c* reduced/15 min/number of cells used in the study, using the extinction coefficient 2.1×10^4 $M^{-1} cm^{-1}$.

Hydroxyl radical detection by HPLC

The detection of hydroxyl radicals based on the formation of 2,3-dihydroxybenzoic and 2,5-dihydroxybenzoic acid was determined using HPLC equipped with a Waters 460 electrochemical detector as previously described by us [14] using 100 μM sodium salicylate as the substrate. The hydroxylated products of salicylic acid after interaction with hydroxyl radicals, 2,3-dihydroxybenzoic and 2,5-dihydroxybenzoic acids, were eluted with a mobile phase containing 0.03 mol/l sodium acetate and 0.03 mol/l of citric acid (pH 3.6) at a flow rate of 1 ml/min. The detection potential was maintained at +0.6 V, employing a glassy carbon working electrode and an Ag/AgCl reference electrode. Retention times for the peaks of 2,3-dihydroxybenzoic and 2,5-dihydroxybenzoic acids were verified by injecting authentic standards.

Lipid peroxidation

Lipid peroxidation was determined as previously described [13], based on the formation of thiobarbituric acid reactive substances (TBARS). Malondialdehyde was used as standard. A molar extinction coefficient of 1.56×10^5 $M^{-1} cm^{-1}$ was used.

DNA fragmentation

Cell or frozen tissue samples were homogenized in lysis buffer (5 mM Tris-HCl, 20 mM EDTA, 0.5% Triton X-100, pH 8.0). Homogenates were centrifuged at 27,000 g for 20 min to separate intact chromatin in the pellets from fragmented DNA in the supernatant fractions. Pellets were resuspended in 0.5 N perchloric acid, and 5.5 N perchloric acid was added to the supernatant fractions to reach a concentration of 0.5 N. Samples were heated at 90°C for 15 min and centrifuged at 1,500 g for 10 min to remove protein. Resulting supernatant fractions were reacted with diphenylamine for 16–20 h at room temperature. Absorbances were measured at 600 nm. DNA fragmentation in control samples is expressed as percent of total DNA appearing in the supernatant fraction. Treatment effects are reported as percent of control fragmentation [14].

DNA ladder analysis by gel electrophoresis

Qualitative analysis of the DNA damage for chromium VI (0–0.60 μM)-treated cells was also assessed by the DNA ladder technique using agarose gel electrophoresis as described by us previously [14]. The effect of various concentrations of sodium dichromate (chromium (VI)) was assessed using calcium activated endonuclease-dependent ladder-like DNA fragmentation by gel electrophoresis. Briefly, the chromium (VI)-treated cells were washed with chilled Tris EDTA sarkosyl (0.1 M Tris HCl/10 mM/ EDTA/0.5 N lauroyl sarcosine, pH 7.8) buffer and digested with proteinase K (50 mg/ml) at 50°C for several hours. After digestion, DNA was extracted three times with phenol/chloroform (1:1) and precipitated with absolute ethanol/2.5 M sodium acetate (pH 4.0). DNA was quantified spectrophotometrically at 260 nm and loaded onto an agarose gel (10 μg DNA/lane), and electrophoresed at 60 V (Heoffer Instruments, CA, USA) at a voltage of 60 V. A Hind III digest of DNA served as a molecular marker. Gels were illuminated with 300 nm UV light and a photographic record of laddering was made with instant Polaroid film No. 667.

Laser scanning confocal microscopy

The overall intracellular oxidized states of the cultured cells following incubation with various concentrations of chromium (VI) were determined at an excitation wavelength of 513 nm using a Meridian ACS 570 Scanning Confocal Laser Microscope (Okemos, MI, USA) and 2,7-DCFD was used as the fluorescent probe. For assays, media was replaced with Hank's solution containing 5 μM DCFD. After 5 min of incubation at room temperature fluorescent intensity was assessed according to the method previously used by us [14]. Signals were quantified by integrating fluorescent intensity over a user-defined area cell number. The relative fluorescent intensity of each cell was calculated using the computer program of the instrument relative to the untreated control cells as the standard. At least twenty randomly selected cells were used for individual time-point determination.

Cell cycle analysis

Chromium (VI)-treated and untreated cells were collected by centrifugation and resuspended in 1 ml of 70% ethanol, and cell cycle analyses were conducted as described by us previously [14]. Briefly, cells were resuspended in Telford's reagent containing EDTA, RNAase, propidium iodide, Triton X-100, and PBS, incubated at 4°C for 24 h, and analyzed using a Becton Dickinson FACSTAR Plus Flow Cytometer (San Jose, CA, USA) equipped with a Cell Quest Software Program. The percentages of cells within the G_0/G_1, S, and G/M phases of the cell cycle were determined by analysis with the software program. The percentages of apoptotic cells were calculated from the forward angle light scatter vs. linear red fluorescence histogram of the Ao population as described by Telford *et al.* [15] and by us [14].

Detection of apoptotic cells

Cells were fixed in 4% buffered paraformaldehyde and embedded in paraffin. The terminal deoxynucleotidyl transferase (TdT)-mediated dUTP biotin nick end labeling method (TUNEL) method, developed by Gavrieli *et al.* [16], was used to visualize DNA fragmentation at the single cell level. In the TUNEL method, 3-OH DNA ends generated by DNA fragmentation are nick end labeled with biotinylated dUTP, introduced by TdT, and counter stained with propidium iodide. A MEBSTAIN Apoptosis Kit from Medical and Biological Laboratories (Nagoya, Japan) was used for fluorescent detection of the dead cells versus live cells.

MTT [3-(4,5-dimethylthiazol-2-yl)-2,5-diphenyltetrazolium bromide] cleavage assay

Respiratory effects of chromium (VI) on exposed J774A.1 cells were assessed by determining the ability of the cells to reduce the tetrazolium dye MTT based on the activation of the enzyme succinate dehydrogenase as described by Mosman [17] and by us [18]. The optical density was read at 560 nm and the amount of formazan produced was calculated by using molar extinction coefficient of 51,000.

Statistical methods

The presence of significant differences between groups was determined using analysis of variance (ANOVA), with Scheffe's S method as the *post hoc* test. Each value is the mean ± S.D. from 4–6 experiments. The level of statistical significance employed in all cases was $p < 0.05$.

Results

Cytochrome c *reduction (superoxide anion production)*

Human peripheral blood mononuclear (HPBM) cells were treated with concentrations of 12.5 and 25 μM sodium dichromate (chromium (VI) or Cr (VI)) for 24 h. 1.4- and 1.55-fold increases were observed in cytochrome *c* reduction at these concentrations of chromium (VI), respectively (Table 1). After treatment of these cells for 48 h, we observed a 1.4- and 1.7-fold increases in cytochrome *c* reduction in these cells treated with 12.5 and 25 μM chromium (VI) (Table 1). Following treatment of the human chronic myelogenous leukemic K562 cells with 12.5 and 25 μM concentrations of chromium (VI) for 24 h, approximately 2.7- and 4.8-fold increases in cytochrome *c* reduction were observed, respectively (Table 2).

Cytochrome *c* reduction was assayed as a marker of superoxide anion production in the liver and brain tissues obtained from both C57BL/6Ntac and p53 deficient C57BL/6TSG p53 mice. Increases in cytochrome *c* reduction were observed in the liver and brain tissues of both strains of mice. Following treatment with an acute oral 0.50 LD_{50} dose of chromium (VI) approximately 6.0- and 4.1-fold increases in cytochrome *c* reduction were observed in the liver and brain tissues of C57BL/6Ntac mice, respectively, while under these same conditions, chromium (VI)-induced 11.1- and 7.7-fold increases in cytochrome *c* reduction in the liver and brain tissues of p53 deficient C57BL/6TSG p53 mice, respectively, as compared to the control untreated mice (Tables 3 and 4). Thus, significantly higher production of superoxide anion was observed in both the liver and brain tissues of p53 deficient C57BL/6TSG p53 mice as compared to the corresponding control animals.

Hydroxyl radical production

Following treatment of the HPBM cells for 24 h with 12.5 and 25 μM concentrations of chromium (VI), approximately 2.4- and 2.5-fold increases in hydroxyl radical production were observed, respectively, and under these same conditions 2.7- and 3.2-fold increases in hydroxyl radical production were observed after 48 h, respectively (Table 1). Approximately 10.7- and 14.1-fold increases in hydroxyl radical production were observed in human K562 cells following treatment with 12.5 and 25 μM concentrations of chromium (VI) for 24 h, respectively, as compared to control untreated cells (Table 2).

Laser scanning confocal microscopy

No significant modulation of intracellular oxidized states in HPBM cells were observed following treatment of these cells

Table 1. Concentration- and time-dependent changes on the enhanced production of superoxide anion and hydroxyl radicals, changes in intracellular oxidized states, DNA fragmentation and apoptotic cell death in human peripheral blood mononuclear (HPBM) cells after treatment with sodium dichromate (chromium (VI) or Cr (VI))

Treatment	Cytochrome *c* reduction (nmoles reduced/15 min/ 1 × 10⁶ cells)	Hydroxyl radical production			Flourescent intensity	DNA fragmentation (% control)	Apoptotic cell death (%)
		2,5-DHBA (nmoles/ml)	2,3-DHBA (nmoles/ml)	Total (nmoles/ml)			
Control 24 h	56.1 ± 8.2[a]	0.12 ± 0.02[a]	0.22 ± 0.03[a]	0.34[a]	89 ± 24[a]	3.16 ± 0.70[a] (100)	0[a]
Control 48 h	59.4 ± 4.3[a]	0.10 ± 0.01[a]	0.25 ± 0.04[a]	0.35[a]	85 ± 15[a]	3.07 ± 0.26[a] (100)	ND
Cr (VI) 12.5 μM 24 h	79.4 ± 13.0[b]	0.21 ± 0.05[b]	0.61 ± 0.05[b]	0.82[b]	87 ± 22[a]	3.19 ± 0.45[a] (101)	0[a]
Cr (VI) 12.5 μM 48 h	82.0 ± 19.1[b]	0.25 ± 0.06[b]	0.65 ± 0.09[b]	0.90[b]	188 ± 56[b]	3.45 ± 0.62[a] (109)	ND
Cr (VI) 25 μM 24 h	86.9 ± 16.8[b]	0.33 ± 0.04[c]	0.50 ± 0.06[b]	0.83[b]	ND	4.04 ± 0.72[b] (128)	0[a]
Cr (VI) 25 μM 48 h	98.4 ± 22.3[c]	0.39 ± 0.06[c]	0.68 ± 0.08[b]	1.07[c]	ND	4.29 ± 0.75[b] (136)	ND

Cultured HPBM cells were treated with 12.5 or 25 μM concentration of sodium dichromate (chromium (VI) or Cr (VI)). See Materials and methods section for details. Each value is the mean ± S.D. of 4–6 experiments. Values with non-identical superscripts are significantly different ($p < 0.05$). ND = Not determined.

with 12.5 μM concentration of chromium (VI) for 24 h. However, treatment of HPBM cells with 12.5 μM concentration of chromium (VI) for 48 h resulted in 2.2-fold increase in fluorescent intensity (Table 1). Approximately 5.7-fold increase in fluorescent intensity was observed following treatment of the chronic myelogenous leukemic K562 cells with 12.5 μM concentration of chromium (VI) for 24 h as compared with the control untreated cells. Due to extensive cell necrosis at 48 h, fluorescent intensity could not be measured at 48 h (Table 2).

Concentration-dependent increases in fluorescent intensity were observed following treatment of the J774A.1 macrophage cells with chromium (VI). Approximately 2.8-, 10.1- and 9.2-fold increases in fluorescent intensities were observed following incubation of the cells with 0.20, 0.40 and 0.60 μM concentrations of chromium (VI), respectively, as compared to the control untreated cells (Table 5).

Reduction of MTT

Concentration- and time-dependent effects were studied on the reductions of tetrazolium dye MTT in the cultured J774A.1 macrophages. The cells were incubated with 0.20, 0.40 and 0.60 μM concentrations of chromium (VI) for 24 and 48 h. This assay measures the activity of succinate dehydrogenase, a marker of the mitochondrial electron transport chain. Treatment of the cells with 0.20 and 0.40 μM concentrations of chromium (VI) for 24 h resulted in no significant changes in succinate dehydrogenase activity. When these cells were treated with 0.60 μM concentration of chromium (VI) for 24 h, a small but non-significant increase in succinate dehydrogenase activity was observed. Incubation of the macrophage cells with the three concentrations of chromium (VI) for 48 h resulted in dose-dependent decreases in succinate dehydrogenase activities (Table 5).

Table 2. Concentration-dependent changes in enhanced production of superoxide anion and hydroxyl radicals, changes in intracellular oxidized states, DNA fragmentation and apoptotic cell death in chronic myelogenous leukemic K562 cells following 24 h treatment with sodium dichromate (chromium (VI) or Cr (VI))

Treatment	Cytochrome *c* reduction (nmoles reduced/15 min/ 1 × 10⁶ cells)	Hydroxyl radical production			Flourescent intensity	DNA fragmentation (% control)	Apoptotic cell death (%)
		2,5-DHBA (nmoles/ml)	2,3-DHBA (nmoles/ml)	Total (nmoles/ml)			
Control	99.0 ± 6.7[a]	0.18 ± 0.05[a]	0.13 ± 0.02[a]	0.31	84 ± 14[a]	8.17 ± 1.80[a] (100)	0[a]
Cr (VI) 12.5 μM	262.6 ± 18.7[b]	0.80 ± 0.21[b]	2.51 ± 0.53[b]	3.31[b]	475 ± 214[a]	18.06 ± 1.12[a] (221)	39.5 ± 6.3[a]
Cr (VI) 25 μM	479.6 ± 41.6[c]	0.89 ± 0.14[c]	3.49 ± 0.66[b]	4.38[c]	ND	24.84 ± 1.86[b] (304)	0[a]

Cultured K562 cells were treated with 12.5 or 25 μM concentration of sodium dichromate (chromium (VI) or Cr (VI)). See Materials and methods section for details. Each value is the mean ± S.D. of 4–6 experiments. Values with non-identical superscripts are significantly different ($p < 0.05$). ND = Not determined.

Table 3. Hepatic lipid peroxidation, DNA fragmentation, superoxide anion production in sodium dichromate (chromium (VI) or Cr (VI))-treated control and p53-deficient mice

Assay	Control (C57BL/6Ntac) mice		p53-deficient (TSG-p53) mice	
	Control	Cr (VI)-treated	Control	Cr (VI)-treated
Lipid peroxidation (nmoles/mg protein)	2.09 ± 0.25[a]	6.84 ± 0.91[b]	1.89 ± 0.26[a]	20.39 ± 3.16[a]
DNA fragmentation (% control)	4.57 ± 0.72[a] (100)	10.06 ± 1.32[b] (220)	3.03 ± 0.46[a] (100)	12.26 ± 1.62[b] (405)
Superoxide anion (nmoles cytochrome *c* reduced/15min/mg/protein)	8.31 ± 1.14[a]	49.86 ± 6.76[b]	6.71 ± 0.90[a]	74.30 ± 10.29[b]

Control C57BL/6Ntac mice and p53-deficient C57BL/6TSG p53 mice were treated with a single oral 0.50 LD_{50} dose of sodium dichromate (chromium (VI)) (95 mg/kg body wt) and sacrificed after 24 h. Each value is the mean ± S.D. of 4 mice. Assays were performed, as described in the Materials and methods section. Values with non-identical superscripts are significantly different ($p < 0.05$).

Table 4. Brain lipid peroxidation, DNA fragmentation, superoxide anion production in sodium dichromate (chromium (VI) or Cr (VI)-treated control and p53-deficient mice

Assay	Control (C57BL/6Ntac) mice		p53-deficient (TSG-p53) mice	
	Control	Cr (VI)-treated	Control	Cr (VI)-treated
Lipid peroxidation (nmoles/mg protein)	1.69 ± 0.19[a]	5.88 ± 0.81[b]	2.02 ± 0.24[a]	14.71 ± 1.69[b]
DNA fragmentation (% control)	2.96 ± 0.40[a] (100)	6.20 ± 0.51[b] (220)	2.76 ± 0.24[a] (100)	12.55 ± 1.22[b] (405)
Superoxide anion (nmoles cytochrome *c* reduced/15min/mg/protein)	5.78 ± 0.41[a]	23.83 ± 5.56[b]	7.34 ± 0.65[a]	56.80 ± 8.94[b]

Control C57BL/6Ntac mice and p53-deficient C57BL/6TSG p53 mice were treated with a single oral 0.50 LD_{50} dose of sodium dichromate (chromium (VI)) (95 mg/kg body wt) and sacrificed after 24 h. Each value is the mean ± S.D. of 4 mice. Assays were performed, as described in the Materials and methods section. Values with non-identical superscripts are significantly different ($p < 0.05$).

Table 5. Concentration-dependent changes in fluorescent intensity, DNA fragmentation and reduction of tetrazolium dye MTT in cultured J774A.1 murine macrophage cells following treatment with sodium dichromate (chromium (VI) or Cr (VI))

Treatment	Fluorescent intensity	DNA fragmentation	Cell viability (%)	
			24 h	48 h
Control	335 ± 40[a]	7.25 ± 0.63[a] (100)	1.07 ± 0.11[a] (100)	1.02 ± 0.16[a] (100)
Cr (VI) 0.20 μM	930 ± 155[b]	13.23 ± 1.89[b] (183)	0.99 ± 0.03[a] (93)	0.71 ± 0.03[b] (70)
Cr (VI) 0.40 μM	3392 ± 617[c]	20.15 ± 0.71[c] (278)	1.14 ± 0.17[a] (107)	0.65 ± 0.06[b,c] (64)
Cr (VI) 0.60 μM	3073 ± 341[c]	21.04 ± 1.08[c] (290)	1.27 ± 0.10[a] (119)	0.58 ± 0.03[c] (57)

Cultured J774A.1 macrophage cells were treated with either 0.20, 0.40, or 0.60 μM sodium dichromate (chromium (VI) or Cr (VI)). See Materials and methods for details. All values are reported as mean ± S.D. from 4–6 replicates. Values with non-identical superscripts are significantly different (*$p < 0.05$).

DNA fragmentation

Following 24 h incubation of the HPBM cells with 12.5 and 25 μM concentrations of chromium (VI), approximately 1.0- and 1.3-fold increases in DNA fragmentation were observed, respectively, and 1.1- and 1.4-fold increases in DNA fragmentation were observed after 48 h, respectively, as compared to the control untreated cells (Table 1). Following treatment of the human K562 cells with 12.5 and 25 μM concentrations of chromium (VI) for 24 h, approximately 2.2- and 3.0-fold increases in DNA fragmentation were observed, respectively (Table 2). Due to extensive cell necrosis at 48 h, DNA fragmentation could not be measured at 48 h. Concentration-dependent increases in DNA fragmentation were observed following incubation of cultured J774A.1 macrophage cells with chromium (VI). Following incubation of J774A.1 macrophage cells with 0.20, 0.40 and 0.60 μM concentrations of chromium (VI) for 24 h approximately 1.8-, 2.8- and 2.9-fold increases in DNA fragmentation were observed, respectively, as compared to control cells (Table 5). The effect of chromium (VI) on ladder-like formation of genomic DNA damage is shown in Fig. 1. Lanes 2–4 demonstrate the concentration-dependent fragmentation of genomic DNA by chromium (VI).

Genomic DNA fragmentation was determined in the liver and brain tissues obtained from both C57BL/6Ntac and p53 deficient C57BL/6TSG p53 mice. Increases in hepatic and brain genomic DNA fragmentation were observed in the both strains of mice. Following treatment with a single oral 0.50 LD_{50} dose of chromium (VI) approximately 2.2- and 2.1-fold increases in DNA fragmentation were observed in the liver and brain tissues of C57BL/6Ntac mice, respectively, as compared to the control untreated mice (Tables 3 and 4). Under these same conditions, approximately 4.1- and 4.6-fold increases in DNA fragmentation were observed in the hepatic and brain tissues of p53 deficient C57BL/6TSG p53 mice, respectively, as compared to the corresponding control animals (Tables 3 and 4). Thus, chromium (VI)-induced more dramatic toxicity in p53-deficient mice.

Flow cytometric analyses

Chromium (VI) didn't induce apoptotic cell death in normal human peripheral blood mononuclear (HPBM) cells at the concentrations and conditions used. On the contrary, following treatment of the human K562 cells with 12.5 μM concentration of chromium (VI) for 24 h approximately 40% apoptotic cell death was observed (Table 2) in conjunction with a significant decrease in the G_2/M phase (data not shown). Due to cell necrosis, no apoptotic cell death was observed in cultured K562 cells following treatment with 25 μM concentration of chromium (VI) for 24 or 48 h, and necrotic cell death was confirmed microscopically (data not shown).

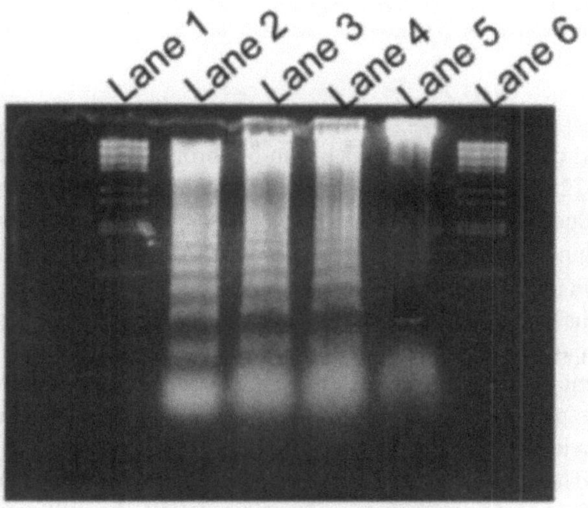

Fig. 1. The electrophoretogram demonstrating sodium dichromate (chromium (VI) or Cr (VI))-induced ladder-like formation of genomic DNA of J774A.1 murine macrophage cells following treatment with 0.20, 0.40 and 0.60 μM concentrations of Cr (VI) for 24 h. Lane 1: Hind III λDNA standard; lane 2: 0.60 μM chromium (VI); lane 3: 0.40 μM chromium (VI); lane 4: 0.20 μM chromium (VI); lane 5: control; lane 6: Hind III λDNA standard.

Chromium (VI)-induced cell morphology and apoptotic cell death in J774A.1 cells

Phase contrast microscopy demonstrated that chromium (VI)-induced concentration-dependent changes in cell morphology. With increasing concentration of chromium (VI), cultured J774A.1 cells lost their adhesion to the culture plates and the cells became typically rounded in appearance. A MEBSTAIN apoptosis kit was used to visualize the effects of 0.20, 0.40 and 0.60 μM concentrations of chromium (VI) on cultured J774A.1 cells. Concentration-dependent effects were observed. An increasing index of apoptotic cell death was evident by the increasing intensity and volume of the green fluorescent color in single cells, demonstrating cytoplasmic condensation with intact membrane and organelle structure (Fig. 2).

Lipid peroxidation

Lipid peroxidation was assayed as a marker of oxidative damage to the lipids in the liver and brain tissues obtained from both C57BL/6Ntac and p53 deficient C57BL/6TSG p53 mice. Increases in lipid peroxidation were observed in the liver and brain tissues of both strains of mice. Following a single acute oral 0.50 LD_{50} dose of chromium (VI) approximately 3.3- and 3.5-fold increases in lipid peroxidation were observed in the liver and brain tissues of C57BL/6Ntac mice, respectively, as compared to the control untreated mice (Tables 3 and 4). Under these identical conditions, chromium (VI)-induced 10.8- and 7.3-fold increases in lipid peroxidation in the liver and brain tissues of p53 deficient C57BL/6TSG p53 mice, respectively (Tables 3 and 4). Thus, significantly higher production of lipid peroxidation was observed in both the liver and brain tissues of p53 deficient C57BL/6TSG p53 mice as compared to the corresponding control animals.

Discussion

A series of *in vitro* and *in vivo* studies were conducted to determine the role reactive oxygen species (ROS) and oxidative stress plays in sodium dichromate (chromium (VI) or Cr (VI))-induced toxicity and carcinogenicity. The relative safety of trivalent chromium has been well documented. Chromium (III) compounds are 1,000-fold less toxic than

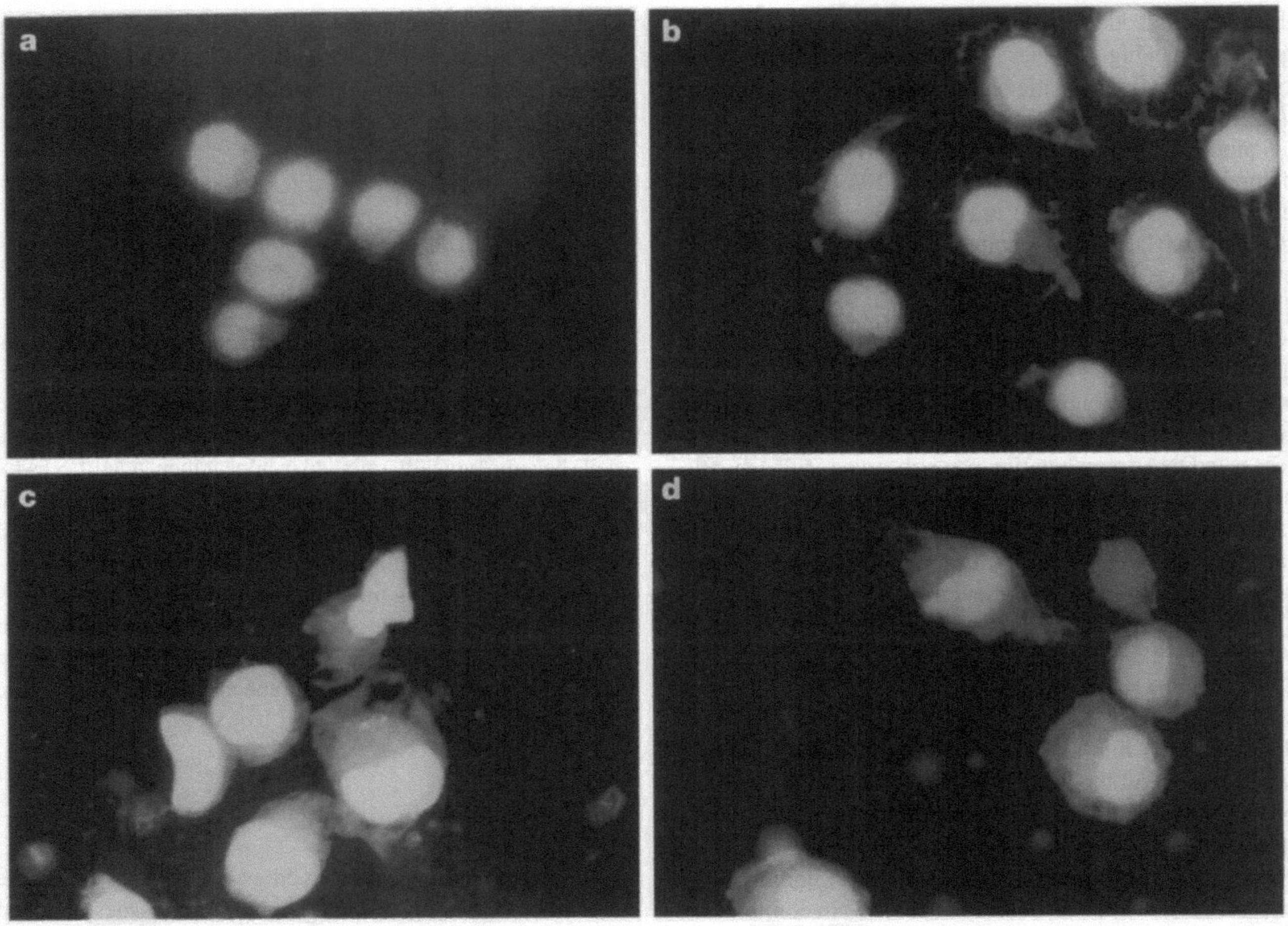

Fig. 2. Sodium dichromate (chromium (VI) or Cr (VI))-induced apoptotic cell death in cultured J774A.1 murine macrophage cells following treatment with 0.20, 0.40 or 0.60 μM concentrations of Cr (VI) for 24 h. The terminal deoxynucleotidyl transferase (TdT) mediateds dUTP biotin nick end labeling (TUNEL) method, developed by Gavrieli *et al.* (1992), was used employing MEBSTAIN apoptosis kit to visualize DNA fragmentation at a single cell level. (a) Control (magnification 40×); (b) 0.20 μM chromium (VI) (magnification 60×); (c) 0.40 μM chromium (VI) (magnification 100×); (d) 0.60 μM chromium (VI) (magnification 100×).

chromium (VI) compounds, but they are still toxic at higher concentrations [2, 3, 6, 13]. It is important to mention that a significant number of studies have demonstrated the adverse toxic effects such as DNA damage, nephrotoxicity and other toxic effects of the nutritional supplement chromium (III) picolonate [19–23]. However, a number of beneficial effects including higher bioavailability have been demonstrated on niacin-bound chromium (III) complex 'chromium polynicotinate' and no adverse or toxic effects have been reported so far [24–27].

Previous studies have demonstrated that chromium (VI) induced an oxidative stress through enhanced production of ROS and nitric oxide, as well as enhanced excretion of oxidative urinary lipid metabolites including malondialdehyde, formaldehyde, acetaldehyde and acetone, activation of protein kinase C in both *in vitro* and *in vivo* models [13, 14, 18, 28–30]. DNA damage can also be induced following acute and chronic oral administration of chromium (VI) [13, 14, 18, 28–30]. Also, at the same concentration chromium (VI) induced more dramatic cytotoxicity as compared to chromium (III) based on increased production of ROS by peritoneal macrophages and hepatic mitochondria and microsomes, enhanced excretion of urinary lipid metabolites, and increased hepatic DNA-single strand breaks [13]. This difference in toxicity is believed to be due to the fact that chromium (VI) can penetrate biological membranes via non-specific anion carriers more readily than chromium (III) [11].

The *in vitro* studies have demonstrated that chromium (VI) induces an oxidative stress in cultured cells. A comparative study was conducted to determine the concentration-dependent effect of chromium (VI) on cultured murine macrophage J774A.1 cells, normal human donor peripheral blood mononuclear cells (HPBM) and human chronic myelogenous leukemic K562 cells. Chromium (VI)-induced a cytotoxic effect of macrophage J774A.1 cells at a much lower dose as compared to other two cells. Modulation of intracellular oxidized states was also observed as assessed by laser scanning confocal microscopy. Chromium (VI)-induced a concentration-dependent effect on macrophage J774A.1 cells with maximal effect observed at 0.40 μM concentration. A differential effect of chromium (VI) was observed in HPBM and human K562 cells. No significant modulation of intracellular oxidized states were observed in HPBM cells following treatment with 12.5 μM concentration of chromium (VI) for 24 h, while under these same conditions a 5.7-fold increase in fluorescent intensity was observed in K562 cells. Approximately 2.2-fold increase in fluorescent intensity was observed following incubation of HPBM cells with 12.5 μM concentration of chromium (VI) for 48 h, while the similar conditions induced cell necrosis in cultured K562 cells. Thus, chromium (VI) was shown to induce more drastic effect on human leukemic cells as compared to human normal cells. Similar results were observed in the production of superoxide anion and hydroxyl radicals. Following incubation of human K562 cells with both 12.5 and 25 μM concentrations of chromium (VI) for either 24 or 48 h induced more dramatic production of superoxide anion and hydroxyl radicals as compared to normal HPBM cells.

Concentration- and time-dependent effects of chromium (VI) were also assessed on the reductions of tetrazolium due MTT in cultured J774A.1 cells. This is a marker of succinate dehydrogenase activity, an index of the mitochondrial electron transport system, which was used to assess cell viability. Only viable cells with intact mitochondria can reduce tetrazolium dye MTT and the amount of MTT reduced is directly proportional to the number of viable cells present. At 24 h time point, chromium (VI) had no effect on the reduction of MTT, suggesting that no loss in cell viability had occurred. However, at the 48 h time point, significant concentration-dependent decreases in MTT reduction were observed. These exposures were accompanied by detachment of dead or dying cells in the macrophage cultures.

The comparative modulation of intracellular oxidized states and enhanced production of superoxide anion (cytochrome *c* reduction) and hydroxyl radicals (HPLC assay) in these cells following incubation with chromium (VI) was further correlated with apoptotic cell death. Induction of apoptotic cell death was demonstrated by three sensitive markers, namely, genomic DNA fragmentation, TUNEL assay and flow cytometry. Apoptosis (programmed cell death) is a widespread phenomenon that plays a crucial role in a myriad of pathophysiological processes, including neoplastic diseases. Apoptosis is accompanied by condensation of cytoplasm, loss of plasma membrane, condensation or fragmentation of nuclei and extensive degradation of chromosomal DNA. The present study demonstrates the toxic effect of chromium (VI) on different cultured cells. Chromium (VI)-induced significant DNA fragmentation in cultured J774A.1 cells and the maximal effect was observed at 0.40 μM concentration of chromium (VI). These genomic DNA fragmentation data were further supported by ladder-like fragmentation of genomic DNA (Fig. 1) and TUNEL assay (Fig. 2). At both concentrations and time points, chromium (VI)-induced more drastic DNA fragmentation of human chronic myelogenous leukemic cells as compared to normal HPBM cells. Similar results were observed in cell cycle analysis. Chromium (VI)-induced no apoptotic cell death on HPBM cells at either concentrations or time points, while 40% apoptotic cell death was observed in human K562 cells treated with 12.5 μM concentration of chromium (VI) for 24 h. Necrotic cell death was observed following treatment of the human K562 cells with 12.5 μM concentration of chromium (VI) for 48 h, or incubation with a 25 μM concentration of chromium (VI) for 24 or 48 h.

In the *in vivo* studies, chromium (VI)-induced more drastic effect of p53-deficient C57BL/6TSG p53 mice as demonstrated by hepatic and brain lipid peroxidation, DNA fragmentation and cytochrome *c* reduction as compared to the control C57BL/6Ntac mice (Tables 3 and 4). The tumor suppressor gene p53 is involved in key responses to genotoxic stress, and Schwartz and Rotter [31] reviewed the known molecular pathways by which the gene controls the cell cycle, with a specific focus on the significance of the p53-mediated checkpoint response for its tumor suppressor function. Post-translational modification of the p53 protein through phosphorylation is believed to be an important regulatory mechanism of p53 function [32]. Furthermore, the p53 tumor suppressor gene plays a major role in the regulation of the cellular stress response, in part, through the transcriptional activation of genes involved in cell cycle control, DNA repair, and apoptosis [33, 34]. Because p53 is activated in response to DNA damage, and because various factors are known to interact to signal and modulate this response, the role of p53 gene was assessed in chromium (VI)-induced toxicity using C57BL/6TSG p53-deficient mice.

In summary, these results in conjunction with our previous studies [13, 14, 18, 28–30] demonstrate that chromium (VI) results in enhanced formation of ROS, including superoxide anion, hydroxyl radicals and nitric oxide, decreased cell viability, increased cellular and genomic hepatic DNA fragmentation, enhanced intracellular oxidized states, membrane damage with leakage of lactate dehydrogenase, activation of protein kinase C, apoptotic and necrotic cell death. Furthermore, these data are consistent with previously published observations regarding the involvement of oxidative stress, apoptotic cell death and modulation of p53 apoptotic regulatory gene in the toxicity and carcinogenicity of chromium (VI).

References

1. Stohs SJ, Bagchi D: Oxidative mechanisms in the toxicity of metal ions. Free Rad Biol Med 18: 321–336, 1995
2. Von Burg R, Liu D: Chromium and hexavalent chromium. J Appl Toxicol 13: 225–230, 1993
3. Barceloux D: Chromium. Clin Toxicol 37:173–194, 1999
4. Patierno SR, Banh D, Landolph JR: Transformation of C3H/10T1/2 mouse embryo cells to focus formation and anchorage independence by insoluble lead chromate but not soluble calcium chromate: Relationship to mutagenesis and internalization of lead chromate particles. Cancer Res 48: 5280–5288, 1988
5. Biedermann KA, Landolph JR: Induction of anchorage independence in human diploid foreskin fibroblasts by carcinogenic metal salts. Cancer Res 47: 3815–3823, 1987
6. Biedermann KA, Landolph JR: Role of valence state and solubility of chromium compounds on induction of cytotoxicity, mutagenesis, and anchorage independence in diploid human fibroblasts. Cancer Res 50: 7835–7842, 1990
7. Shi X, Dalal NS: Chromium (V) and hydroxyl radical formation during the glutathione reductase-catalyzed reduction of chromium (VI). Biochem Biophys Res Commun 163: 627–634, 1989
8. Shi X, Dalal NS: On the hydroxyl radical formation in the reaction between hydrogen peroxide and biologically generated chromium (V) species. Arch Biochem Biophys 277: 342–350, 1990
9. Jones P, Kortenkamp A, O'Brien P, Wang G, Yang G: Evidence for the generation of hydroxyl radicals from a chromium (V) intermediate isolated from the reaction of chromate and glutathione. Arch Biochem Biophys 286: 652–655, 1991
10. Kawanishi S, Inoue S, Sano S: Mechanism of DNA cleavage induced by sodium chromate (VI) in the presence of hydrogen peroxide. J Biol Chem 261: 5952–5958, 1986
11. Danielsson BRG, Hassoun E, Dencker L: Embryo toxicity of chromium: Distribution in pregnant mice and effects on embryonic cells *in vitro* Arch Toxicol 51: 233–245, 1982
12. Babior BM, Kipnes RS, Curnutte JT: Biological defense mechanisms. The production by leukocytes of superoxide, a potential bactericidal agent. J Clin Invest 52: 741–744, 1973
13. Bagchi D, Hassoun E, Bagchi M, Stohs S: Chromium-induced excretion of urinary lipid metabolites, DNA damage, nitric oxide production, and generation of reactive oxygen species in Sprague-Dawley rats. Comp Biochem Physiol 110C: 177–187, 1995
14. Bagchi D, Joshi SS, Bagchi M, Balmoori J, Benner EJ, Kuszynski CA, Stohs SJ: Cadmium- and chromium-induced oxidative stress, DNA damage, and apoptotic cell death in cultured human chronic myelogenous leukemic K562 cells, promyelocytic leukemic HL-60 cells, and Norman human peripheral blood mononuclear cells. J Biochem Mol Toxicol 14: 33–41, 2000
15. Telford W, King L, Fraker P: Evaluation of glucocorticoid-induced DNA fragmentation in mouse thymocyte by flow cytometry. Cell Prolif 24: 447–459, 1991
16. Gavrieli Y, Sherman Y, Ben-Sassen SA: Identification of programmed cell death *in situ* via specific labeling of nuclear DNA fragmentation. J Cell Biol 119: 493–501, 1992
17. Mosmann T: Rapid colorimetric assay for cellular growth and survival: Application to proliferation and cytotoxicity assays. J Immunol Meth 65: 55–63, 1983
18. Bagchi D, Tran MX, Newton S, Bagchi M, Ray SD, Kuszynski CA, Stohs SJ: Chromium and cadmium induced oxidative stress and apoptosis in cultured J774A.1 macrophage cells. *In Vitro* Mol Toxicol 11: 171–181, 1998
19. Fowler JF Jr: Systemic contact dermatitis caused by oral chromium picolinate. Cutis 65: 116, 2000
20. Young PC, Turiansky GW, Bonner MW, Benson PM: Acute generalized exanthematous pustulosis induced by chromium picolinate. J Am Acad Dermatol 41: 820–823, 1999
21. Cerulli J, Grabe DW, Gauthier I, Malone M, McGoldrick MD: Chromium picolinate toxicity. Ann Pharmacother 32: 428–431, 1998
22. Speetjens JK, Collins RA, Vincent JB, Woski SA: The nutritional supplement chromium (III) tris(picolinate) cleaves DNA. Chem Res Toxicol 12: 483–487, 1999
23. Stearns DM, Wise JP Sr, Patierno SR, Wetterhahn KE: Chromium (III) picolinate produces chromosome damage in Chinese hamster ovary cells. FASEB J 9: 1643–1648, 1995
24. Grant KE, Chandler RM, Castle AL, Ivy JL: Chromium and exercise training: Effect on obese women. Med Sci Sports Exer 29: 992–998, 1997
25. Lefavi RG, Wilson GD, Keith RE, Anderson RA, Blessing DL, Hames CG, McMillan JL: Lipid-lowering effect of a dietary chromium (III) nicotinic acid complex in male athletes. Nutr Res 13: 239–249, 1993
26. Crawford V, Scheckenbach R, Preuss HG: Effects of niacin-bound chromium supplementation on body composition in overweight African-American women. Diab Obes Metab 1: 331–337, 1999
27. Olin KL, Stearns DM, Armstrong WH, Keen CL: Comparative retention/ absorption of ^{51}chromium (^{51}Cr) from ^{51}Cr chloride, ^{51}Cr nicotinate and ^{51}Cr picolinate in a rat model. Tr Elem Electr 11: 182–186, 1994
28. Bagchi D, Hassoun EA, Bagchi M, Muldoon D, Stohs SJ: Oxidative stress induced by chronic administration of sodium dichromate (Cr (VI)) to rats. Comp Biochem Physiol 110C: 281–287, 1995
29. Bagchi D, Vuchetich PJ, Bagchi M, Hassoun EA, Tran MX, Tang L, Stohs SJ: Induction of oxidative stress by chronic administration of sodium dichromate (chromium VI) and cadmium chloride (cadmium II) to rats. Free Rad Biol Med 22: 471–478, 1997
30. Bagchi D, Bagchi M, Tang L, Stohs SJ: Comparative *in vitro* and *in vivo* protein kinase C activation by selected pesticides and transition metal salts. Toxicol Lett 91: 31–39, 1997
31. Schwartz D, Rotter V: p53-dependent cell cycle control: Response to genotoxic stress. Cancer Biol 8: 326–336, 1998
32. Milczarek GJ, Martinez J, Bowden GT: p53 phosphorylation: Biochemical and functional consequences. Life Sci 60: 1–11, 1997
33. Amundson SA, Myers TG, Fornace AJ Jr: Roles of p53 in growth arrest and apoptosis: putting on the brakes after genotoxic stress. Oncogene 17: 3287–3299, 1998
34. Singh J, Carlisle DL, Pritchard DE, Patierno SR: Chromium-induced genotoxicity and apoptosis: Relationship to chromium carcinogenesis. Oncol Rep 5: 1307–1318(review), 1998

Molecular and Cellular Biochemistry **222:** 159–171, 2001.

Review

Carcinogenic metals and NF-κB activation

Fei Chen, Min Ding, Vince Castranova and Xianglin Shi
Health Effects Laboratory Division, National Institute for Occupational Safety and Health, Morgantown, WV, USA

Abstract

Epidemiological and animal studies suggest that several metals and metal-containing compounds are potent mutagens and carcinogens. These metals include chromium, arsenic, vanadium, and nickel. During the last two decades, chemical and cellular studies have contributed enormously to our understanding of the mechanisms of metal-induced pathophysiological processes. Although each of these metals is unique in its mechanism of action, some common signaling molecules, such as reactive oxygen species (ROS), may be shared by many of these carcinogenic metals. New techniques are now available to reveal the mechanisms of carcinogenesis in precise molecular terms. In this review, we focused our attentions on metal-induced signal transduction pathways leading to the activation of NF-κB, a transcription factor governing the expression of most early response genes involved in a number of human diseases. (Mol Cell Biochem **222**: 159–171, 2001)

Key words: NF-κB, IKK, metals, signal transduction, ROS

Introduction

There is increasing evidence that most of the toxic metals or metal-containing particles from either environmental or occupational sources are human carcinogens [1–4]. However, a detailed molecular mechanism of metal-induced malignant transformation and cancer remains elusive. Emerging evidence indicates that cellular transformation and tumorigenesis in humans is a multistep process that requires both non-genetic and genetic alterations that promote the transformation of normal human cells into highly malignant tumors [5]. However, it is still not clear which step or steps are targeted by these metals. For example, how the metals affect critical carcinogenic transformation events, such as NF-κB or apoptosis, is poorly understood. It is generally believed that oxidative stress, resulting from metal-induced generation of reactive oxygen species (ROS), is a critical mediator for the malignant transformation. However, ROS-independent effects of metals on cellular signaling pathways and genomic stability may also play a significant role [6, 7].

This review will focus on pathways leading to the activation of the transcription factor, NF-κB, and its possible roles in cell apoptosis and cell cycle regulation induced by metals, especially arsenic, vanadium and chromium.

Signal transduction of NF-κB activation

A wide range of signals, which typically include cytokines, mitogens, environmental and occupational particles or metals, intracellular stresses, viral and bacterial products, and UV light, induce the expression of early response genes through the NF-κB family of transcription factors [8–12]. In resting cells, NF-κB is sequestered in the cytoplasm of most cells where it is bound to one of a number of inhibitory molecules, including IκBα, IκBβ, IκBε, p105, and p100. Activation of the NF-κB signaling cascade results in a complete degradation of IκB or partial degradation of the carboxy termini of p105 and p100 precursors, allowing nuclear translocation of the NF-κB complexes (Fig. 1). Activated NF-κB binds to specific DNA sequences in target genes, designated as κB-elements, and regulates transcription of genes mediating inflammation, carcinogenesis and anti-apoptotic reactions.

Three IκB proteins, IκBα, IκBβ and IκBε, have been identified, among which IκBα is the most abundant inhibitory protein for NF-κB [9]. All IκB proteins contain two conserved serine residues within their N-terminal domain. Phosphorylation of these conserved serine residues in response to inducers, leads to the immediate polyubiquitination of IκB proteins by the SCF-β-TrCP complex, a step which has been

Address for offprints: X. Shi, National Institute for Occupational Safety and Health, 1095 Willowdale Road, Morgantown, WV 26505, USA

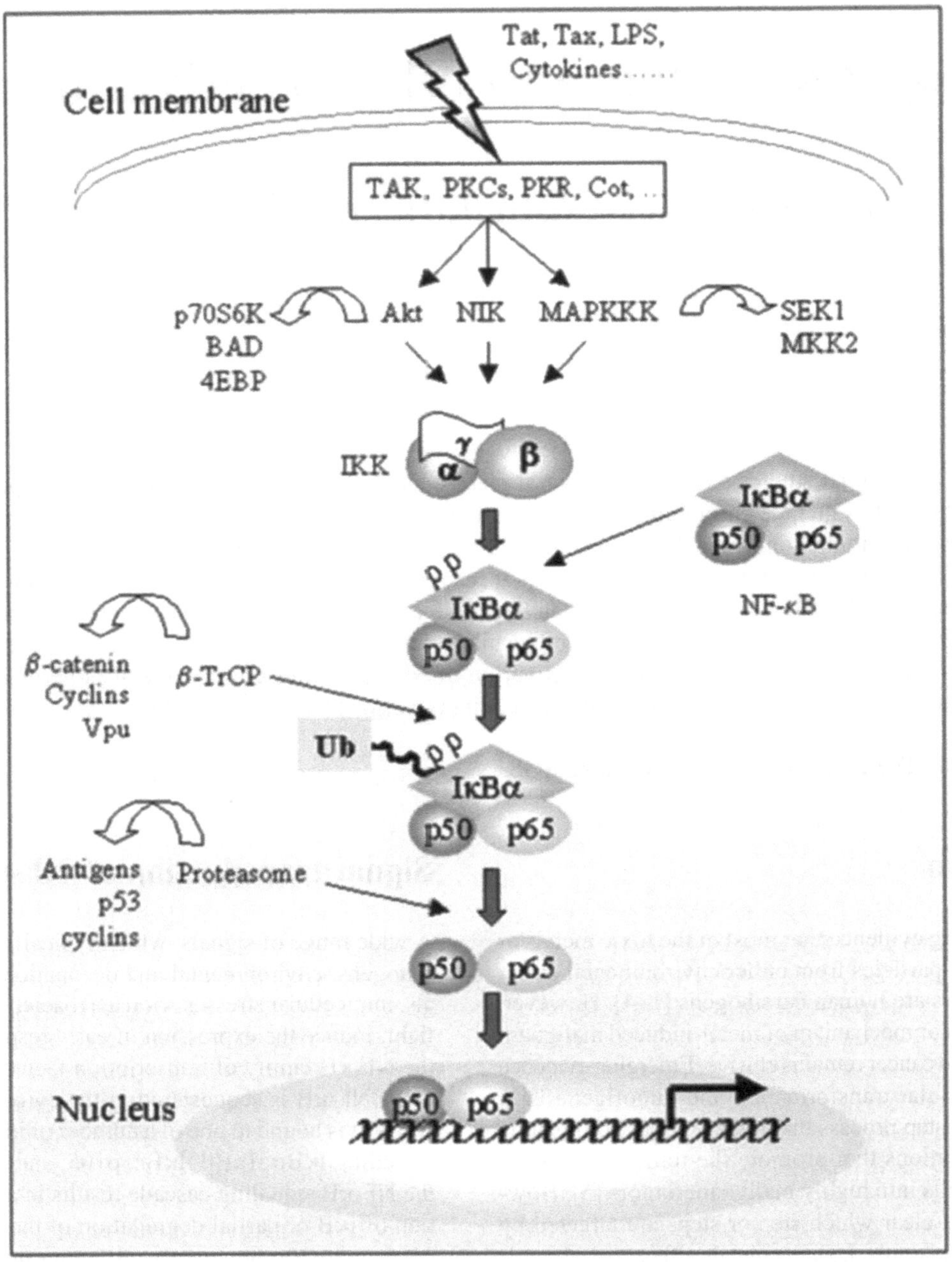

Fig. 1. Signaling pathways activating NF-κB. Extracellular inducers, including cytokines, viral and bacterial products and ROS, activate IKK through upstream kinases directly or indirectly. Activated IKK phosphorylates N-terminal S32 and S36 residues of IκBα which is associated with the NF-κB p50 and p65 heterodimer. The β-TrCP complex recognizes phosphorylated IκBα and modifies IκBα with polyubiquitin chains. This is followed by proteasome-mediated degradation of IκBα. After degradation of IκBα, the activated NF-κB translocates into the nucleus and binds to the κB-sites of gene promoters or enhancers to up-regulate target gene expression. Line arrows and filled arrows denote the NF-κB signaling pathways; non-filled arrows denote the involvement in other signaling pathways.

shown recently to be inhibited by the nonpathogenic *Salmonella* bacteria in gut epithelial cells [13]. This modification subsequently targets IκB proteins for rapid degradation by the 26S proteasome [14]. The phosphorylation is accomplished by a specific IκB kinase (IKK) complex containing two catalytic subunits, IKKα and IKKβ, and a structural component named NEMO/IKKγ/IKKAP [15, 16]. An earlier report by Cohen *et al.* [17] suggested the presence of a scaffold protein named IKK complex-associated protein (IKAP) in the IKK complex, which could not be confirmed in other studies. A late study by the group, who originally identified IKAP, indicated that the observed association of IKAP with

IKK was due to a non-strict elusion condition of chromatographic extracts during the purification of IKK [18]. Recently, two groups independently reported the identification of a novel protein, CIKS/Act1, associated with the IKK complex and suggested that CIKS/Act1 functions as an anchoring protein in the assembly of the IKK complex and in providing a possible connection between IKK and JNK signaling [19, 20]. IKKα and IKKβ share 50% sequence similarity. Both proteins contain an amino terminal kinase domain, a carboxy terminal region with a leucine zipper, and a helix-loop-helix domain. *In vitro* or *in vivo* studies indicate that although both IKKα and IKKβ are capable of phosphorylating IκBα on ser32 and ser36, IKKβ is more potent in IκBα phosphorylation induced by proinflammatory stimuli. Several groups have recently indicated the existence of an additional IKK-like kinase complex in T cells, named IKKi/ε, which shares 27% homology with IKKα and IKKβ and possibly mediates NF-κB activating kinase (NAK) signaling and PMA/PKCε-induced S36 phosphorylation of IκBα and NF-κB activation [21, 22].

Diverse stimuli, with the possible exception of UV, activate NF-κB through activation of the kinase activity of IKK that phosphorylates IκBs, a critical step required for subsequent ubiquitination and degradation. UV-induced NF-κB activation may be through a mechanism that does not depend on IKK or N-terminal phosphorylation of IκB [23, 24]. It is known that the activity of IKK can be stimulated by TNFα, IL-1, LPS, Tax protein, ionizing radiation, vanadate, or double-stranded RNA. However, how these diverse stimuli activate IKK remains unresolved. A number of kinases, mostly based on transient overexpression in which wild-type forms activate IKK and catalytically inactive forms inhibit IKK, have been suggested to be able to phosphorylate and activate IKK. These kinases include MEKK1 [25], NIK [15], TBK1/NAK [22, 26], Akt [27, 28], PKC-θ [29, 30], PKCξ [31], Cot kinase [32], PKR [33], and MLK3 [34]. It is unclear whether each of these upstream kinases represents one of the many routes to IKK activation by specific stimuli, or represents a ubiquitous pathway leading to the activation of IKK induced by diverse stimuli.

MEKK1

MEKK1 is a mammalian serine/threonine kinase in the mitogen-activated protein kinase kinase kinase (MAPKKK) group. It was found that MEKK1 was a far more important activator for JNK signaling than for ERK signaling as proposed originally [35, 36]. The first evidence, indicating the involvement of MEKK1 in signal-induced IKK activation, was provided by Lee *et al.* [25, 37]. In their studies, they reported that addition of the recombinant catalytic domain of MEKK1 (MEKK1Δ) to a partially enriched fraction of non-stimulated HeLa cells stimulated an IKK-like kinase activity that phosphorylated IκBα at serines 32 and 36 and induced subsequent ubiquitination and degradation of IκBα. It was subsequently demonstrated that an overexpression of MEKK1 stimulated NF-κB-dependent transcriptional reporter [38, 39]. The activation of NF-κB by HTLV Tax protein was shown to involve MEKK1 [40]. MEKK1 may also contribute to Toll- and IL-1 receptor-mediated IKK activation, since it was demonstrated that an adaptor protein named evolutionarily conserved signaling intermediate in Toll pathways (ECSIT) could promote the proteolytic activation of MEKK1 and subsequent activation of NF-κB [41]. Further studies by Mercurio *et al.* [16] indicate the presence of a protein that is recognized by anti-MEKK1 anti-serum in the IKK complex. Both IKKα and IKKβ contain a MAPKK activation loop, 176/177 S-X-X-X-S 180/181, (where X is any amino acid), required for their activation. Substitution of these serines with alanine residues inactivates both kinases, while phosphomimetic glutamic acid substitutions at these positions result in constitutively active kinases. It remains to be confirmed whether MEKK1 is a physiologically important activator of IKK in cells in response to various stimuli. Indeed, a recent study by Xia *et al.* [42] demonstrated that inactivation of MEKK1 does not result in any impairment of NF-κB activation in response to TNFα, IL-1, LPS or dsRNA. In addition, several studies indicated that a modest expression of MEKK1, that was sufficient for JNK activation, failed to stimulate IKK [43]. Although some studies, using a dominant negative mutant of MEKK1, showed an inhibition of IKK, these results should be interpreted with caution since many dominant negatives can bind to their upstream partners upon overexpression and function as potent general inhibitors. The best approach to determine whether MEKK1 is a direct upstream kinase to phosphorylate and activate IKK is the use of MEKK1-deficient mice to examine whether cells or tissues from these mice display defects in activation of IKK or NF-κB in response to cytokines or other inflammatory stimuli.

NF-κB-inducing kinase (NIK)

NIK, a member of MAPKKK family, was originally identified as a TNFα receptor associating factor 2 (TRAF2)-interacting kinase whose overexpression results in potent NF-κB activation without any considerable effect on MAPKs [44, 45]. A study using yeast two-hybrid screens identified an interaction between NIK and IKK [15, 46], suggesting that NIK might be a direct upstream activator of IKK. Transient transfection of NIK into human embryonic kidney 293 cells indicated that IKKα was more responsive to NIK, whereas IKKβ was slightly more responsive to MEKK1 [47]. Most studies show that when the abilities of MEKK1 and NIK to activate total IKK kinase activity are compared, NIK is a much stronger activator of the NF-κB transcriptional reporter than MEKK1 is [45, 48]. NIK can preferentially phosphorylate IKKα on ser176 in the activation loop, leading to the activation of IKKα kinase activity. In contrast, MEKK1 was found to preferentially phosphorylate the corresponding serine residue, ser177, in the activation loop of IKKβ. A dominant negative mutant of NIK blocked NF-κB activation by TNFα, IL-1, Fas [44], Toll-like receptors 2 and 4 [49, 50], LMP-1 [51], and CD3/CD28 stimulation [32]. Thus, NIK appears to be a general kinase mediating IKK activation induced by diverse stimuli. However, a recent analysis using a NIK-mutant mouse strain, *alymphoplasia (aly)*, contradicted this assumption. The *Alymohoplasia* mouse strain failed to develop lymphoid organs, such as lymph node, Peyer's patches, due to a point mutation in the NIK locus [52]. The mutation of NIK locus results in a disruption of interactions between NIK and IKKα or TRAF proteins. Further analysis indicated that the *aly* mutation did not affect TNFα-induced activation of NF-κB, but only blocked lymphotoxin-mediated activation of NF-κB. Similarly, studies using cells derived form NIK-deficient mice have indicated that NIK appears to be dispensable in IKK activation induced by TNFα or IL-1. It raises the possibility that NIK may be specifically involved in IKK activation induced by lymphotoxin but not other stimuli.

NF-κB activating kinase (NAK)

Three groups independently identified a novel serine/threonine kinase possibly activating IKK through direct phosphorylation in cells stimulated with PMA [22, 26, Tularik Inc., US patent 5,776,717 (1998)]. This novel kinase was named NAK, TANK-binding kinase 1 (TBK1) and T2K, respectively. Pomerantz and Baltimore [26] cloned NAK by a yeast two-hybrid screen using a N-terminal stimulatory domain of TANK 1-190 fused to GAL4 as bait and a human B-cell library fused to the GAL4 activation domain. The same kinase was also identified by PCR using degenerate primers based on sequences common to IKKα and IKKβ [22]. Amino acid sequence analysis indicated that NAK protein contains a kinase domain at its N-terminus that exhibits about 30% identity to the corresponding kinase domains of IKKα and IKKβ and more than 60% identity to the corresponding kinase domain of IKKε/i. Whereas the report by Pomerantz and Baltimore [26] showed that NAK might form a ternary complex with TANK and TRAF2, suggesting that NAK functions at distal upstream of signal cascade leading to IKK activation, *in vitro* kinase activation assay by Tojima *et al.* [22] demonstrated that NAK was a direct upstream kinase phosphorylating IKKβ. Intriguingly, activation of endogenous NAK resulted in only ser36, but not ser32, phosphorylation of IκBα, a phenomenon similarly observed in recombinant IKKε/i-mediated IκBα phosphorylation [53]. Since both IKKα and IKKβ are able to phosphorylate both ser32 and ser36 of IκBα protein, it is unclear whether IKKε/i or a novel IKK isozyme functions as a downstream kinase of NAK to induce ser36 phosphorylation of IκBα. Transient transfection studies showed that dominant negative NAK inhibited NF-κB transcriptional reporter activity induced by PMA, PKCε and PDGF, but not by TNFα, IL-1β, LPS, or ionizing radiation. These results, therefore, suggest that NAK is likely a downstream kinase of PKCε or related isozymes and an upstream kinase of IKK in the signaling pathway through which growth factors, such as PDGF, stimulate NF-κB activity.

Akt

The pro-survival function of Akt has been well documented. The kinase activity of Akt is activated via the phosphoinositide-3-OH kinase (PI3K) and PI3K-dependent kinase 1/2 (PDK1/2) signaling pathway [54]. Overexpression or constitutive activation of Akt has been associated with tumourigenesis in a number of studies. As a serine/threonine kinase, Akt is able to phosphorylate pro-apoptotic protein Bad, anti-apoptotic protein Bcl-x, apoptotic protease caspase-9, Forkhead transcription factors, and eNOS [54]. However, considerable controversies remain regarding the involvement of Akt in signal-induced IKK activation. Studies by Ozes *et al.* [27] and Xie *et al.* [55] indicate that Akt was required for TNFα- or G protein activator-induced NF-κB activation by directly phosphorylating and activating IKKα in 293, HeLa and ME-180 cells. A putative Atk phosphorylation site at amino acids 18–23 in both IKKα and IKKβ was identified. Akt induced Thr23 phosphorylation of IKKα both *in vitro* and *in vivo*. Mutation of Thr23 significantly decreased Akt-induced IKKα phosphorylation and TNFα-induced NF-κB activation in 293, HeLa and ME-180 cells [27]. In contrast, Romashkova *et al.* [28] demonstrated that Akt was involved in PDGF-mediated, but not in TNFα- or PMA-mediated, NF-κB activation in human or rat fibroblasts. In this study, the authors suggested that upon PDGF stimulation, Akt could transiently associate with IKK and induce IKK activation, especially IKKβ activation. Several other studies, however, suggested that the effects of Akt on NF-κB did not occur at the level of IKK activation in several cell types. Studies by Delhase *et al.* [56] indicate that Akt activation induced by IGF-1 failed to activate IKKα kinase activity, IκBα phosphorylation and degradation, and NF-κB DNA binding in HeLa cells, the same cell line used by Ozes *et al.* [27]. Similarly, several recent studies showed that Akt was not involved in TNFα-induced NF-κB activation in human vascular smooth muscle cells, skin fibroblasts, or endothelial cells [57, 58]. Rather, Akt might enhance the ability of the p65 (RelA) transactivation to induce transcription [59, 60]. In Jurkat T-cells, Akt alone failed to activate NF-κB, but it was capable of potentiating NF-κB activation induced by PMA, partially by enhancing IκBβ degradation [61]. Thus, the questions remain to be answered are: (1) does Akt phosphorylate IKK in a cell context- and stimulation-dependent manner? and (2) do upstream kinases of Akt, such as PDK1 and PDK2, also activate IKK, since both IKKα and IKKβ contain a putative PDK1 phosphorylation site (S-F-X-G-T-X-X-Y-X-A-P-E) directly juxtaposed to MAPKKK phosphorylation site [62, 63]?

Mixed-lineage kinase 3 (MLK3)

MLK3, another member of MAPKKK family, contains a N-terminal SH3 domain, followed by the catalytic domain and two tandem leucine/isoleucine zippers, a basic region, a Cdc43/Rac binding motif, and a proline-rich C terminus [64]. Based on these structural characteristics, MLK3 may associate with a variety of protein modules. Studies by Hehner *et al.* [34] suggested that MLK3 could directly associate with IKK complex through its leucine zipper domain and phosphorylate Ser176 of IKKα and Ser177 and Ser181 of IKKβ. Transfection of Jurkat T cells with a kinase mutated form of MLK3 blocked the CD3-CD28 signal- and PMA-induced NF-κB transcriptional reporter. No significant influence of this mutated MLK3 was observed on either TNFα- or IL-1β-induced NF-κB activation. These results suggest that MLK3 may be important in mediating T cell costimulation-induced activation of IKK and consequent NF-κB-dependent transcription. MLK3 has also been shown to form a complex with a JNK scaffold protein, JIP, and stimulate JNK activation. Thus, MLK3 may function as integral molecule between the signaling pathways leading to the activation of NF-κB and JNK and provide a molecular explanation as to why many stimuli induce activation of NF-κB and JNK simultaneously under certain circumstances.

A variety of other kinases have been shown to function upstream of IKK. Because direct association of these kinases with IKK upon activation or specific phosphorylating site(s) of these kinases on IKK has not yet been demonstrated, whether these kinases are direct upstream kinases phosphorylating and activating IKK or far more distal kinases indirectly activating IKK remains unsolved. These kinases include Cot [32], PKCθ [29, 30], PKCα, PKCξ [31], TAK1 [65, 66], PKR [33], etc. Due to the fact that a variety of kinases either related or unrelated can activate IKK, it seems likely that different cell types and stimuli may utilize distinct upstream kinases for the activation of IKK. An example to support this notion is provided by the observation that PKCθ and Cot kinase participate in CD3-CD28 costimulation signal-induced but not TNFα-induced activation of NF-κB [30, 32].

Metals and NF-κB activation

A number of reports during the last few years indicate that some metals are able to affect the activation or activity of NF-κB transcription factors [67]. However, a detailed mechanistic understanding of metal-induced activation or inactivation of NF-κB is still lacking. In addition, certain results from different groups have been contradictory [68–71]. A possible explanation for this controversy is that previous studies have utilized different doses of metal ions and have examined the cellular response in different types of cells. These previous studies have not been conducted over a wide dose range and with the simultaneous evaluation of both bioactivity and cytotoxicity. Since it is possible that different doses of metals may either stimulate the activation of NF-κB or inhibit the activity of NF-κB as a result of their outright ability to initiate cell death, it will be important to examine the relationship between the dose level and the type of metal on the activation of NF-κB.

Using mouse JB6 fibroblast cells stably expressing both the NF-κB-dependent chloramphenicol acetyltransferase (CAT) reporter gene and AP-1-dependent luciferase reporter gene, we recently investigated the potency of individual metals and some metal-containing particles, such as silica and residual oil fly ashes (ROFA), to activate NF-κB. All of these metals produced an optimal dose-dependent induction of NF-κB-dependent and AP-1-dependent reporter gene activity (Fig. 2). Under the optimal dose range, cobalt and copper showed the strongest induction of NF-κB reporter gene activity. Arsenic (As (III) and As (V)) and vanadium (V) (V (V)) exhibited a similar mild induction of NF-κB, whereas chromium (VI) (Cr (VI)) and nickel were relatively weak in their ability to induce NF-κB reporter gene activity. It is interesting to note that very low doses of Cr (VI) as compared to other metals, i.e., from 4–100 ng/ml, are sufficient for the induction of NF-κB. Higher doses of Cr (VI), (> 200 ng/ml), showed considerable cytotoxicity on this type of cells. In human bronchial epithelial cell line, the optimal dose range of Cr (VI) required for stimulating NF-κB activity is between 0.125–2 μg/ml (data not shown). Thus, NF-κB induction by metals appears to be both dose-dependent and cell type-dependent.

Figure 2 also showed that arsenite, arsenate, copper, vanadate, and silica induced NF-κB and AP-1 with a similar dose-response curve, whereas both cobalt and zinc preferentially induce NF-κB but not AP-1. Although both the original ROFA and the washing buffer of ROFA exhibited a similar dose-response induction of both NF-κB and AP-1, the washed ROFA could only induce NF-κB, indicating that the AP-1 induction by ROFA was dependent upon water soluble components of this particle.

Previous studies by several groups indicate that, at a non-cytotoxic dose range, Cr (VI), As (III) or V (V) is capable of

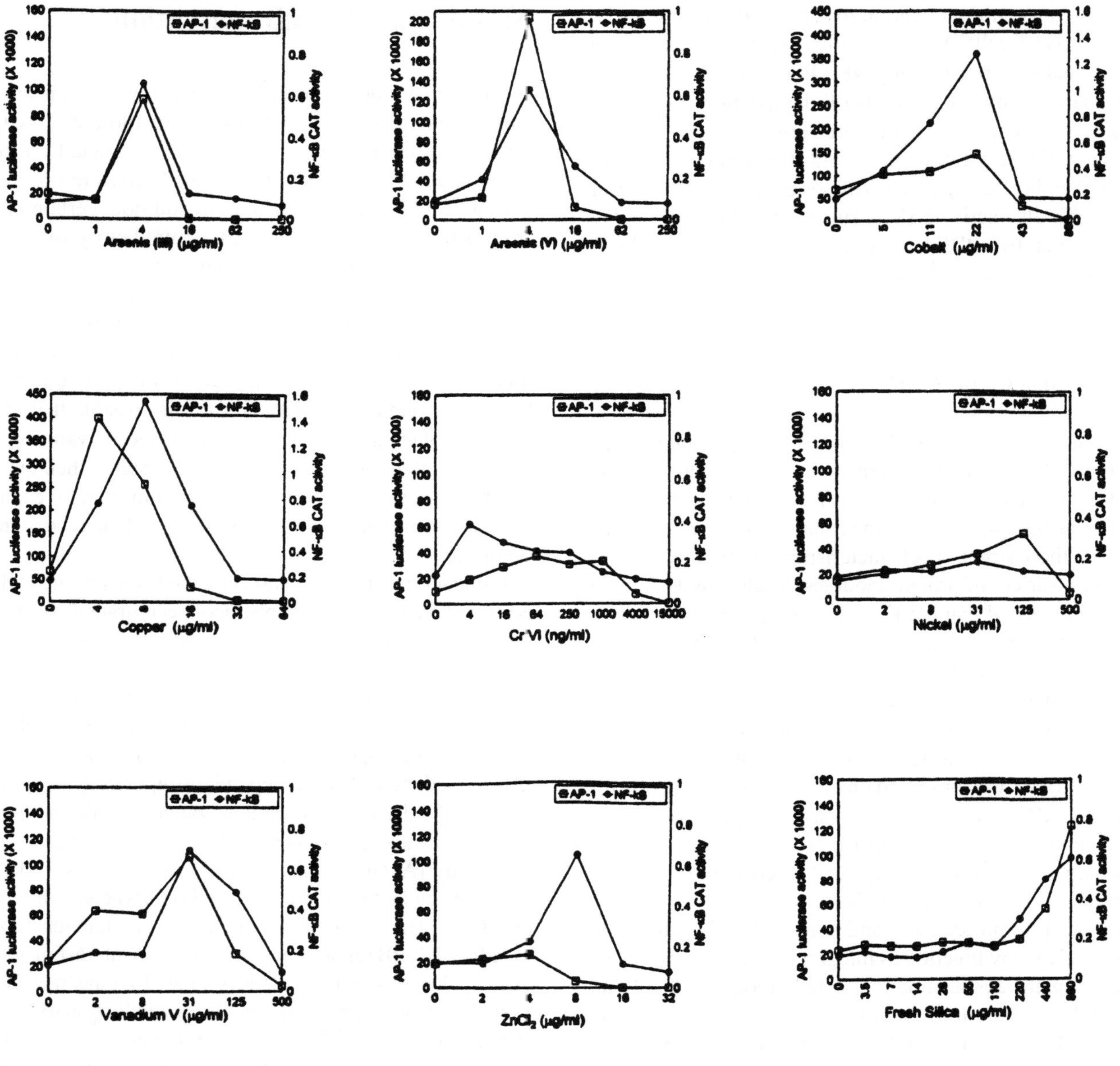

Fig. 2. Metals or metal-containing particles induce NF-κB-dependent CAT activity and AP-1 dependent luciferase activity. JB6 mouse epidermal cells stably transfected with NF-κB- and AP-1-dependent reporter genes were cultured with 0.1% FBS for 24 h in 24-well tissue culture plates. Cells were then treated with various doses of metals or metal-containing particles as indicated for an additional 12 h. Total cell lysates were prepared and subjected to CAT and luciferase activity assay, respectively.

activating NF-κB as determined by either a gel shift assay, which reflects the activation and nuclear translocation of NF-κB, or NF-κB-dependent reporter gene assay, an indicator of NF-κB activity. In contrast, Cr (VI), As (III), and other metals have been reported to inhibit NF-κB activation through interfering with IKK kinase, DNA binding, or the interactions with nuclear co-factor, cAMP-responsive element-binding protein (CREB)-binding protein (CBP) [70, 71]. These results raise the mechanistic question of how metals affect NF-κB activation and activity. Although some metals showed similar capabilities on either the activation or inhibition of NF-κB, different metals may use different signal transduction pathways to affect NF-κB.

Arsenic

The pioneer study of arsenic-induced NF-κB activation was carried out by Barchowsky *et al.* [72]. This study suggested that lower concentrations of As (III) activated NF-κB through increasing intracellular oxidant levels in vascular endothelial cells. Later studies by Hamilton *et al.* [6] indicated that the induction of NF-κB by As (III) is cell-type dependent. As (III) is able to induce NF-κB in MDA epithelial-like cells, whereas As is unable to induce NF-κB in H4IIE rat hepatoma cells. Mechanistic studies by Jaspers *et al.* [73] indicated a possible non-classic signaling pathway involved in mediating arsenite-induced NF-κB activation in airway epithelial cells. In contrast to other NF-κB inducers such as vanadyl sulfate, inducible IκBα degradation or p65 nuclear translocation did not occur in As (III)-induced NF-κB reporter gene activity. At higher concentrations, As (III) appears to exhibit an inhibitory effect on NF-κB by interfering with the DNA binding of NF-κB, possibly through a direct modification of As (III) on the sulfhydryls of NF-κB proteins [74]. If that is the case, there are a number of intracellular proteins, either within NF-κB signaling pathways or other pathways, which contain exterior sulfhydryls and would be targeted by As (III). Indeed, a direct inhibition of IKK kinase activity by As (III) has been demonstrated recently [70, 75]. In Hela cells and HEK293 cells, As (III) has been shown to be able to bind to cysteine 179 of IKKβ and inhibit IKK activity induced by TNFα, IL-1 and PMA [75]. These latest findings provided a new insight into the role of As (III) on NF-κB signaling, but still did not solve the dilemma of the stimulation of NF-κB by As (III) at lower concentrations. One may speculate that the activation of NF-κB by lower concentrations of As (III) is possibly through the regulation of the redox status of cells as suggested by the earlier studies [72], while the inhibition of NF-κB by lower or higher concentrations of As (III) is through direct binding of As (III) to either IKK or NF-κB proteins. However, a more difficult question to answer is how redox changes participate in the signaling pathways leading to the activation of NF-κB. Obviously, IKK kinase is unlikely to be activated directly by changes in cellular redox status, since, similar to the modification by As (III), the oxidation of cysteine 179 in the activation loop of IKKβ will inactivate the activity of this kinase.

Vanadium

Vanadium is a transitional element that exists in oxidation states ranging from –1 to +5. The most common forms of vanadium include V (IV) and V (V). The first evidence indicating the induction of NF-κB by V (V) was provided by Schieven *et al.* [76] who demonstrated that V (V) potentiated ROS-induced NF-κB activation in a tyrosine kinase-dependent fashion [76]. Further studies consistently showed that V (V) or V (IV) could activate NF-κB in almost all types of cells tested, such as B-lymphocyte precursors, T-lymphocytes [77], monocytes [78], macrophages [79], epidermal cells [68] and epithelial cells [73]. In human myeloid ML-1a cells, V (V) itself exhibited a mild induction of NF-κB, but it attenuated the activation of NF-κB induced by TNFα [80]. The mechanism involved in vanadium- or vanadium compound-induced NF-κB activation is a matter of debate. Early study by Imbert *et al.* [77] indicated that the activation of NF-κB by V (V) occurred independently of IκBα degradation. However, this observation could not be reproduced by several recent studies, suggesting that V (V) did induce degradation of IκBα following the phosphorylation of serine or tyrosine [79, 81]. In mouse macrophage RAW264.7 cells, V (V)-induced IκBα degradation occurred within 10–20 min with a peak degradation at 40 min [79]. In human myeloid U937 cells or epithelial cells, V (V)-induced IκBα degradation occurred at 30 min and reached a maximum at 240 min [81]. A similar result was obtained in Jurkat E6.1 cells, Ramos, a human B cell lymphoma line, and normal human bronchial epithelial (NHBE) cells [73, 82, 83]. In contrast to the earlier report that no resynthesis of IκBα occurred after V (V) treatment [77], several subsequent studies have indicated that the resynthesis of IκBα indeed occurs at 80–180 min after V (V) treatment [79, 81]. Phosphorylation on either tyrosine 42 or serine 32/36 sites of IκBα has been demonstrated in cells treated with V (V). The phosphorylation of IκBα on these sites may contribute to the subsequent degradation of this protein. At present, there are limited data available regarding which kinase or how many kinases contribute to the phosphorylation of IκBα. Since IKK is required for LPS- and inflammatory cytokine-induced IκBα phosphorylation and NF-κB activation, IKK may also contribute to vanadium-induced NF-κB activation. Indeed, in mouse macrophage RAW264.7 cells, the activation of IKKβ kinase by V (V) is both dose- and time-dependent [79]. In comparison to cytokine- or LPS-induced IKKβ kinase activation, V (V) induces IKKβ activation in a relatively persistent manner. In addition, blockage of the JNK signaling pathway resulted in a partial inhibition of IKK kinase activity induced by V (V). In the case of V (IV)-induced NF-κB activation, studies by Jaspers *et al.* [84] suggest that the p38 MAPK pathway may contribute to the NF-κB activation induced by V (IV). The phosphorylation of Tyr42 of IκBα might be through a passive enhancement of tyrosine kinase activity due to the inhibition of tyrosine kinase phosphatases by V (V). A new model has been proposed for the V (V)-induced, tyrosine phosphorylation-dependent NF-κB activation [85]. In this model, p85α, the regulatory subunit of PI3-kinase, had been suggested as a sequester of IκBα through specific association of its Src homology 2 domains with tyrosine-phosphorylated IκBα *in vitro* and *in vivo* after stimulation of T cells with pervanadate.

Chromium

Chromium occurs primarily in the trivalent state (III), which is the most stable form, or in the hexavalent state (VI), which is a strong oxidizing agent and carcinogen [86, 87]. NF-κB activation by Cr (VI) was observed in T cells [88], macrophages [67], bronchial epithelial cells (Chen *et al.*, unpublished), and human breast cancer cells [69]. The ability of Cr (VI) to activate NF-κB is also dose-dependent. It has been frequently observed that at higher concentrations, Cr (VI) exhibits an inhibitory effect on NF-κB activation or activity due to its cytotoxicity, interference with NF-κB DNA binding, and/or disruption of the interaction between NF-κB p65 protein and CEBP-binding protein [71, 74]. Our recent observations suggest that the stimulation or inhibition of Cr (VI) with a given dose-range on NF-κB is also cell density-dependent. The activation of NF-κB by Cr (VI) could only be detected at a relatively higher cell density in the presence of 1 to 5% of serum. We are currently testing the possibility that Cr (VI)-induced NF-κB activation is through a secondary cell-damaging effect in which damaged cells induced by Cr (VI) elicit an activation process for NF-κB in the adjacent undamaged cells. It has been suggested that Cr (VI) is one of the most potent metals which can elicit an oxidative stress response in cells through the generation of ROS, including superoxide anion, hydroxy radical, nitric oxide, and the by-product of superoxide, H_2O_2 [89–92]. Thus, under certain circumstances, this Cr (VI)-induced ROS generation may be accountable for the activation of NF-κB.

ROS generation in metal-induced NF-κB

A number of excellent review articles suggesting a central role of ROS in signal-induced NF-κB activation have been published and well cited during the last few years [93–99]. A high amount of ROS generated from chronic and acute inflammatory responses or environmental stresses is cytotoxic. It is known that limited production of ROS, as a consequence of electron transfer reaction in cytosol, peroxisomes and mitochondria, is buffered or scavenged by both enzymatic and nonenzymatic antioxidant systems inside the cells [99]. The mechanisms of ROS generation by metals under biologically relevant conditions may involve Fenton/Haber–Weiss chemistry and autoxidation [92]. In macrophages or other phagocytes, metal-induced ROS generation may also involve the activation of hypochlorous acid, lipid peroxides and nucleoside hydroperoxides [100]. Studies using of ROS scavengers or antioxidants have suggested the involvement of cellular redox regulation with certain kinases, leading to the activation of NF-κB or other transcription factors [97]. In an early model, ROS-induced oxidative stress was proposed as a universal mechanism for NF-κB activation by diverse agents [101]. An unanswered and difficult question is which point in the activation pathway of NF-κB is targeted by ROS. The signal transduction pathways, such as the upstream and proximal kinases (e.g., IKK, for NF-κB activation induced by TNF, LPS or CD28) have recently been identified [8, 21]. However, no evidence has been presented to indicate that these kinase cascades would be oxidant-responsive or redox-regulated. Evidence provided by Li and Karin [102] demonstrated that the ROS scavenger, N-acetyl-L-cysteine (NAC), reduced TNFα-induced IκBα degradation and NF-κB DNA-binding activity, but failed to affect TNFα-induced IKK kinase activity in HeLa cells. These results raise the possibility that ROS may not target the activation pathway of IKK, but rather may interfere with the steps of ubiquination and degradation of IκB. The model proposed by Roederer *et al.* [103] suggests that NF-κB activation is controlled by intracellular thiol levels, and that NF-κB inducers somehow potentiate oxidative stress by depleting glutathione levels. However, this model contradicts several later observations, which showed that a glutathione-oxidizing agent could inhibit NF-κB through interference with its DNA binding [104–106]. Several studies indicate that cysteine 62 of the p50 subunit of NF-κB is essential for NF-κB DNA binding and that oxidation of cysteine 62 inhibits NF-κB DNA binding activity. Thioredoxin, a ubiquitous dithiol-reducing enzyme, can reduce cysteine 62 and restore DNA-binding activity of NF-κB [104]. It thus appears that the observed NF-κB activation by metal-induced ROS may depend on some alternative pathways, such as MAP kinases, Ras and Rac1, with the potential to stimulate the IKKs or affect the transcriptional activity of NF-κB. Indeed, JNK inhibition by a transient transfection of macrophages with a dominant negative SEK1, an upstream kinase of JNK, decreased V (V)-induced IκBα degradation [79].

ROS and some transition metals, including V (V) and As (III), are potent inhibitors of protein tyrosine phosphatases

(PTP) [107, 108]. All PTPs contain a signature active site characterized by the sequence of His-Cys-X-X-Gly-X-X-Arg-Ser/Thr, where X is any amino acid [109]. Oxidation of the cysteine residue in this signature motif that is essential for the achievement of phosphatase activity will inactivate PTPs. Protein serine/threonine phosphatases, such as PP1, PP2A, PP2B and PP2C, may also be subject to redox regulation through the oxidative formation of a disulfide bond between a conserved pair of cysteine residues [110]. Because the extent of protein phosphorylation, such as IκBα phosphorylation, reflects the balance between the opposing actions of protein kinases and phosphatases, changes in either can consequently shift the extent of phosphorylation (Fig. 3). It has been suggested that hypoxia and reoxygenation induce NF-

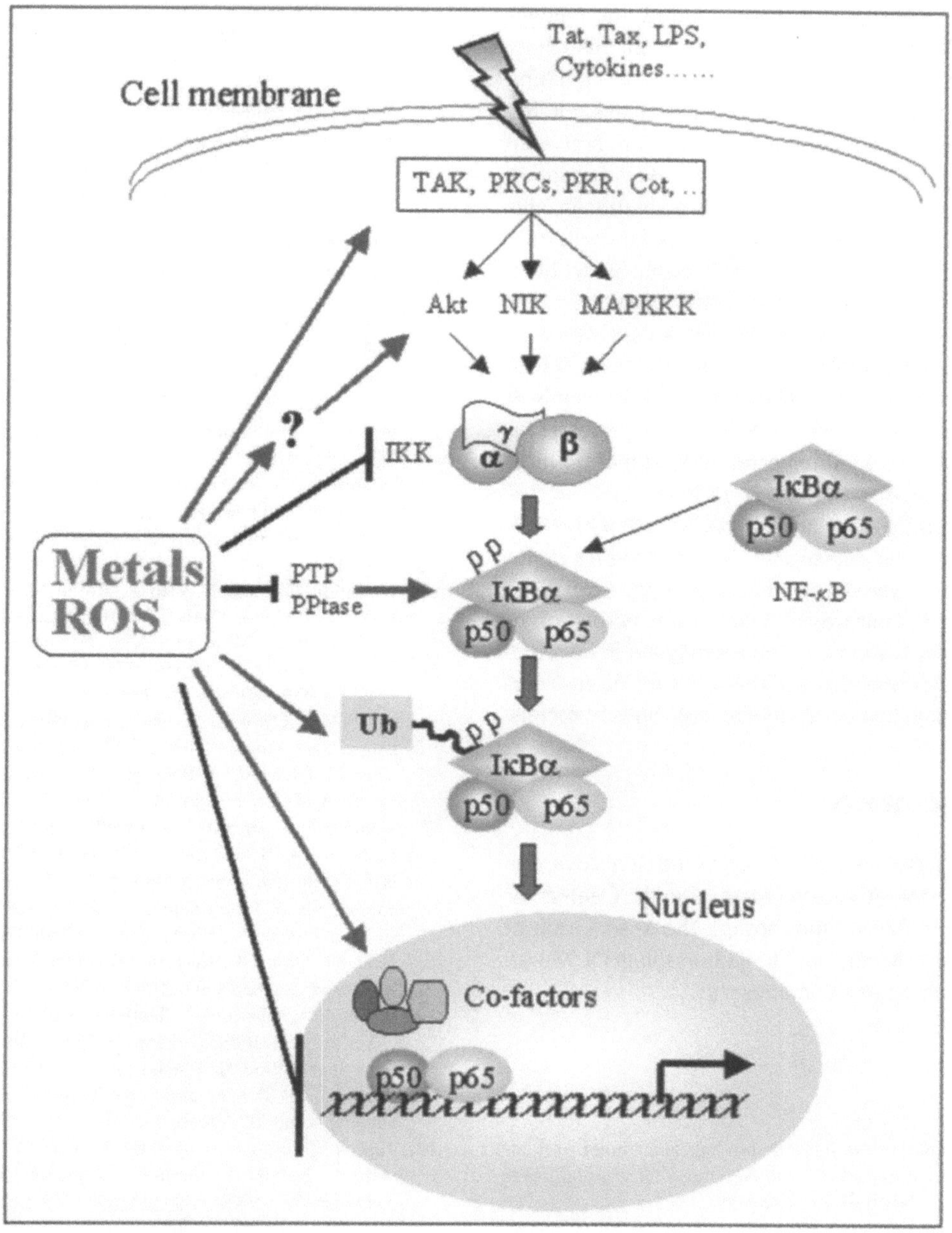

Fig. 3. Conflicting effects of metals or their ROS derivates on NF-κB signaling. Metals or ROS can possibly activate or potentiate the activation of up-stream kinases leading to the activation of IKK kinase complex. Meanwhile, metals or ROS can also inhibit IKK kinase activity through a direct modification of cysteine residues in the activation loop of IKK kinases. The phosphorylation of IκBα may be potentiated by metals or ROS due to the direct inhibition of tyrosine phosphatase (PTP) or serine/threonine phosphatases (PPtase) by metals or ROS. In the nucleus, metals or ROS may activate co-factors required for NF-κB-mediated transcriptional regulation on target genes, whereas they may also inhibit DNA binding activity of NF-κB by direct oxidation of critical cysteine residues in the DNA binding domain of NF-κB p50 or p65 proteins. →: activation; ⊣: inhibition.

κB activation through the induction of phosphorylation of tyrosine at position 42 on IκBα. Although this notion remains controversial, it will be important to determine whether tyrosine kinases are required for metal-induced NF-κB activation.

Summary

One of the major challenges in understanding the mechanisms of metal-induced NF-κB activation is to elucidate how signal transduction pathways are activated and how signaling cross-talk and specificity are achieved when several signaling pathways eliciting different cellular responses are activated at the same time by metals. For instance, why does activation of the NF-κB coincide with obvious MAPK activation in cells treated with metals under many circumstances, whereas only activation of NF-κB or activation of MAPK occurs under other circumstances? Since metals or their ROS derivatives are highly reactive but nonspecific molecules, an activation of only one specific signaling molecule, for example, an up-stream kinase, is hard to achieve in the cells in response to metals. It is a common scenario that metals or ROS can in principle induce conflicting signals by interfering with signaling molecules at different levels (Fig. 3). A good example is that oxidative stress amplifies or potentiates NF-κB activation, whereas at the same time oxidation of NF-κB proteins inhibits its NF-κB function. Translating the knowledge gained by studying the signaling pathways leading to the activation of NF-κB in the cellular response to metals may aid in developing novel preventive measures and therapies for diseases related to environmental and occupational exposures to metals.

Acknowledgements

Fei Chen was supported by a Career Development Award in Genetics under a cooperative agreement from the Centers for Disease Control and Prevention through the Association of Teachers of Preventive Medicine. The authors thank Dr. Murali Rao for critical reading of the manuscript.

References

1. Sarkar B: Metal replacement in DNA-binding zinc finger proteins and its relevance to mutagenicity and carcinogenicity through free radical generation. Nutrition 11: 646–649, 1995
2. Pollan M, Gustavsson P: High-risk occupations for breast cancer in the Swedish female working population. Am J Public Health 89: 875–881, 1999
3. Maier H, Tisch M: Epidemiology of laryngeal cancer: Results of the Heidelberg case-control study. Acta Otolaryngol Suppl 527: 160–164, 1997
4. Snow ET: Metal carcinogenesis: Mechanistic implications. Pharmacol Ther 53: 31–65, 1992
5. Hanahan D, Weinberg RA: The hallmarks of cancer. Cell 100: 57–70, 2000
6. Hamilton JW, Kaltreider RC, Bajenova OV, Ihnat MA, McCaffrey J, Turpie BW, Rowell EE, Oh J, Nemeth MJ, Pesce CA, Lariviere JP: Molecular basis for effects of carcinogenic heavy metals on inducible gene expression. Environ Health Perspect 106 (suppl 4): 1005–1015, 1998
7. Bartsch H, Nair J: New DNA-based biomarkers for oxidative stress and cancer chemoprevention studies. Eur J Cancer 36: 1229–1234, 2000
8. Karin M, Ben-Neriah Y: Phosphorylation meets ubiquitination: The control of NF-[kappa]B activity. Annu Rev Immunol 18: 621–663, 2000
9. Chen F, Castranova V, Shi X, Demers LM: New insights into the role of nuclear factor-kappaB, a ubiquitous transcription factor in the initiation of diseases. Clin Chem 45: 7–17, 1999
10. Pahl HL: Activators and target genes of Rel/NF-kappaB transcription factors. Oncogene 18: 6853–6866, 1999
11. Gilmore TD: The Rel/NF-kappaB signal transduction pathway: Introduction. Oncogene 18: 6842–6844, 1999
12. Sun SC, Ballard DW: Persistent activation of NF-kappaB by the tax transforming protein of HTLV-1: Hijacking cellular IkappaB kinases. Oncogene 18: 6948–6958, 1999
13. Neish AS, Gewirtz AT, Zeng H, Young AN, Hobert ME, Karmali V, Rao AS, Madara JL: Prokaryotic regulation of epithelial responses by inhibition of IkappaB-alpha ubiquitination. Science 289: 1560–1563, 2000
14. Palombella VJ, Rando OJ, Goldberg AL, Maniatis T: The ubiquitin-proteasome pathway is required for processing the NF-kappa B1 precursor protein and the activation of NF-kappa B. Cell 78: 773–785, 1994
15. Woronicz JD, Gao X, Cao Z, Rothe M, Goeddel DV: IkappaB kinase-beta: NF-kappaB activation and complex formation with IkappaB kinase-alpha and NIK. Science 278: 866–869, 1997
16. Mercurio F, Zhu H, Murray BW, Shevchenko A, Bennett BL, Li J, Young DB, Barbosa M, Mann M, Manning A, Rao A: IKK-1 and IKK-2: Cytokine-activated IkappaB kinases essential for NF-kappaB activation. Science 278: 860–866, 1997
17. Cohen L, Henzel WJ, Baeuerle PA: IKAP is a scaffold protein of the IkappaB kinase complex. Nature 395: 292–296, 1998
18. Krappmann D, Hatada EN, Tegethoff S, Li J, Klippel A, Giese K, Baeuerle PA, Scheidereit C: The ikappa B kinase (IKK) complex is tripartite and contains IKKgamma but not IKAP as a regular component. J Biol Chem 275: 29779–29787, 2000
19. Li X, Commane M, Nie H, Hua X, Chatterjee-Kishore M, Wald D, Haag M, Stark GR: Act1, an NF-kappa B-activating protein. Proc Natl Acad Sci USA 97: 10489–10493, 2000
20. Leonardi A, Chariot A, Claudio E, Cunningham K, Siebenlist U: CIKS, a connection to Ikappa B kinase and stress-activated protein kinase. Proc Natl Acad Sci USA 97: 10494–10499, 2000
21. Israel A: The IKK complex: An integrator of all signals that activate NF-kappaB? Trends Cell Biol 10: 129–133, 2000
22. Tojima Y, Fujimoto A, Delhase M, Chen Y, Hatakeyama S, Nakayama K, Kaneko Y, Nimura Y, Motoyama N, Ikeda K, Karin M, Nakanishi M: NAK is an IkappaB kinase-activating kinase. Nature 404: 778–782, 2000
23. Li N, Karin M: Ionizing radiation and short wavelength UV activate NF-kappaB through two distinct mechanisms. Proc Natl Acad Sci USA 95: 13012–13017, 1998
24. Bender K, Gottlicher M, Whiteside S, Rahmsdorf HJ, Herrlich P: Sequential DNA damage-independent and -dependent activation of NF-kappaB by UV. Embo J 17: 5170–5181, 1998
25. Lee FS, Hagler J, Chen ZJ, Maniatis T: Activation of the IkappaB

alpha kinase complex by MEKK1, a kinase of the JNK pathway. Cell 88: 213–222, 1997

26. Pomerantz JL, Baltimore D: NF-kappaB activation by a signaling complex containing TRAF2, TANK and TBK1, a novel IKK-related kinase. Embo J 18: 6694–6704, 1999
27. Ozes ON, Mayo LD, Gustin JA, Pfeffer SR, Pfeffer LM, Donner DB: NF-kappaB activation by tumour necrosis factor requires the Akt serine-threonine kinase. Nature 401: 82–85, 1999
28. Romashkova JA, Makarov SS: NF-kappaB is a target of AKT in anti-apoptotic PDGF signalling. Nature 401: 86–90, 1999
29. Sun Z, Arendt CW, Ellmeier W, Schaeffer EM, Sunshine MJ, Gandhi L, Annes J, Petrzilka D, Kupfer A, Schwartzberg PL, Littman DR: PKC-theta is required for TCR-induced NF-kappaB activation in mature but not immature T lymphocytes. Nature 404: 402–407, 2000
30. Lin X, O'Mahony A, Mu Y, Geleziunas R, Greene WC: Protein kinase C-theta participates in NF-kappaB activation induced by CD3-CD28 costimulation through selective activation of IkappaB kinase beta. Mol Cell Biol 20: 2933–2940, 2000
31. Lallena MJ, Diaz-Meco MT, Bren G, Paya CV, Moscat J: Activation of IkappaB kinase beta by protein kinase C isoforms. Mol Cell Biol 19: 2180–2188, 1999
32. Lin X, Cunningham ET, Jr., Mu Y, Geleziunas R, Greene WC: The proto-oncogene Cot kinase participates in CD3/CD28 induction of NF-kappaB acting through the NF-kappaB-inducing kinase and IkappaB kinases. Immunity 10: 271–280, 1999
33. Zamanian-Daryoush M, Mogensen TH, DiDonato JA, Williams BR: NF-kappaB activation by double-stranded-RNA-activated protein kinase (PKR) is mediated through NF-kappaB-inducing kinase and IkappaB kinase. Mol Cell Biol 20: 1278–1290, 2000
34. Hehner SP, Hofmann TG, Ushmorov A, Dienz O, Wing-Lan Leung I, Lassam N, Scheidereit C, Droge W, Schmitz ML: Mixed-lineage kinase 3 delivers CD3/CD28-derived signals into the IkappaB kinase complex. Mol Cell Biol 20: 2556–2568, 2000
35. Minden A, Lin A, McMahon M, Lange-Carter C, Derijard B, Davis RJ, Johnson GL, Karin M: Differential activation of ERK and JNK mitogen-activated protein kinases by Raf-1 and MEKK. Science 266: 1719–1723, 1994
36. Yan M, Dai T, Deak JC, Kyriakis JM, Zon LI, Woodgett JR, Templeton DJ: Activation of stress-activated protein kinase by MEKK1 phosphorylation of its activator SEK1. Nature 372: 798–800, 1994
37. Lee FS, Peters RT, Dang LC, Maniatis T: MEKK1 activates both IkappaB kinase alpha and IkappaB kinase beta. Proc Natl Acad Sci USA 95: 9319–9324, 1998
38. Meyer CF, Wang X, Chang C, Templeton D, Tan TH: Interaction between c-Rel and the mitogen-activated protein kinase kinase kinase 1 signaling cascade in mediating kappaB enhancer activation. J Biol Chem 271: 8971–8976, 1996
39. Hirano M, Osada S, Aoki T, Hirai S, Hosaka M, Inoue J, Ohno S: MEK kinase is involved in tumor necrosis factor alpha-induced NF-kappaB activation and degradation of IkappaB-alpha. J Biol Chem 271: 13234–13238, 1996
40. Yin MJ, Christerson LB, Yamamoto Y, Kwak YT, Xu S, Mercurio F, Barbosa M, Cobb MH, Gaynor RB: HTLV-I Tax protein binds to MEKK1 to stimulate IkappaB kinase activity and NF-kappaB activation. Cell 93: 875–884, 1998
41. Kopp E, Medzhitov R, Carothers J, Xiao C, Douglas I, Janeway CA, Ghosh S: ECSIT is an evolutionarily conserved intermediate in the Toll/IL-1 signal transduction pathway. Genes Dev 13: 2059–2071, 1999
42. Xia Y, Makris C, Su B, Li E, Yang J, Nemerow GR, Karin M: MEK kinase 1 is critically required for c-Jun N-terminal kinase activation by proinflammatory stimuli and growth factor-induced cell migration. Proc Natl Acad Sci USA 97: 5243–5248, 2000
43. DiDonato JA, Hayakawa M, Rothwarf DM, Zandi E, Karin M: A cytokine-responsive IkappaB kinase that activates the transcription factor NF-kappaB. Nature 388: 548–554, 1997
44. Malinin NL, Boldin MP, Kovalenko AV, Wallach D: MAP3K-related kinase involved in NF-kappaB induction by TNF, CD95 and IL-1. Nature 385: 540–544, 1997
45. Song HY, Regnier CH, Kirschning CJ, Goeddel DV, Rothe M: Tumor necrosis factor (TNF)-mediated kinase cascades: Bifurcation of nuclear factor-kappaB and c-jun N-terminal kinase (JNK/SAPK) pathways at TNF receptor-associated factor 2. Proc Natl Acad Sci USA 94: 9792–9796, 1997
46. Regnier CH, Song HY, Gao X, Goeddel DV, Cao Z, Rothe M: Identification and characterization of an IkappaB kinase. Cell 90: 373–383, 1997
47. Nakano H, Shindo M, Sakon S, Nishinaka S, Mihara M, Yagita H, Okumura K: Differential regulation of IkappaB kinase alpha and beta by two upstream kinases, NF-kappaB-inducing kinase and mitogen-activated protein kinase/ERK kinase kinase-1. Proc Natl Acad Sci USA 95: 3537–3542, 1998
48. Natoli G, Costanzo A, Moretti F, Fulco M, Balsano C, Levrero M: Tumor necrosis factor (TNF) receptor 1 signaling downstream of TNF receptor-associated factor 2. Nuclear factor kappaB (NFkappaB)-inducing kinase requirement for activation of activating protein 1 and NFkappaB but not of c-Jun N-terminal kinase/stress-activated protein kinase. J Biol Chem 272: 26079–26082, 1997
49. Kirschning CJ, Wesche H, Merrill Ayres T, Rothe M: Human toll-like receptor 2 confers responsiveness to bacterial lipopolysaccharide. J Exp Med 188: 2091–2097, 1998
50. Muzio M, Natoli G, Saccani S, Levrero M, Mantovani A: The human toll signaling pathway: Divergence of nuclear factor kappaB and JNK/SAPK activation upstream of tumor necrosis factor receptor-associated factor 6 (TRAF6). J Exp Med 187: 2097–2101, 1998
51. Sylla BS, Hung SC, Davidson DM, Hatzivassiliou E, Malinin NL, Wallach D, Gilmore TD, Kieff E, Mosialos G: Epstein–Barr virus-transforming protein latent infection membrane protein 1 activates transcription factor NF-kappaB through a pathway that includes the NF-kappaB-inducing kinase and the IkappaB kinases IKKalpha and IKKbeta. Proc Natl Acad Sci USA 95: 10106–10111, 1998
52. Shinkura R, Kitada K, Matsuda F, Tashiro K, Ikuta K, Suzuki M, Kogishi K, Serikawa T, Honjo T: Alymphoplasia is caused by a point mutation in the mouse gene encoding Nf-kappa b-inducing kinase. Nat Genet 22: 74–77, 1999
53. Peters RT, Liao SM, Maniatis T: IKKepsilon is part of a novel PMA-inducible IkappaB kinase complex. Mol Cell 5: 513–522, 2000
54. Khwaja A: Akt is more than just a Bad kinase. Nature 401: 33–34, 1999
55. Xie P, Browning DD, Hay N, Mackman N, Ye RD: Activation of NF-kappa B by bradykinin through a Galpha (q)- and Gbeta gamma-dependent pathway that involves phosphoinositide 3-kinase and Akt. J Biol Chem 275: 24907–24914, 2000
56. Delhase M, Li N, Karin M: Kinase regulation in inflammatory response. Nature 406: 367–368, 2000
57. Madge LA, Pober JS: A phosphatidylinositol 3-kinase/Akt pathway, activated by tumor necrosis factor or interleukin-1, inhibits apoptosis but does not activate NFkappaB in human endothelial cells. J Biol Chem 275: 15458–15465, 2000
58. Rauch BH, Weber A, Braun M, Zimmermann N, Schror K: PDGF-induced Akt phosphorylation does not activate NF-kappa B in

human vascular smooth muscle cells and fibroblasts. FEBS Lett 481: 3–7, 2000
59. Sizemore N, Leung S, Stark GR: Activation of phosphatidylinositol 3-kinase in response to interleukin-1 leads to phosphorylation and activation of the NF-kappaB p65/RelA subunit. Mol Cell Biol 19: 4798–4805, 1999
60. Madrid LV, Wang CY, Guttridge DC, Schottelius AJ, Baldwin AS Jr, Mayo MW: Akt suppresses apoptosis by stimulating the transactivation potential of the RelA/p65 subunit of NF-kappaB. Mol Cell Biol 20: 1626–1638, 2000
61. Kane LP, Shapiro VS, Stokoe D, Weiss A: Induction of NF-kappaB by the Akt/PKB kinase. Curr Biol 9: 601–604, 1999
62. Pullen N, Dennis PB, Andjelkovic M, Dufner A, Kozma SC, Hemmings BA, Thomas G: Phosphorylation and activation of p70s6k by PDK1. Science 279: 707–710, 1998
63. Peterson RT, Schreiber SL: Kinase phosphorylation: Keeping it all in the family. Curr Biol 9: R521–524, 1999
64. Teramoto H, Coso OA, Miyata H, Igishi T, Miki T, Gutkind JS: Signaling from the small GTP-binding proteins Rac1 and Cdc42 to the c-Jun N-terminal kinase/stress-activated protein kinase pathway. A role for mixed lineage kinase 3/protein-tyrosine kinase 1, a novel member of the mixed lineage kinase family. J Biol Chem 271: 27225–27228, 1996
65. Sakurai H, Miyoshi H, Toriumi W, Sugita T: Functional interactions of transforming growth factor beta-activated kinase 1 with IkappaB kinases to stimulate NF-kappaB activation. J Biol Chem 274: 10641–10648, 1999
66. Ninomiya-Tsuji J, Kishimoto K, Hiyama A, Inoue J, Cao Z, Matsumoto K: The kinase TAK1 can activate the NIK-I kappaB as well as the MAP kinase cascade in the IL-1 signalling pathway. Nature 398: 252–256, 1999
67. Chen F, Ding M, Lu Y, Leonard SS, Vallyathan V, Castranova V, Shi X: Participation of MAP kinase p38 and IkappaB kinase in chromium (VI)-induced NF-kappaB and AP-1 activation. J Environ Pathol Toxicol Oncol 19: 231–238, 2000
68. Huang C, Chen N, Ma WY, Dong Z: Vanadium induces AP-1- and NFkappB-dependent transcription activity. Int J Oncol 13: 711–715, 1998
69. Kaltreider RC, Pesce CA, Ihnat MA, Lariviere JP, Hamilton JW: Differential effects of arsenic (III) and chromium (VI) on nuclear transcription factor binding. Mol Carcinog 25: 219–229, 1999
70. Roussel RR, Barchowsky A: Arsenic inhibits NF-kappaB-mediated gene transcription by blocking IkappaB kinase activity and IkappaBalpha phosphorylation and degradation. Arch Biochem Biophys 377: 204–212, 2000
71. Shumilla JA, Broderick RJ, Wang Y, Barchowsky A: Chromium (VI) inhibits the transcriptional activity of nuclear factor-kappaB by decreasing the interaction of p65 with cAMP-responsive element-binding protein-binding protein. J Biol Chem 274: 36207–36212, 1999
72. Barchowsky A, Dudek EJ, Treadwell MD, Wetterhahn KE: Arsenic induces oxidant stress and NF-kappa B activation in cultured aortic endothelial cells. Free Radic Biol Med 21: 783–790, 1996
73. Jaspers I, Samet JM, Reed W: Arsenite exposure of cultured airway epithelial cells activates kappaB-dependent interleukin-8 gene expression in the absence of nuclear factor-kappaB nuclear translocation. J Biol Chem 274: 31025–31033, 1999
74. Shumilla JA, Wetterhahn KE, Barchowsky A: Inhibition of NF-kappa B binding to DNA by chromium, cadmium, mercury, zinc, and arsenite *in vitro*: Evidence of a thiol mechanism. Arch Biochem Biophys 349: 356–362, 1998
75. Kapahi P, Takahashi T, Natoli G, Adams SR, Chen Y, Tsien RY, Karin M: Inhibition of NF-kappa B activation by arsenite through reaction with a critical cysteine in the activation loop of Ikappa B kinase. J Biol Chem 275: 36062–36066, 2000
76. Schieven GL, Kirihara JM, Myers DE, Ledbetter JA, Uckun FM: Reactive oxygen intermediates activate NF-kappa B in a tyrosine kinase-dependent mechanism and in combination with vanadate activate the p56lck and p59fyn tyrosine kinases in human lymphocytes. Blood 82: 1212–1220, 1993
77. Imbert V, Rupec RA, Livolsi A, Pahl HL, Traenckner EB, Mueller-Dieckmann C, Farahifar D, Rossi B, Auberger P, Baeuerle PA, Peyron JF: Tyrosine phosphorylation of I kappa B-alpha activates NF-kappa B without proteolytic degradation of I kappa B-alpha. Cell 86: 787–798, 1996
78. Klebanoff SJ, Watts DH, Mehlin C, Headley CM: Lactobacilli and vaginal host defense: Activation of the human immunodeficiency virus type 1 long terminal repeat, cytokine production, and NF-kappaB. J Infect Dis 179: 653–660, 1999
79. Chen F, Demers LM, Vallyathan V, Ding M, Lu Y, Castranova V, Shi X: Vanadate induction of NF-kappaB involves IkappaB kinase beta and SAPK/ERK kinase 1 in macrophages. J Biol Chem 274: 20307–20312, 1999
80. Singh S, Aggarwal BB: Protein-tyrosine phosphatase inhibitors block tumor necrosis factor-dependent activation of the nuclear transcription factor NF-kappa B. J Biol Chem 270: 10631–10639, 1995
81. Mukhopadhyay A, Manna SK, Aggarwal BB: Pervanadate-induced nuclear factor-kappaB activation requires tyrosine phosphorylation and degradation of IkappaBalpha. Comparison with tumor necrosis factor-alpha. J Biol Chem 275: 8549–8555, 2000
82. Krejsa CM, Nadler SG, Esselstyn JM, Kavanagh TJ, Ledbetter JA, Schieven GL: Role of oxidative stress in the action of vanadium phosphotyrosine phosphatase inhibitors. Redox independent activation of NF-kappaB. J Biol Chem 272: 11541–11549, 1997
83. Barbeau B, Bernier R, Dumais N, Briand G, Olivier M, Faure R, Posner BI, Tremblay M: Activation of HIV-1 long terminal repeat transcription and virus replication via NF-kappaB-dependent and -independent pathways by potent phosphotyrosine phosphatase inhibitors, the peroxovanadium compounds. J Biol Chem 272: 12968–12977, 1997
84. Jaspers I, Samet JM, Erzurum S, Reed W: Vanadium-induced kappaB-dependent transcription depends upon peroxide-induced activation of the p38 mitogen-activated protein kinase. Am J Respir Cell Mol Biol 23: 95–102, 2000
85. Beraud C, Henzel WJ, Baeuerle PA: Involvement of regulatory and catalytic subunits of phosphoinositide 3-kinase in NF-kappaB activation. Proc Natl Acad Sci USA 96: 429–434, 1999
86. Barceloux DG: Chromium. J Toxicol Clin Toxicol 37: 173–194, 1999
87. Alexander BH, Checkoway H, Wechsler L, Heyer NJ, Muhm JM, O'Keeffe TP: Lung cancer in chromate-exposed aerospace workers. J Occup Environ Med 38: 1253–1258, 1996
88. Ye J, Zhang X, Young HA, Mao Y, Shi X: Chromium (VI)-induced nuclear factor-kappa B activation in intact cells via free radical reactions. Carcinogenesis 16: 2401–2405, 1995
89. Shi X, Chiu A, Chen CT, Halliwell B, Castranova V, Vallyathan V: Reduction of chromium (VI) and its relationship to carcinogenesis. J Toxicol Environ Health B Crit Rev 2: 87–104, 1999
90. Shi XL, Dalal NS: Chromium (V) and hydroxyl radical formation during the glutathione reductase-catalyzed reduction of chromium (VI). Biochem Biophys Res Commun 163: 627–634, 1989
91. Shi XL, Dalal NS: NADPH-dependent flavoenzymes catalyze one electron reduction of metal ions and molecular oxygen and generate hydroxyl radicals. FEBS Lett 276: 189–191, 1990

92. Shi XL, Dalal NS: The role of superoxide radical in chromium (VI)-generated hydroxyl radical: The Cr (VI) Haber–Weiss cycle. Arch Biochem Biophys 292: 323–327, 1992
93. Janssen-Heininger YM, Poynter ME, Baeuerle PA: Recent advances towards understanding redox mechanisms in the activation of nuclear factor kappaB. Free Radic Biol Med 28: 1317–1327, 2000
94. Flohe L, Brigelius-Flohe R, Saliou C, Traber MG, Packer L: Redox regulation of NF-kappa B activation. Free Radic Biol Med 22: 1115–1126, 1997
95. Ginn-Pease ME, Whisler RL: Redox signals and NF-kappaB activation in T cells. Free Radic Biol Med 25: 346–361, 1998
96. Bonizzi G, Piette J, Merville MP, Bours V: Cell type-specific role for reactive oxygen species in nuclear factor-kappaB activation by interleukin-1. Biochem Pharmacol 59: 7–11, 2000
97. Bowie A, O'Neill LA: Oxidative stress and nuclear factor-kappaB activation: A reassessment of the evidence in the light of recent discoveries. Biochem Pharmacol 59: 13–23, 2000
98. Dalton TP, Shertzer HG, Puga A: Regulation of gene expression by reactive oxygen. Annu Rev Pharmacol Toxicol 39: 67–101, 1999
99. Davies KJ: Oxidative stress: The paradox of aerobic life. Biochem Soc Symp 61: 1–31, 1995
100. Buzard GS, Kasprzak KS: Possible roles of nitric oxide and redox cell signaling in metal-induced toxicity and carcinogenesis: A review. J Environ Pathol Toxicol Oncol 19: 179–199, 2000
101. Schreck R, Meier B, Mannel DN, Droge W, Baeuerle PA: Dithiocarbamates as potent inhibitors of nuclear factor kappa B activation in intact cells. J Exp Med 175: 1181–1194, 1992
102. Li N, Karin M: Is NF-kappaB the sensor of oxidative stress? Faseb J 13: 1137–1143, 1999
103. Roederer M, Staal FJ, Raju PA, Ela SW, Herzenberg LA: Cytokine-stimulated human immunodeficiency virus replication is inhibited by N-acetyl-L-cysteine. Proc Natl Acad Sci USA 87: 4884–4888, 1990
104. Matthews JR, Wakasugi N, Virelizier JL, Yodoi J, Hay RT: Thioredoxin regulates the DNA binding activity of NF-kappa B by reduction of a disulphide bond involving cysteine 62. Nucleic Acids Res 20: 3821–3830, 1992
105. Matthews JR, Watson E, Buckley S, Hay RT: Interaction of the C-terminal region of p105 with the nuclear localisation signal of p50 is required for inhibition of NF-kappa B DNA binding activity. Nucleic Acids Res 21: 4516–4523, 1993
106. Matthews JR, Kaszubska W, Turcatti G, Wells TN, Hay RT: Role of cysteine62 in DNA recognition by the P50 subunit of NF-kappa B. Nucleic Acids Res 21: 1727–1734, 1993
107. Cavigelli M, Li WW, Lin A, Su B, Yoshioka K, Karin M: The tumor promoter arsenite stimulates AP-1 activity by inhibiting a JNK phosphatase. Embo J 15: 6269–6279, 1996
108. Pugazhenthi S, Tanha F, Dahl B, Khandelwal RL: Inhibition of a Src homology 2 domain containing protein tyrosine phosphatase by vanadate in the primary culture of hepatocytes. Arch Biochem Biophys 335: 273–282, 1996
109. Denu JM, Dixon JE: Protein tyrosine phosphatases: Mechanisms of catalysis and regulation. Curr Opin Chem Biol 2: 633–641, 1998
110. Fetrow JS, Siew N, Skolnick J: Structure-based functional motif identifies a potential disulfide oxidoreductase active site in the serine/threonine protein phosphatase-1 subfamily. FASEB J 13: 1866–1874, 1999

Molecular and Cellular Biochemistry **222**: 173–182, 2001.

Effects of glutathione on chromium-induced DNA crosslinking and DNA polymerase arrest

Travis O'Brien, Jian Xu and Steven R. Patierno

Department of Pharmacology, The George Washington University Medical Center, Washington DC, USA

Abstract

Hexavalent chromium (Cr (VI)) is reduced intracellularly to Cr (V), Cr (IV) and Cr (III) by ascorbate (Asc), cysteine and glutathione (GSH). These metabolites induce a spectrum of genomic DNA damage resulting in the inhibition of DNA replication. Our previous studies have shown that treatment of DNA with Cr (III) or Cr (VI) plus Asc results in the formation of DNA-Cr-DNA crosslinks (Cr-DDC) and guanine-specific arrests of both prokaryotic and mammalian DNA polymerases. GSH not only acts as a reductant of Cr (VI) but also becomes crosslinked to DNA by Cr, thus, the focus of the present study was to examine the role of GSH in Cr-induced DNA damage and polymerase arrests. Co-incubation of Cr (III) with plasmid DNA in the presence of GSH led to the crosslinking of GSH to DNA. GSH co-treatment with Cr (III) also led to a decrease in the degree of Cr-induced DNA interstrand crosslinks relative to Cr (III) alone, without affecting total Cr DNA binding. DNA polymerase arrests were observed following treatment of DNA with Cr (III) alone, but were markedly reduced when GSH was added to the reaction mixture. Pre-formed polymerase-arresting lesions (Cr-DDC) were not removed by subsequent addition of GSH. Treatment of DNA with Cr (VI), in the presence of GSH, resulted in crosslinking of GSH to DNA, but failed to produce detectable DNA interstrand crosslinks or polymerase arrests. The inhibitory effect of GSH on Cr-induced polymerase arrest was further confirmed in human genomic DNA using quantitative PCR (QPCR) analysis. Treatment of genomic DNA with Cr (III) resulted in a marked inhibition of the amplification of a 1.6 kb target fragment of the p53 gene by *Taq* polymerase. This was almost completely prevented by co-treatment with GSH and Cr (III). These results indicate that Cr-induced DNA interstrand crosslinks, and not DNA-Cr-GSH crosslinks, are the principal lesions responsible for blocking DNA replication. Moreover, the formation of DNA-Cr-GSH crosslinks may actually preclude the formation of the polymerase arresting lesions. (Mol Cell Biochem **222**: 173–182, 2001)

Key words: chromium, glutathione, DNA polymerase arrest, DNA interstrand crosslinks, QPCR

Introduction

The carcinogenicity of certain hexavalent chromium (Cr (VI)) compounds has been demonstrated by extensive epidemiological, animal, and tissue culture studies [1]. However, the cellular and molecular mechanisms of Cr (VI)-carcinogenesis remain unclear. Many Cr (VI) compounds exist as chromate oxyanions (CrO_4^{2-}) in solution which readily enter cells through relatively non-selective anion channels for SO_4^{2-} [2]. Once inside cells, Cr (VI) is reduced to Cr (III) and other reductive intermediates which react with nucleic acids, resulting in a spectrum of DNA lesions [3–13] and mutations [14–19]. Cr-induced DNA damage is preferentially formed in and repaired from nuclear matrix-associated DNA [13], the site of DNA replication and transcription in cells. At least one Cr-induced DNA lesion, the DNA-Cr-DNA interstrand crosslink (Cr-DDC), interferes with the processivity of both prokaryotic and mammalian DNA polymerases [10, 21–24]; and RNA polymerases (Xu and Patierno, in preparation) leading to guanine-specific arrests [22, 23]. Cr (VI) has been dem-

Present address: J. Xu, Merck and Company, West Point, PA, USA
Address for offprints: S.R. Patierno, Department of Pharmacology, The George Washington University Medical Center, 2300 I Street N.W., Washington DC 20037, USA

onstrated to inhibit both DNA replication [25, 26] and the expression of inducible genes [10, 27]. Exposure of cells to Cr (VI)-containing compounds results in a variety of cellular effects including the arrest of cell growth and division [20, 28–33] and the activation of the p53-dependent apoptotic cell death pathway [34].

The genetic effects of Cr are the result of its reduction within the cell. The main intracellular reductants of Cr (VI) include ascorbate (Asc), cysteine and glutathione (GSH) [35–37]. GSH, a cysteine-containing tripeptide, is present at high concentrations in cells (up to 10 mM) suggesting that this compound is a physiologically relevant intracellular reductant of Cr-containing compounds. The reaction of Cr (VI) with GSH results in the formation of Cr-thiolate complexes that may react further to yield Cr (III), Cr (V), Cr (IV), hydroxyl radicals and GSH thiyl radicals [38–42]. During this process, GSH may become reactive and bind to DNA [43, 44]. Indeed, Cr (VI)-induced DNA-protein crosslinks increase in proportion to increasing GSH concentrations in cultured cells and cell-free systems [43, 45, 46]. When H_2O_2 is present, the reduction of Cr (VI) by GSH also produces O_2• and •OH, apurinic/apyrimidinic (AP) sites and DNA single-strand breaks [5, 47–50]. Likewise, the depletion of GSH decreases the formation of Cr (VI)-induced DNA single-strand breaks and lowers Cr (V) levels in the cultured cells [40, 51, 52], but has no effect on Cr-DNA binding [53]. Some evidence suggests that the formation of GSH-Cr (III)-DNA crosslinks may also lead to single-base substitutions in DNA [54].

GSH is one of the main cellular peptides crosslinked to DNA by Cr in cells [44] and may play a critical role in Cr carcinogenesis as both a reductant and a crosslinking substrate. Previous studies have not investigated the seemingly important relationship between GSH and Cr-induced DNA damage and inhibition of replication. The present study was aimed at investigating the role that GSH plays in formation of Cr-induced DNA-DNA crosslinks, DNA adducts, DNA-Cr-GSH crosslinks and the impact that these lesions have on the function of DNA polymerase on plasmid and human DNA. The results suggest that DNA-Cr-GSH crosslinks, unlike DNA interstrand crosslinks, are not polymerase arresting lesions. Moreover, formation of Cr-GSH complexes precludes the formation of Cr-induced replication-blocking lesions.

Materials and methods

Cell culture and reagents

Human lung fibroblasts (HLF) (ATCC LL-24 cells, number 151-CCL) were grown in F12K medium (Life Technologies, Gaithersburg, MD, USA) containing 15% fetal bovine serum (FBS) and supplemented with 0.5% penicillin-streptomycin in a humidified incubator under 5% CO_2 at 37°C. Sodium chromate (J.T. Baker Chemical, Phillipsburg, NJ, USA), ascorbic acid (Fisher Scientific, Fair Lawn, NJ, USA) and chromium chloride (Sigma Chemical, St. Louis, MO, USA) were freshly prepared in sterile double-distilled, de-ionized water prior to each experiment.

Preparation of DNA

The preparation of the closed circular plasmid pGEM-3Z-TS, used as a DNA replication template in this work, was carried out according to the standard procedure described by Sambrook [55]. This plasmid carries a target sequence (TS) used previously to determine the base-specificity of polymerase arrest [21–23]. The plasmid DNA was quantitated using a spectrophotometer at OD_{260} and stored at –20°C until use. HLF cells were harvested by scraping and genomic DNA was isolated using the Wizard Genomic DNA Purification Kit (Promega Corp., Madison, WI, USA) per the manufacturer's protocol and was digested with *Eco*R1 prior to treatment.

Measurement of Cr (VI) reduction

Cr (VI) reduction by Asc and GSH, under the same conditions as in polymerase arrest assays, was monitored by absorption spectroscopy as described [40]. Absorbance at a wavelength of 372 nm (OD_{372} the λmax of Cr (VI)), reflects Cr (VI) concentration in a linear dose-dependent manner but does not correlate with the other forms of Cr such as Cr (V), Cr (IV), or Cr (III). Na_2CrO_4 (Cr (VI)) (60 μM) was incubated with either Asc or GSH in 1 ml of final volume of 100 mM HEPES, pH 7.4 for 2 h at 37°C in the absence of DNA. OD_{372} was measured and the data were expressed as the percent of Cr (VI) remaining.

Cr treatment of DNA

The closed circular pGEM-3Z-TS DNA was linearized by using 2 units/μg DNA of *Not*I restriction endonuclease (Promega, Madison, WI, USA) which cut the pGEM-3Z-TS DNA at the unique site near the 3′-end of the TS and left an 83-base DNA template beyond the T7 promoter. One microgram of the linearized pGEM-3Z-TS DNA (approximately 1.5 nmol DNA-bp), or genomic DNA (0.04 μg/μl), was treated with 0 or 60 μM $CrCl_3$ (Cr (III)) (Sigma, St. Louis, MO, USA) or Cr (VI) in the presence or absence of GSH or Asc in a final volume of 25–100 μl of 100 mM HEPES, pH 7.4 at 37°C for 2 h. To assess the effect of GSH on preformed Cr-induced DNA lesions, 1 μg of the linearized pGEM-3Z-TS DNA was treated with 0 or 60 μM of $CrCl_3$ in a final volume of 20 μl of

100 mM HEPES, pH 7.4 at 37°C for 18 h and then added with 0–6 mM GSH for an additional 2 h. Following the treatment, the pGEM-3Z-TS DNA was dialyzed with a MicroDialysis membrane (10 kD of MW cutoff; Spectrum, Houston, TX, USA) against 200 ml of 100 mM HEPES, pH 7.4 at room temperature for 2 h to remove unreacted Cr and GSH. Genomic DNA was filter dialyzed against sterile, deionized water using microcon filter columns (10,000 MWCO) (Millipore, Bedford, MA, USA).

Measurement of Cr-induced DNA-GSH crosslinks

One µg of the linearized pGEM-3Z-TS DNA was treated with 0 or 60 µM of Cr (III) or Cr (VI) in the presence of [^{3}H]-labeled GSH stock (6.25 mCi/mmol), which were prepared by mixing unlabeled GSH with radiolabeled [glycine-2-^{3}H]-GSH (47 Ci/mmol; NEN, Wilmington, DE, USA), in a final volume of 25 µl of 100 mM HEPES, pH 7.4 for 2 h at 37°C. Following treatment, 0.5 ml of ice-cold 5% trichloroacetic acid (TCA) was added and incubated in an ice bath for at least 30 min. The entire volume was loaded onto a 5% TCA pre-wetted Whatman GF/A glass micro fiber filter in a Millipore filtration apparatus under vacuum. The filter (containing GSH bound to DNA) was sequentially washed twice with 5 ml of 5% TCA and once with 2 ml of acetone and air-dried completely. DNA-[^{3}H]-GSH adduct levels were determined by liquid scintillation counting. A 10-fold dilution series of the [^{3}H]-labeled GSH stock was applied to the filters, which were air-dried and counted. All scintillation vials contained 3 ml of EcoLite® scintillation fluid. The Cr-induced DNA-GSH crosslinks were calculated and expressed as GSH/kbp (mmol GSH/mol DNA kilobase pairs) according to the [^{3}H]-labeled GSH stock.

DNA polymerase arrest assay on Cr-treated plasmid DNA

The end-labeling of T7 promoter primer (Promega, Madison, WI, USA), dideoxy DNA sequencing, and DNA polymerase arrest assay were performed as previously described by our laboratory [21–23]. Each reaction (0.5–2 µl) was resolved on a 6% polyacrylamide/8 M urea sequencing gel with the adjacent lanes loaded with a dideoxy DNA sequencing reaction of an identical template using the same T7 promoter primer.

Measurement of Cr-DNA adducts

Cr-DNA adducts were detected in pGEM-3Z-TS DNA under the same conditions used in the polymerase arrest assay. One microgram of pGEM-3Z-TS DNA was treated 2 h at 37°C using [^{51}Cr]-labeled $CrCl_3$ (1.35 Ci/mmol) or Na_2CrO_4 (1.2 Ci/mmol). Radiolabeled stocks were prepared by mixing unlabeled Cr (III) or Cr (VI) with radioactive $^{51}CrCl_3$ (265 mCi/mg Cr; ICN, Costa Mesa, CA, USA) or $Na_2{}^{51}CrO_4$ (229 mCi/mg Cr; ICN), respectively. Following the treatment, the pGEM-3Z-TS DNA reaction mixture was combined with 500 µl of ice-cold 5% trichloroacetic acid (TCA), incubated in an ice bath for at least 30 min to allow for precipitation and applied onto a 5% TCA pre-wetted Whatman GF/A glass micro fiber filter in a Millipore filtration apparatus under vacuum. The filter was washed twice with 5 ml of 5% TCA, once with 2 ml of acetone and dried. The amount of ^{51}Cr remained on the filter was counted in 3 ml of EcoLite® scintillation cocktail (ICN, Irvine, CA, USA). The Cr-DNA adducts were calculated and expressed as Cr/kbp (mmol Cr/mol DNA) kilobase pairs based on the [^{51}Cr]-labeled Cr (III) or Cr (VI) stock.

Renaturing Agarose Gel Electrophoresis (RAGE)

Analysis of Cr-DDC was conducted essentially as described previously by our laboratory [22]. The presence of DNA interstrand crosslinks is characterized by DNA that migrates as double stranded following heat denaturation on an agarose gel. Five microliter (200 ng DNA) aliquots of Cr-treated pGEM-3Z-TS DNA were diluted 50-fold with 100 mM HEPES, pH 7.4 following the treatment. A 25 µl (20 ng) aliquot of the diluted pGEM-3Z-TS DNA was denatured by heating at 96°C for 10 min and rapidly chilled in an ice bath. The other aliquot was stored in an ice bath and was not denatured. Both aliquots were admixed with 0.2 volume of 6X gel-loading buffer, resolved on 1% agarose gels at 35 V for 18 h in 0.5X TBE containing 0.5 µg/ml ethidium bromide. The DNA in the gel was transferred to a Biotrace HP charge-modified polysulfone binding matrix membrane (Gelman, Ann Arbor, MI, USA) by capillary force in 0.4 N NaOH according to the manufacturer's protocol. The *Not*I linearized pGEM-3Z-TS DNA was radiolabeled with [α-^{32}P]-dCTP (3000 Ci/mmol, ICN) using a method described [56]. Then, the membrane was hybridized with the radiolabeled probe as previously described by our laboratory [10].

Quantitative PCR analysis of Cr-treated human DNA

The p53 gene (exons 5–8, 1.6 kb) was used as the target sequence for quantitative PCR analysis (QPCR) from chromium treated HLF DNA. QPCR was carried out using previously published primers ($5' \rightarrow 3'$: sense-TGCCCTGACTTTCAACTCTGTCTCC and antisense-CTTGGTCTCCTCCACCGCTTCTTG) [57, 58]. The concentration of template was optimized to be the limiting factor for PCR-mediated amplification. Briefly, one hundred nanograms of HLF DNA was amplified for 22-34 cycles to determine the range (28 cycles) at which the amount of amplified product was proportional to the amount of template. A titration curve with increasing amounts of template (25–200 ng) was produced to confirm that the amount of PCR product was linearly related to template concentration in the reactions after 28 cycles of amplification. A typical PCR reaction (50 μl) consisted of 5 μl 10X Thermophillic Buffer (500 mM KCl, 100 mM Tris-HCl (pH 9.0), 1.0% Triton® X-100), 3 mM $MgCl_2$, 0.2 mM each dNTP, 2 μCi $\alpha^{32}P$-dCTP (3000 Ci/mmol) (ICN, Costa Mesa, CA, USA) 0.5 ml *Taq* polymerase (2.5 units) (Promega Life Science, Madison, WI, USA) and 100 ng *Eco*R1-digested HLF DNA. The reactions were initially heated to 94°C before primer (10 pmol each) addition. Reactions were denatured at 94°C for 1 min, annealed at 65°C for 2 min and extended at 72°C for 3 min. A final extension 72°C for 7 min was included at the end of the reaction. Reactions (12.5 μl) were admixed with 6X Loading buffer and resolved on a 5% nondenaturing acrylamide gel at 200 V for ~ 2 h. Gels were dried and exposed to film for 3 h to overnight at –70°C.

Results

Reduction Cr (VI) by GSH

The reduction of Cr (VI) is necessary for Cr (VI)-induced DNA damage because Cr (VI) itself does not react with DNA. Cr (VI) is reduced by GSH and Asc to Cr (III), which remains within the cell achieving millimolar concentrations [59]. Therefore, we compared GSH and Asc-mediated reduction of a biologically relevant intracellular concentration of Cr (VI) (60 μM) by absorption spectrometry (OD_{372}). Figure 1 shows the dose-effect curve of GSH and Asc-induced Cr (VI) reduction (2 h at 37°C). The GSH results paralleled those obtained for Asc, however, Asc was more effective at reducing Cr (VI) at relatively lower concentrations than GSH. At 6–10 mM GSH, greater than 80% of Cr (VI) was reduced after a 2 h incubation.

Binding of GSH to pGEM-3Z-TS DNA during Cr reduction

GSH can become crosslinked to DNA during the process of chromium reduction and is one of the major DNA-protein crosslinks formed by Cr in cultured cells [44]. To measure the extent of Cr-induced GSH-DNA crosslinks and to evaluate the effects of these lesions on DNA replication pGEM-3Z-TS was treated with Cr in the presence of 60 μM of Cr (III) or Cr (VI) (1 Cr/DNA-bp treatment ratio) in the presence of various concentrations of [^{3}H]-GSH for 2 h at 37°C. As shown in Fig. 2, there was an association between increas-

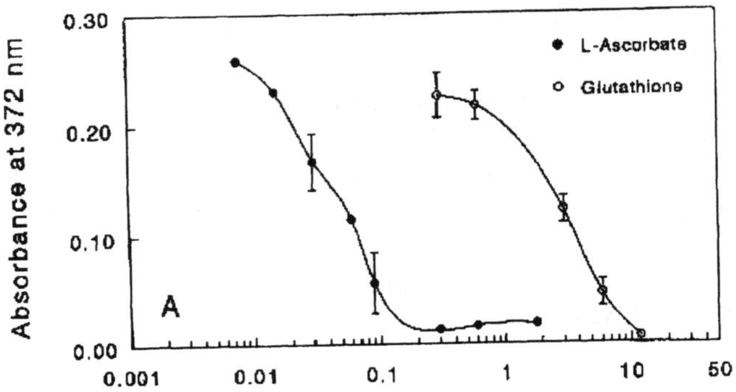

Fig. 1. Cr (VI) reduction by Asc and GSH. The relative reduction of Cr (VI) was measured by a shift in absorbance at 372 nm in a spectrophotometer. The figure shows the percent of Cr (VI) remaining after incubation of 60 μM Cr (VI) with the indicated concentrations of Asc or glutathione in 100 mM HEPES (pH 7.4) for 2 h at 37°C.

ing GSH concentration and GSH-Cr-DNA crosslinks induced by 60 μM Cr (III) achieving 6, 13, and 146 GSH/kbp (mmol GSH/mole DNA kilobase pairs) at 0.3, 0.6, and 6 mM of GSH, respectively. Cr (VI) led to a similarly apparent association producing 10, 12 and 107 GSH/kbp at 0.3, 0.6, and 6 mM of GSH, respectively.

Effects of GSH on formation of DNA polymerase arresting lesions by Cr in pGEM-3Z-TS DNA

To determine the role of GSH and Cr-induced GSH-DNA crosslinks on induction of Cr-induced DNA polymerase arrests, a DNA polymerase arrest assay was performed using pGEM-3Z-TS incubated with Cr (III) and GSH (2 h, Fig. 3) or Cr (III) followed by GSH (18 h Cr/2 h GSH, Fig. 4). In both untreated and GSH-treated controls, most initiated replication continued to the end of the template (*Not*I site) or one base beyond, as previously reported [22, 23]. Guanine-specific DNA polymerase arrests were observed in the 60 μM Cr (III)-treated positive controls (Figs 3 and 4). Co-treatment of Cr (III) (60 μM) with even low concentrations of GSH (0.06–0.6 mM) for 2 h resulted in a dose-dependent decrease in the intensity of the arrest fragments (Fig. 3). No detectable DNA polymerase arrest occurred following treatment with 60 μM Cr (VI) in the presence or absence of GSH for 2 h (Fig. 4), even though the presence of GSH resulted in both reduction of Cr (VI) and crosslinking of GSH to DNA. Therefore, neither Cr (VI) alone nor Cr (VI) plus GSH under these experimental conditions could produce enough Cr-DDC to cause detectable DNA polymerase arrests. GSH was unable to reverse the inhibition of DNA polymerase activity when the DNA was pre-incubated with Cr (III) for 18 h (preformed Cr-DDC) before treatment with GSH for 2 h (Fig. 4).

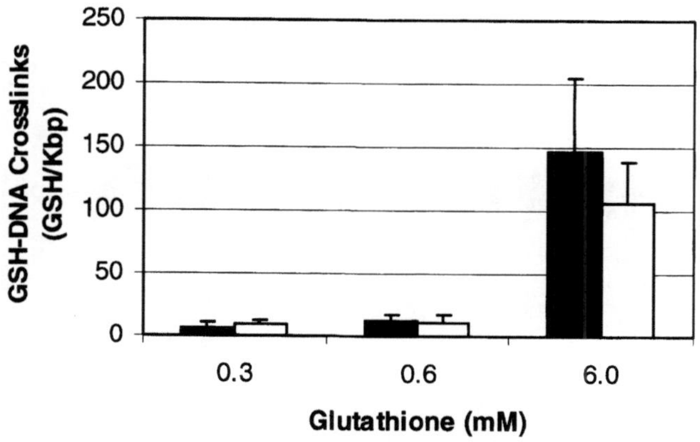

Fig. 2. Formation of GSH-DNA crosslinks in the presence of Cr. pGEM-3Z-TS DNA (1 μg) was treated with 60 μM of Cr (III) (closed bars) or 60 μM Cr (VI) (open bars) in the presence of the indicated concentrations of [glycine-2-^{3}H]-GSH for 2 h. TCA precipitation was then performed and amount of the ^{3}H that remained on the filter was counted in scintillation cocktail. Cr-induced GSH-DNA crosslinks were calculated and expressed as mmol GSH/mol DNA kilobase pairs (GSH/kbp). Error bars represent the mean ± S.E. of 3–4 separate experiments.

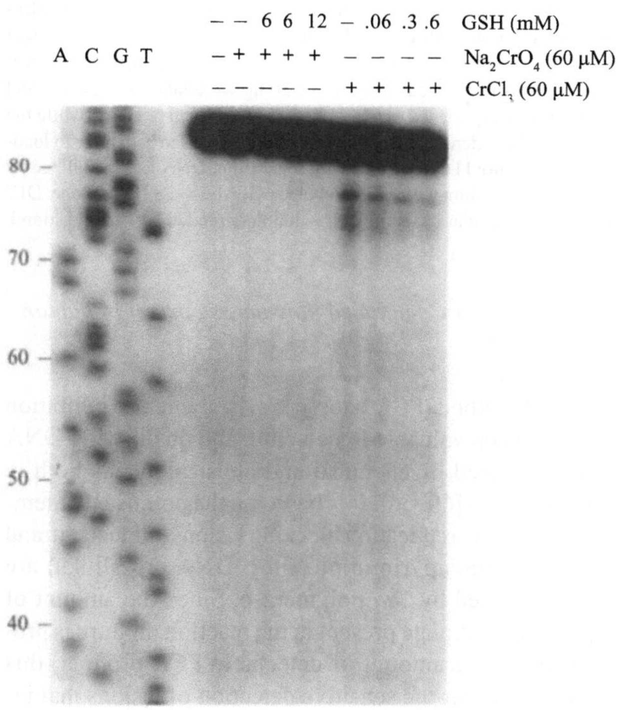

Fig. 3. DNA polymerase arrest assay of pGEM-3Z-TS treated with Cr in the presence of GSH. The pGEM-3Z-TS DNA (1 μg) was treated with 60 μM of Cr (III) or 60 μM Cr (VI) in the presence of the indicated concentrations of GSH for 2 h at 37°C. After treatment pGEM-3Z-TS DNA was dialyzed to remove free Cr and GSH and the DNA polymerase arrest assay was conducted using a [^{32}P]-end-labeled T7 promoter primer (see Materials and methods). Shown is a typical autoradiogram of the DNA polymerase arrest assay. On the left, adjacent lanes (A, C, G, and T) are the sequence of the pGEM-3Z-TS DNA. The numbers on the left side (40–80) indicate the position in the target sequence of plasmid pGEM-3Z-TS DNA.

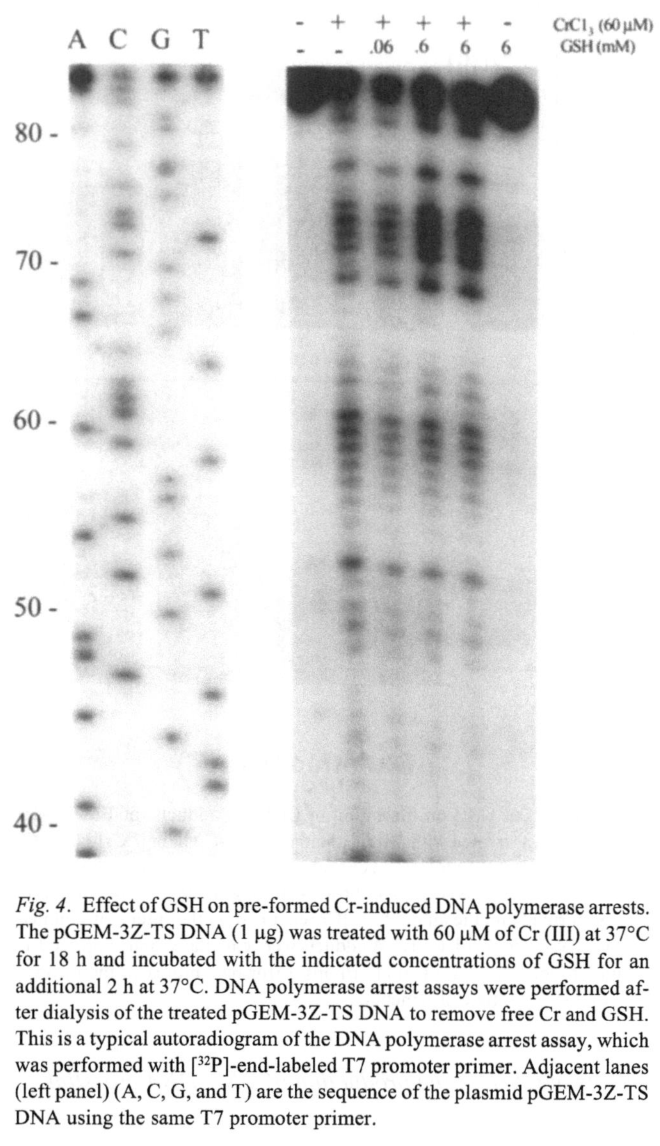

Fig. 4. Effect of GSH on pre-formed Cr-induced DNA polymerase arrests. The pGEM-3Z-TS DNA (1 μg) was treated with 60 μM of Cr (III) at 37°C for 18 h and incubated with the indicated concentrations of GSH for an additional 2 h at 37°C. DNA polymerase arrest assays were performed after dialysis of the treated pGEM-3Z-TS DNA to remove free Cr and GSH. This is a typical autoradiogram of the DNA polymerase arrest assay, which was performed with [^{32}P]-end-labeled T7 promoter primer. Adjacent lanes (left panel) (A, C, G, and T) are the sequence of the plasmid pGEM-3Z-TS DNA using the same T7 promoter primer.

Binding of Cr to pGEM-3Z-TS DNA in the presence of GSH

To determine the effect of GSH on the binding of Cr to DNA, pGEM-3Z-TS DNA was treated with 60 μM (the treatment ratio was 1 Cr per DNA-bp) of [^{51}Cr]-labeled Cr (III) or Cr (VI) in the presence of GSH (0.06–12 mM) at a GSH/Cr treatment ratio of 1–200 for 2 h at 37°C. As shown in Fig. 5, Cr (III) readily becomes adducted to DNA and GSH had little impact on the formation of Cr (III)-induced Cr-DNA adducts. Cr (VI), alone, does not form DNA adducts in the absence of GSH. However, addition of GSH to the reaction mixture resulted in Cr-DNA adduct formation at levels previously shown to result in polymerase arrests when induced by either Cr (III) or Asc reduction of Cr (VI) [22, 23].

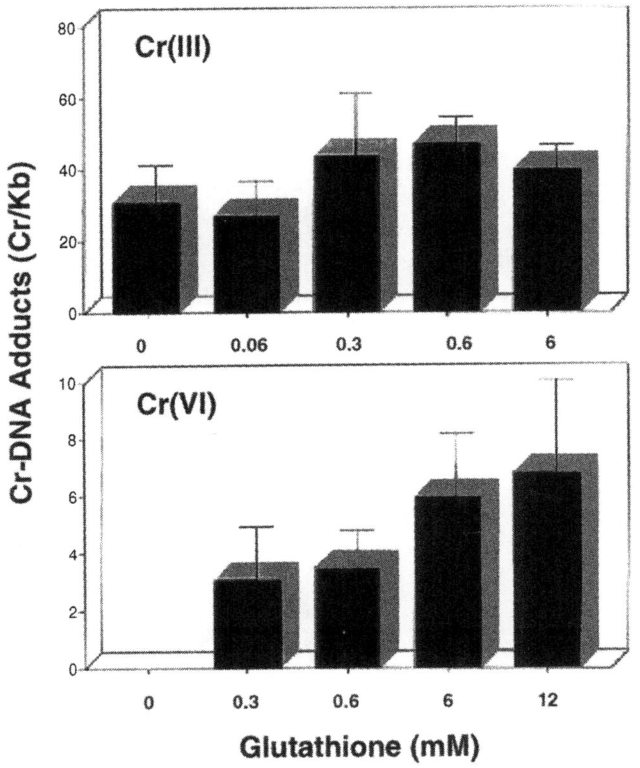

Fig. 5. Effect of GSH on formation of Cr-DNA adducts. pGEM-3Z-TS DNA (1 μg) was treated with 60 μM of either [^{51}Cr]-$CrCl_3$ or [^{51}Cr]-Na_2CrO_4 in the presence of the indicated concentrations of GSH as described in the Materials and methods section. Following the treatment, TCA precipitation was performed and the amount of ^{51}Cr remained on the filter was counted in scintillation cocktail. Cr-DNA adducts were calculated and expressed as mmol Cr/mol DNA base pairs (Cr/kbp). Error bars represent the mean ± S.E. of 3–5 separate experiments.

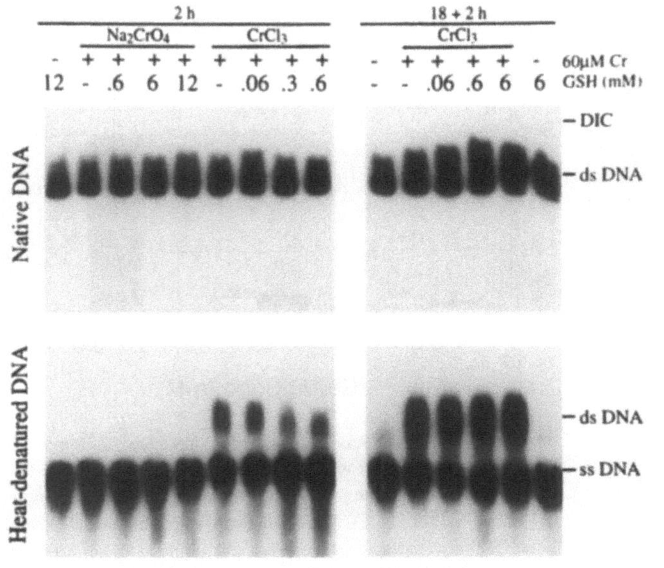

Fig. 6. Renaturing agarose gel electrophoresis analysis of GSH on formation of Cr-induced DNA-DNA crosslinks. The linearized pGEM-3Z-TS was either co-treated with 60 μM of Cr (VI) or Cr (III) and the indicated concentrations of glutathione for 2 h (left panel) or pre-treated with 60 μM of Cr (III) for 18 h, then added and incubated with the indicated concentrations of GSH for an additional 2 h (right panel). Following the treatment and dialysis, samples were diluted 50-fold and renaturing agarose gel electrophoresis was performed. Aliquots of 20 ng were taken from each, and one set was directly loaded onto a 1% agarose gel (Native DNA) while the other set was heat-denatured and rapidly chilled in an ice bath prior to loading (Heat-denatured DNA). Following separation, DNA was visualized by Southern hybridization with a radiolabeled pGEM-3Z-TS DNA probe. DIC – DNA intermolecular crosslinks; ds – double-stranded; ss – single strand.

Effects of GSH on the production of DNA-DNA crosslinks by Cr

Cr-induced DNA interstrand crosslinks have been shown to be a DNA polymerase arresting lesion [21–23]. Renaturing Agarose Gel Electrophoresis (RAGE) was employed to examine the effect of GSH on the induction of Cr-induced DNA interstrand crosslinks. pGEM-3Z-TS DNA was exposed to 60 μM of Cr (III) or Cr (VI) in the presence, or absence, of GSH for 2 h at 37°C. The left panel in Fig. 6 shows that increasing the ratio of GSH to Cr (III) resulted in a dose-dependent decrease of the intensity of the double-stranded DNA band at micromolar concentrations of GSH. However, neither Cr (VI) alone nor Cr (VI) plus GSH produced any detectable DNA crosslinks. Figure 6 (right panel) also shows that DNA-DNA crosslinks were produced by treatment with 60 μM Cr (III) for 18 h and that these preformed lesions could not be removed by GSH, even at the 6 μM concentration. These results were consistent with the polymerase arrest data shown in Fig. 1.

Effect of GSH on Cr-induced polymerase arrest in human genomic DNA

The QPCR method (PCR 'stop' assay, polymerase inhibition assay) has been valuable in detecting and quantifying DNA lesions produced by chemical and physical agents both *in vitro* and *in vivo* [60–66]. It is based on the premise that templates containing replication blocking lesions (adducts, strand breaks, apurinic/apyrimidinic sites, DNA crosslinks) are poorly amplified by *Taq* polymerase. Since the amount of 'amplifiable' template present in the reaction mixture is proportionate to the amount of detectable PCR product, this method allows for the sensitive detection of lesions that inhibit *Taq* polymerase processivity on a single target sequence in genomic DNA. In order to confirm the data obtained in the plasmid-based polymerase arrest and RAGE assays and extend the results to genomic DNA, we examined the effect of physiological concentrations of GSH on the formation of arresting lesions in Cr (III) and Cr (VI)-treated HLF DNA using QPCR analysis of the p53 gene (exons 5–8, 1.6 kb) under reaction conditions identical to those employed in the RAGE and sequencing-based assays [58]. Consistent with the

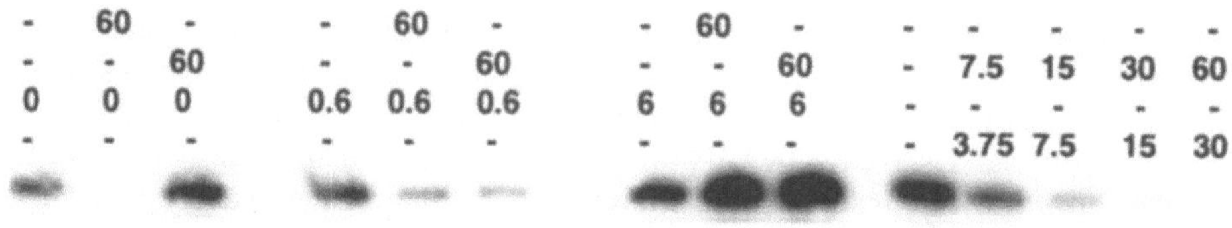

Fig. 7. Effect of GSH on the formation Cr-induced polymerase-arresting lesions in HLF DNA by QPCR. HLF DNA (0.04 μg/μl) was reacted with either 60 μM Cr (III) or 60 μM Cr (VI) in the presence or absence of GSH (mM), processed and a fragment of the p53 gene (exons 5–8) was amplified for 28 cycles. PCR products (1.6 kb) were separated on a 5% non-denaturing acrylamide gel for 2 h. A representative gel is shown. The lanes on the right display the inhibition of *Taq* processivity following the reduction of Cr (VI) by Asc as a reference.

RAGE and polymerase arrest data, Cr (III) (60 μM) led to an inhibition of *Taq*-mediated amplification of the target sequence (Fig. 7). Similarly, co-treatment with GSH abrogated Cr (III)-mediated polymerase arrest in a concentration-dependent manner beginning at 0.6 mM and completely prevented Cr (III)-induced replication blockage at 6.0 mM GSH. While Cr (VI) alone had no effect, GSH (0.6 mM) co-incubated with Cr (VI) (60 μM) led to a slight inhibition of amplification indicating that some polymerase-arresting lesions were being formed as a consequence of the reduction of Cr (VI). However, at 6.0 mM, GSH completely prevented the formation of Cr (VI)-induced polymerase arresting lesions even though Cr (VI) was being reduced (Fig. 1) and GSH could be found extensively crosslinked to DNA (Fig. 2). In contrast, reduction of Cr (VI) with Asc resulted in extensive polymerase arrest (far right lanes).

Discussion

The intracellular reduction of Cr (VI) is crucial for the genotoxicity of chromate because Cr (VI) alone, does not react with DNA. GSH is one of most abundant intracellular reductants of Cr (VI), approaching millimolar concentrations within the cell [51]. Although Asc has been demonstrated to be a principle reductant of Cr (VI) [36, 67], GSH also facilitates the reduction of Cr (VI) resulting in DNA damage [37]. Unlike Asc, GSH can be crosslinked to DNA by Cr, *in vivo* and *in vitro*, leading to DNA-Cr-GSH crosslinks [43, 44, 46]. Moreover, GSH is one of the most common peptides crosslinked to DNA by Cr [44]. Our previous studies using Asc as the reductant, established that Cr-induced DNA-DNA crosslinks, but not monoadducts, caused base-specific DNA polymerase arrest [22]. Given the potential bulky nature of a DNA-Cr-GSH cross-link, we sought to determine whether this complex was also a polymerase arresting lesion. The results from the present study showed that DNA-GSH crosslinks are not polymerase arresting lesions. Furthermore, the presence of GSH prevented the formation of Cr-mediated DNA interstrand crosslinks and attenuated the Cr-induced blockage of DNA polymerase chain extension.

Cr (III) contains six coordinate sites and is capable of forming complexes with many ligands such as single amino acids, proteins, and DNA [68, 69]. The reactivity of Cr (III) towards DNA can be modified by the nature of associated ligands [68]. Cr (III)-amino acid complexes have been demonstrated to exhibit a lower level of DNA binding than the corresponding aquo complexes [69]. In our studies, reduction of 60 μM Cr (VI) with 6 mM GSH resulted in approximately 6 adducts per 1000 base pairs (Fig. 5) whereas an equivalent adduct level can be produced by the reaction of 7.5 μM Cr (VI) with 3.25 μM Asc [22]. This indicates that the formation of Cr-DNA adducts is markedly decreased when Cr (VI) is reduced by GSH compared to reduction by Asc. GSH is capable of reacting with Cr (VI) to yield Cr (V), Cr (IV), and GSH thiyl radicals [39–42]. The potential to both reduce Cr (VI) and interact with Cr affords GSH a unique opportunity to form complexes with Cr (III) in cells. GSH often occupies three coordination sites of Cr (III) and forms stable tridentate Cr (III)-GSH complexes [69], however, it is plausible that incomplete Cr (III)-GSH complexes are still capable of binding to DNA resulting in DNA-Cr-GSH complexes [7, 44]. Like Cr DNA-binding, the Cr-GSH complex primarily binds to the phosphate groups present in DNA [68, 71–73]. In addition, Cr-GSH complexes can directly interact with guanine-containing polynucleotides presumably through the guanine-*N7* position [43, 74].

An interesting result from the present study was that the presence of GSH, leading to the formation of DNA-Cr-GSH crosslinks, actually attenuated the formation of both DNA interstrand crosslinks and polymerase arrests when DNA was exposed to Cr (III). This effect was observed even when the GSH:Cr (III) treatment ratio was lower than 10. However, GSH could not alter levels of the pre-formed DNA interstrand crosslinks and polymerase-arresting lesions. This study also showed that GSH did not significantly alter the total levels of Cr binding to DNA induced by Cr (III) *in vitro* suggesting that the total Cr-DNA binding includes Cr-DNA monoadducts, DNA-Cr-GSH crosslinks, and DNA interstrand crosslinks. Thus, a probable explanation for the decreased DNA interstrand crosslinks while Cr-DNA adduct levels remained the same, is that formation of the Cr-GSH complexes and DNA-Cr-GSH crosslinks decreased the availability of free Cr coordination sites that could participate in the formation of DNA interstrand crosslinks.

Although GSH is capable of reducing Cr (VI), the rate of this reaction is slower than that observed with Asc [40, 75]. Our data re-confirmed this showing that GSH had a lower reductive potency than Asc at physiological pH. The process of Cr (VI) reduction by GSH results in binding of GSH to the reductive intermediates of Cr, forming Cr-GSH complexes, which have a lower affinity for DNA than free Cr (III). Formation of such complexes during reduction is the likely explanation for our finding that the reduction of Cr (VI) by GSH led to a lower level of Cr-DNA binding relative to reduction by Asc (and relative to equivalent concentrations of Cr (III)). For example, treatment of DNA with either 60 μM Cr (III) or 60 μM Cr (VI) plus Asc led to adduct levels around 30 adducts/kbp. In contrast, 60 μM Cr (VI) plus 6 mM GSH yielded adduct levels of around 6 adducts/kb. At approximately 6 Cr-adducts per 1000 bases, our prior sequencing-based techniques were capable of detecting a 10% inhibition of primer extension of an 88-bp target sequence nested in a 2.8 kb plasmid [22, 23]. In the present study, QPCR analysis conducted at this adduct level detected an ~ 50% inhibition of amplification of a 1.6 kb fragment. This suggests that QPCR yielded a 5-fold increased sensitivity for detecting polymerase arresting lesions than sequencing. Both techniques showed that when this adduct level is primarily composed of GSH-Cr-DNA adducts, rather than DNA-Cr-DNA crosslinks, both polymerase arrest and DNA-DNA interstrand crosslinks were absent.

An additional focus of this work was to extend our observations from our plasmid-based assays to genomic DNA. We utilized a PCR based-assay (QPCR) to detect Cr-induced replication-blocking lesions because of the increased sensitivity and selectivity that such methodology affords. QPCR has been used to examine the *in vitro* and *in vivo* formation and repair of *Taq* polymerase arresting lesions, within a single gene, induced by drugs [61–64], chemical carcinogens [58, 65, 66] and UV irradiation [60, 63, 66]. Under conditions identical to those employed in the pGEM-3Z-TS experiments, which produced Cr-monoadducts and DNA-Cr-GSH crosslinks, we observed that physiological concentrations of GSH effectively blocked the formation of both Cr (III) and Cr (VI)-induced polymerase arresting lesions. However, it is interesting that, in contrast to the sequencing-based polymerase arrest data using pGEM-3Z-TS, QPCR of a fragment of genomic DNA actually detected a low level of polymerase arrest when Cr [VI] was co-incubated with 0.6 mM GSH. This is probably because QPCR is much more sensitive in identifying polymerase arresting lesions than the Sequenase®-based polymerase assay [66] and was able to detect a low level of DNA-interstrand crosslinking probably formed at a low rate by mass exchange between the GSH ligand and an additional DNA strand. This is not surprising since the former exponentially amplifies undamaged template DNA via PCR whereas the latter involves only a single linear extension reaction. Indeed, QPCR has been demonstrated to be more sensitive than other methods of DNA damage detection including Southern analysis [60, 63].

In conclusion, the levels of polymerase-arresting lesions were proportionally related to Cr-mediated DNA interstrand crosslinks levels, but did not correlate with the levels of either Cr-DNA binding or GSH-Cr-DNA crosslinks. The results from this study indicate that (i) the formation of DNA-Cr-GSH crosslinks prevents the coordinating Cr atom from participating in a DNA interstrand crosslink, (ii) the DNA-Cr-GSH complexes, alone, were not capable of causing DNA polymerase arrest and (iii) Cr-mediated covalent DNA interstrand crosslinks are the major polymerase arresting DNA lesion produced by Cr.

Acknowledgements

This work was supported by NIH grant R01-ES-05304 and an Elaine H. Snyder Cancer Research Award to SRP.

References

1. IARC: Chromium, nickel, and welding. In: International Agency for Research on Cancer Monographs on the Evaluation of the Carcinogenic Risk of Chemicals to Human, vol. 49. IARC, Lyon, France, 1990, pp 1–214
2. Clarkson TW: Molecular ionic mimicry of toxic metals. Annu Rev Pharmacol Toxicol 32: 545–571, 1993
3. DeFlora S, Wetterhahn KE: Mechanisms of chromium metabolism and genotoxicity. Life Chem Rep 7: 169–244, 1989
4. DeFlora S, Bagnasco M, Serra D, Zanacchi P: Genotoxicity of chromium compounds. A review. Mutat Res 238: 99–172, 1990
5. Jones P, Kortenkamp A, O'Brien P, Wang G, Yang G: Evidence for the generation of hydroxyl radicals from a chromium (V) intermediate isolated from the reaction of chromate with glutathione. Arch Biochem Biophys 286: 652–655, 1991
6. Kasprzak KS: The role of oxidative damage in metal carcinogenicity. Chem Res Toxicol 4: 604–615, 1991
7. Kortenkamp A, O'Brien P, Beyersmann D: The reduction of chromate is a prerequisite of chromium binding to cell nuclei. Carcinogenesis 12: 1143–1144, 1991
8. Léonard A: Mechanisms in metal genotoxicity: The significance of *in vitro* approaches. Mutat Res 198: 321–326, 1988
9. Levis AG, Majone F: Cytotoxic and clastogenic effects of soluble and insoluble compounds containing hexavalent and trivalent chromium. Br J Cancer 44: 219–235, 1981
10. Manning FCR, Xu J, Patierno SR: Transcription inhibition by carcinogenic chromate: Relationship to DNA damage. Mol Carcinogen 6: 270–279, 1992
11. Sugiyama M, Ando A, Nakao K, Ueta H, Hidaka T, Ogura R: Influence of vitamin B2 on formation of chromium (V), alkali-labile sites, and lethality of sodium chromate (VI) in Chinese hamster V-79 cells. Cancer Res 49: 6180–6184, 1989

12. Xu J, Wise JP, Patierno SR: DNA damage induced by carcinogenic lead chromate particles in cultured mammalian cells. Mutat Res 280: 129–136, 1992
13. Xu J, Manning FCR, Patierno SR: Preferential formation and repair of chromium-induced DNA damage in nuclear matrix DNA. Carcinogenesis 15: 1443–1450, 1994
14. Bennicelli C, Camoirano A, Petruzzeli S, Zanacchi P, DeFlora S: High sensitivity of Salmonella TA 102 in detecting hexavalent chromium mutagenicity and its reversal by liver and lung preparations. Mutat Res 122: 1–5, 1983
15. Chen J, Thilly WG: Mutation spectrum of chromium (VI) in human cells. Mutat Res 323: 21–27, 1994
16. Cheng L, Liu S, Dixon K: Analysis and repair and mutagenesis of chromium-induced DNA damage in yeast, mammalian cells, and transgenic mice. Environ Health Perspect 106(suppl 4): 1027–1032, 1998
17. Liu S, Medvedovic M, Dixon K: Mutational specificity in a shuttle vector replication in chromium (VI)-treated mammalian cells. Environ Mol Mutagen 33: 313–319, 1999
18. Liu S, Dixon K: Induction of mutagenic DNA damage by chromium (VI) and glutathione. Environ Mol Mutagen 28: 71–79, 1996
19. Yang JL, Hsieh YC, Wu CW, Lee TC: Mutational specificity of chromium (VI) compounds at the *hprt* locus of Chinese hamster ovary-K1 cells. Carcinogenesis 13: 2053–2057, 1992
20. Bakke O, Jakobsen K, Eik-Nes KB: Concentration-dependent effects of potassium dichromate on the cell cycle. Cytometry 5: 482–486, 1984
21. Bridgewater LC, Manning FC, Patierno SR: Arrest of replication by mammalian DNA polymerases alpha and beta caused by chromium-DNA lesions. Mol Carcinogen 23: 201–6, 1998
22. Bridgewater LC, Manning FCR, Patierno SR: Base-specific arrest of *in vitro* DNA replication by carcinogenic chromium: Relationship to DNA intrastrand crosslinks. Carcinogenesis 15: 2421–2427, 1994
23. Bridgewater LC, Manning FCR, Patierno SR: DNA polymerase arrest by adducted trivalent chromium. Mol Carcinogen 9: 122–133, 1994
24. Xu J, Bubley GJ, Detrick B, Blankenship LJ, Patierno SR: Chromium (VI) treatment of normal human lung cells results in guanine-specific DNA polymerase arrest and blockade of cell cycle and DNA synthesis. Carcinogenesis 17: 1511–1517, 1996
25. Bianchi V, Dal Toso R, Debetto P, Levis AG, Luciani S, Majone F, Tamino G: Mechanisms of chromium toxicity in mammalian cell cultures. Toxicology 17: 219–224, 1980
26. Levis AG, Buttignol M, Bianchi V, Sponza G: Effects of potassium dichromate on nucleic acid and protein syntheses and on precursor uptake in BHK fibroblasts. Cancer Res 38: 110–116, 1978
27. Hamilton JW, Wetterhahn KE: Differential effects of chromium (VI) in constitutive and inducible gene expression in chick embryo liver *in vivo* and correlation with chromium (VI)-induced DNA damage. Mol Carcinogen 2: 274–286, 1989
28. Jakobsen K, Bakke O, Østgaard K, White LR, Eik-Nes KB: Effects of potassium dichromate on the cell cycle of an established human cell line (NHIK 3025). Toxicology 24: 281–292, 1982
29. Blankenship LJ, Carlisle DL, Wise JP, Orenstein JM, Dye LE III, Patierno SR: Induction of apoptotic cell death by particulate lead chromate: Differential effects of vitamins C and E on genotoxicity and survival. Toxicol Appl Pharmacol 146: 270–280, 1997
30. Blankenship LJ, Manning FCR, Orenstein JM, Patierno SR: Apoptosis is the mode of cell death caused by carcinogenic chromium. Toxicol Appl Pharmacol 126: 75–83, 1994
31. Singh J, Pritchard DE, Carlisle DL, Mclean JA, Montaser A, Orenstein J, Patierno SR: Internalization of carcinogenic lead chromate particles by cultured normal human lung epithelial cells: formation of lead-inclusion bodies and induction of apoptosis. Toxicol Appl Pharmacol 161: 240–248, 1999
32. Singh J, Carlisle DL, Pritchard DE, Patierno SR: Chromium-induced apoptosis: relationship to chromium carcinogenesis. Oncology Rep 5: 1307–1318(review), 1998
33. Carlisle DL, Pritchard DE, Singh J, Owens BM, Blankenship LJ, Orenstein JM, Patierno SR: Apoptosis and p53 induction in human fibroblasts exposed to chromium (VI): Effects of ascorbate and tocopherol. Toxicol Sci 55: 60–68, 2000
34. Carlisle DL, Pritchard DE, Singh J, Patierno SR: Chromium (VI) induces p53 dependent apoptosis in diploid human lung and mouse dermal fibroblasts. Mol Carcinogen 28: 111–118, 2000
35. Arslan P, Beltrame M, Tomasi A: Intracellular chromium reduction. Biochim Biophys Acta 931: 10–15, 1987
36. Standeven AM, Wetterhahn KE: Ascorbate is the principal reductant of chromium (VI) in rat liver and kidney ultrafiltrates. Carcinogenesis 12: 1733–1737, 1991
37. Wiegand HJ, Ottenwalder H, Bolt HM: The reduction of chromium (VI) to chromium (III) by glutathione: An intracellular redox pathway in the metabolism of the carcinogen chromate. Toxicology 33: 341–348, 1984
38. Liu KJ, Shi X, Dalal NS: Synthesis of Cr (IV)-GSH, its identification and its free hydroxyl radical generation: A model compound for Cr (VI) carcinogenicity. Biochem Biophys Res Commun 235: 54–58, 1997
39. Aiyar J, Berkovits HJ, Floyd RA, Wetterhahn KE: Reaction of chromium (VI) with glutathione or with hydrogen peroxide: Identification of reactive intermediates and their role in chromium (VI)-induced DNA damage. Environ Health Perspect 92: 53–62, 1991
40. Borges KM, Boswell JS, Liebross RH, Wetterhahn KE: Activation of chromium (VI) by thiols results in chromium (V) formation, chromium binding to DNA and altered DNA conformation. Carcinogenesis 12: 551–561, 1991
41. O'Brien P, Kortenkamp A: Chemical models important in understanding the ways in which chromate can damage DNA. Environ Health Perspect 102(suppl 3): 3–10, 1994
42. Shi XL, Dalal NS: On the mechanism of the chromate reduction by glutathione: Evidence for the glutathionyl radical and an insoluble Cr (V) intermediate. Biochem Biophys Res Commun 156: 137–142, 1988
43. Borges KM, Wetterhahn KE: Chromium crosslinks glutathione and cysteine to DNA. Carcinogenesis 10: 2165–2168, 1989
44. Zhitkovich A, Voitkun V, Costa M: Glutathione and free amino acids form stable complexes with DNA following exposure of intact mammalian cells to chromate. Carcinogenesis 16: 907–913, 1995
45. Capellmann M, Mikalsen A, Hindrum M, Alexander J: Influence of reducing compounds on the formation of DNA-protein crosslinks in HL-60 cells induced by hexavalent chromium. Carcinogenesis 16: 1135–1139, 1995
46. Lin X, Sugiyama M, Costa M: Differences in the effect of vitamin E on nickel sulfoxide or nickel chloride-induced chromosomal aberrations in mammalian cells. Mutat Res 260: 159–164, 1991
47. Casadevall M, Kortenkamp A: The generation of apurinic/apyrimidinic sites in isolated DNA during the reduction of chromate by glutathione. Carcinogenesis 15: 407–409, 1994
48. Casadevall M, Kortenkamp A: The formation of both apurinic/apyrimidinic sites and single-strand breaks by chromate and glutathione arises from attack by the same single reactive species and is dependent on molecular oxygen. Carcinogenesis 16: 805–809, 1995
49. Kawanishi S, Inoue S, Sano S: Mechanism of DNA cleavage induced by sodium chromate (VI) in the presence of hydrogen peroxide. J Biol Chem 261: 5952–5958, 1986
50. Shi XL, Dalal NS: One-electron reduction of chromate by NADPH-dependent glutathione reductase. J Inorg Biochem 40: 1–12, 1990
51. Cupo DY, Wetterhahn KE: Modification of chromium (VI)-induced DNA damage by glutathione and cytochrome P-450 in chicken embryo hepatocytes. Proc Natl Acad Sci USA 82: 6755–6759, 1985

52. Sugiyama M, Tsuzuki K: Effects of glutathione depletion on formation of paramagnetic chromium in Chinese hamster V-79 cells. FEBS Lett 341: 273–276, 1994
53. Misra M, Alcedo JA, Wetterhahn KE: Two pathways for chromium (VI)-induced DNA damage in 14 day chick embryos: Cr-DNA binding in liver and 8-oxo-2′-deoxyguanosine in red blood cells. Carcinogenesis 15: 2911–2917, 1994
54. Voitkun V, Zhitkovich A, Costa M: Cr (III)-mediated crosslinks of glutathione or amino acids to the DNA phosphate backbone are mutagenic in human cells. Nucleic Acids Res 26: 2024–2030, 1998
55. Sambrook J, Fritsch EF, Maniatis T: In: Molecular Cloning, A Laboratory Manual, 2nd edn. Cold Spring Harbor Laboratory Press, Cold Spring Harbor, New York, 1989
56. Feinberg A, Vogelstein B: Addendum to 'A technique for radiolabeling DNA restriction endonuclease fragments to high specific activity.' Anal Biochem 137: 266–267, 1984
57. Tornaletti S, Rozek D, Pfeifer GP: The distribution of UV photoproducts along the human p53 gene and its relation to mutations in skin cancer. Oncogene 8: 2051–2057, 1993
58. Denissenko MF, Cahill J, Koudriakova TB, Gerber N, Pfeifer GP: Quantitation and mapping of aflatoxin B1-induced DNA damage in genomic DNA using aflatoxin B1-8,9-epoxide and microsomal activation systems. Mutat Res 425: 205–211, 1999
59. Wise JP, Orenstein JM, Patierno SR: Inhibition of lead chromate clastogenesis by ascorbate: Relationship to particle dissolution and uptake. Carcinogenesis 14: 429–434, 1993
60. McCarthy MJ, Rosenblatt JI, Lloyd RS: A modified quantitative polymerase chain reaction assay for measuring gene-specific repair of UV photoproducts in human cells. Mutat Res 363: 57–66, 1996
61. Grimaldi KA, Bingham JP, Souhami RL, Hartley JA: DNA damage by anticancer agents and its repair: Mapping in cells at the subgene level with quantitative polymerase chain reaction. Anal Biochem 222: 236–242, 1994
62. Oshita F, Saijo N: Rapid polymerase chain reaction assay to detect variation in the extent of gene-specific damage between cisplatin- or VP16-resistant and sensitive lung cancer cell lines. Jpn J Cancer Res 85: 669–673, 1994
63. Kalinowski DP, Illenye S, Van Houten B: Analysis of DNA damage and repair in murine leukemia L1210 cells using a quantitative polymerase chain reaction assay. Nucleic Acids Res 20: 3485–3494, 1992
64. Jennerwein MM, Eastman A: A polymerase chain reaction-based method to detect cisplatin adducts in specific genes. Nucleic Acids Res 19: 6209–6214, 1991
65. Govan HL, Valles-Ayoub Y, Braun J: Fine-mapping of DNA damage and repair in specific genomic segments. Nucleic Acids Res 18: 3823–3830, 1990
66. Jenkins GJ, Burlinson B, Parry JM: The polymerase inhibition assay: A methodology for the identification of DNA-damaging agents. Mol Carcinogen 27: 289–297, 2000
67. Standeven AM, Wetterhahn KE: Ascorbate is the principal reductant of chromium (VI) in rat lung ultrafiltrates and cytosol, and mediates chromium-DNA binding *in vitro*. Carcinogenesis 13: 1319–1324, 1992
68. Larkworthy LF, Nolan KB, O'Brien EC: Chromium. In: G. Wilkinson, R.D. Gillard, J.A. McCleverty (eds). Comprehensive Co-ordination Chemistry. Pergamon Press, Oxford, 1988, pp 699–969
69. Hneihen AS, Standeven AM, Wetterhahn KE: Differential binding of chromium (VI) and chromium (III) complexes to salmon sperm nuclei and nuclear DNA and isolated calf thymus DNA. Carcinogenesis 14: 1795–1803, 1993
70. Abdullah N, Barrett J, O'Brien P: Synthesis and characterization of some chromium (III) complexes with glutathione. J Chem Soc Dalton Trans 5: 2085–2089, 1985
71. Fiol JJ, Terron A, Moreno V: Chromium (III) interactions with nucleotides. Inorg Chim Acta 83: 69–73, 1984
72. Campromar JA, Fiol JJ, Terron A: Chromium (III) interactions with nucleotides. II. Inorg Chim Acta 124: 75–81, 1986
73. Krishnamoorthy CR, Harris GM: An equilibrium and rate study of the interaction of aqueous chromium (III) ion with adenine. J Coord Chem 10: 69–73, 1980
74. Mikulski CM, Mattucci L, Smith Y, Tran TB, Karayannis NM: Guanine complexes with first transition metal perchlorates. Inorg Chim Acta 80: 127–133, 1983
75. Suzuki Y, Fukuda K: Reduction of hexavalent chromium by ascorbic acid and glutathione with special reference to the rat lung. Arch Toxicol 64: 169–176, 1990

Molecular and Cellular Biochemistry **222**: 183–188, 2001.

Mini review

Cell apoptosis induced by carcinogenic metals

Fei Chen, Val Vallyathan, Vince Castranova and Xianglin Shi
Health Effects Laboratory Division, National Institute for Occupational Safety and Health, Morgantown, WV, USA

Abstract

Well-documented evidence suggests that environmental and occupational exposure of toxic metals or metal-containing compounds can cause a number of human diseases, including inflammation and cancer, through DNA damage, protein modifications, or lipid peroxidation. This mini-review addresses the mechanisms of cell death induced by some carcinogenic metals, including arsenic (III), chromium (VI) and vanadium (V). A possible contribution of reactive oxygen species to metal-induced cell death is also discussed. (Mol Cell Biochem **222**: 183–188, 2001)

Key words: apoptosis, metal, ROS, DNA damage, carcinogenesis

Apoptotic pathways

Programmed cell death or apoptosis is a process in which cell death is initiated and completed in an orderly fashion through the activation of various apoptotic pathways [1, 2]. During the last decade, there has been overwhelming interest in apoptosis and elucidation of mechanisms controlling this process. Apoptosis is an essential process required for development, morphogenesis, immune regulation, tissue remodeling, and some pathological reactions. Most apoptotic cells are characterized by unique morphological features, such as membrane blebbing, cell shrinking, cytosolic and nuclear condensation, and breakdown of chromosomal DNA. Depending on the use of different initiating caspases, signal-induced apoptosis can be roughly divided into two categories: receptor-mediated 'extrinsic' apoptosis and mitochondria-mediated 'intrinsic' apoptosis [1, 2]. Whereas procaspase 8 is activated by receptors for Fas ligand and TNF through the recruitment of intracellular death domain-containing proteins, procaspase 9 is initiated by cytochrome c released from damaged mitochondria (Fig. 1). Both activated caspase 8 and caspase 9 use the same executive caspases, mainly caspase 3, to complete an apoptotic process. All caspases are expressed as inactive precursors and are activated by cleavage at specific peptide bonds. There are a number of cellular proteins which act as roadblocks in the activation cascade of caspases. These proteins act either by counteracting the effect of caspases, such as in the case of cIAP1, cIAP2 and XIAP, or stabilizing the outer membrane of mitochondria, such as in the case of Bcl-x and Bcl-2 [3, 4]. The apoptosis mediated by mitochondria may be more important and relevant in metal-induced cell death.

ROS in apoptosis

Under normal conditions, mitochondria may be the main source for the generation of basic and homeostatic reactive oxygen species (ROS) through the chain reaction of electron transport. Paradoxically, mitochondria are also very vulnerable to the adverse effects of excessive ROS generation. Excessive ROS that can not be neutralized by the limited reducing capacity of the mitochondria and cytosol not only increase mitochondrial membrane permeability, but also diffuse into the mitochondrial matrix to damage the respiratory chain, and consequently amplify the generation of ROS further due to the less efficient transferring of electrons to oxygen (O_2) [5]. Mitochondria with increased membrane permeability release several apoptosis-promoting factors including cytochrome c, apoptosis inducing factor (AIF), and Diablo/Smac [6–9]. Cytochrome c forms a complex with a cytosolic protein, named apoptosis protease-activating factor-1 (Apaf-1), to activate caspase-9. AIF is a flavoprotein with homology to plant ascorbate reductases and bacterial NADH oxidases [9]. After its release from the intermembrane space of damaged

Address for offprints: X. Shi, National Institute for Occupational Safety and Health, 1095 Willowdale Road, Morgantown, WV 26505, USA

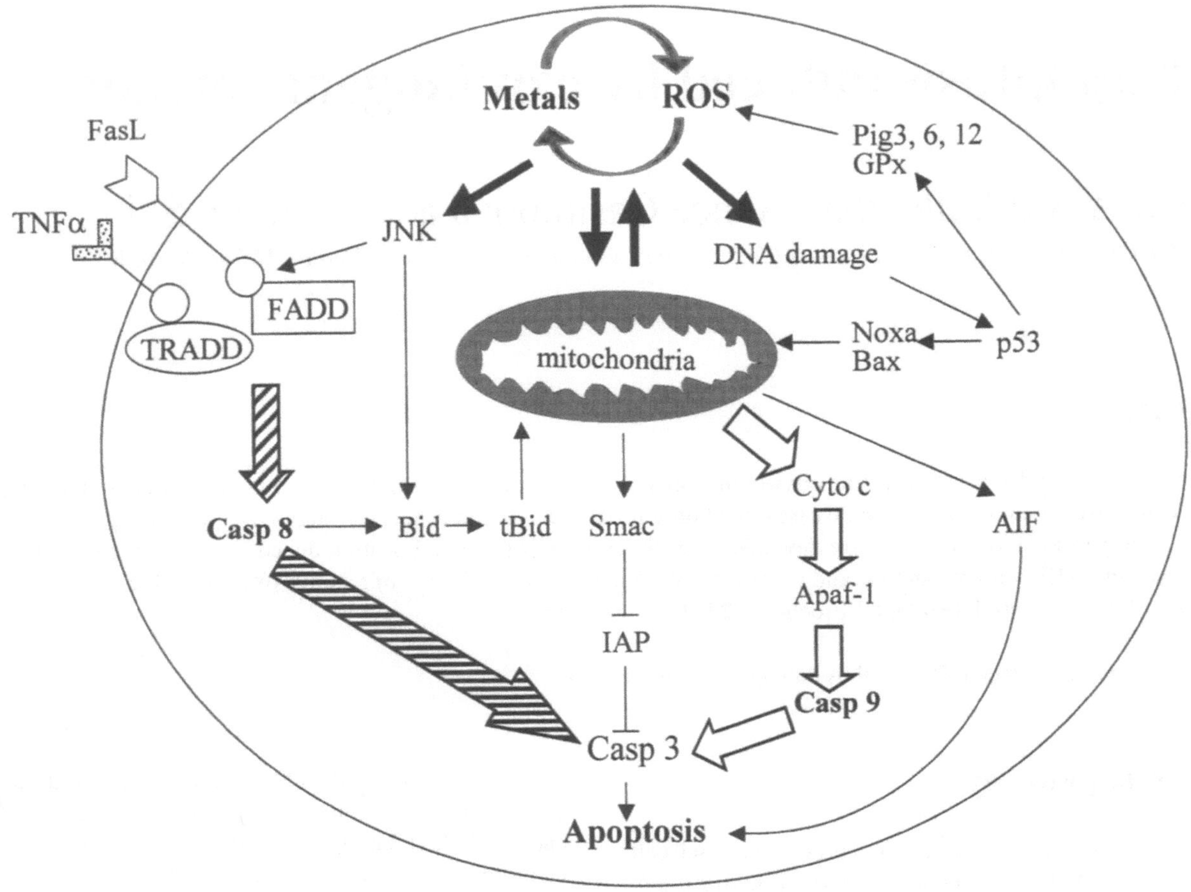

Fig. 1. Apoptosis signaling pathways mediated by metals. Metals or ROS generated by metals directly initiate the intrinsic apoptotic pathway at the mitochondria (unfilled block arrows). Metals or ROS can also indirectly initiate this pathway by activating p53 as a result of metal- or ROS-induced DNA damage. The activation of p53 triggers the expression of Bax and Noxa, two apoptotic proteins which induce an increase in mitochondrial permeability. Cytochrome c is released into the cytosol from damaged mitochondria and binds to Apaf1, resulting in the activation of caspase-9. Damaged mitochondria can also release AIF, which induces apoptosis in a caspase-independent manner, and Smac which antagonizes the antiapoptotic effect of IAP proteins. Metals or ROS may also potentiate the extrinsic apoptotic pathway (shadow block arrows) through MAP kinase-, especially, JNK-mediated induction of FasL or Fas expression. Recruitment of death domain containing proteins, such as FADD and TRADD to these receptor complexes, activates caspase-8. The activation of caspase-8 and caspase-9 leads to the activation of the effector caspases, including caspase-3 and caspase-7.

mitochondria, AIF exhibits a caspase-independent apoptotic function. Recently, another mitochondrial intermembrane protein, Smac/Diablo, has been identified [7, 8]. Once released from the mitochondria to cytoplasm, Smac/Diablo binds to and antagonizes cIAP1 and cIAP2, allowing the activation of caspases.

Mechanisms of metal-induced apoptosis

Cr (VI)

Since apoptosis is an important process influencing the malignant transformation of cells, the regulation of cell apoptosis may be critical in metal-induced carcinogenesis. The earliest report revealing apoptotic induction by chromium (VI) (Cr (VI)) was provided by Blankenship *et al.* [10, 11], who demonstrated that cells treated with Cr (VI) exhibited apoptotic features. Evidence provided by Singh *et al.* [12] indicated that depending on the rate, magnitude and spectrum of genotoxicity and mitochondrial damage, cells exposed to Cr (VI) are fated to undergo either terminal growth arrest or p53-dependent apoptosis. It has been suggested that Cr (VI) was able to damage DNA by forming DNA-Cr-DNA cross links [13]. Cr (VI) itself was found to be unable to react with macromolecules, such as DNA, RNA, proteins and lipids. Instead, Cr (V) and Cr (III), intermediates of Cr (VI) reduction form covalent interactions with DNA and other macromolecules [12], a process which activates DNA-dependent protein kinases (DNA-PK) and induces subsequent p53 activation and cell apoptosis. The DNA damaging effect of Cr (VI) might be also through ROS generated by Cr (VI) reduction. Supporting this notion is the evidence by our laboratory showing that antioxidants, including pyrroline dithiocarbamate (PDTC) and aspirin, protected DNA from damage and p53 from activation induced by Cr (VI) [14, 15].

Activation of p53 tumor suppressor protein is considered to be one of the critical steps in the induction of apoptosis by Cr (VI) [10, 15, 16], although p53-independent apoptosis induced by Cr (VI) has also been reported [16]. Several mechanisms are involved in the Cr (VI)-induced p53 activation. First, direct DNA damage by Cr (VI) or ROS generated during cellular Cr (VI) reduction activates upstream kinases, including DNA-PK, ATM, ATR and others, for p53 phosphorylation, stabilization/activation [17, 18]. Second, Cr (VI)-induced MAP kinase activation, especially, c-Jun-N-terminal kinase and p38, which is independent of DNA damage, modulates the activation and activity of p53 [19–21]. Third, the p53 protein contains several redox-sensitive cysteines critical for the DNA binding activity of p53 [22, 23]. As a transcription factor, p53 is able to up-regulate the expression of genes involved in either ROS production or metabolism, including quinone oxidoreductase (Pig3), proline oxidase (Pig6) homologues, glutathione transferase (Pig12), and glutathione peroxidase (GPx) [24]. Moreover, p53 also activates the expression of several genes that directly control or regulate the process of apoptosis. These genes include Bax, Fas, Fas ligand, IGF-BP3, PAG608 [25], ei24 (Pig8) [26], and Noxa [27]. The strongest evidence supporting the notion that p53 is involved in Cr (VI)-induced apoptosis was demonstrated through the use of fibroblasts from both wild type mice and p53-deficient mice [16]. Cr (VI) induced a 2- to 3-fold greater increase in apoptosis in wild type fibroblasts than in p53-deficient fibroblasts. Obviously, the activation of p53 by Cr (VI) will lead to the increased expression of p53-targeted apoptotic genes, such as Bax, Fas, Fas ligand, ei24, and Noxa. Several recent studies from our group and others have highlighted the importance of H_2O_2 and hydroxyl radical in Cr (VI)-induced p53 activation and cell death [15, 28–30]. As discussed earlier, excessive ROS are toxic to the mitochondria. Therefore, mitochondrial damage, resulting either from the direct effect of Cr (VI) or indirect effect mediated by Cr (VI)-induced activation of p53, appears to be a critical step in Cr (VI)-induced cell apoptosis. Indeed, studies by Carlisle *et al.* [31] indicated that cyclosporin A, which prevents the pre-apoptotic release of cytochrome c from mitochondria, blocked Cr (VI)-induced apoptosis and increased the survival rate of replication-competent but genetically damaged cells.

As (III)

Chronic exposure to As (III), another important toxic submetal widely found both in the environment and in occupational settings, can cause a number of human diseases, especially cancer [32, 33]. The apoptotic effect of As (III) was originally observed in human acute promyelocytic leukemia cells [34]. Later studies suggested that As (III) might be a direct toxic agent for mitochondria in this type of cell. As (III) (1 μM) induced a condensation of the mitochondrial matrix and caused disruption of the mitochondrial transmembrane potential leading to the release of cytochrome c and AIF [35, 36]. As (III) can also induce apoptosis in other types of cells. Although the mechanism remains unclear, it has been consistently shown that p53 activation may not be involved in As (III)-induced cell apoptosis under many circumstances [34, 37]. Based on studies using several antioxidants or ROS scavengers in Chinese hamster ovary cells, two reports have suggested an oxidative stress model for As (III)-induced apoptosis [38, 39]. In mouse skin epidermal cells, the apoptosis induced by arsenic is possibly through protein kinase C and MAP kinases pathways [40]. MAP kinases, especially c-Jun-N-terminal kinase (JNK) and p38, have been proposed as important mediators in signal-induced apoptosis [40]. Both JNK- and p38-mediated cell death may be due to the rapid induction of FasL as observed in Jurkat cells [41], microglial cells [42] and hepatoma cells [43]. In addition, it has been demonstrated that JNK is also required for UV-induced caspase-independent proteolytic activation of Bid, a pro-apoptotic BH3-only member of the Bcl2 group which translocates to the mitochondria after activation and induces cytochrome c release [44, 45]. Our recent studies have found that JNK activation is required for arsenic-induced gene expression of GADD45, a cell cycle checkpoint protein which arrests cells in G2/M phase transition, and for the possible induction of apoptosis [46].

V (V)

Emerging evidence suggests that various forms of vanadium or vanadium containing particles from environmental and occupational sources are able to trigger or potentiate cell apoptosis. It was believed that among the various oxidation states of vanadium, the pentavalent form, vanadium (V) (V (V)), one of the most common forms of vanadium found in nature and in tissues or cells after exposure, is the most toxic [47]. Short-term exposure to V (V) causes local irritation to the eyes and upper respiratory tract rather than systemic toxicity. There is limited information concerning the possible neoplastic transforming activity of V (V) following long-term exposure [47–49]. V (V) might cause apoptosis in some types of cells, but may be anti-apoptotic in other types of cells. In lymphoid cell lines, V (V) treatment caused the activation of caspases 3, 8 and 9, the induction of mitochondrial permeability transition, the release of cytochrome c and DNA fragmentation. Further studies suggested that there was no involvement of protein tyrosine kinase $p56^{lck}$ or phosphatase CD45 in this apoptotic effect of V (V) in lymphoid cell lines [50]. V (V)-induced apoptosis in mouse epidermal JB6 cells was thought to involve the generation of ROS and the activation of p53. Pre-treatment of the cells with antioxidants, including NAC and catalase, blunted the apoptotic effects of V (V) on these cells [51]. In other types of cells, V (V) might elicit an anti-apoptotic effect through the activation of Akt/PKB or the inactivation of tyrosine phosphatase. Support for this comes from the work of Chin *et al.* [52], who showed that V (V) protected malignant glioma cells from apoptosis. The addition of wortmannin, a phosphatidylinositol 3-kinase (PI3K) inhibitor, abolished the protective effect of V (V) on these cells, indicating that PI3K and Atk/PKB may be responsible for the V (V)-mediated protective effect. In intestinal epithelial cells, studies by Scheving *et al.* [53] demonstrated a correlation between a decline in protein phosphatase activity and inhibition of cell apoptosis in V (V)-treated cells. It is not clear why V (V) is anti-apoptotic in some types of cells, but is pro-apoptotic in other types of cells. The explanation for this may be the use of different forms of V (V). It was noted that some studies used sodium vanadate, while others used peroxovanadate or pervanadate. The later form is a reactive product of V (V) in the presence of H_2O_2. Many reports indicated that sodium vanadate and peroxovanade exhibit different effects on the induction of cell apoptosis, inactivation of protein phosphatase and generation of ROS. An additional explanation for the contradictory effects of V (V) on cell apoptosis may be due to the use of different doses of V (V). It has been frequently observed that lower concentrations of V (V) protect cells from signal-induced apoptosis, whereas higher concentrations of V (V) are either cytotoxic or pro-apoptotic. The final explanation for the varying effects of V (V) on cell apoptosis is the types of cells used in each experiment. It is well-known that cells, originating from different tissues, exhibit different capacities for ROS generation and respond differently to metal or exogenous ROS stimulation.

Summary

Evidence indicates that many toxic metals, such as Cr (VI), As (III) and V (V), are carcinogenic. It is difficult to reconcile the pro-apoptotic effects of metals with the observed carcinogenic potential of metals in humans. Many studies addressed the pro-apoptotic effect of metals on selected cell types, but failed to delineate how this process would contribute to cell transformation and carcinogenesis. Cell apoptosis was originally viewed as a normal physiologic process by which correct functional cellular population dynamics are maintained through the apoptotic loss of cell populations carrying abnormal genetic information [4]. It is known that metals under certain circumstances are apoptotic, but it is not known whether this apoptotic process induced by metals is a perfect or an imperfect process. An imperfect apoptotic process might result in the escape of cells that would be potentially carcinogenic. Thus, increased apoptosis under the conditions of chronic metal exposure could possibly coincide with an increase in the number of cells carrying damaged but replication-competent genetic information. Conversely, chronic and lower dose exposure of cells or tissues to metals may perturb or even inhibit appropriate apoptosis, leading to the accumulation of cells with carcinogenic potential. Among metals discussed above, it is likely that Cr (VI) is not only the most toxic metal, but also the most carcinogenic metal toward cells [13, 21]. The genotoxicity of Cr (VI), resulting from either direct DNA binding of Cr (VI) derivatives or generation of ROS during cellular Cr (VI) reduction, may create a cell sub-population that may be predisposed to mutagenesis and capable of averting apoptosis. Considering the fact that Cr (VI)-induced apoptosis is p53 dependent under many circumstances, mutations in the p53 gene would undoubtedly facilitate the development of apoptosis resistant and potentially carcinogenic cells. Moreover, although Cr (VI) itself and ROS generated from Cr (VI) reduction intracellularly are able to activate upstream signals leading to the accumulation and activation of p53 in the early phase of cells in response to Cr (VI), neither Cr (VI) nor ROS can persistently activate the function of p53 protein in the later phase. Instead, due to the oxidation of cysteine residues in the DNA binding domain of p53 protein [22, 23], Cr (VI) may actually inactivate the tumor suppressing function of p53 by impairing the DNA binding activity of p53.

Acknowledgements

Fei Chen was supported by a Career Development Award in Genetics under a cooperative agreement from the Centers for Disease Control and Prevention through the Association of Teachers of Preventive Medicine.

References

1. Green DR, Reed JC: Mitochondria and apoptosis. Science 281: 1309–1312, 1998
2. Ashkenazi A, Dixit VM: Death receptors: Signaling and modulation. Science 281: 1305–1308, 1998
3. LaCasse EC, Baird S, Korneluk RG, MacKenzie AE: The inhibitors of apoptosis (IAPs) and their emerging role in cancer. Oncogene 17: 3247–3259, 1998
4. Reed JC: Bcl-2 family proteins. Oncogene 17: 3225–3236, 1998
5. Beckman KB, Ames BN: Mitochondrial aging: Open questions. Ann NY Acad Sci 854: 118–127, 1998
6. De Laurenzi V, Melino G: Apoptosis. The little devil of death. Nature 406: 135–136, 2000

7. Du C, Fang M, Li Y, Li L, Wang X: Smac, a mitochondrial protein that promotes cytochrome c-dependent caspase activation by eliminating IAP inhibition. Cell 102: 33–42, 2000
8. Verhagen AM, Ekert PG, Pakusch M, Silke J, Connolly LM, Reid GE, Moritz RL, Simpson RJ, Vaux DL: Identification of DIABLO, a mammalian protein that promotes apoptosis by binding to and antagonizing IAP proteins. Cell 102: 43–53, 2000
9. Susin SA, Lorenzo HK, Zamzami N, Marzo I, Snow BE, Brothers GM, Mangion J, Jacotot E, Costantini P, Loeffler M, Larochette N, Goodlett DR, Aebersold R, Siderovski DP, Penninger JM, Kroemer G: Molecular characterization of mitochondrial apoptosis-inducing factor. Nature 397: 441–446, 1999
10. Blankenship LJ, Manning FC, Orenstein JM, Patierno SR: Apoptosis is the mode of cell death caused by carcinogenic chromium. Toxicol Appl Pharmacol 126: 75–83, 1994
11. Manning FC, Blankenship LJ, Wise JP, Xu J, Bridgewater LC, Patierno SR: Induction of internucleosomal DNA fragmentation by carcinogenic chromate: Relationship to DNA damage, genotoxicity, and inhibition of macromolecular synthesis. Environ Health Perspect 102 (suppl 3): 159–167, 1994
12. Singh J, McLean JA, Pritchard DE, Montaser A, Patierno SR: Sensitive quantitation of chromium-DNA adducts by inductively coupled plasma mass spectrometry with a direct injection high-efficiency nebulizer. Toxicol Sci 46: 260–265, 1998
13. Hamilton JW, Kaltreider RC, Bajenova OV, Ihnat MA, McCaffrey J, Turpie BW, Rowell EE, Oh J, Nemeth MJ, Pesce CA, Lariviere JP: Molecular basis for effects of carcinogenic heavy metals on inducible gene expression. Environ Health Perspect 106 (suppl 4): 1005–1015, 1998
14. Shi X, Leonard SS, Wang S, Ding M: Antioxidant properties of pyrrolidine dithiocarbamate and its protection against Cr (VI)-induced DNA strand breakage. Ann Clin Lab Sci 30: 209–216, 2000
15. Wang S, Leonard SS, Ye J, Ding M, Shi X: The role of hydroxyl radical as a messenger in Cr (VI)-induced p53 activation. Am J Physiol Cell Physiol 279: C868–875, 2000
16. Carlisle DL, Pritchard DE, Singh J, Patierno SR: Chromium (VI) induces p53-dependent apoptosis in diploid human lung and mouse dermal fibroblasts. Mol Carcinog 28: 111–118, 2000
17. Jimenez GS, Khan SH, Stommel JM, Wahl GM: p53 regulation by post-translational modification and nuclear retention in response to diverse stresses. Oncogene 18: 7656–7665, 1999
18. Ashcroft M, Kubbutat MH, Vousden KH: Regulation of p53 function and stability by phosphorylation. Mol Cell Biol 19: 1751–1758, 1999
19. Milne DM, Campbell DG, Caudwell FB, Meek DW: Phosphorylation of the tumor suppressor protein p53 by mitogen-activated protein kinases. J Biol Chem 269: 9253–9260, 1994
20. Milne DM, Campbell LE, Campbell DG, Meek DW: p53 is phosphorylated *in vitro* and *in vivo* by an ultraviolet radiation-induced protein kinase characteristic of the c-Jun kinase, JNK1. J Biol Chem 270: 5511–5518, 1995
21. Chen F, Ding M, Lu Y, Leonard SS, Vallyathan V, Castranova V, Shi X: Participation of MAP kinase p38 and IkappaB kinase in chromium (VI)-induced NF-kappaB and AP-1 activation. J Environ Pathol Toxicol Oncol 19: 231–238, 2000
22. Hainaut P, Milner J: Redox modulation of p53 conformation and sequence-specific DNA binding *in vitro*. Cancer Res 53: 4469–4473, 1993
23. Meplan C, Richard MJ, Hainaut P: Redox signalling and transition metals in the control of the p53 pathway. Biochem Pharmacol 59: 25–33, 2000
24. Polyak K, Xia Y, Zweier JL, Kinzler KW, Vogelstein B: A model for p53-induced apoptosis. Nature 389: 300–305, 1997
25. Amundson SA, Myers TG, Fornace AJ Jr: Roles for p53 in growth arrest and apoptosis: Putting on the brakes after genotoxic stress. Oncogene 17: 3287–3299, 1998
26. Gu Z, Flemington C, Chittenden T, Zambetti GP: ei24, a p53 response gene involved in growth suppression and apoptosis. Mol Cell Biol 20: 233–241, 2000
27. Oda E, Ohki R, Murasawa H, Nemoto J, Shibue T, Yamashita T, Tokino T, Taniguchi T, Tanaka N: Noxa, a BH3-only member of the Bcl-2 family and candidate mediator of p53-induced apoptosis. Science 288: 1053–1058, 2000
28. Shi XL, Dalal NS: The role of superoxide radical in chromium (VI)-generated hydroxyl radical: The Cr (VI) Haber–Weiss cycle. Arch Biochem Biophys 292: 323–327, 1992
29. Stohs SJ, Bagchi D, Hassoun E, Bagchi M: Oxidative mechanisms in the toxicity of chromium and cadmium ions. J Environ Pathol Toxicol Oncol 19: 201–213, 2000
30. Buzard GS, Kasprzak KS: Possible roles of nitric oxide and redox cell signaling in metal-induced toxicity and carcinogenesis: A review. J Environ Pathol Toxicol Oncol 19: 179–199, 2000
31. Carlisle DL, Pritchard DE, Singh J, Owens BM, Blankenship LJ, Orenstein JM, Patierno SR: Apoptosis and p53 induction in human lung fibroblasts exposed to chromium (VI): Effect of ascorbate and tocopherol. Toxicol Sci 55: 60–68, 2000
32. Iyengar GV, Nair PP: Global outlook on nutrition and the environment: Meeting the challenges of the next millennium. Sci Total Environ 249: 331–346, 2000
33. Simeonova PP, Wang S, Toriuma W, Kommineni V, Matheson J, Unimye N, Kayama F, Harki D, Ding M, Vallyathan V, Luster MI: Arsenic mediates cell proliferation and gene expression in the bladder epithelium: Association with activating protein-1 transactivation. Cancer Res 60: 3445–3453, 2000
34. Chen GQ, Zhu J, Shi XG, Ni JH, Zhong HJ, Si GY, Jin XL, Tang W, Li XS, Xong SM, Shen ZX, Sun GL, Ma J, Zhang P, Zhang TD, Gazin C, Naoe T, Chen SJ, Wang ZY, Chen Z: *In vitro* studies on cellular and molecular mechanisms of arsenic trioxide (As2O3) in the treatment of acute promyelocytic leukemia: As2O3 induces NB4 cell apoptosis with downregulation of Bcl-2 expression and modulation of PML-RAR alpha/PML proteins. Blood 88: 1052–1061, 1996
35. Cai X, Shen YL, Zhu Q, Jia PM, Yu Y, Zhou L, Huang Y, Zhang JW, Xiong SM, Chen SJ, Wang ZY, Chen Z, Chen GQ: Arsenic trioxide-induced apoptosis and differentiation are associated respectively with mitochondrial transmembrane potential collapse and retinoic acid signaling pathways in acute promyelocytic leukemia. Leukemia 14: 262–270, 2000
36. Larochette N, Decaudin D, Jacotot E, Brenner C, Marzo I, Susin SA, Zamzami N, Xie Z, Reed J, Kroemer G: Arsenite induces apoptosis via a direct effect on the mitochondrial permeability transition pore. Exp Cell Res 249: 413–421, 1999
37. Huang C, Ma WY, Li J, Dong Z: Arsenic induces apoptosis through a c-Jun NH2-terminal kinase-dependent, p53-independent pathway. Cancer Res 59: 3053–3058, 1999
38. Wang TS, Kuo CF, Jan KY, Huang H: Arsenite induces apoptosis in Chinese hamster ovary cells by generation of reactive oxygen species. J Cell Physiol 169: 256–268, 1996
39. Chen YC, Lin-Shiau SY, Lin JK: Involvement of reactive oxygen species and caspase 3 activation in arsenite-induced apoptosis. J Cell Physiol 177: 324–333, 1998
40. Chen NY, Ma WY, Yang CS, Dong Z: Inhibition of arsenite-induced apoptosis and AP-1 activity by epigallocatechin-3-gallate and theaflavins. J Environ Pathol Toxicol Oncol 19: 287–295, 2000
41. Bauer MKA, Vogt M, Los M, Siegel J, Wesselborg S, Schulze-Osthoff K: Role of reactive oxygen intermediates in activation-induced CD95 (APO-1/Fas) ligand expression. J Biol Chem 273: 8048–8055, 1998

42. Vogt M, Bauer MK, Ferrari D, Schulze-Osthoff K: Oxidative stress and hypoxia/reoxygenation trigger CD95 (APO-1/Fas) ligand expression in microglial cells. FEBS Lett 429: 67–72, 1998
43. Hug H, Strand S, Grambihler A, Galle J, Hack V, Stremmel W, Krammer PH, Galle PR: Reactive oxygen intermediates are involved in the induction of CD95 ligand mRNA expression by cytostatic drugs in hepatoma cells. J Biol Chem 272: 28191–28193, 1997
44. Bossy-Wetzel E, Newmeyer DD, Green DR: Mitochondrial cytochrome c release in apoptosis occurs upstream of DEVD-specific caspase activation and independently of mitochondrial transmembrane depolarization. Embo J 17: 37–49, 1998
45. Bessho R, Matsubara K, Kubota M, Kuwakado K, Hirota H, Wakazono Y, Lin YW, Okuda A, Kawai M, Nishikomori R, Heike T: Pyrrolidine dithiocarbamate, a potent inhibitor of nuclear factor kappa B (NF-kappa B) activation, prevents apoptosis in human promyelocytic leukemia HL-60 cells and thymocytes. Biochem Pharmacol 48: 1883–1889, 1994
46. Chen F, Lu Y, Zhang Z, Vallyathan V, Ding M, Castranova V, Shi X: Opposite effect of NF-kB and c-Jun-N-terminal kinase on p53-independent GADD45 induction by arsenite. J Biol Chem 276: 11414–11419, 2001
47. Barceloux DG: Vanadium. J Toxicol Clin Toxicol 37: 265–278, 1999
48. Afshari CA, Kodama S, Bivins HM, Willard TB, Fujiki H, Barrett JC: Induction of neoplastic progression in Syrian hamster embryo cells treated with protein phosphatase inhibitors. Cancer Res 53: 1777–1782, 1993
49. Sabbioni E, Pozzi G, Devos S, Pintar A, Casella L, Fischbach M: The intensity of vanadium (V)-induced cytotoxicity and morphological transformation in BALB/3T3 cells is dependent on glutathione-mediated bioreduction to vanadium (IV). Carcinogenesis 14: 2565–2568, 1993
50. Hehner SP, Hofmann TG, Droge W, Schmitz ML: Inhibition of tyrosine phosphatases induces apoptosis independent from the CD95 system. Cell Death Differ 6: 833–841, 1999
51. Huang C, Zhang Z, Ding M, Li J, Ye J, Leonard SS, Shen HM, Butterworth L, Lu Y, Costa M, Rojanasakul Y, Castranova V, Vallyathan V, Shi X: Vanadate induces p53 transactivation through hydrogen peroxide and causes apoptosis. J Biol Chem 275: 32516–32522, 2000
52. Chin LS, Murray SF, Harter DH, Doherty PF, Singh SK: Sodium vanadate inhibits apoptosis in malignant glioma cells: A role for Akt/PKB. J Biomed Sci 6: 213–218, 1999
53. Scheving LA, Thomas JR, Zhang L: Regulation of intestinal tyrosine phosphorylation and programmed cell death by peroxovanadate. Am J Physiol 277: C572–579, 1999

Molecular and Cellular Biochemistry **222:** 189–197, 2001.

Gene expression profile in response to chromium-induced cell stress in A549 cells

Jianping Ye and Xianglin Shi

Pathology and Physiology Research Branch, Health Effects Laboratory Division, National Institute for Occupational Safety and Health, Morgantown, WV, USA

Abstract

Chromium (Cr) is a trace element required for life. Biological activities of Cr are complicated and remain to be fully investigated. It is known that the valence of Cr plays an important role in the biological activities of Cr. For example, Cr (VI) is classified as a metal carcinogen [1], but Cr (III) is widely used as a nutritional supplement [2, 3]. Establishment of a gene expression profile for Cr-induced cellular response is necessary to facilitate investigation of the biological activities of Cr. In the present study, a large-scale gene expression analysis was conducted using RNA of human lung epithelial cells after *in vitro* exposure to Cr (VI). Utilizing high-density oligonucleotide arrays representing 2400 genes, we observed that expression of 150 genes was up-regulated, and that of 70 genes were down-regulated by Cr (VI). Functional analysis of these responsive genes led to an outline of potential biological activities of Cr in six aspects. The gene expression profile reveals that Cr may involves in redox stress, calcium mobilization, energy metabolism, protein synthesis, cell cycle regulation and carcinogenesis in the cell. The results provide a critical clue for understanding molecular mechanisms of the biological activities of Cr. (Mol Cell Biochem **222**: 189–197, 2001)

Key words: chromium (VI), gene chip, redox stress, cell cycle regulation, carcinogenesis

Introduction

Chromium is one of the trace elements in human body that have double properties. It is a carcinogen when its valence becomes VI. It is an essential nutritional element if its valence is in III. Chromate (Cr (VI)) compounds are widely used in industry for chrome plating, chrome pigment, leather tanning, and stainless steel production. It has been proposed that chromium-induced DNA damage, such as chromosome aberration, DNA breakage, and gene mutation, is an important mechanism through which Cr (VI) induces cancer. Studies from our laboratory and other groups have demonstrated that reactive oxygen species (ROS) can mediate the biological activity of Cr (VI) [1]. Cr (VI) is able to generate ROS both intra-cellularly and extra-cellularly. ROS generation is required for several biological activities of Cr (VI), such as induction of apoptosis [4], and activation of NF-κB [5].

Cr (III) compounds are nontoxic and non-carcinogenic as being demonstrated in experimental studies. Chromium picolinate, a Cr (III) compound, is able to improve glucose metabolism, serum lipid parameters, and to reduce body fat [2, 3]. As a result, chromium picolinate has been promoted widely as a health aid for the general population. Unlike Cr (VI), trivalent Cr has a very large safety range and is not toxic at levels up to 1 mg per day in nutritional studies. It remains unclear how Cr (III) exerts its biological effects in the body [2, 6]. One barie in Cr (III) study is that Cr (III) can not enter the cell in cell culture. Since Cr (VI) is able to enter the cell and to be reduced to Cr (III) in the cell, a study of the cellular response to Cr (VI) may facilitate our understanding of nutritional mechanism of chromium picolinate. In this respect, establishment of a gene expression profile for Cr (VI)-induced cellular responses will provide critical guidance for the study of chromium in both carcinogenic and nutritional aspects.

Address for offprints: X. Shi, NIOSH, 1095 Willowdale Road, Morgantown, WV 26505, USA

Advances in high-throughput screening technology has made it possible to analyze expression of thousands of genes at a time. Oligonucleotide-based micro-array represents a state of the art approach in large-scale screening technology. This technology has been used to study HIV polymorphisms [7], aging [8] and mitochondrial function. In this study, a genechip of 2400 genes was used to analyze differential gene expression in response to Cr (VI)-induced cellular stress.

Materials and methods

The human lung type II epithelial cell line (A549) was used as a model in the study. Potassium dichromate was utilized as a source of Cr (VI). The cells were plated in a 100 mm cell culture dish in 10 ml culture medium, which is formed by RPMI-1640 supplemented with 5% fetal bovine serum and 2 mM glutamine. After a 2 h treatment with 300 μM potassium dichromate *in vitro*, the cells were harvested for RNA extraction using Trizole. Rest of the experiment was conducted through a contract service at Alpha Gene, Inc. (Woburn, MA, USA). The messenger RNA (mRNA) was purified from total RNA (400 μg) with an affinity column. Then, the mRNA (4 μg) was reverse transcribed into cDNA that was labeled with fluorescence, Cy3 and Cy5. The fluorescence-labeled cDNA was used to hybridize the genechip that represents 2388 human genes and 12 negative controls (Alphagene, Woburn, MA, USA). The fluorescence data were collected using ScanArray 3000 and analyzed with program ImaGene (BioDiscovery, Los Angeles, CA, USA). The data presented in this study represent mean values of 2 individual hybridizations. Only those genes that exhibited at least a 2 fold change were evaluated in this study.

Results

Induction of redox gene expression

The study reveals that Cr-induced oxidative stress occurs at the gene expression level (Table 1).

Copper-zinc superoxide dismutase (Cu/ZnSOD) and glutathione peroxidase (GPx) were induced by Cr (VI). Cr (VI)-induced oxidative stress was reported previously, but molecular events in the response to oxidative stress remain unclear [4, 5, 9]. Cu/ZnSOD is located in the cytosolic fraction of the cell. Its function is to specifically catalyze conversion of O^-_2 into H_2O_2. A protective role of SOD against oxidative stress has been observed in many biological processes including aging [10]. SOD expression was induced 2.45 fold by Cr (VI), indicating that Cr (VI) may induce O^-_2 production in the cell. The induced SOD expression may be responsible for removal of the extra O^-_2. GPx is a major enzyme responsible for conversion of H_2O_2 into H_2O through consumption of GSH. GPx is located in the cytosolic fraction of the cell. The result in Table 1 shows that GPx is induced 4.33 fold by Cr (VI), indicating that GPx may take care of H_2O_2 induced by Cr (VI). These data provide another piece of evidence that Cr (VI) induces ROS in the cells. Increased expression of both SOD and GPx indicates that the cellular antioxidant systems are activated in response to Cr (VI) to remove extra O^-_2 and H_2O_2.

Expression of metallothionein IIA (MT-IIA) and metal-response transcription factor 1 (MTF-1) were induced by Cr (VI). Metallothionein (MT) is a small protein whose primary function is to protect cells against toxic metals [11]. Other functional roles of MT include maintaining zinc homeostasis and protecting cells from ROS. A hallmark of MT-I and MT-II genes is their rapid transcriptional induction by metals, such as zinc and cadmium. Transcriptional induction of MT is mediated by MTF-1 at the metal response elements (MREs) in the gene promoter [12]. In the human body, MT-IIA is the major MT molecule. In this study, both MT-IIA (3.99 fold) and MTF-1 (3.04 fold) were induced by Cr (VI), suggesting that MTF-1 may be responsible for the increased expression of MT-II.

Photolyase is a group of enzymes responsible for DNA repair in response to ultraviolet light (UV) [13]. UV-induced DNA damage is associated with the activation of photolyases which repair damaged-DNA in a light-dependent manner. Two types of photolyases are known, one specific for cyclobutane pyrimidine dimers (CPD photolyase) and another specific to pyrimidine (6-4) pyrimidone photoproducts ((6-4) photolyase) [14]. Data in Table 1 show that the human homolog of the Drosophila (6-4) photolyase gene was induced 5.17 fold by chromium. This is the first evidence of induction of photolyase in response to agents other than UV light.

In a previous study, we observed that cells undergo programed-cell death (apoptosis) and p53 activation in response to Cr (VI) in the same experimental system [4]. This study further supports apoptosis and p53 activation induced by chromium [15–19]. An apoptosis-related gene, hSIAH (the human homologue of the Drosophila seven in absentia gene), was induced 7.69 fold by Cr (VI) (Table 1). hSIAH is activated during apoptosis in the intestinal epithelium [20]. This gene serves as a target of the p53 tumor suppressor, and it transduces the signaling of p53 during development of apoptosis [21].

Other stress-related genes that were induced by Cr (VI) include heat shock protein 75 (hsp75) and activating transcription factor 3 (ATF-3). Hsp75 belongs to the Hsp90 family whose function is to protect newly synthesized proteins, and regulate their functions through protein folding [22, 23]. Data presented in Table 1 show that hsp75 was induced 3.63 fold by Cr (VI). Association of ATF3 with stress responses has been reviewed [24]. The ATF3 mRNA increases greatly when the cells are exposed to stress signals. It has been sug-

Table 1. Stress response genes

	Acccession #	Gene	Function	Fold ↑	% ↓
1	U76247	hSIAH1	Apotosis	7.69	–
2	D83702	Photolyase	DNA repair	5.17	–
3	X58295	Glutathione peroxidase	Removal of hydrogen peroxide	4.33	–
4	X76717	MT-II	Protection of cells from metal toxicity and oxidative stress	3.99	–
5	AF043254	Heat shock protein 75 (hsp75)	Involved in protein transportation and comformation change	3.63	–
6	X78710	Metal-regulatory transcription factor 1(MTF-1)	Control gene expression of MT-II	3.04	–
7	L19871	Activating transcription factor 3 (ATF3)	Stress response	2.97	–
8	X02317	Cu/Zn superoxide dismutase (SOD)	Catalyzing conversion of superoxide into hydrogen peroxide	2.45	–

gested that the JNK/SAPK pathway is involved in the induction of ATF3 by stress signals. Data in Table 1 shows that ATF3 expression was increased 2.45 fold in response to Cr (VI).

Calcium mobilization

The present study indicates a role of Cr (VI) in the regulation of calcium signaling. Alterations in the expression of calcium-related genes are presented in Table 2. Calcineurin A [25] and calcineurin B were induced by Cr (VI). Calcineurin (also known as protein phosphatase 2B, PP2B) is a serine/threonine phosphatase whose function is to hydrolyze phosphorylated serine and threonine residues in the target proteins, such as nuclear factor of activated T cells (NFAT). This phosphatase contains two subunits: calcinuerin A (catalytic subunit) and calcineurin B (regulatory subunit). Calcinuerin is responsible for activating NFAT through dephosphorylation of serine or threonine in NFAT protein. Activity of calcineurin is regulated by calmodulin, which is a cytosolic protein serving as an intracellular receptor for calcium. The results indicate that expression of calcineurin A2 (an isoform of calcineurin A) was induced 76.91 fold, which is the maximal induction in this study. Calcinuerin B was also induced 2.43 fold by Cr (VI). Expression of both subunits of calcinuerin were activated by Cr (VI), indicating the induction of calcium mobilization in response to Cr (VI). Although it has been reported that other metals, such as cadmium (Cd^{2+}), can induce calcium mobilization, this appears to be the first evidence that Cr (VI) may regulate calcium signaling pathways. It is possible that Cr (VI) induces calcium mobilization through induction of oxidative stress, since oxidants can initiate calcium mobilization.

In accordance with the increased expression of calcineurin genes, cyclophilin (immunophilin) [26], an inhibitor of the calcineurin, was also induced. Cyclophilin is a member of prolyl isomerases, which serves as an intracellular receptor for cyclosporin A (CsA). CsA binds to cyclophilin and the resultant complex binds to calcineurin with a high affinity [27]. This interaction leads to inhibition of calcineurin activity and finally results in immunosuppression. Data in Table 2 indicate that chromium induced the expression of cyclophilin by 3.54 fold.

As shown in Table 2, three other calcium-related genes were induced by Cr (VI): caldesmon (18.76 fold), sorcin (6.27 fold) and PEP-19 (2.37 fold). Activities of these gene products are controlled by calcium.

Calcium-ATPase, encoded by gene ATP2C1, is an ATP-powered calcium pump that sequesters calcium into the Golgi in the cytosol [28]. Calcium-ATPase mutation leads to Hailey-Hailey disease (HHD) due to disregulation of cytoplasmic calcium in the keratinocytes [28]. As shown in Table 2, calcium pump expression was reduced by 59% after Cr (VI) treatment. This change might be responsible for a calcium increase in the cytoplasmic in response to Cr (VI).

Energy metabolism

Cr may play a significant role in the regulation of energy metabolism. Several energy metabolism-related genes, including the obese gene, were induced by Cr (VI) treatment

Table 2. Calcium signaling genes

	Acccession #	Gene	Function	Fold ↑	% ↓
1	M29551	Calcineurin A2	Activating NFAT in response to calcium mobilization	76.91	–
2	M83216	Caldesmon	Involved in control of cell morphology in response to calcium mobilization	18.76	–
3	L12387	Sorcin (SRI)	Related to drug resistance	6.27	–
4	Y00052	Cyclophilin	Cyclosporin A (CsA) receptor	3.54	–
5	M30773	Calcineurin B	Regulating calcinurine activity	2.43	–
6	X93349	PEP-19	Regulating cell movement in response to calcium mobilization	2.37	–
7	M23114	Calcium-ATPase (HK1)	Calcium pump	–	59.00

(Table 3). The major energy resources in the body are glucose and fatty acid. The energy-related gene expression profile supports the hypothesis that Cr induces fatty acid degradation, but reduces glucose consumption. Obesity is a common and complex disease that increases the risk of hypertension and diabetes. Both genetic and environmental factors play a role in the development and progression of human obesity [29, 30]. One of the genes that regulate fatty acid accumulation in the mouse is the obese (ob) gene [31, 32]. The ob gene product, leptin, functions as a hormone to regulate the body fat deposition. Animal experiments demonstrated that mutation of ob gene resulted in profound obesity and type II diabetes [31]. As shown in Table 3, chromium induced ob gene expression by 4.1 fold, indicating that chromium may reduce fat deposition and control body weight through the activation of ob gene expression. This mechanism may explain how Cr compounds act to control body weight.

Enhancement of fatty acid consumption by Cr is further supported by the increased expression of a fat metabolism enzyme, medium-chain acyl-CoA dehydrogenase (MCAD) (Table 3). MCAD is one of the three similar enzymes that catalyze the initial step of fatty acid beta-oxidation [33]. In the fatty acid oxidation process, the first step is the enzymatic dehydrogenation of saturated fatty acid to form a double bond in the carbon chain. MCAD is the enzyme responsible for catalysis of this step. In Cr-treated cells, MCAD was increased by 4.39 fold (Table 3), indicating an increase in the metabolism of saturated fatty acids.

On the other side, it is indicated that glucose consumption may be reduced by Cr (Table 3). Pyruvate dehydrogenase kinase 2 (PDK2) is a major enzyme responsible for the regulation of ATP generation from glucose metabolism. The key function of PDK is to reduce ATP generation from glucose by inactivating pyruvate dehydrogenase (PD). Data in Table 3 show that expression of PDK2 was induced 7.11 fold in Cr (VI) treated cells. This suggests that Cr (VI) may enhance PDK activity and lead to an inhibition of glucose degradation in the cell through Cr (III) which may promote glucose uptake in cells by enhancing insulin signaling [3, 6].

ATP synthesis may be enhanced by Cr because expression of several ATPases was enhanced by Cr (VI) (Table 3). These ATPases include: beta-1 subunit of sodium, potassium- transporting ATPase (Na, K-ATPase) (6.19 fold), the 115 kDa subunit of H^+-ATPase (4.18 fold), medium-chain acyl-CoA dehydrogenase (MCAD) (4.39 fold) and subunit c of ATP synthase (2.71 fold). Taken together, these data suggest that Cr (VI) enhances ATP synthesis from consumption of fatty acid, not glucose. Since Cr (VI) can be easily reduced to Cr (III) in cells, we propose that a net effect of chromium on energy metabolism is to increase the degradation of fatty acid, and at the same time promote synthesis of glycogen. This effect may be due to the combined biological activities of Cr (VI) and Cr (III).

Carcinogenesis

Several important molecules in the intracellular kinase cascades responded to Cr (VI). These include G-proteins, Src kinases and MAP kinase (Mitogen-activated protein) (Table 4). G-proteins and Src-kinase are very important proteins mediating receptor signals in the cell. In mammals, both G-proteins and Src-kinases are involved in the regulation of cell proliferation and differentiation. Crosstalk between these two families has become a new area of interest [34, 35]. G proteins are able to mediate signals from the receptors of hormones, neurotransmitters, and local mediators. A putative G protein was induced 3.33 fold by Cr (VI) [36]. Src kinases mediate receptor signals from cytokine, antigen or extracellular matrix receptors. Two Src-related genes were induced by Cr (VI). They were consensus tyrosine-lacking kinase (Cyl, 6.64 fold) [37] and hematopoietic consensus tyrosine-lacking kinase (HYL, 6.55 fold) [38]. These results indicate that Cr (VI) might be able to regulate cell growth or differentiation through G protein and Src kinases.

G-protein activation often leads to an increase in intracellular cyclic AMP (cAMP) by activating adenylyl cyclase that synthesizes cAMP from ATP. cAMP is an intracellular signaling molecule in all procaryotic and animal cells that

Table 3. Energy metabolism genes

	Acccession #	Gene	Function	Fold ↑	% ↓
1	L42451	Pyruvate dehydrogenase kinase isoenzyme 2 (PDK2)	Inhibition of ATP synthesis from glucose degradation	7.11	–
2	U16799	Na,K-ATPase beta-1 subunit	ATP-dependent Na, K pump	6.19	–
3	D12775	AMP deaminase	Energy generation	5.30	–
4	M16827	Medium-chain acyl-CoA dehydrogenase (MCAD)	Promoting degradation of saturated fatty acid	4.39	–
5	U73006	H(+) ATPase (115 kDa subunit)	Energy transfer in mitochondria	4.18	–
6	U43653	Obese gene (ob)	Enhancement of fatty acid degradation	4.10	–
7	X56549	Muscle fatty-acid-binding protein (FABP)	Fatty-acid metabolism	3.20	–
8	J03620	Dihydrolipoamide dehydrogenase	Glucose degradation	3.18	–
9	D13118	ATP synthase subunit C	Biosynthesis of ATP	2.71	–

Table 4. Carcinogenesis and kinase genes

	Acccession #	Gene	Function	Fold ↑	% ↓
1	X60114	Cytoplasmic tyrosine kinase (Cyl)	Src-related kinase	6.64	–
2	X77278	HYL tyrosine kinase	Src-related kinase	6.55	–
3	AF001434	NEN-1	Tumor suppressor	6.03	–
4	X94612	Type II cGMP-dependent protein kinase	Ion transport and fluid secretion	5.99	–
5	U02882	cAMP phosphodiesterase	cAMP regulation	5.74	–
6	U62800	Cystatin M (CST6)	Inhibition of metastasis	4.24	–
7	Y14391	Putative GTP-binding protein	G-protein	3.33	–
8	U09578	MAPKAP kinase (3pK)	MAP kinase p38 substrate	3.27	–
9	X51345	Jun-B	Nuclear protein	3.27	–
10	S80864	Cytochrome c-like polypeptide	Tumor marker	2.79	–
11	Z71621	Wnt-13	Tumor marker	2.73	–
12	M33011	Carcinoma-associated antigen GA733-2	Tumor marker	2.70	–
13	X03484	Raf oncogene	Kinase	2.63	–
14	U68723	Checkpoint suppressor 1	Tumor marker	2.57	–
15	U24576	Breast tumor autoantigen	Tumor marker	2.18	–
16	M84820	Retinoid X receptor beta (RXR-beta)	Tumor suppressor	–	79.00

have been studied. To maintain cell homeostasis, cAMP is continuously destroyed by one or more cAMP phosphodiesterases, which hydrolyze cyclic AMP to adenosine 5′-monophosphate (5′-AMP). In response to the increase in cAMP, cAMP phosphodiesterases activity should be enhanced. As shown in Table 4, cAMP phosphodiesterase was induced 5.74 fold by Cr (VI) [39], indicating a role of Cr (VI) in the regulation of cAMP signaling pathways.

G-proteins and Src-kinases regulate many signal pathways, including MAP kinases (through Ras and Raf). Raf is the immediate downstream substrate of Ras in the Ras/Raf signaling pathway. After activation, Raf passes a signal to its downstream mediators, such as MAP kinases. Table 4 shows that Raf expression was induced 2.63 fold by Cr (VI). MAP kinase p38 signal is mediated by its downstream kinase, such as MAPKAP2 (MAP kinase-activated protein kinase 2) [40]. Like MAPKAP2, 3pK (Mitogen-activated protein kinase-activated protein kinase 3) is another downstream substrate of p38 [41]. Table 4 shows that 3pK expression was induced 3.27 fold by Cr (VI). In addition, MAP kinase p38 is able to induce expression of c-jun and c-fos, two protooncogenes. Data in Table 4 demonstrate that one of the c-Jun family members, Jun B, was induced 3.27 fold by Cr (VI). Taken together, these data suggest that Cr (VI) may regulate MAP kinase signaling at multiple levels, including G protein, Src and Raf.

Activation of oncogenes by Cr (VI) may lead to carcinogenesis as several tumor-related genes were up-regulated (Table 4). They are: NEN-1 (6.03 fold), Cystatin M (4.24 fold), carcinoma-associated antigen GA733-2 (2.7 fold), breast tumor autoantigen (2.18 fold), checkpoint suppressor 1 (2.57 fold), Wnt-13 (2.73 fold), and cytochrome c-like polypeptide (2.79 fold). In addition, the anticancer molecule, retinoid X receptors beta (RXR beta), was reduced by 79% after Cr (VI) treatment, further implying a possible carcinogenic role of Cr (VI).

Biosynthesis

Metal-induced cell stresses often result in damage to a variety of proteins such as enzymes. This type of protein damage can lead to inactivation of an enzyme. Stressed cells respond to damage with an increase in protein synthesis in order to replace the inactivated or damaged protein, and an enhancement in protein-recycling process to eliminate the damaged proteins. Data in Table 5 provide a global view of genes that are associated with these processes.

New protein synthesis is carried out by ribosomes that are composed of two subunits, the large subunit (60S subunit) and small subunit (40S subunit). Each ribosome subunit contains two types of components: ribosomal protein and ribosomal RNA (rRNA). The eukaryotic ribosomes contain over 70 different proteins. The results show that the gene expression of key proteins in both the large ribosomal subunit and small ribosomal subunit were induced (Table 5). The induced-large subunit ribosomal proteins include: L21 (3.47 fold), L34 (2.8 fold), L38 (5.13 fold), and L41 (2.86 fold). The induced-small subunit ribosomal proteins include S4 (3.12 fold) [42], S17 (2.56 fold), S20 (2.25 fold) and S24 (2.39 fold). The increased expression of ribosomal proteins indicates that Cr-treated cells are initiating a reconstruction process through up-regulation of protein synthesis. In line with this hypothesis, expression of the eukaryotic translation initiation factor 5 (eIF-5) [43] and threonyl-tRNA synthetase [44] were induced 4.34 and 2.73 fold, respectively, in Cr-treated cells.

The ubiquitin-dependent proteolytic pathways are the major mechanisms responsible for removal of damaged pro-

Table 5. Biosynthesis and metabolism genes

	Acccession #	Gene	Function	Fold ↑	% ↓
1	U18244	Excitatory amino acid transporter 4	Glutamate transporter	16.75	–
2	X63237	Ubiquitin 80	Degradation of protein	17.47	–
3	L38961	Putative transmembrane protein precursor (B5)	Glycosylation of protein	6.52	–
4	Z26876	Ribosomal protein L38	Protein synthesis	5.13	–
5	U49436	Translation initiation factor 5 (eIF5)	Promote protein synthesis	4.34	–
6	U49260	Mevalonate pyrophosphate decarboxylase (MPD)	Biosynthesis of steroid	3.82	–
7	X89401	Ribosomal protein L21	Protein synthesis	3.47	–
8	M58458	Ribosomal protein S4 (RPS4X) isoform	Protein synthesis	3.12	–
9	AF026844	Ribosomal protein L41	Protein synthesis	2.86	–
10	L38941	Ribosomal protein L34 (RPL34)	Protein synthesis	2.80	–
11	M63180	Threonyl-tRNA synthetase	Activation of amino acid for protein synthesis	2.73	–
12	M13932	Ribosomal protein S17	Protein synthesis	2.56	–
13	M31520	Ribosomal protein S24	Protein synthesis	2.39	–
14	L06498	Ribosomal protein S20 (RPS20)	Protein synthesis	2.25	–

teins in the stress response. This proteolytic pathway contains two components: proteasome and ubiquitin. The proteasome is a large protein complex whose function is to degrade ubiquinized proteins. Ubiquitin is a small protein whose function is to mark the unwanted proteins and bring them to the proteasome for degradation. Ubiquitin 80 is an oxidation response protein [45, 46], which covalently attaches to oxidatively damaged protein and activates ubiquitin-dependent proteolytic pathways [45]. As shown in Table 5, the expression of ubiquitin 80 was increased 17.47 fold by Cr (VI), indicating an increased activity in protein recycling.

Cholesterol is an important component of cell membranes and plasma lipoproteins. It is also the precursor of many other biologically important steroids, such as various steroid hormones. Biosynthesis of cholesterol may be increased in the response to chromium (Table 5). This is supported by an up-regulated expression of cholesterol enzyme. Mevalonate pyrophosphate decarboxylase (MPD) is a unique enzyme in the cholesterol biosynthetic pathway [47]. This enzyme was induced 3.82 fold in response to Cr (VI). The increased expression of this enzyme may reflect a cellular mechanism for the repairing of cell membranes that are damaged by stress-induced lipid oxidation.

Cell cycle regulation

Our other experiments demonstrated that Cr (VI) inhibited cell proliferation, and the inhibitory effect of Cr (VI) was dose-dependent (data not shown). In a previous study, we observed that Cr (VI) elevates p53 protein levels in the cytoplasm [4], indicating a molecular mechanism of the cell growth inhibition. The microarray data provides more molecular evidence for cell growth regulation by chromium. First, Cr (VI) induced expression of an inhibitor of cyclin-dependent kinase (CDK), INK4 p19 (Table 6). INK4 p19 is a CDK inhibitor [48, 49]. An increased INK4 activity leads to cell cycle arrest. INK4 p19 selectively inhibits CDK4 and CDK6 and results in G1 cell cycle arrest [49]. Expression of INK4 p19 was induced 2.65 fold by Cr (VI). Second, Cr (VI) induced expression of a cell growth inhibitor, retinoblastoma binding protein (Rbp), which is a cofactor of retinoblastoma (Rb) [50]. Rbp acts as a suppressor of the transcription factor E2F and leads to cell growth inhibition. Deficiency of Rbp 2 is a feature of sporadic human melanoma skin cancer [51]. Rbp 2 was induced 3.45 fold by Cr (VI) (Table 6). Third, as shown in Table 6, Cr (VI) reduced expression of several cell cycle or DNA replication mediators, such as CDC25B , CDC2, CDC47 and casein kinase II by 87, 98, 73 and 52%, respectively.

CDC25 is a phosphatase and a potential human oncogene [52]. CDC25 is responsible for activation of CDKs. CDC25 phosphatase removes inhibitory phosphate from tyrosine and threonine residues of CDKs and activates kinase activities of CDKs. In the human cells, CDC25 proteins are encoded by a multigene family, consisting of CDC25A, CDC25B, and CDC25C. While CDC25A plays a crucial role at the G1/S phase transition, CDC25C is responsible for the activation of the mitotic kinase, CDC2/cyclinB. Loss of CDC2/cyclinB activity leads to G2 arrest. CDC25B phosphatase is essential for the G2/M phase transition in human cells [53, 54]. CDC25B is overexpressed in 32 percent of human primary breast cancers tested. Expression of both CDC2 and CDC25B were reduced by 98 and 87%, respectively, by Cr (VI) (Table 6).

CDC47 and casein kinase II (CK2) are two kinases closely related to DNA replication. CDC47 is a member of the MCM protein family [55], which has been implicated in the regulatory machinery of DNA replication in the S phase. Expression of CDC47 mRNA is decreased in quiescent cells, but increased at the late G1 or S phase by growth factor stimulation [56]. CDC47 was reduced 73% by Cr (VI) (Table 6).

Table 6. Cell growth and cell cycle genes

	Acccession #	Gene	Function	Fold ↑	% ↓
1	AJ000414	Cdc42-interacting protein 4 (CIP4)	Cell growth	11.86	–
2	X67951	Proliferation-associated gene (pag)	Cell proliferation	11.77	–
3	M24398	Parathyrmosin	Zinc binding	6.90	–
4	L07758	IEF SSP 9502	Cell growth	6.75	–
5	U39360	CROC-1A	Cell growth	6.47	–
6	M76979	Pigment epithelium-differentiation factor (PEDF)	Cell differentiation	6.11	–
7	AF026547	Neurocan (CSPG3)	Cell motility	5.63	–
8	U32315	Syntaxin 3	Protein secretion	5.49	–
9	S66431	Retinoblastoma binding protein 2 (RBP2)	Tumor suppressor	3.45	–
10	U34051	Cdk 5 activator isoform p39	Cell growth	3.00	–
11	U40343	INK4 p19	Induce G1 cell cycle arrest	2.65	–
12	X74979	trk E	NGF receptor kinase	2.68	–
13	X66360	p34 CDC2	Cell growth	–	98.00
14	L42611	Keratin 6 isoform K6e (KRT6E)	Growth marker	–	88.00
15	Z68092	CDC25B tyrosine phosphatase	Promote cell growth	–	87.00
16	D55716	CDC47	DNA replication	–	73.00
17	J02853	Casein kinase II alpha subunit	DNA replication	–	52.00

CK2 is a serine/threonine kinase overexpressed in many human tumors and rapidly proliferating tissues. CK2 is composed of two catalytic subunits (alpha) and two regulatory subunits (beta). CK2 is required for DNA replication and over-expression of CK2 induces cell transformation. CK2 alpha, a catalytic subunit of CK2 [57], was reduced by 52% after Cr treatment.

On the other hand, some of the cell cycle promoting kinases were induced by Cr (VI). They are CIP4 (CDK42-interacting protein 4, 11.86 fold), and CDK5 (3 fold). In line with these changes, some cell proliferation markers were also induced by Cr (VI). They are PAG (proliferation-associated gene, 11.77 fold), IEF SSP 9502 (a human homologue of the yeast PWP1 protein, 6.75 fold), IEF SSP 9502 (6.75 fold), CROC-1A (6.47 fold), and trk (2.68 fold). It is not clear why these genes were induced by Cr (VI). Their expression might be a part of the cellular stress response.

Discussion

Chromium has dual effects on the human body. The molecular mechanisms of the biological effects of chromium has been dominated by the free radical theory [1, 16, 58]. However, there has been no report on how Cr regulates gene expressions as a result of generation of ROS. In this study, genechip technology was used to explore differential expressions of more than two thousand genes after Cr treatment. The results show that many oxidative stress-related genes are induced in response to Cr treatment. These include Cu/Zn SOD, GPx and MT-II. Expression of these ROS-related genes provides a new evidence in favor of the free radical theory. Calcium mobilization is associated oxidative stress [59]. Although there have been many reports about oxidative stress in response to chromium, it is not clear whether calcium mobilization is involved in biological mechanisms of chromium. This study provides evidence about a role of calcium mobilization in chromium-induced signal transduction. Several calcium-related genes were induced by chromium. Products of these genes play different roles in calcium signal transduction pathways. Since calcium signal is coupled with ROS generation, induction of calcium-related genes become an indirect evidence that chromium induces oxidative stress in the cells.

It is believed that ROS is involved in many metabolism pathways [60–62]. The gene expression profiles from this genechip study provide a detail map for the chromium-responsive pathways. The study suggests that Cr-induced ROS may regulated energy metabolism through induction of gene expression. The induced-genes may enhance energy metabolism. The energy resource will mainly be fatty acid as several genes induced by chromium can promote degradation of fatty acid, and at the same time reduce glucose degradation. This effect will lead to a control of body fat deposition, and therefore may explain why chromium can reduce body fat as well as enhance muscle strength [3, 63]. Intracellular signaling components are targets of ROS [60–62]. These signaling components may extend ROS activities in carcinogenesis and cell cycle regulation. This study suggests that signaling molecules including G-protein, Src kinase, MAP kinases and cyclin-dependent kinases are activated by Cr-induced ROS. These signaling molecules are distributed all over in the signaling cascades from cell membrane to nucleus. These ROS-responsive genes are responsible for ROS responsiveness of different signal transduction pathways. In addition, many cell cycle-related genes are regulated by chromium. Some of them are enhanced by chromium, such as INK p19 and Rb binding protein 2. Some of them are decreased by chromium, such as CDC25B and CDC47. Changes in these genes expression may explain the cell cycle rest activity of chromium [16, 17, 64]. These results together may provide a new basis for hypothesis of chromium carcinogenesis.

ROS oxidize proteins and leads to protein damage [62]. This study shows that the stressed cells may enhance protein synthesis to replace the damaged proteins. At the same time, the stressed cells may accelerate recycling process to remove the damaged proteins. Chromium exposure resulted in expression of genes responsible for both processes. This suggests that biosynthesis is activated in response to Cr-induced ROS.

In summary, this study provides a global view of differential gene expression in response to chromium treatment. Cr regulates expression of genes involved in oxidative stress, calcium signaling, fatty acid degradation, carcinogenesis and cell growth. This information may prove useful not only in understanding molecular mechanisms in Cr toxicity, but also in advancing our knowledge in a number of fields, such as neurology, toxicology, immunology, physiology, biochemistry and oncology.

Acknowledgement

We thank Drs. Vincent Castranova, Murali Rao and Val Vallyathan for their suggestions to improvement of the manuscript.

References

1. Shi X, Chiu A, Chen CT, Halliwell B, Castranova V, Vallyathan V: Reduction of chromium (VI) and its relationship to carcinogenesis. J Toxicol Environ Health, Part B, Crit Rev 2: 87–104, 1999
2. Anderson RA: Chromium as an essential nutrient for humans. Reg Toxicol Pharmacol 26: S35–S41, 1997
3. Jeejeebhoy KN: The role of chromium in nutrition and therapeutics and as a potential toxin. Nutr Rev 57: 329–335, 1999
4. Ye J, Wang S, Leonard SS, Sun Y, Butterworth L, Antonini J, Ding M, Rojanasakul Y, Vallyathan V, Castranova V, Shi X: Role of reactive oxygen species and p53 in chromium (VI)-induced apoptosis. J Biol Chem 274: 34974–34980, 1999
5. Ye J, Zhang X, Young HA, Mao Y, Shi X: Chromium (VI)-induced nuclear factor-kappa B activation in intact cells via free radical reactions. Carcinogenesis 16: 2401–2405, 1995
6. Anderson RA: Chromium in the prevention and control of diabetes. Diabetes Metab 26: 22–27, 2000
7. Kozal MJ, Shah N, Shen N, Yang R, Fucini R, Merigan TC, Richman DD, Morris D, Hubbell E, Chee M, Gingeras TR: Extensive polymorphisms observed in HIV-1 clade B protease gene using high-density oligonucleotide arrays. Nature Med 2: 753–759, 1996
8. Lee CK, Klopp RG, Weindruch R, Prolla TA: Gene expression profile of aging and its retardation by caloric restriction. Science 285: 1390–1393, 1999
9. Friedman J, Shabtai F, Levy LS, Djaldetti M: Chromium chloride induces chromosomal aberrations in human lymphocytes via indirect action. Mutat Res 191: 207–210, 1987
10. Orr WC, Sohal RS: Extension of life-span by overexpression of superoxide dismutase and catalase in *Drosophila melanogaster*. Science 263: 1128–1130, 1994
11. Andrews GK: Regulation of metallothionein gene expression. Prog Food Nutr Sci 14: 193–258, 1990
12. Radtke F, Heuchel R, Georgiev O, Hergersberg M, Gariglio M, Dembic Z, Schaffner W: Cloned transcription factor MTF-1 activates the mouse metallothionein I promoter. Embo J 12: 1355–1362, 1993
13. Thoma F: Light and dark in chromatin repair: Repair of UV-induced DNA lesions by photolyase and nucleotide excision repair. Embo J 18: 6585–6598, 1999
14. Todo T, Ryo H, Yamamoto K, Toh H, Inui T, Ayaki H, Nomura T, Ikenaga M: Similarity among the Drosophila (6-4) photolyase, a human photolyase homolog, and the DNA photolyase-blue-light photoreceptor family. Science 272: 109–112, 1996
15. Shimada H, Shiao YH, Shibata M, Waalkes MP: Cadmium suppresses apoptosis induced by chromium. J Toxicol Environ Health 54: 159–168, 1998
16. Singh J, Carlisle DL, Pritchard DE, Patierno SR: Chromium-induced genotoxicity and apoptosis: Relationship to chromium carcinogenesis. Oncol Rep 5: 1307– 1318 (review), 1998
17. Bagchi D, Joshi SS, Bagchi M, Balmoori J, Benner EJ, Kuszynski CA, Stohs SJ: Cadmium- and chromium-induced oxidative stress, DNA damage, and apoptotic cell death in cultured human chronic myelogenous leukemic K562 cells, promyelocytic leukemic HL-60 cells, and normal human peripheral blood mononuclear cells. J Biochem Mol Toxicol 14: 33–41, 2000
18. Carlisle DL, Pritchard DE, Singh J, Patierno SR: Chromium (VI) induces p53-dependent apoptosis in diploid human lung and mouse dermal fibroblasts. Mol Carcinogen 28: 111–118, 2000
19. Blankenship LJ, Manning FC, Orenstein JM, Patierno SR: Apoptosis is the mode of cell death caused by carcinogenic chromium. Toxicol Appl Pharmacol 126: 75–83, 1994
20. Nemani M, Linares-Cruz G, Bruzzoni-Giovanelli H, Roperch JP, Tuynder M, Bougueleret L, Cherif D, Medhioub M, Pasturaud P, Alvaro V, der Sarkissan H, Cazes L, Le Paslier D, Le Gall I, Israeli D, Dausset J, Sigaux F, Chumakov I, Oren M, Calvo F, Amson RB, Cohen D, Telerman A: Activation of the human homologue of the Drosophila sina gene in apoptosis and tumor suppression. Proc Nat Acad Sci USA 93: 9039–9042, 1996
21. Matsuzawa S, Takayama S, Froesch BA, Zapata JM, Reed JC: p53-inducible human homologue of Drosophila seven in absentia (Siah) inhibits cell growth: Suppression by BAG-1. Embo J 17: 2736–2747, 1998
22. Schulte TW, Akinaga S, Murakata T, Agatsuma T, Sugimoto S, Nakano H, Lee YS, Simen BB, Argon Y, Felts S, Toft DO, Neckers LM, Sharma SV: Interaction of radicicol with members of the heat shock protein 90 family of molecular chaperones. Mol Endocrinol 13: 1435–1448, 1999
23. Chen CF, Chen Y, Dai K, Chen PL, Riley DJ, Lee WH: A new member of the hsp90 family of molecular chaperones interacts with the retinoblastoma protein during mitosis and after heat shock. Mol Cell Biol 16: 4691–4699, 1996
24. Hai T, Wolfgang CD, Marsee DK, Allen AE, Sivaprasad U: ATF3 and stress responses. Gene Exp 7: 321–335, 1999
25. Guerini D, Klee CB: Cloning of human calcineurin A: Evidence for two isozymes and identification of a polyproline structural domain. Proc Natl Acad Sci USA 86: 9183–9187, 1989
26. Haendler B, Hofer-Warbinek R, Hofer E: Complementary DNA for human T-cell cyclophilin. Embo J 6: 947–950, 1987
27. Liu J, Albers MW, Wandless TJ, Luan S, Alberg DG, Belshaw PJ, Cohen P, MacKintosh C, Klee CB, Schreiber SL: Inhibition of T cell

signaling by immunophilin-ligand complexes correlates with loss of calcineurin phosphatase activity. Biochemistry 31: 3896–3901, 1992
28. Hu Z, Bonifas JM, Beech J, Bench G, Shigihara T, Ogawa H, Ikeda S, Mauro T, Epstein EH Jr: Mutations in ATP2C1, encoding a calcium pump, cause Hailey-Hailey disease. Nat Genet 24: 61–65, 2000
29. Bouchard C, Despres JP, Tremblay A: Genetics of obesity and human energy metabolism. Proc Nutr Soc 50: 139–147, 1991
30. Boulanger BR, Milzman DP, Rodriguez A: Obesity. Crit Care Clin 10: 613–622, 1994
31. Zhang Y, Proenca R, Maffei M, Barone M, Leopold L, Friedman JM: Positional cloning of the mouse obese gene and its human homologue. Nature 372: 425–432, 1994
32. Gong DW, Bi S, Pratley RE, Weintraub BD: Genomic structure and promoter analysis of the human obese gene. J Biol Chem 271: 3971–3974, 1996
33. Kelly DP, Kim JJ, Billadello JJ, Hainline BE, Chu TW, Strauss AW: Nucleotide sequence of medium-chain acyl-CoA dehydrogenase mRNA and its expression in enzyme-deficient human tissue. Proc Natl Acad Sci USA 84: 4068–4072, 1987
34. Ptasznik A, Gewirtz AM: Crosstalk between G protein-coupled receptors and tyrosine kinase signaling: Src take centre stage. Arch Immunol Ther Exp 48: 27–30, 2000
35. Diverse-Pierluissi M, Remmers AE, Neubig RR, Dunlap K: Novel form of crosstalk between G protein and tyrosine kinase pathways. Proc Nat Acad Sci USA 94: 5417–5421, 1997
36. Gianfrancesco F, Esposito T, Montanini L, Ciccodicola A, Mumm S, Mazzarella R, Rao E, Giglio S, Rappold G, Forabosco A: A novel pseudoautosomal gene encoding a putative GTP-binding protein resides in the vicinity of the Xp/Yp telomere. Hum Mol Genet 7: 407–414, 1998
37. Partanen J, Armstrong E, Bergman M, Makela TP, Hirvonen H, Huebner K, Alitalo K: cyl encodes a putative cytoplasmic tyrosine kinase lacking the conserved tyrosine autophosphorylation site (Y416src). Oncogene 6: 2013–2018, 1991
38. Sakano S, Iwama A, Inazawa J, Ariyama T, Ohno M, Suda T: Molecular cloning of a novel non-receptor tyrosine kinase, HYL (hematopoietic consensus tyrosine-lacking kinase). Oncogene 9: 1155–1161, 1994
39. Livi GP, Kmetz P, McHale MM, Cieslinski LB, Sathe GM, Taylor DP, Davis RL, Torphy TJ, Balcarek JM: Cloning and expression of cDNA for a human low-Km, rolipram-sensitive cyclic AMP phosphodiesterase. Mol Cell Biol 10: 2678–2686, 1990
40. Winzen R, Kracht M, Ritter B, Wilhelm A, Chen CY, Shyu AB, Muller M, Gaestel M, Resch K, Holtmann H: The p38 MAP kinase pathway signals for cytokine-induced mRNA stabilization via MAP kinase-activated protein kinase 2 and an AU-rich region-targeted mechanism. Embo J 18: 4969–4980, 1999
41. Uddin S, Majchrzak B, Woodson J, Arunkumar P, Alsayed Y, Pine R, Young PR, Fish EN, Platanias LC: Activation of the p38 mitogen-activated protein kinase by type I interferons. J Biol Chem 274: 30127–30131, 1999
42. Fisher EM, Beer-Romero P, Brown LG, Ridley A, McNeil JA, Lawrence JB, Willard HF, Bieber FR, Page DC: Homologous ribosomal protein genes on the human X and Y chromosomes: Escape from X inactivation and possible implications for Turner syndrome. Cell 63: 1205–1218, 1990
43. Si K, Das K, Maitra U: Characterization of multiple mRNAs that encode mammalian translation initiation factor 5 (eIF-5). J Biol Chem 271: 16934–16938, 1996
44. Cruzen ME, Arfin SM: Nucleotide and deduced amino acid sequence of human threonyl-tRNA synthetase reveals extensive homology to the *Escherichia coli* and yeast enzymes. J Biol Chem 266: 9919–9923, 1991
45. Ishii T, Itoh K, Sato H, Bannai S: Oxidative stress-inducible proteins in macrophages. Free Rad Res 31: 351–355, 1999
46. Pancre V, Pierce RJ, Fournier F, Mehtali M, Delanoye A, Capron A, Auriault C: Effect of ubiquitin on platelet functions: Possible identity with platelet activity suppressive lymphokine (PASL). Eur J Immunol 21: 2735–2741, 1991
47. Toth MJ, Huwyler L: Molecular cloning and expression of the cDNAs encoding human and yeast mevalonate pyrophosphate decarboxylase. J Biol Chem 271: 7895–7898, 1996
48. Guan KL, Jenkins CW, Li Y, O'Keefe CL, Noh S, Wu X, Zariwala M, Matera AG, Xiong Y: Isolation and characterization of p19INK4d, a p16-related inhibitor specific to CDK6 and CDK4. Mol Biol Cell 7: 57–70, 1996
49. Hirai H, Roussel MF, Kato JY, Ashmun RA ,Sherr CJ: Novel INK4 proteins, p19 and p18, are specific inhibitors of the cyclin D-dependent kinases CDK4 and CDK6. Mol Cell Biol 15: 2672–2681, 1995
50. Fattaey AR, Helin K, Dembski MS, Dyson N, Harlow E, Vuocolo GA, Hanobik MG, Haskell KM, Oliff A, Defeo-Jones D, Jones ARE: Characterization of the retinoblastoma binding proteins RBP1 and RBP2. Oncogene 8: 3149–3156, 1993
51. Vogt T, Kroiss M, McClelland M, Gruss C, Becker B, Bosserhoff AK, Rumpler G, Bogenrieder T, Landthaler M, Stolz W: Deficiency of a novel retinoblastoma binding protein 2-homolog is a consistent feature of sporadic human melanoma skin cancer. Lab Invest 79: 1615–1627, 1999
52. Galaktionov K, Lee AK, Eckstein J, Draetta G, Meckler J, Loda M, Beach D: CDC25 phosphatases as potential human oncogenes. Science 269: 1575–1577, 1995
53. Lammer C, Wagerer S, Saffrich R, Mertens D, Ansorge W, Hoffmann I: The cdc25B phosphatase is essential for the G2/M phase transition in human cells. J Cell Sci 111: 2445–2453, 1998
54. Forrest AR, McCormack AK, DeSouza CP, Sinnamon JM, Tonks ID, Hayward NK, Ellem KA, Gabrielli BG: Multiple splicing variants of cdc25B regulate G2/M progression. Biochem Biophys Res Commun 260: 510–515, 1999
55. Kiyono T, Fujita M, Hayashi Y, Ishibashi M: Cloning of a cDNA encoding a human homologue of CDC47, a member of the MCM family. Biochim Biophys Acta 1307: 31–34, 1996
56. Fujita M, Kiyono T, Hayashi Y, Ishibashi M: hCDC47, a human member of the MCM family. Dissociation of the nucleus-bound form during S phase. J Biol Chem 271: 4349–4354, 1996
57. Meisner H, Heller-Harrison R, Buxton J, Czech MP: Molecular cloning of the human casein kinase II alpha subunit. Biochemistry 28: 4072–4076, 1989
58. Stohs SJ, Bagchi D: Oxidative mechanisms in the toxicity of metal ions. Free Rad Biol Med 18: 321–336, 1995
59. Hoyal CR, Giron-Calle J, Forman HJ: The alveolar macrophage as a model of calcium signaling in oxidative stress. J Toxicol Environ Health. Part B, Crit Rev 1: 117–134, 1998
60. Vallyathan V, Shi X: The role of oxygen free radicals in occupational and environmental lung diseases. Environ Health Persp 105(suppl 1): 165–177, 1997
61. Buzard GS, Kasprzak KS: Possible roles of nitric oxide and redox cell signaling in metal-induced toxicity and carcinogenesis: A review. J Environ Pathol Toxicol Oncol 19: 179–199, 2000
62. Stadtman ER, Berlett BS: Reactive oxygen-mediated protein oxidation in aging and disease. Drug Metab Rev 30: 225–243, 1998
63. Anderson RA: Effects of chromium on body composition and weight loss. Nutr Rev 56: 266–270, 1998
64. Xu J, Bubley GJ, Detrick B, Blankenship LJ, Patierno SR: Chromium (VI) treatment of normal human lung cells results in guanine-specific DNA polymerase arrest, DNA-DNA cross-links and S-phase blockade of cell cycle. Carcinogenesis 17: 1511–1517, 1996

Molecular and Cellular Biochemistry **222:** 199–204, 2001.

Cr (VI) increases tyrosine phosphorylation through reactive oxygen species-mediated reactions

Yong Qian,[1] Bing-Hua Jiang,[1] Daniel C. Flynn,[1] Stephen S. Leonard,[2] Suiwei Wang,[2] Zhuo Zhang,[2] Jianping Ye,[2] Fei Chen,[2] Liying Wang[2] and Xianglin Shi[2]

[1]Cancer Center and Department of Microbiology and Immunology, School of Medicine, West Virginia University, Morgantown, WV; [2]Pathology and Physiology Research Branch, Health Effects Laboratory Division, National Institute for Occupational Safety and Health, Morgantown, WV, USA

Abstract

While Cr (VI)-containing compounds are well established carcinogens, the mechanisms of their action remain to be investigated. In this study we show that Cr (VI) causes increased tyrosine phosphorylation in human lung epithelial A549 cells in a time-dependent manner. N-acetyl-cysteine (NAC), a general antioxidant, inhibited Cr (VI)-induced tyrosine phosphorylation. Catalase, a scavenger of H_2O_2, sodium formate and aspirin, scavengers of hydroxyl radical ($^{\bullet}OH$), also inhibited the increased tyrosine phosphorylation induced by Cr (VI). SOD, an inhibitor of superoxide radical ($O_2^{\bullet-}$), caused less inhibition. ESR study shows that incubation of Cr (VI) with the A549 cells generates $^{\bullet}OH$ radical. The generation of radical was decreased by addition of catalase and sodium formate, while SOD did not have any inhibitory effect. Oxygen consumption measurements show that addition of Cr (VI) to A549 cells resulted in enhanced molecular oxygen consumption. These results indicate that Cr (VI) can induce an increase in tyrosine phosphorylation. H_2O_2 and $^{\bullet}OH$ radicals generated during the process are responsible for the increased tyrosine phosphorylation induced by Cr (VI). (Mol Cell Biochem **222**: 199–204, 2001)

Key words: Cr (VI), tyrosine phosphorylation, A549 cells

Introduction

Cr (VI)-containing compounds, used widely in industry, are established carcinogens [1]. These compounds have been found to exert toxic and carcinogenic effects in humans and animals and to cause DNA damage and mutations in bacteria and transformation of mammalian cells [2–7]. Epidemiological studies show that workers occupationally exposed to Cr (VI) had a higher incidence of respiratory cancers than the general population [8, 9]. Although the mechanism of Cr (VI)-induced carcinogenesis remains to be investigated, it is generally believed that reduction of Cr (VI) to lower oxidation states by cellular reductants is an important step [3, 10, 11]. During the Cr (VI) reduction, reactive oxygen species (ROS) are generated [2]. Through these species, Cr (VI) is able to cause apoptosis [12] and activation of nuclear transcription factors, such as NF-κB [13], AP-1 [14] and p53 [15]. Several studies have focused on the signal transduction pathways leading to metal-induced carcinogenesis [12–18]. Tyrosine phosphorylation is an important step in the regulation of many key cellular functions. It is involved in control of cell proliferation, differentiation, cell-cycle, cell signal transduction, metabolism, transcription, morphology, adhesion and ion channels [18]. There are two major types of tyrosine kinases: receptor tyrosine kinases (RTKs) and nonreceptor tyrosine kinases (NTKs). The former family is represented by proteins such as the EGF receptor (EGFr) or PDGF receptor (PDGFr) and become activated upon binding to lig-

Address for offprints: D.C. Flynn, Cancer Center and Department of Microbiology and Immunology, School of Medicine, West Virginia University, Morgantown, WV 26506, USA and or by receptor cross-linking [19]. Activation of the EGFr is followed by autophosphorylation on tyrosine residues within its carboxy terminus, followed by binding to cellular substrates via SH2 interactions. NTKs, such as Src family kinases, are tightly controlled by intramolecular interactions that keep Src in a repressed, closed conformation [20]. This repressed state is mediated by an intramolecular interaction between a carboxy terminal phosphotyrosine residue and Src's own SH2 domain. Upon dephosphorylation of the carboxy terminal phosphotyrosine, the structure of Src becomes relaxed and the ATP binding site becomes available for binding ATP. Activated Src can then phosphorylate a specific number of cellular substrates on tyrosine. Tyrosine phosphorylation is also controlled by the activity of tyrosine phosphatases, which dephosphorylate tyrosine residues. In this way, RTK or NTK signaling can be fine-turned by a balance between tyrosine kinases and tyrosine phosphatases. Many cellular proteins are phosphorylated by tyrosine kinases *in vivo*, which modulate different cellular processes. Glenney [21] categoried those proteins into five groups: (1) membrane proteins: connexin-43, T-cell receptor, caveolin and coated pits; (2) serine/threonine kinases: CDC2, MAP kinase and Raf; (3) SH2 containing proteins: phospholipase Cγ, PI-3 kinase. GAP and tensin; (4) cytoskeleton proteins: AFAP-110, vinculin, talin, paxillin, tensin, integrin, annexins I and II, and Ezrin; and (5) other proteins: pp15, IRS1, LDH, Enolase and pp120. Uncontrolled tyrosine phosphorylation, as represented by constitutively high tyrosine phosphorylation of cellular substrates, can lead to aberrant signaling and uncontrolled growth of cells in culture.

Although tyrosine phosphorylation has been considered to be important in signal transduction and in cancer development, the possible relationship of metal carcinogenesis and tyrosine phosphorylation has not been reported. In the present study, the following questions will be answered. (i) Is Cr (VI) capable of causing tyrosine phosphorylation? (ii) Do ROS play a role? (iii) If they do, which species are responsible for Cr (VI)-induced tyrosine phosphorylation? Our results show that Cr (VI) is indeed able to cause tyrosine phosphorylation through $H_2O_2^-$ and ·OH-mediated reactions. We hypothesize that Cr (VI)-induced tyrosine phosphorylation may be important in the signal transduction pathways of Cr (VI)-induced carcinogenesis.

Materials and methods

Reagents

Potassium dichromate ($K_2Cr_2O_7$), 5,5-dimethyl-1-pyrroline *N*-oxide (DMPO), superoxide dismutase (SOD), catalase, aspirin, N-acetyl-cysteine (NAC), and sodium formate were purchased from Sigma (St. Louis, MO, USA). The spin trap, DMPO, was purified by charcoal decolorization. DMPO thus purified did not contain any ESR detectable impurities.

Cell culture

The human lung epithelial cell line (A549) were purchased from the American Type Culture Collection (ATCC, Manassas, VA, USA). The cells were cultured to form a confluent monolayer at 37°C in a 5% CO_2 incubator with RPMI 1640 cell culture medium that was supplemented with 5% fatal calf serum, 2 mM L-glutamine, 100 units/ml penicillin and 100 μg/ml streptomycin. Trypsin EDTA solution (0.25%) was used to detach the cells from the culture plates to pass the cells.

Western blot analysis

The A549 cells were lysated with 1% NP-40 buffer containing 1 mg/ml aprotinin, 200 mM sodium vanadate, 200 mM PMSF, 0.5 M EGTA and 2 μg/ml leupeptin for 5 min at 37°C. The cell lysates were centrifuged at 12,000 g for 10 min at 4°C. The supernatants were saved and standardized to 1 μg/μl total protein concentration. Each sample (80 μl) was applied to an 8% SDS-PAGE. After electrophoresis, protein was transferred to a PVDF membrane. The membrane was washed with TBST (Tris-buffered saline containing 0.1% Tween 80) and then was blocked with 1% BSA in TBST at 4°C overnight. The membrane was first probed with 1:1000 dilution of antiphosphotyrosine antibody (Transduction Laboratories, San Diego, CA, USA) in TBST for 1 h at room temperature, and then probed with 1:1000 dilution of horseradish peroxidase-conjugated donkey anti-rabbit IgG in TBST for 40 min at room temperature. Finally, the tyrosine phosphorylated proteins were visualized using the enhanced chemiluminescence Western blotting detection reagents (Amersham Pharmacia Biotech, Piscataway, NJ, USA).

Electron spin resonance (ESR) measurements

ESR spin trapping was used to examine free radical generation. This technique involves an addition-type reaction of a short-lived radical with a diamagnetic compound (spin trap) to form a relatively long-lived free radical product, the so-called spin adduct, which can be studied by conventional ESR. The intensity of the spin adduct signal corresponds to the amount of short-lived radicals trapped, and the hyperfine splittings of the spin adduct are generally characteristic of the original, short-lived, trapped radical. All measurements were conducted using a Varian E9 ESR spectrometer and a flat cell assembly as described previously [13, 15]. Briefly, hyperfine couplings were measured directly from magnetic field separation using potassium tetraperoxochromate (K_3CrO_8) and 1,1-diphenyl-2-picrylhydrazyl (DPPH) as reference standards. The relative radical concentration was calculated by multiplying half of the peak height by $(\Delta H_{pp})^2$, where $_{Dhpp}$ represents peak-to-peak width. A549 cells were mixed with DMPO to a total final volume of 0.5 ml. The reaction mixture was transferred to a flat cell for measurement.

Oxygen consumption measurements

Oxygen consumption measurements were carried out with a Gilson oxygraph equipped with a Clark electrode (Gilson Medical Electronics, Middleton, WI, USA). These measurements were made at 37°C from mixtures containing 1.0×10^6/ml cells in PBS and various treatments in a total volume of 1.5 ml. The oxygraph was calibrated with media equilibrated with oxygen of known concentrations.

Results

Cr (VI) increases tyrosine phosphorylation in a time-dependent manner

Human lung epithelial A549 cells were used to study Cr (VI)-induced tyrosine phosphorylation. A549 cells were treated with 0.3 mM Cr (VI) at 37°C for different periods of times as indicated in Fig. 1. The western blot in tyrosine phosphorylated proteins shown in Fig. 1 demonstrates that substantial increase of tyrosine phosphorylation occurred after 30 min of Cr (VI) treatment. Cr (VI)-induced tyrosine phosphorylation continued to increase further through 1 h of Cr (VI) treatment. Tyrosine phosphorylation returned to the basal level after prolonged treatment with Cr (VI) (data not shown). There was one distinctive 125 kDa protein whose tyrosine phosphorylation level increases after 30 min of treatment with Cr (VI). In normal cells, anti-phosphotyrosine antibodies will detect pp125Fak, or focal adhesion kinase, a known Src substrate. It is possible that the 125 kDa protein seen in Fig. 1 is pp125Fak, which is consistent with the data reported by Gozin [18].

Role of ROS in Cr (VI)-induced tyrosine phosphorylation

Various antioxidants were used to study the role of ROS in Cr (VI)-induced tyrosine phosphorylation. The results are shown in Fig. 2. It may be noted from this figure that both aspirin and NAC abolished the Cr (VI)-induced tyrosine phosphorylation, indicating that ROS play a key role. Catalase, a specific scavenger of H_2O_2, also abolished the Cr (VI)-induced tyrosine phosphorylation. Since H_2O_2 is a precursor of $^{\bullet}OH$ radical generation, it is possible that $^{\bullet}OH$ radicals are also involved in the Cr (VI)-induced tyrosine phosphorylation. Indeed, sodium formate, a specific $^{\bullet}OH$ radical scavenger inhibited the Cr (VI)-induced tyrosine phosphorylation. In contrast, SOD, a specific $O_2^{\bullet-}$ scavenger only had a slight inhibitory effect. These results suggest that, among ROS, H_2O_2 and $^{\bullet}OH$ are responsible for Cr (VI)-induced tyrosine phosphorylation.

Detection of Cr (VI)-mediated ROS generation

ESR spin trapping was used to detect free radical generation from Cr (VI)-stimulated cells. Figure 3A shows the spectrum recorded from a mixture containing Cr (VI), A549 cells, and DMPO (a spin trapping agent) in PBS. This spectrum consists of a 1:2:2:1 quartet with hyperfine splittings of $a_H = a_N = 14.9$ G, where a_H and a_N denote hyperfine splittings of the α-hydrogen and the nitroxyl nitrogen, respectively. On the basis of these splitting constants [19], the 1:2:2:1 quartet spectrum was assigned to a DMPO/$^{\bullet}OH$ adduct. Addition of catalase (5,000 units/ml) inhibited the $^{\bullet}OH$ generation by 88% (Fig. 3B), and addition of sodium formate (50 mM) decreased $^{\bullet}OH$ generation by 57% (Fig. 3C). The results presented in this figure thus show that both H_2O_2 and $^{\bullet}OH$ are generated in Cr (VI)-stimulated cells and H_2O_2 functioned as a precursor of $^{\bullet}OH$ generation.

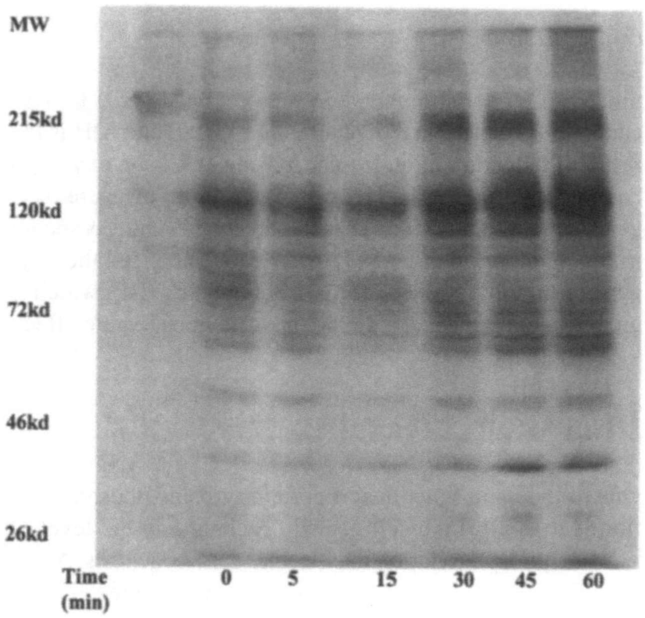

Fig. 1. Cr (VI) increases tyrosine phosphorylation in human lung epithelial A549 cells in a time-dependent manner. The human lung epithelial A549 cells (1.6×10^6 cells/ml in PBS) were treated with 0.3 mM Cr (VI) for different periods of time as indicated. After Cr (VI) treatment, the cells were lysed and resolved by 8% SDS-PAGE. The protein bands were detected using rabbit anti-phosphotyrosine antibody.

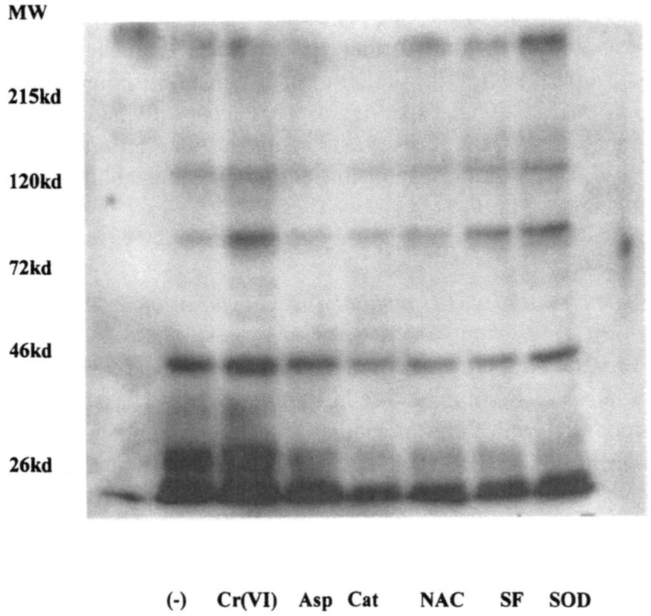

Fig. 2. Free radical scavengers and antioxidants can inhibit Cr (VI)- induced tyrosine phosphorylation. The human lung epithelial A549 cells (1.6 × 10^6 cells/ml in PBS) were treated with Asp (Aspirin, 2.5 mM), Cat (catalase, 5000 units/ml), NAC (N-acetyl-cysteine, 20 mM), SF (sodium formate, 5 mM) and SOD (superoxide dismutase, 800 units/ml) for 1 h at 37°C, and then were incubated with 0.3 mM Cr (VI) for 45 min. After the treatments, the cells were lysed and resolved by 8% SDS-PAGE. The protein bands were detected using rabbit anti-phosphotyrosine antibody.

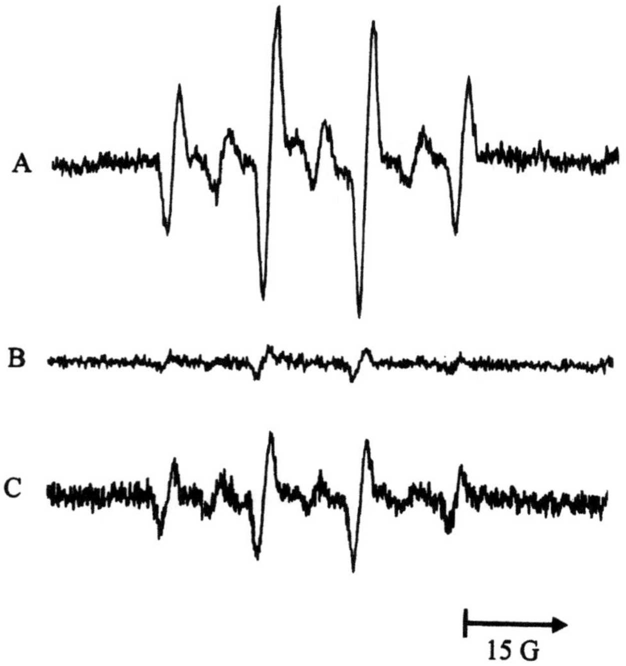

Fig. 3. Cr (VI)-induced ·OH generation. (A) ESR spectrum was recorded 2 min after mixing 1.0 × 10^6 A549 cells/ml with 1 mM Cr (VI) and 100 mM DMSO at a pH 7.4 in PBS. (B) Same as (A) but with 5,000 units/ml catalase added. (C) Same as (A) but with 50 mM sodium formate added.

Oxygen consumption

Since in Cr (VI)-stimulated cells, no exogenous H_2O_2 was added, it is likely that the generation of H_2O_2 and ·OH is the result of reduction of molecular oxygen. A Gilson oxygraph equipped with a Clark electrode was used to measure oxygen consumption in Cr (VI)-stimulated A549 cell. As shown in Fig. 4, addition of Cr (VI) A549 cells enhanced the oxygen consumption by 1.75 fold. In contrast, Cr (III), which is unable to enter the cells, did not have any observable effect.

Discussion

One of the main functions of protein tyrosine phosphorylation is to control the cell growth. A change in the level of tyrosine phosphorylation may lead to carcinogenesis. A role for RTKs and NTKs in cancer has been documented [19, 23, 24]. Indeed, tyrosine phosphorylation has been implicated in malignancies, such as leukemia, lymphoma, multiple endocrine neoplasia type 2, small lung cancer, breast cancer, colon cancer, etc.

There is increasing evidence that tyrosine phosphorylation plays an important role in regulating the activation of transcriptional factors. A good example is Ras-Map kinase transcytoplasmic signaling pathway. Tyrosine phosphorylation is involved in recruiting and activating Ras, Raf, MAP kinase and ERK, which convey signals from plasma membrane to activate the binding of transcription factors to DNA [25] Tyrosine phosphorylation has also been found to have functional roles in the Jak-Stat transcytoplasmic signaling pathway [26]. Recently, it was found that NF-κB can be activated via tyrosine phosphorylation of IκB without subsequent proteolytic degradation of IκB [27].

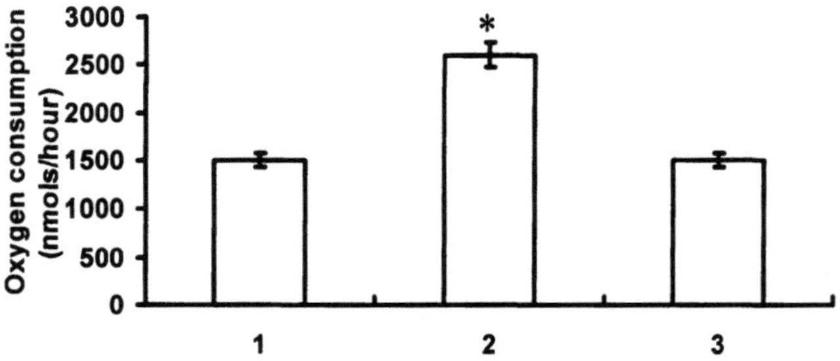

Fig. 4. Cr (VI) treated human lung epithelial A549 cells show increased consumption of molecular oxygen. The human lung epithelial A549 cells (1.6 × 10^6 cells/ml) were incubated in PBS at 37°C (1), or treated with 0.3 mM Cr (VI) (2) or Cr (III) (3) for 20 min. After the treatment, A549 cells were assayed for oxygen consumption at 37°C ($p < 0.05$, n = 3). *indicates a significant increase from control.

The present studies show that Cr (VI) is able to induce tyrosine phosphorylation in the epithelial cell line A549. H_2O_2 and ˙OH play a key role in the Cr (VI)-induced tyrosine phosphorylation. Previous studies have demonstrated that reduction of Cr (VI) by cellular reductants, such as glutathione reductase in the presence of NAD(P)H, generates Cr (V) [5, 15]. During the reduction process, molecular oxygen is reduced to $O_2^{\bullet-}$ radical, which produces H_2O_2 via dismutation. Cr (V) can react with H_2O_2 to generate ˙OH radical via a Fenton-like reaction (Cr (V) + H_2O_2 $\rightarrow$ Cr (VI) + ˙OH + OH^-). Using ESR spin trapping, the present study has demonstrated that Cr (VI)-stimulated cells generate a whole spectrum of ROS and molecular oxygen is the source of these ROS.

Among the ROS generated by Cr (VI)-stimulated cells, both H_2O_2 and ˙OH appears to be the key species responsible for tyrosine phosphorylation. The following experimental observation support the above conclusion. (a) Sodium formate, an ˙OH radical scavenger, decreased but did not abolish, tyrosine phosphorylation. (b) SOD, a scavenger of $O_2^{\bullet-}$ only slightly decreased tyrosine phosphorylation. (c) Catalase, a scavenger of H_2O_2, abolished the tyrosine phosphorylation. It may be noted that H_2O_2 is a precursor of ˙OH radical generation. Elimination of H_2O_2 eliminated ˙OH generation as shown by ESR spin trapping measurements. As for $O_2^{\bullet-}$, while it is not directly involved in tyrosine phosphorylation, it is an intermediate for the generation of both H_2O_2 and ˙OH.

It has been reported that ROS are involved in various physiological and pathological processes [29]. Biological systems generate ROS naturally. They are also produced in these system from exposure to exogenous substances [30]. Emerging evidence points to the involvement of ROS as signaling intermediates [31]. ROS are involved in the activation of several nuclear transcription factors, such as AP-1, NF-κB, and p53 and appear to regulate apoptosis, cell growth regulation, cytokine production, and oncogene expression [32]. Recently, it has been shown that ROS help Ras transmit growth-stimulating messages to the cell interior [33]. This report adds to the growing evidence that ROS, traditionally viewed as toxins, can also play a role in the cell's normal signaling pathways. Although it is believed that ROS function as signaling intermediates, their mechanism of action remains to be investigated. For example, how ROS can turn certain signals on and off. ROS-mediated tyrosine phosphorylation may be involved in this process. Thus, elucidation of mechanisms involved in Cr (VI)-induced tyrosine phosphorylation may not only increase our understanding of Cr (VI)-induced carcinogenesis, but may also help in understanding ROS-mediated signal pathways in general.

It may also be noted that many other metal ions, mineral particles and chemical carcinogens, such as cobalt, nickel, vanadium, asbestos, and silica, have been reported to be capable of generating ROS. It is possible that these agents may function in the same way as Cr (VI); i.e. they may cause tyrosine phosphorylation through H_2O_2 and ˙OH mediated common messages. Because tyrosine phosphorylation is important in signal transduction, ROS could play a critical role in various signal pathways in both physiological and pathological processes.

In conclusion, (a) Cr (VI) is able to induce increased tyrosine phosphorylation in human lung epithelial A549 cells; (b) Cr (VI)-stimulated A549 cells generate ROS, including $O_2^{\bullet-}$, H_2O_2 and ˙OH , via reduction of molecular oxygen; and (c) the generated H_2O_2 and ˙OH are responsible for the increased tyrosine phosphorylation induced by Cr (VI). Therefore, ROS may play a critical role in various signal pathways under both physiological and pathological conditions via their ability to cause tyrosine phosphorylation.

Acknowledgements

The authors are grateful to Drs. Vince Castranova, Murali Rao, and Val Vallyathan for helpful comments and suggestions during the preparation of this manuscript.

References

1. De Flora S, Bagnasco M, Serra D, Zanacchi P: Genotoxicity of chromium compounds: a review. Mutat Res 238: 99–172, 1990
2. Shi X, Chiu A, Chen CT, Halliwell B, Castranova V, Vallyathan V: Reduction of chromium (VI) and its relationship to carcinogenesis. J Toxicol Environ Health 2: 87–104, 1999
3. De Flora S, Wetterhahn KE: Mechanism of chromium metabolism and genotoxicity. Life Chem Rep 7: 169–244, 1989
4. Xu J, Wise JP, Patierno SR: DNA damage induced by carcinogenic lead chromate particles in cultured mammalian cells. Mutat Res 280: 129–136, 1992
5. Leonard S, Wang S, Zang L, Castranova V, Vallyathan V, Shi X: Role of molecular oxygen in the generation of hydroxyl and superoxide anion radicals during enzymatic Cr (VI) reduction and its implication to Cr (VI)-induced carcinogenesis. J Environ Pathol Toxicol Oncol 19: 49–60, 2000
6. Xu J, Wise JP, Patierno SR: DNA damage induced by carcinogenic lead chromate particles in cultured mammalian cells. Mutat Res 280: 129–136, 1992
7. Kortenkamp A, Ozolins Z, Beyersmann D, O'Brien P: Generation of PM2 DNA breaks in the course of reduction of chromium (VI) by glutathione. Mutat Res 216: 19–26, 1989
8. Hayes RB: Review of occupational epidemiology of chromium chemicals and respiratory cancer. Sci Total Environ 71: 331–339, 1988
9. Langard S: One hundred years of chromium and cancer: A review of epidemiological evidence and selected case reports. Am J Ind Med 17: 189–215, 1990
10. Connett PH, Wetterhahn KE: Metabolism of the carcinogenic chromate by cellular constituents. Struct Bond 54: 93–124, 1983
11. Cohen MD, Kargacin B, Klein CB, Costa M: Mechanism of chromium carcinogenicity and toxicity. Crit Rev Toxicol 23: 255–281, 1983
12. Ye J, Wang S, Leonard SS, Sun Y, Butterworth, L, Antonini J, Ding M, Yongyut Rojanasakul, Vallyathan V, Castranova V, Shi X: Role of

reactive oxygen species and p53 in chromium (VI)-induced apoptosis. J Biol Chem 274: 34974–34980, 1999
13. Ye J, Zhang X, Young HA, Mao Y, Shi X: Chromium (VI)-induced nuclear factor-κB activation in intact cells via free radical reactions. Carcinogenesis 16: 2401–2405, 1995
14. Chen F, Ding M, Lu Y, Leonard SS, Vallyathan V, Castranova V, Shi X: Participation of MAP kinase p38 and IκB kinase in chromium (VI)-induced NF-κB and AP-1 activation. J Environ Pathol Toxicol Oncol 19: 231–238, 2000
15. Wang S, Leonard SS, Ye J, Ding M, Shi X: The role of hydroxyl radical as a messenger in Cr (VI)-induced p53 activation. Am J Physiol 279: C868–C875, 2000
16. Chuang SM, Liou GY, Yang JL: Activation of JNK, p38, and ERK mitogen-activated protein kinase by chromium (VI) is mediated through oxidative stress but does not affect cytotoxicity. Carcinogenesis 21: 1491–1500, 2000
17. Chuang SM, Wang IC, Yang JL: Roles of JNK, p38 and ERK mitogen-activated protein kinase in the growth inhibition and apoptosis induced by cadmium. Carcinogenesis 21: 1423–1432, 2000
18. Gozin A, Franzini E, Andrieu V, Da Costa L, Rollet-Labelle E, Pasquier C: Reactive oxygen species activate focal adhesion kinase, paxillin and tyrosine phosphorylation in endothelial cells. Free Rad Biol Med 9: 1021–1032, 1998
19. Hunter T: The Croonian Lecture 1997. The phosphorylation of proteins on tyrosine: Its role in cell growth and disease. Phil Trans R Soc Lond B Biol Sci 353: 583–605, 1998
20. Jiang G, Hunter T: Receptor signaling: When dimerization is not enough. Curr Biol 9: R568–R571, 1999
21. Brown MT, Cooper JA: Regulation, substrates and functions of src. Biochim Biophys Acta 1287: 121–149, 1996
22. Glenney JR Jr: Tyrosine-phosphorylated proteins: Mediators of signal transduction from the tyrosine kinases. Biochim Biophys Acta 1134: 113–127, 1992
23. Kolibaba KS, Druker BJ: Protein tyrosine kinases and cancer. Biochim Biophys Acta. 1333: F217–F248, 1997
24. Biscardi JS, Tice DA, Parsons SJ: c-Src, receptor tyrosine kinases, and human cancer. Adv Cancer Res 76: 61–119, 1999
25. Hill CS, Treisman R: Transcriptional regulation by extracellular signals: Mechanisms and specificity. Cell 80: 199–211, 1995
26. Darnell JE Jr, Kerr IM, Stark GR: Jak-STAT pathways and transcriptional activation in response to IFNs and other extracellular signaling proteins. Science 264: 1415–1421, 1994
27. Imbert V, Rupec RA, Livolsi A, Pahl HL, Traenckner EB, Mueller-Dieckmann C, Farahifar D, Rossi B, Auberger P, Baeuerle PA, Peyron JF: Tyrosine phosphorylation of IκB-alpha activates NF-κB without proteolytic degradation of IκB alpha. Cell 86: 787–798, 1996
28. Buettner GR: ESR parameters of spin adducts. Free Rad Biol Med 3: 259–303, 1987
29. Halliwell B, Gutteridge JMC: In: Free Radicals in Biology and Medicine. Oxford University Press, Oxford, UK, 2000, pp 1–540
30. Vallyathan V, Shi X: The role of free radicals in occupational and environmental lung diseases. Environ Health Perspect 105: 195–177, 2000
31. Pennisi E: Superoxides relay Ras protein's oncogenic message. Science 1567–1568, 1997
32. Buzard GS, Kasprzak KS: Possible role of nitroxide and redox cell signaling in metal-induced toxicity and carcinogenesis: A review. J Environ Pathol Toxicol Oncol 19: 179–199, 2000
33. Irani K, Xia Y, Zweier JL, Sollott SJ, Der CJ, Fearon ER, Sundaresan M, Finkel T, Goldschmidt-Clermont PJ: Mitogenic signaling mediated by oxidants in Ras-transformed fibroblasts. Science 275: 1649–1652, 1997

Molecular and Cellular Biochemistry **222:** 205–211, 2001.

Molecular biology of nickel carcinogenesis

Max Costa, Jessica E. Sutherland, Wu Peng, Konstantin Salnikow, Limor Broday and Thomas Kluz

Department of Environmental Medicine, New York University School of Medicine, NY, USA

Abstract

A review of the molecular mechanisms of nickel carcinogenesis has been compiled. This work is based upon approximately 20 years of research conducted in my laboratory. Molecular mechanisms of nickel carcinogenesis are considered from the point-of-view of the uptake of nickel, both soluble and insoluble particles in cells, its dissolution and its effects on heterochromatin. Molecular mechanisms by which nickel induces gene silencing in cells by DNA hypermethylation in mammalian cells and by inhibiting histone acetylation in yeast cells are also discussed. (Mol Cell Biochem **222**: 205–211, 2001)

Key words: nickel carcinogenesis, molecular mechanisms, gene silencing

Epidemiological studies have confirmed that water-soluble nickel compounds represent a substantial hazard in the induction of human lung and nasal cancers [1–3]. In experimental animals, water-insoluble nickel compounds are also very potent carcinogens. They are able to induce a very high incidence of cancers at the site of administration. For example, in a recent inhalation study conducted by the National Toxicology Progra, 'NTP Bioassay of Carcinogenic Activity of Nickel Subsulfide', the subsulfide was found to induce lung cancer in rats, whereas the water-soluble nickel sulfate did not cause a statistically higher incidence of lung cancer in this species [4].

Figure 1 summarizes work that has been conducted in our laboratory illustrating why the water-insoluble nickel compounds are such potent carcinogens. Using tissue culture cell systems, such as Syrian hamster embryo cells, as well as a variety of other fibroblast cell lines, we were able to demonstrate that the crystalline nickel sulfide and nickel subsulfide compounds were phagocytized by cells [5, 6] (Fig. 1). In contrast, a non-carcinogenic nickel compound, amorphous nickel sulfide was not phagocytized by cells (Fig. 1). Water-soluble nickel salts are taken up poorly by cells. Shown in this cartoon is the fact that the surface charge of the particles influences phagocytosis [7]. The crystalline nickel sulfide particles have a negative surface charge but amorphous nickel sulfide particles have a high positive surface charge (Fig. 1). For this reason, it is not phagocytized by cells. Water-soluble nickel salts penetrate cells poorly. Thus, it is the delivery of the nickel inside the cell that appears to determine the carcinogenic potencies of these different nickel compounds (Fig. 1). Following phagocytosis, the particles are contained in vacuoles that become highly acidified and this greatly enhances the dissolution of soluble nickel from the particles and increases the intracellular levels of nickel, that can subsequently attack chromatin and produce effects on chromatin which are related to carcinogenic activity.

Table 1 shows the influence of surface charge in regulating the phagocytosis and transforming activity of nickel sulfide compounds [7]. Amorphous nickel sulfide has a +9 mv potential and is phagocytized to a lesser degree than all the compounds here and produces a low incidence of transformation. Treatment of amorphous nickel sulfide with lithium aluminium hydride makes these particles considerably less positive, and now they are phagocytized and induce a higher incidence of transformation [8]. The basal surface charge of crystalline nickel sulfide is negative with –27 mv potential. Its level of phagocytosis is high and its transforming activity is also high [7–9] (Table 1). Furthering the negativity of the surface charge by treating crystalline nickel sulfide particles with lithium aluminium hydride enhances their phago-

Address for offprints: M. Costa, Department of Environmental Medicine, New York University School of Medicine, NY 10016, USA

Table 1. The influence of surface charge in regulating phagocytosis and transformation induction by nickel sulfide compounds

Treatment compound (1–2.5 μm particle size)	Surface charge	Phagocytosis[a]	Transformation[b] (%)
Amorphous NiS	+9 mV Zeta potential	4.8	0.49
$LiALH_A$ reduced amorphous NiS	Considerably less positive	27.7*	3.27*
Crystalline NiS	–27 mV Zeta potential	29.2	3.67
$LiALH_4$ reduced crystalline NiS	Considerably more negative	53.1*	7.49*

*$p < 0.001$ vs. untreated particles, χ^2 test; [a]number of cells with phagocytized metal/total number examined (10 μg/ml, 24 h, Chinese hamster ovary cells); [b]number of transformed colonies/total number of surviving colonies (10 μg/ml, 4 days, Syrian hamster embryo cells).

cytosis and further enhances the transforming activity of the nickel compounds. These studies demonstrate that the transforming activity of these nickel compounds is proportional to cellular uptake which is related to the degree of negative surface charge.

Table 2 demonstrates the importance of this phagocytosis process in delivering high concentrations of nickel inside the cell [7, 10, 11]. We show calculations performed on a 1.45 micron particle that should it totally dissolve in the cell, it would produce a potential nickel concentration of 250 mM. Should a 4 micron particle dissolve inside the cell, the concentration of nickel would be 4.75 M. Thus, it can be seen that the phagocytosis and dissolution of these particles can yield very high concentrations of nickel inside the cell. Similarly, Table 3 shows the effect of soluble and insoluble nickel compounds on cytotoxicity in cells as determined by plating efficiency comparing doses in μM, as well as μg per cm^2 [12]. What can be seen in this table is that at a concentration of approximately 20 μM of nickel subsulfide, the cytotoxicity is equivalent to 600 μM of soluble nickel chloride. Thus, the delivery of the nickel as particles by phagocytosis represents a higher degree of cytotoxic and transforming response than the delivery of soluble Ni^{2+} to cells.

Nickel selectively effects heterochromatin (Fig. 2). Shown is a chromosome spread from Chinese hamster ovary (CHO) cells treated with crystalline nickel sulfide compounds [13, 14]. There is selective damage on the entirely heterochromatic long-arm of the X chromosome of CHO cells. Figure 3 shows the selectivity on the X chromosome, heterochromatic long-arm vs. the short-arm, which is euchromatic [15]. The long-arm but not the short-arm was damaged by crystalline nickel sulfide. Subsequent studies have examined the effect of crystalline nickel sulfide on gene silencing and it was found that nickel could silence tumor suppressor and senes-

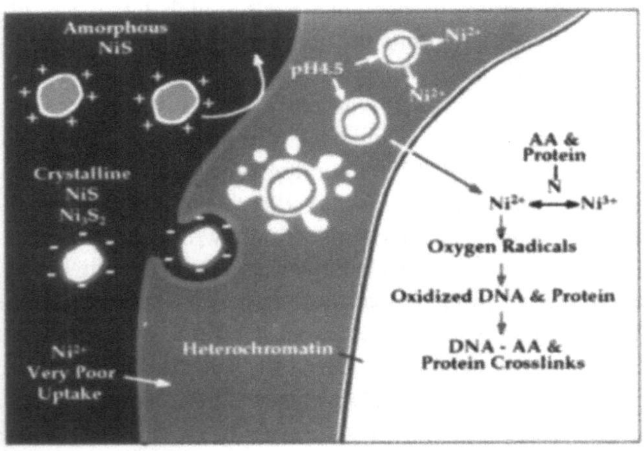

Fig. 1. A model of the molecular mechanisms by which particle nickel compounds are phagocytized by cells. Both crystalline nickel sulfide and subsulfide have a negative surface charge and are actively phagocytized in cells and are subsequently contained in acidic vacuoles which enhances the dissolution of nickel. In contrast, amorphous NiS has a positive charge and is not phagocytized by cells while soluble nickel salts enter cells poorly.

Table 2. Potential intracellular concentration of a phagocytized crystalline NiS particle[a]

Mean particle diameter used in calculation (μm)	Approximate NiS cellular concentration[b] (M)
1.45	0.25
4.00	4.75

[a]Cell volume was determined in CHO cells with a Coulter counter-particle size analyzer and log range expander; [b]cell volume, 393.5 μm^3; density of NiS, 5.5 g/cm^3; particles assumed spherical.

Table 3. Effects of soluble vs. insoluble nickel compounds on survival of CH V79 (G12)

Nickel compound	Exposure concentration	Survival (plating efficiency % of control)
Ni_3S_2	0.1 $\mu g/cm^2$ (4.1 μM)	90
	0.3 $\mu g/cm^2$ (12.3 μM)	60
	0.5 $\mu g/cm^2$ (20.5 μM)	20
NiS	0.1 $\mu g/cm^2$ (10.8 μM)	80
	0.3 $\mu g/cm^2$ (32.4 μM)	50
	0.5 $\mu g/cm^2$ (54.0 μM)	15
$NiCl_2$	200 μM	100
	400 μM	64
	600 μM	30

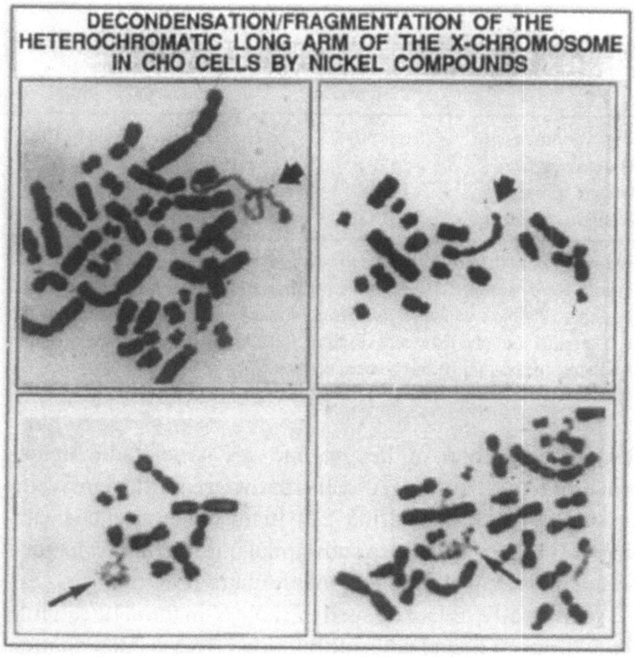

Fig. 2. The effect of nickel sulfide on the heterochromatic long-arm of the X-chromosome.

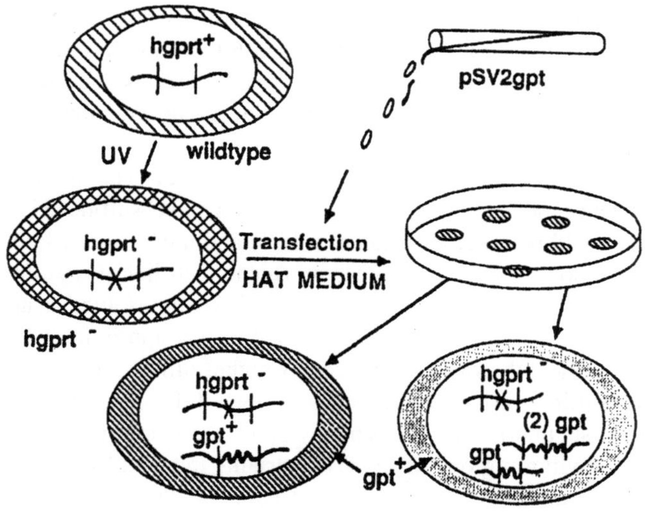

Fig. 4. Cartoon illustrating the preparation of the transgenic G12 and G10 cell lines. The endogenous *HPGRT* gene in V79 cells was inactivated by a mutation caused by UV light. A bacterial *gpt* gene under the control of the SV40 promoter was transfected into these cells. A single copy of the bacterial *gpt* gene was inserted in each cell line at different positions.

cence genes by enhancing DNA hypermethylation on the X chromosome in CHO cells [16]. We used a model system to investigate the mechanisms by which nickel was inducing gene silencing and the involvement of DNA methylation in this silencing. This system utilizes a transgenic cell line where the endogenous *HPGRT* gene has ben inactivated by a permanent mutation induced by UV light and a bacterial *gpt* gene under the SV40 promoter control has been transfected into this cell line [17, 18] (Fig. 4).

Inactivation of the *gpt* gene enables cells to grow in 10 μg/ml 6-thioguanine. Figure 5 shows a fluorescent *in situ* hybridization showing the position of the *gpt* gene in the G12 cell line, a cell line that was responsive to crystalline nickel sulfide induced gene inactivation [19]. Figure 6 shows the effect of crystalline nickel sulfide and nickel sulfate on 6-thioguanine resistance in the G12 cell line [20, 21]. Note that nickel sulfate was not effective in producing 6-thioguanine resistant colonies in the G12 cell line regardless of the dose utilized, either toxic or non-toxic. In contrast, crystal-

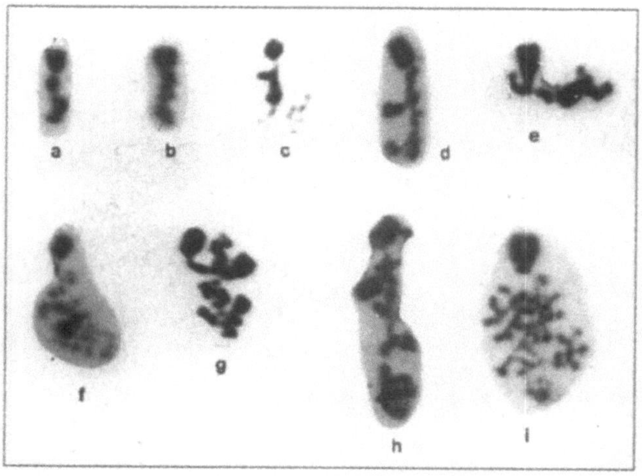

Fig. 3. The X-chromosome was isolated from Chinese hamster ovary cells treated with nickel sulfide. Shown is the selective damage produced in heterochromatin (bottom portion) compared to the lack of damage in euchromatin (top portion).

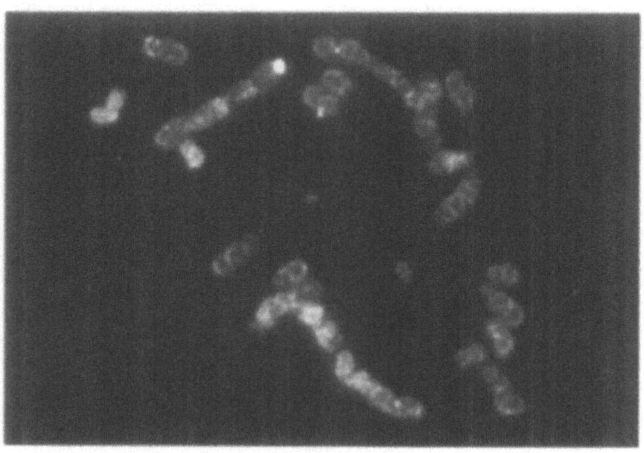

Fig. 5. In situ hybridization showing the position of the *gpt* gene in chromosome 1 in the G12 cell line.

line nickel sulfide induced a 1:1000 incidence of 6-thioguanine resistance at doses that caused 60–60% cytotoxicity (Fig. 6). This incidence of 6-thioguanine resistance was much too high a frequency to have been caused solely by mutations.

Table 4 shows the analysis of the G12 6-thioguanine resistance induced by nickel subsulfide and nickel sulfide [21]. It should be noted that nickel sulfide and nickel subsulfide were not effective in another transgenic cell line (G10). In this cell line, the *gpt* gene was located in chromosome 6 near a euchromatic region. This table shows that approximately 80–90% of the variants induced by nickel sulfide and nickel subsulfide were negative for transcripts of the *gpt* gene. Table 5 summarizes the DNA methylation status in the promoter and coding regions of the G12 variants induced by nickel sulfide and nickel subsulfide, and again, 80–90% if the variants were hypermethylated in the promoter or coding region of the gene [21].

Figure 7 shows the sensitivity of the variants to DNase I digestion [21]. In the wild-type, the *gpt* gene is degraded by increasing concentration of DNase I, whereas in the *gpt*-silenced variants induced by nickel sulfide and nickel subsulfide (e.g. Ni-2, Ni-3, Ni-5, and Ni-9) the *gpt* gene was resistant to DNase I degradation. Ni-8 is a variant that was not hypermethylated and it showed similar sensitivity to the wild-type cells. We next examined the position of the *gpt* gene in biochemically fractionated heterochromatin. As shown in Fig. 8, the G12 cells, more of the *gpt* gene was associated with heterochromatin than in G10 cells that were not responsive to crystalline nickel subsulfide [22]. In the nickel-induced variants of G12 (Ni-2, Ni-3), a substantial portion of the *gpt* gene was associated with the heterochromatic fraction.

Table 4. Analysis of *gpt* transcripts of $6TG^R$ variants

	$6TG^R$ variants (% of proportion)		
	NiS-induced	Ni_3S_2-induced	Total
Total variant tested	16 (100)	15 (100)	31 (100)
Deleted variant	2 (13)	0 (0)	2 (6.5)
Positive transcripts	1 (6)	1 (7)	2 (6.5)
Negative transcripts	13 (81)*	14 (93)*	27 (87)

*No significant difference between two groups (p = 0.1515). A total of 31 variants were examined for deletion, positive and negative transcripts of the *gpt* gene. The base of nickel sufide and nickel subsulfide, 80–90% of the G12 variants induced that were resistant to 6-thioguanine became resistant as a result of the loss of the *gpt* transcripts.

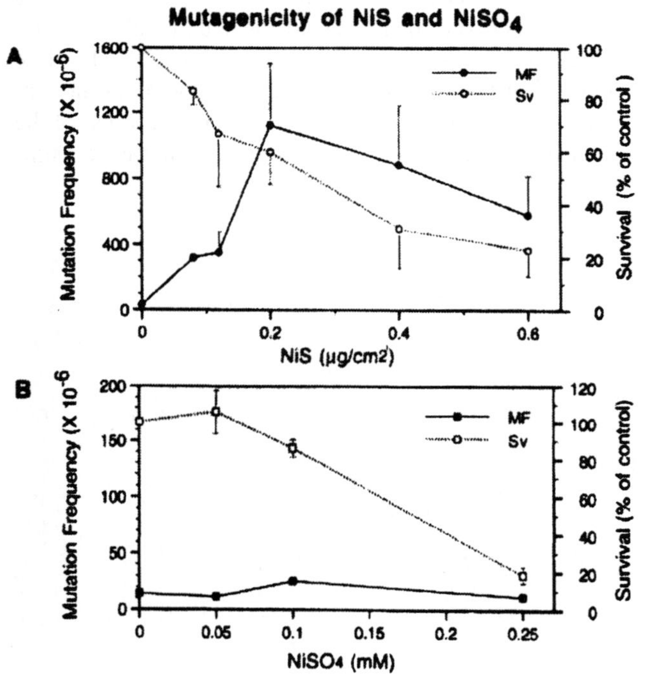

Fig. 6. The effect of NiS and $NiSO_4$ on 6-thioguanine resistance in G12 cells.

Additionally, as can be seen in Table 6, in the G12 cell line the addition of nickel chloride inhibited DNase I degradation of *gpt*, whereas in G10 cell lines, addition of nickel chloride did not cause the *gpt* gene to become resistant to DNase I degradation [22]. We attribute this finding to the fact that in the G12 cell line, the *gpt* gene is located near the telomere of chromosome 1 and near a heterochromatic fragment, whereas in the G10 cell lines, the *gpt* gene is located on chromosome 6, away from any heterochromatic or other silencing elements.

We developed a model of how nickel was inducing gene silencing [22]. This model is shown in Fig. 9 where nucleosomes are pictured in a non-condensed state and in a

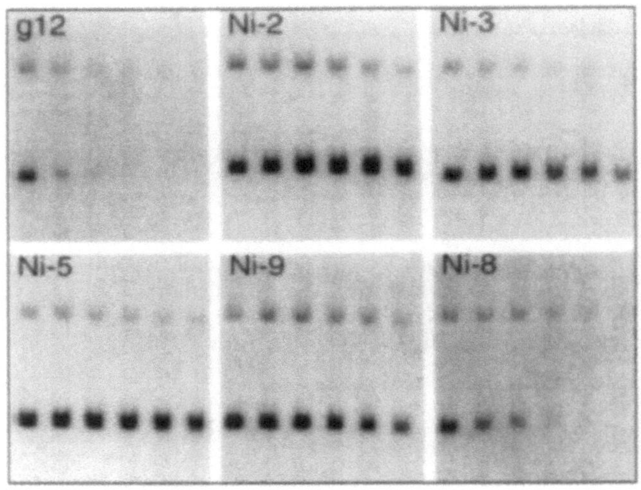

Fig. 7. A Southern blot showing the effects of increasing concentration of DNase I on the degradation of the *gpt* gene in wild-type and nickel-induced variants that were hypermethylated, Ni-1, Ni-3, Ni-5, and Ni-9. It also shows a nickel-induced variant that did not exhibit hypermethylation (Ni-8).

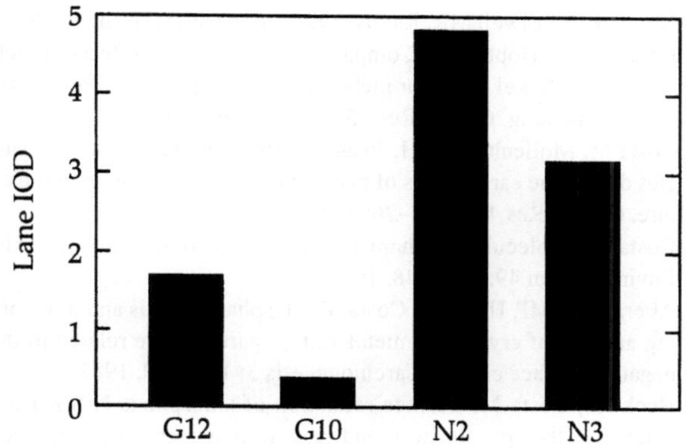

Fig. 8. Heterochromatin fraction was isolated and the density of the *gpt* gene in the heterochromatin fraction was examined. Note that in the nickel-induced variants, Ni-2 and Ni-3, a high intensity of *gpt* gene was found in the heterochromatic fraction.

Table 5. Summary of methylation studies of $6TG^R$ variants

	$6TG^R$ variants (% of proportion)		
	Spontaneous	NiS-induced[2]	Ni_3S_2-induced[a]
Total variant tested	10 (100)	16 (100)	15 (100)
Non-deleted variant	8	14	15
Deleted variant	2 (20)	2 (13)	0 (0)
Non-methylated	7 (70)	1 (6)	1 (7)
Hypermethylated	1 (10)	13 (81)	14 (93)

Examination of spontaneous nickel sulfide and nickel subsulfide induced variants for methylation in the promoter or coding region of the *gpt* gene. Note that in the case of the spontaneous variants, about 10% of these had hypermethylation in the promoter or coding region, whereas 80–90% of the nickel sulfide and nickel subsulfide-induced variants had methylation.

Table 6. Immediate effects of $NiCl_2$ on *gpt*-specific chromatin condensation *in vitro*

		% of initial density	
Cell line	Treatment	0.125 DNase I (units)	0.5 DNase I (units)
G12	Untreated	73.9 ± 11.4	38.6 ± 1.8
	$NiCl_2$	82.1 ± 14.9	70.5 ± 4.3
G10	Untreated	82.0 ± 2.9	57.4 ± 4.0
	$NiCl_2$	89.2 ± 15.8	54.0 ± 2.3

higher order chromatin state where the DNA is highly methylated. When nickel is present, it binds to the phosphate backbone of DNA and increases chromatin condensation pulling in tumor suppressor genes that are near heterochromatin into the heterochromatic fraction. The condensation of genes into the heterochromatic fraction in the absence of methylation, we believe triggered *de novo* methylation of these genes that would then cause these genes to be inherited in the heterochromatic fraction, even in the absence of nickel. We believe that this may be a model mechanism by which nickel may be inducing gene silencing of tumor suppressor and senescence genes located near heterochromatin.

Yeast do not use DNA methylation to silence genes but use histone acetylation to activate genes and loss of histone acetylation to inactivate genes. We were interested in the effect of nickel chloride on gene silencing in yeast and we used a telomere silencing element which had a *URA3* marker gene placed 1.3 kb from the telomere silencing element. As shown in Fig. 10, nickel chloride was very effective at silencing the gene at doses that were not toxic to cells [23]. It should be noted that if the gene was placed 2.0 kb from the telomere silencing element, nickel was not effective at silencing this gene. Other metals, such as cobalt, cadmium, and copper were not as effective as nickel at silencing this telomere silencing element [23] (Fig. 11).

We next examined the effect of nickel on histone acetylation in the yeast system. Figure 12 shows that nickel chloride has substantial effects in inhibiting histone acetylation

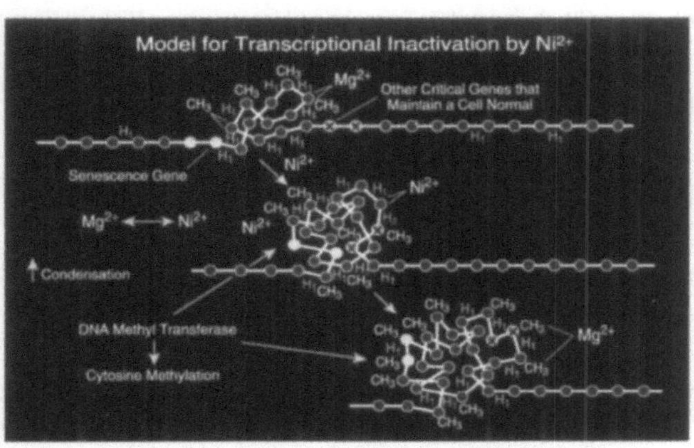

Fig. 9. Cartoon of model for Ni-induced silencing.

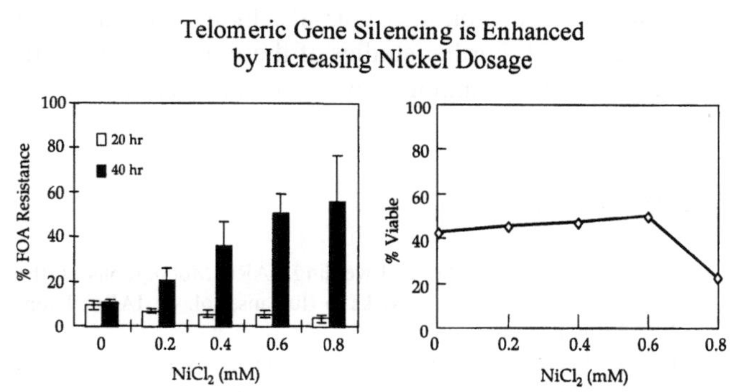

Fig. 10. The effect of $NiCl_2$ on silencing of the *URA3* gene inserted 1.3 kb from a telomere silencing element in yeast. This is shown as the toxicity of $NiCl_2$ in this yeast system.

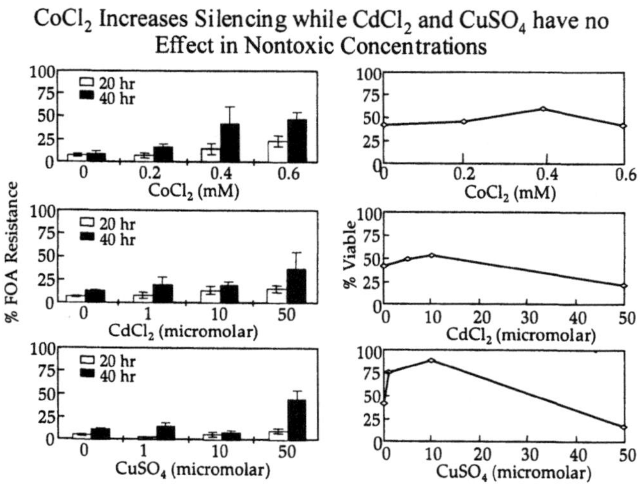

Fig. 11. The effect of $CoCl_2$, $CdCl_2$ and $CuSO_4$ on silencing of the *URA3* gene in yeast and the toxicity of these metals.

[24]. This inhibition of histone acetylation may be responsible for the observed gene silencing in the yeast system. In fact, we have recently shown that pretreatment of yeast cells with trichostatin A, a histone deacetylation inhibitor, which would hyperacetylate the histones, inhibited the ability of nickel to silence genes. We attribute the inhibition of histone acetylation by nickel to the possible binding of nickel to the N-terminal tail of histone H4 which has a metal anchoring histidine site at position 18.

Conclusions

We believe that an important event in nickel-induced carcinogenesis is gene silencing. The loss of expression of tumor suppressor and senescence genes is a critical event in chemical carcinogenesis. In fact, fusion of a normal cell with a cancer cell yields a normal cell, suggesting that cancer cells lose genetic information since normal cells are dominant over cancer cells [25]. Further studies of these mechanisms will enhance our understanding of nickel carcinogenesis.

References

1. IARC: Chromium, nickel and welding. IARC Monographs on the Evaluation of Carcinogenic Risks to Humans, vol. 49. IARC, Lyon: 1–677, 1990
2. Doll R, Morgan LG, Speizer FE: Cancers of the lung and nasal sinuses in nickel workers. Br J Cancer 24: 623–632, 1970
3. Doll R, Mathews JD, Morgan LG: Cancers of the lung and nasal sinuses in nickel workers: A reassessment of the period of risk. Br J Ind Med 34: 102–105, 1977
4. Dunnick JK, Elwell MR, Radovsky AE, Benson JM, Hahn FF, Nikula KJ, Barr EB, Hobbs CH: Comparative carcinogenic effects of nickel subsulfide, nickel oxide, or nickel sulfate hexahydrate chronic exposures in the lung. Cancer Res 55: 5251–5256, 1995
5. Costa M, Mollenhauer HH: Phagocytosis of nickel subsulfide particles during the early stages of neoplastic transformation in tissue culture. Cancer Res 40: 2688–2694, 1980
6. Costa M: Molecular mechanisms of nickel carcinogenesis. Toxicol Environ Chem 49: 145–148, 1995
7. Abbracchio MP, Heck JD, Costa M: The phagocytosis and transforming activity of crystalline metal sulfide particles are related to their negative surface charge. Carcinogenesis 3: 175–180, 1982
8. Heck JD, Costa M: Surface reduction of amorphous NiS particles potentiates their phagocytosis and subsequent induction of morphological transformation in Syrian hamster embryo cells. Cancer Lett 15: 19–26, 1982
9. Costa M: Sequential events in the induction of transformation in cell culture by specific nickel compounds. Biol Trace Element Res 5: 285–295, 1983
10. Abbracchio MP, Simmons-Hansen J, Costa M: Cytoplasmic dissolution of phagocytized crystalline nickel sulfide particles: A prerequisite for nuclear uptake of nickel. J Toxicol Environ Health 9: 663–676, 1982
11. Costa M, Simmons-Hansen J, Bedrossian CW, Bonura J, Caprioli RM: Phagocytosis, cellular distribution, and carcinogenic activity of particulate nickel compounds in tissue culture. Cancer Res 41: 2868–2876, 1981
12. Kargacin B, Klein CB, Costa M: Mutagenic responses of nickel oxides and nickel sulfides in Chinese hamster V79 cell lines at the xanthine-guanine phosphoribosyl transferase locus. Mutat Res 300: 63–72, 1993
13. Conway K, Wang XW, Xu LS, Costa M: Effect of magnesium on nickel-induced genotoxicity and cell transformation. Carcinogenesis 8: 1115–1121, 1987
14. Sen P, Costa M: Induction of chromosomal damage in Chinese hamster ovary cells by soluble and particulate nickel compounds: Preferential fragmentation of the heterochromatic long arm of the X-chromosome by carcinogenic crystalline NiS particles. Cancer Res 45: 2320–2325, 1985
15. Sen P, Costa M: Pathway of nickel uptake influences its interaction with heterochromatic DNA. Toxicol Appl Pharmacol 84: 278–285, 1986
16. Klein CB, Conway K, Wang XW, Bhamra RK, Lin XH, Cohen MD, Annab L, Barrett JC, Costa M: Senescence of nickel-transformed cells by an X chromosome: Possible epigenetic control. Science 251: 796–799, 1991
17. Klein CB, Rossman TG: Transgenic Chinese hamster V79 cell lines which exhibit variable levels of gpt mutagenesis. Environ Mol Mutagen 16: 1–12, 1990
18. Klein CB: The genotoxicity and mutagenic mechanisms of Beta-propriolactone in Chinese hamster and bacterial cells. Ph.D. Thesis. New York University, New York, 1988
19. Klein C, Snow E: Localization of the gpt sequence in transgenic G12 cells by fluorescent *in situ* hybridization. Environ Mol Mutagen 21: 35, 1993
20. Lee YW, Pons C, Tummolo DM, Klein CB, Rossman TG, Christie NT: Mutagenicity of soluble and insoluble nickel compounds at the gpt locus in G12 Chinese hamster cells. Environ Mol Mutagen 21: 365–371, 1993
21. Lee Y-W: Molecular mechanisms of 6-thioguanine resistance induced by carcinogenic nickel compounds in the transgenic G12 Chinese hamster cells: Relevance to an epigenetic mechanism of nickel carcinogenesis. Ph.D. Thesis. New York University, New York, 1995

22. Lee YW, Klein CB, Kargacin B, Salnikow K, Kitahara J, Dowjat K, Zhitkovich A, Christie NT, Costa M: Carcinogenic nickel silences gene expression by chromatin condensation and DNA methylation: A new model for epigenetic carcinogens. Mol Cell Biol 15: 2547–2557, 1995
23. Broday L, Cai J, Costa M: Nickel enhanced telomeric silencing in *Saccharomyces cerevisiae*. Mutat Res 440: 121–130, 1999
24. Broday L, Peng W, Kuo M-H, Salnikow K, Zoroddu M, Costa M: Nickel compounds are novel inhibitors of histone H4 acetylation. Cancer Res 60: 238–241, 2000
25. Johnson TL, Moyer MP: Normal human colon cells suppress malignancy when fused with colon cancer cells. *In Vitro* Cell Dev Biol 26: 1095–1100, 1990

Molecular and Cellular Biochemistry **222:** 213–219, 2001.

Model reactions of Cr (VI) with DNA mediated by thiol species

Dmitriy Krepkiy,[1] William E. Antholine,[2] Charles Myers[3] and David H. Petering[1]

[1]*Department of Chemistry, University of Wisconsin-Milwaukee, Milwaukee, WI;* [2]*Department of Radiation Biology and Biophysics, National Biomedical ESR Center, Medical College of Wisconsin, Milwaukee, WI;* [3]*Department of Pharmacology and Toxicology, Medical College of Wisconsin, Milwaukee, WI, USA*

Abstract

Model reactions were devised to investigate the capacity of physiologically interesting thiol compounds to mediate reactions between CrO_4^{2-} (Cr (VI)) and DNA. The sulfhydryl containing reagents included cysteine, glutathione, apo-metallothionein (apoMT). Zinc finger 3 of transcription factor IIIA (Zn-F_3) of *Xenopus laevis* was investigated as a potential redox active site of reaction of Cr (VI) and thiol compounds. The DNA samples were calf thymus DNA and two oligomers, one of them specific for binding Zn-F_3. Results showed that in the presence of Cr (VI) apoMT readily participated in damaging DNA in a reaction that appeared to be hydroxyl radical dependent. It also became cross-linked to oligomer and native DNA samples. In comparison, the other two thiol donors were largely inactive in these assays even though they, like apoMT, were able to reduce Cr (VI) to Cr (III) under the conditions of the experiments. Direct attempts to cross link thiols with DNA in the presence of Cr^{3+} were unsuccessful at pH 7.4. Together, the results indicate that apoMT can effectively collaborate with Cr (VI) in reactions that are deleterious to DNA. (Mol Cell Biochem **222**: 213–219, 2001)

Key words: chromate, DNA, thiols, metallothionein, glutathione

Introduction

Chromium is a metal used extensively for plating of steel, to formulate paint pigments, and in the tanning industry. As CrO_4^{2-} or Cr (VI), it is mutagenic and carcinogenic [1]. These effects are thought to arise as a result of intracellular reduction coupled with modification of DNA [2]. Thus, *in vitro* and *in vivo* studies show that exposure to Cr (VI) under reductive conditions results in alteration of gene expression, DNA strand breaks, guanine hydroxylation, generation of abasic sites, and formation of Cr (III) mediated cross links between DNA and other cellular nucleophiles [3].

The features of reduction of Cr (VI) to Cr (III) have been investigated using a variety of reducing agents including ascorbate, thiols, and NAD(P)H via redox enzymes such as cytochrome P-450 reductase coupled to cytochrome b_5 [4–6]. Several *in vitro* model systems have also been used to characterize oxidative DNA damage during the reductive activation of Cr (VI) [7–10]. Nevertheless, much remains to be resolved about the underlying chemistry of these reactions.

In this context, experiments were carried out in order to evaluate model reactions for Cr (VI) dependent DNA damage involving thiol compounds as reducing agents. In this work, we used small oligomers of defined sequence, proteins that hypothetically might directly bind to chromium species, and thiol compounds that might reasonably be present in biological systems.

These included apometallothionein, the metal free form of the thiol rich metal binding protein. In addition, we explored the reactivity of Cr^{3+} with DNA and sulfhydryl compounds (RSH) as a basis to consider how RS-Cr (III)-DNA cross-linking might occur in cells.

Address for offprints: D.H. Petering, Department of Chemistry, University of Wisconsin-Milwaukee, Milwaukee, WI 53201, USA

Metallothionein (MT) was investigated because it has been implicated in a variety of reactions involving toxic metals and oxidant species [11]. Zn-MT is relatively unreactive with Cr (VI), but apoMT serves as an effective reductant [12]. Originally thought to be unstable in cells, apoMT has now been identified in a variety of tumors and tumor cell lines [13]. The extent of its presence and role in normal cells is a matter of ongoing study.

The interaction of Cr (VI) with a zinc finger peptide-oligomer adduct under reductive conditions was also examined. It has been hypothesized that zinc fingers which are natural metal binding sites for toxic metal ions and are bound to DNA might be preferentially reactive with toxic metals leading to modifications of DNA- protein interactions [14]. Finger 3 of transcription factor IIIA of *Xenopus laevis* was chosen for this study as a model of a zinc finger binding motif that contains two cysteine and two imidazole metal binding site.

Materials and methods

Synthesis of zinc finger 3 of transcription factor IIIA from Xenopus laevis

Zn finger 3 (Zn-F_3) was made and purified by peptide synthesis as previously described [15]. Its integrity was confirmed by amino acid analysis and electrospray mass spectroscopy. A one to one ratio of Zn^{2+} to apo-peptide sulfhydryl groups was used to constitute the holoprotein.

Preparation of DNA oligomers and the HPLC analysis of the products of oligonucleotide degradation

The 10-mer self-complementary oligonucleotide, d(GGAAGCTTCC)$_2$, and the 5S ribosomal RNA C box, d(3′-ACCTACCCTCTGG-5′, 5′-TGGATGGGAGACC-3′), that is part of its internal control region were synthesized using a DNA synthesizer (Millipore Cyclone Plus DNA synthesizer) and purified with HPLC and G-15 gel-filtration chromatography based on a published procedure [16]. In order to investigate oligonucleotide damage, d(GGAAGCTTCC)$_2$ was incubated with thiol and Cr (VI) over time. Some precipitation occurred rapidly after mixing the oligonucleotide with apoMT and Cr (VI). The supernatant was loaded onto an ion exchange column (Vydac 301 VHP375 DEAE column, 7.5 mm × 10 cm) and eluted with a linear gradient of (a) 10 mM sodium phosphate, pH 7.4, and (b) 10 mM sodium phosphate, 0.5 M NaCl, pH 7.4. The peaks were detected at 260 nm.

Radio-labeling of the oligonucleotide and PAGE experiments

d(GGAAGCTTCC)$_2$ was radio-labeled with ^{32}P from ATP using T4 polynucleotide kinase (Promega) as previously described [16]. The unincorporated ^{32}P γ ATP was separated by repeated precipitation of the oligomer in 70% EtOH. The oligonucleotide was resuspended in 20 μL of 0.18 M Tris/Cl buffer, pH 7.4, and the radioactivity counted by liquid scintillation analysis (Packard), using Bio-Safe II scintillation fluid (Research Products).

The products of the reaction of oligonucleotide with Cr (VI) and apoMT or other thiols were loaded on the 8% denaturing polyacrylamide gel with 7 M urea and submitted to electrophoresis at 200 V for 1.5 h in 1X TB (Tris-borate) buffer. At its completion, the gel was frozen at –80°C and then exposed to X-ray (Fuji) film. The film was then developed and the bands analyzed with the Kodak Digital Science 1 software. In the electrophoretic mobility shift assay experiments, the complex between the C box portion of the Internal Control Region (ICR) of the 5S rRNA gene and Zn-F_3, the third zinc finger of transcription factor IIIA, was formed in the 1X binding buffer in the presence of 0.7 μg poly[d(I-C)]. The bound and unbound oligonucleotide were separated in 8% non-denaturing PAGE after electrophoresis for 1.5 hour in 1X TB (Tris-borate) buffer at 200 V. The gel was then dried and X-ray film (Fuji) was exposed to the gel.

Formation of the calf thymus DNA-thiolate cross-links

The formation of the cross-links of apo-MT with DNA was investigated by another method [17]. Calf thymus DNA (100 μL of 2 mg/mL) was incubated with CrO_4^{2-} of different concentrations of GSH or apoMT for 2 h at 37°C at pH 7.5 in the volume of 300 μL. In order to precipitate DNA, 700 μL of EtOH was added and the mixture was left overnight at –20°C. Under this condition, thiol compounds not bound to DNA remained in the supernatant. The resultant suspension was centrifuged at 13,000 g for 30 min and its supernatant discarded. The pellet was hydrolyzed after the addition of 40 μL of 0.2 M NaOH during a 2 h incubation at 37°C. A 20 μL aliquot was loaded on the C18 reverse phase analytical HPLC column (Bio-Rad R.- 318 250 mm × 4.6 mm) and eluted with a linear H_2O-MeOH gradient. The eluate was analyzed by the absorbance at 215 nm, using authentic cysteine and apoMT hydrolysate to define the elution time of the cysteine peak. The amount of cysteine was then calculated from the calibration curve after integration of the cysteine peak.

NMR spectroscopy

NMR data were acquired on a Bruker DRX500 NMR spectrometer. All samples contained 10% D_2O (Sigma) and spectra were recorded in 100 mM NaCl that was self-buffered at pH 7.4 in the experiments with 3,3′D2Cys. Two-dimensional COSY, TOCSY, and NOESY spectra were obtained in the presence of water suppression and analyzed as described previously to determine whether the oligonucleotide remained intact after reaction with Cr (VI) and cysteine [18].

ESR spectroscopy of the complex formation of Cr (III) with different ligands

X-band EPR spectra of Cr (III) were recorded at 9°K in liquid helium on the Varian E4 spectrometer with the following parameters: microwave power 20 dB, frequency 9.2469 GHz, time constant 0.064, scan time 4 min, modulation amplitude 5G, modulation frequency 100 KHz, receiver gain 2000. The g values were calculated based on a calibration with 1,1-diphenyl-2-picrylhydrazyl (g = 2.0037).

Results

Oxidative DNA modification during the reaction of Cr (VI) with thiols

Cr (VI) is considered to be carcinogenic upon reduction by cellular reductants, possibly including thiols such as cysteine and glutathione. Although DNA is considered the target of these reactions, the underlying chemistry remains to be established. The reactions of a small oligonucleotide, $d(GGAAGCTTCC)_2$ with Cr (VI) and either cysteine (Cys), glutathione (GSH), or apoMT were examined using anion exchange HPLC to analyze the integrity of DNA. According to Fig. 1a, incubation of the 10-mer with Cr (VI) with either 1–5 mM Cys or GSH failed to alter its chromatographic profile over time. In contrast, substitution of an equivalent concentration of thiol in the form of apoMT (20 sulfhydryl groups per molecule) progressively altered the oligomer. Initially, there seemed to be specificity to the reaction since only 1 major product was detected after 12 h of incubation (Fig. 1b). Later, a number of products were detected (Fig. 1c).

Strand cleavage or possibly cross-linking might have occurred in this reaction by direct reaction between a chromium species and DNA or indirectly after generation of a reactive oxygen species such as hydroxyl radical. In order to discriminate between these possibilities, a potent hydroxyl radical scavenger, 10% (v/v) dimethyl sulfoxide, was included in the reaction mixture. In its presence, detectable DNA damage was almost completely eliminated (Fig. 1d). Thus, it was concluded that apoMT dependent DNA damage involved the generation of OH$^{\bullet}$, probably through a Fenton-like reaction of an intermediate, reduced species of Cr with H_2O_2. It is hypothesized that frank DNA strand cleavage occurred in this reaction.

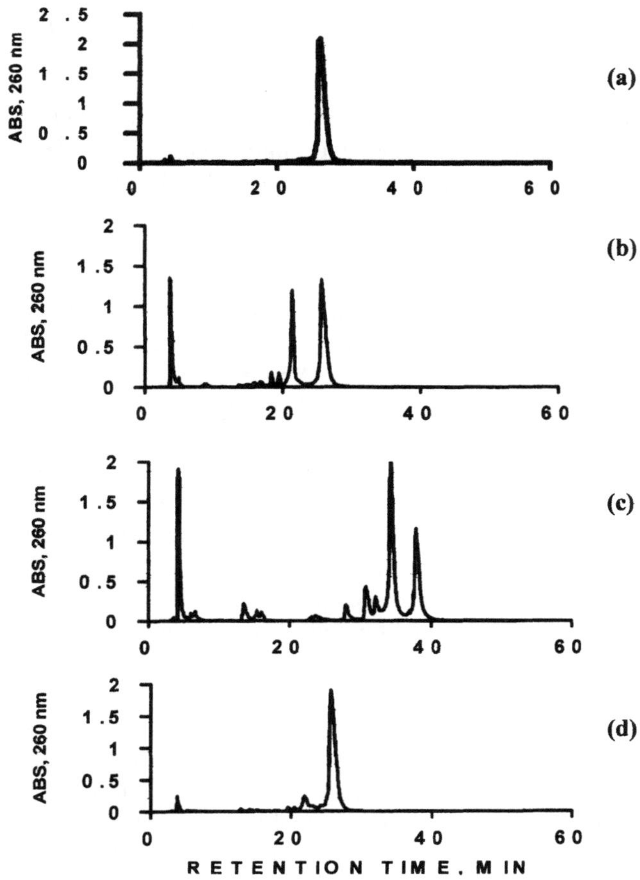

Fig. 1. Reaction of $d(GGAAGCTTCC)_2$ with apoMT and Cr (VI). Condition: 250 µM $d(GGAAGCTTCC)_2$, 50 µM apoMT, and 250 µM (Cr (VI), in 2.5 mM phosphate buffer, pH 7.4. (a) Native oligonucleotide or oligonucleotide + Cr (VI) + GSH or Cys; (b) oligonucleotide + Cr (VI) + apoMT, 12 h incubation; (c) oligonucleotide + Cr (VI) + apoMT, 24 h incubation; (d) oligonucleotide + Cr (VI) + apoMT + 10% DMSO, 24 h incubation.

It was possible that Cr (VI) and cysteine or glutathione caused other DNA lesions, which left the backbone intact. The one-dimensional ^1H NMR spectra of the reactions mixtures of $d(GGAAGCTTCC)_2$ with Cr (VI) and either of these thiols were obtained. Figure 2 shows the example of cysteine. It was evident with both thiol reductants, that the only changes in the NMR spectrum of the reaction mixture were due to resonances attributed to the reducing agent. In a more refined examination of the oligomer by 2-dimensional NMR spectroscopy as described in the Materials and methods, there was no indication of modification of the structure. Therefore, it was concluded that the native oligomer remained unchanged after the reduction of Cr (VI) by cysteine or glutathione.

DNA cross-linking to thiol species during the reaction of Cr (VI) with thiols

Costa and coworkers showed that in cells exposed to Cr (VI) DNA became cross-linked with amino acids [17]. Cysteine was the predominant amino acid bound to DNA. Similar reactions were carried out as above, except that smaller concentrations of d(GGAAGCTTCC)$_2$, Cr (VI) and thiol compounds were used. In order to observe DNA damage under these conditions, an electrophoretic mobility shift assay was employed. The striking finding was the evidence for apoMT-DNA cross-linking in Fig. 3. DNA migration was substantially slowed upon reaction with Cr (VI) and apoMT. Interestingly, as the concentration of apoMT increased, it appeared that more than one MT protein molecule became bound to the DNA structure. Thus, these two sets of conditions led to either strand scission or cross-linking. Experiments are underway to solidify the identification of the products and understand the key variables that favor each pathway.

Cysteine, substituted for apoMT, caused no indication of DNA strand scission or retardation of oligonucleotide movement. If cysteine-DNA cross-linking occurred, the adduct may have been too small to noticeably slow the migration of the DNA sample. In fact, when glutathione was substituted for cysteine a very small retardation seemed to occur.

Finally, reactions of Cr (VI), apoMT and calf thymus DNA were carried out with the intent to see whether the protein became cross-linked to DNA. According to Fig. 4, a substantial elevation in cysteine bound to DNA was detected after these reactions. Under these conditions which included thorough washing of the product to remove unbound protein, the extent of association of cysteine with DNA was dependent upon the apoMT but not the Cr (VI) concentration.

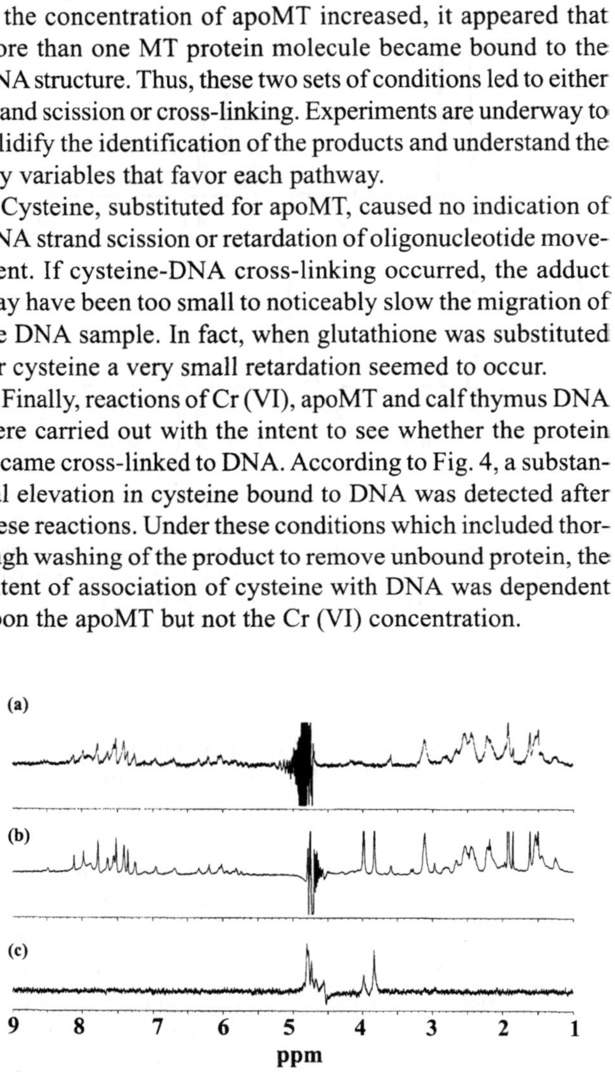

Fig. 2. One-dimensional ^{1}H NMR spectra of reaction of d(GGAAGCTT-CC)$_2$ with cysteine and Cr (VI). Five hundred MHz ^{1}H NMR spectra of 150 μM d(GGAAGCTTCC)$_2$ in 0.1 M NaCl, 20 mM phosphate buffer, pH 7.4, treated with 30 mM deuterated Cys (3, 3′D2) and 5 mM Cr (VI). (a) 5 mM Cr (VI) + 150 M oligo, 1024 scans; (b) 5 mM Cr (VI) + 30 mM Cys (D$_2$) + 150 M oligo after 24 h of incubation, 2048 scans; (c) 5 mM Cr (VI) + 30 mM Cys (D$_2$), 0 h, 1024 scans.

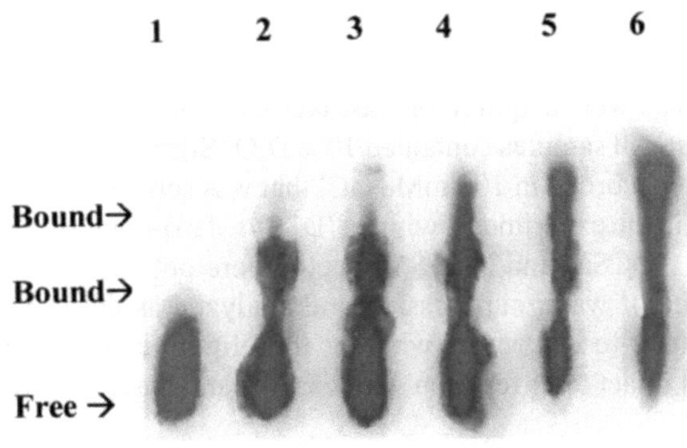

Fig. 3. Formation of the DNA-protein cross-links during the reaction of d(GGAAGCTTCC)$_2$ with apoMT with Cr (VI). Conditions: 5 μM ^{32}P labeled oligonucleotide, 2 mM Cr (VI), and various concentrations of apoMT, self-buffered at pH 7.4 for 2 h at 37°C. Lines 1–8: 1: no apoMT; 2: 1 μM; 3: 2.5 μM; 4: 5 μM; 5: 10 μM; 6: 20 μM apoMT.

Reaction of Cr (VI) with thiols and a model zinc finger-DNA adduct

The nine zinc finger protein, transcription factor IIIA, binds tightly to the internal control region (ICR) of the 5S rRNA gene, with the first 6 zinc finger domains making the key recognition and stabilizing interactions with DNA [19]. Synthetic zinc finger 3 (F_3) forms a specific complex with the C box which displays enough stability to be measured according to the electrophoretic mobility shift assay (Fig. 5:1). This binding was dependent upon the presence of Zn^{2+} in Zn-F_3 as shown by the ability of the zinc chelator, zincon, to inhibit binding (Fig. 5:2). Although oxidized or reduced glutathione alone had no effect on the complex (Fig. 5:3–5), adduct formation was abolished when glutathione and Cr (VI) were jointly present (Fig. 5:6). The reaction was dependent upon the concentration of Cr (VI) as well as glutathione (data not

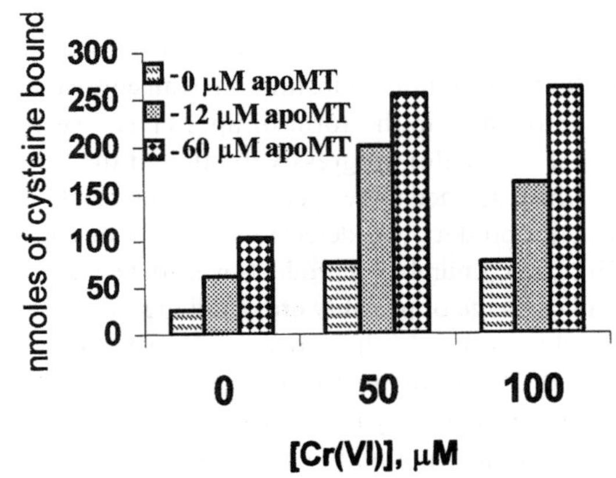

Fig. 4. Calf thymus DNA cross linking with apoMT. Conditions: calf thymus DNA (100 μL of 2 mg/mL) was incubated with CrO_4^{2-} and apoMT for 2 h at 37°C at pH 7.5 in 300 μL.

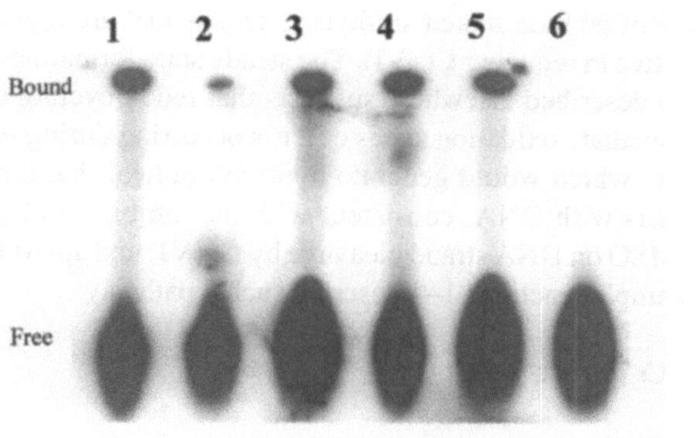

Fig. 5. Reaction of Zn-F_3-C box complex with glutathione (GSH) and Cr (VI). Conditions: Preformation of complex between 2.5 μM oligonucleotide and 30 μM F_3 in the 1X binding buffer in the presence of 0.7 μg poly[d(I-C)]. Lanes: (1) no GSH; (2) 5 mM Zincon; (3) 30 mM GSSG; (4) 15 mM GSH; (5) 30 mM GSH; (6) 30 mM GSH + 15 mM Cr (VI); 2 h incubation.

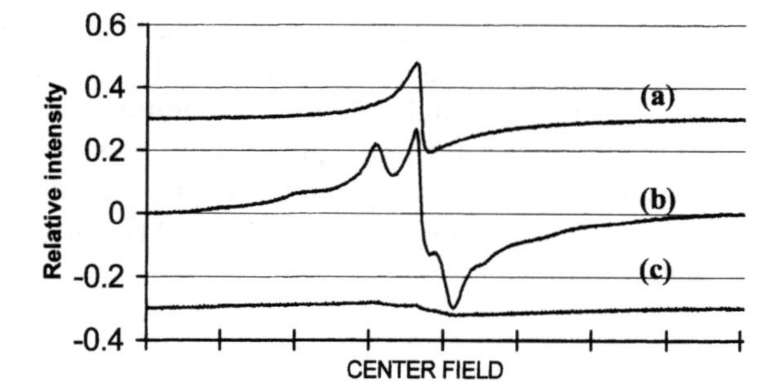

Fig. 6. ESR spectra of Cr^{3+} in the absence of presence of d(GGAAGC-TTCC)$_2$. Conditions: (a) 1 mM Cr^{3+} as $CrCl_3$, (b) 1 mM Cr^{3+} + 30 μM oligonucleotide in 0.1 M NaCl, 0.2 mM phosphate buffer, pH 5.5. ESR parameters: T = 12°K, field set 3400 G, scan range 2000 G, microwave power 20 dB, frequency 9.2469 GHz, time constant 0.064, scan time 4 min, modulation amplitude 5 G, modulation frequency 100 KHz, receiver gain 2000.

shown). Cysteine had no effect on the mobility of the C box or its complex with by Zn-F_3 (data not shown). Thus, under thiol reducing conditions Cr either disrupted the ability of Zn-F_3 to bind to the oligomer or altered the oligomer so that it no longer could interact with Zn-F_3. ApoMT, itself, readily removes Zn from Zn-F_3, so it could not be employed in this type of assay system to assess effects upon zinc finger-DNA binding.

Interaction of Cr^{3+} and Cr (III)-thiolate complexes with DNA

Experiments were done to explore the interaction of Cr (III) species with DNA, using d(GGAAGCTTCC)$_2$ as a target oligonucleotide. Cr^{3+} precipitates from solution at pH 7.4, so experiments with the aquated metal ion were conducted at pH 5.5. Its ESR spectrum is shown in Fig. 6a. Upon addition of a substoichiometric concentration of DNA, the original Cr (III) spectrum changed, indicative of an interaction between the metal ion and the oligomer (Fig. 6b). At a larger 10-mer concentration, only a broad signal was detected that is thought to represent binding of Cr^{3+} to the phosphodiester linkages (Fig. 6c). Assuming that Cr^{3+} favors this interaction, then the ESR signal in Fig. 6b may represent Cr (III)-base interactions formed under the condition of saturation of backbone sites with Cr (III). When the pH of the reaction mixtures was raised to 7.4, precipitate formed slowly and the ESR signal of the supernatant was lost, consistent with the disappearance of Cr (III) from solution.

According to Fig. 7, the one-dimensional ^1H NMR titration of the DNA 10-mer with Cr^{3+} at pH 5.5 led to the changes in intensity of the degenerate imino protons of the bases but not to their splitting. The onset of these effects was detected at 100–150 μM metal ion or at a ratio of approximately 1–10 Cr^{3+} to base pairs. According to the spectral perturbations, the Cr (III)-DNA interaction was fast exchange on the NMR time scale and non-specific in nature such that chemical shift differences of the imino protons were not induced. Upon raising the pH to 7.4, a precipitate slowly formed that was thought to be polymeric Cr (III).

In order to overcome the insolubility of Cr^{3+} at neutral pH, it was first reacted with cysteine at pH 5.5. An octahedral complex was formed in the presence of a large excess of cysteine as indicated by the ESR spectrum of the product (Fig. 8a). Cr (III)-(cysteine)$_n$ remained soluble at pH 7.4 for hours before precipitating and displayed a new ESR signature that represent several Cr (III) species (Fig. 8b). At this pH, the mixture of Cr (III)-thiolate species was unreactive with d(GGAAGCTTCC)$_2$ according to an NMR titration experiment similar to that in Fig. 7.

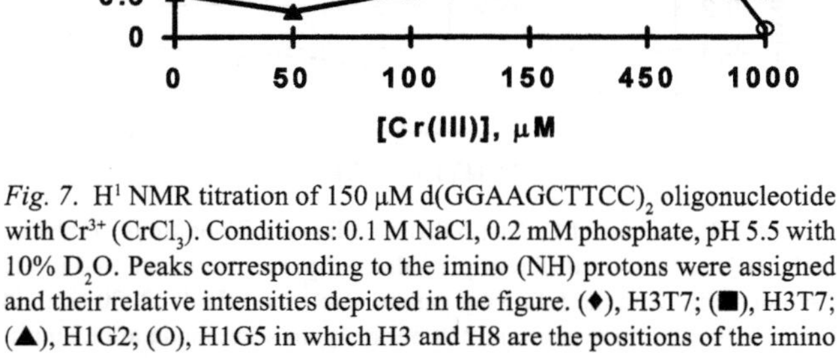

Fig. 7. H^1 NMR titration of 150 μM d(GGAAGCTTCC)$_2$ oligonucleotide with Cr^{3+} ($CrCl_3$). Conditions: 0.1 M NaCl, 0.2 mM phosphate, pH 5.5 with 10% D_2O. Peaks corresponding to the imino (NH) protons were assigned and their relative intensities depicted in the figure. (♦), H3T7; (■), H3T7; (▲), H1G2; (O), H1G5 in which H3 and H8 are the positions of the imino protons on the heterocycle and T7, G2, and G5 are the sequence numbers of the bases from the 5′ end.

Cr^{3+} also binds to apoMT at pH 5.5 and 7.4 (data not shown). With this multidentate thiolate structure, Cr (III) is stable indefinitely at either pH under anaerobic conditions. In the presence of air at pH 7.4, Cr (III) precipitates, presumably as the metallothionein sulfhydryl groups oxidize. There was no reaction between the oligomer and the Cr-MT complex either at pH 5.5 or 7.4.

Discussion

The results described here show that DNA damage as well as cross-linking of DNA with thiols by Cr can be demonstrated in simple model systems consisting of Cr (VI), a defined oligomer in the absence or presence of a DNA binding peptide, and a thiol reductant. In these experiments, DNA was unreactive with Cr (VI) or Cr (III) in the presence of cysteine acting either as a thiol reductant or metal binding agent. Nevertheless, under the conditions of these experiments, Cr (VI) reduction does take place, showing that the details of the reductive pathway are important in determining whether DNA becomes a target for reaction.

Glutathione only displayed unequivocal reactivity in the Zn-F_3-DNA system, where in the presence of Cr (VI) it inhibited the reaction binding of the Zn-peptide and cognate DNA oligomer (Fig. 5). Conceivably, the redox reaction between Cr (VI) and GSH led to the oxidation of the two cysteines at the Zn binding site of F_3, liberating Zn^{2+} and unfolding the peptide so that it did not recognize the oligomer. Alternatively, a reduced species of Cr (VI) such as Cr (III) might have directly competed with Zn for the F_3 metal binding site and perturbed the conformation of the peptide such that it could no longer bind specifically to the DNA sample.

ApoMT was much more reactive in these model systems. In every one, it either collaborated with Cr (VI) to produce DNA strand cleavage or a DNA-protein cross-linked species (Figs 1, 3 and 4). Its reactivity is attributed to the high density of peptide linked sulfhydryl groups that are highly effective in reducing Cr (VI). The steady state formation of Cr (V) described elsewhere suggests that redox cycling of intermediate oxidation states of Cr is occurring during reduction, which would generate hydroxyl radical that directly reacts with DNA, consistent with the inhibitory effect of DMSO on DNA strand cleavage by Cr (VI) and apoMT. For example reactions 1–4 describe such a pathway:

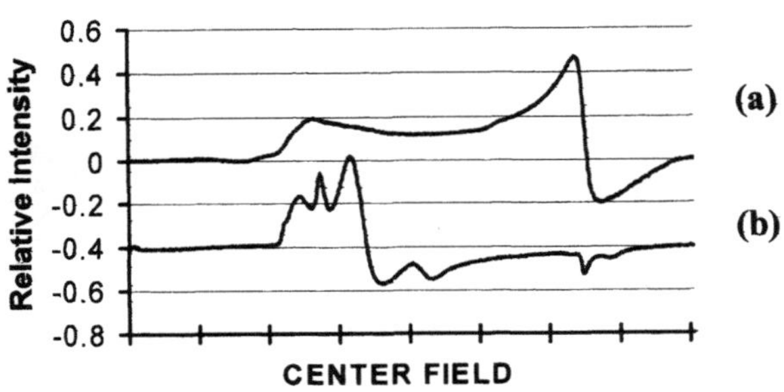

Fig. 8. ESR spectra of Cr^{3+} with different ligands. Conditions: 5 mM Cr (III), 100 mM Cys at different pH values). Spectra were recorded in the helium cavity with the following parameters: center field – 2000 G, field range – 4000 G, time constant 250 msec, 4 min scan, modulation amplitude 5G, modulation frequency 100 KHz, receiver gain – 2000, T = 8°K, power – 13 dB, frequency – 9.2722 GHz. (a) pH 5.5; (b) pH 7.4.

$$Cr\ (V) + e^- \rightarrow Cr\ (IV) \quad (1)$$

$$Cr\ (IV) + O_2 \rightarrow Cr\ (V) + O_2^{\cdot -} \quad (2)$$

$$2\ O_2^{\cdot -} + 2H^+ \rightarrow H_2O_2 + O_2 \quad (3)$$

$$Cr\ (IV) + H_2O_2 \rightarrow Cr\ (V) + OH^{\cdot} + OH^- \quad (4)$$

There was also electrophoretic evidence that apoMT became cross-linked with a DNA oligonucleotide during reaction with Cr (VI) (Fig. 3). The generation of stable MT-DNA cross links was also supported by the direct observation of cysteine in hydrolysates of calf thymus DNA isolated after reaction with apoMT and Cr (VI) (Fig. 4). In each case, a first hypothesis would be that an apoMT-Cr (III)-DNA linked complex had formed in which Cr (III) coordinates both thiolate and DNA phosphate or base ligands [20]. It does not seem likely that this occurred at pH 7.4 through the direct reaction of the terminally reduced species, Cr^{3+}, with apoMT and DNA:

$$Cr^{3+} + apoMT\text{-}SH_m + DNA \rightarrow MT\text{-}S_m\text{-}Cr\ (III)\text{-}DNA \quad (5)$$

The reasons for skepticism are the insolubility of Cr^{3+} at pH 7.4, the aerobic instability of Cr (III)-MT at pH 7.4, and the lack its reaction with DNA under this condition, presumably, because of the formation of a more stable Cr (III)-MT species.

These concerns extend to other plausible thiol donors in reaction 5 such as cysteine and glutathione. According to experiments designed to examine this reaction with these sulfhydryl reagents, Cr (III)-thiolate complexes are even less stable at neutral pH. Thus, at present an alternative hypothesis is extended to explain cross-linking. An intermediate oxidation state of Cr binds to DNA as well as to MT and this intermediate reaches the final cross-linked state through an internal oxidation-reduction reaction:

$$MT\text{-}S_p\text{-}Cr\ (IV,V)\text{-}DNA \rightarrow MT\text{-}S_r\text{-}Cr\ (III)\text{-}DNA \quad (6)$$

According to this reaction, the pathway may dictate a different product than observed in reaction 5, which is kinetically stable at pH 7.4. A caution about this interpretation is that measurement of the presence of Cr in the final adduct has not been undertaken.

These experiments point to some features of chromium reactions with thiols and DNA that leave doubt about their relevance to the observed toxic effects of Cr (VI). Notably, large concentrations of Cr (VI) in relation to DNA and long incubation times are used to achieve the observed reactions. What these systems do show is that Cr (VI) is not a highly reactive species *in vitro* and that more work will be necessary to link chemical studies to the pathological effects of Cr (VI) *in vivo*. The present experiments provide new opportunities to explore mechanisms by which Cr (VI) reaction with DNA can be achieved.

Acknowledgments

The authors acknowledge the support of NIH grants ES-04026, ES-04184, and RR01008 (National Biomedical ESR Center).

References

1. O'Brien P, Kortenkamp A: Chemical models important in understanding the ways in which chromate can damage DNA. Environ Health Perspect 102: 3–10, 1994
2. Levina A, Barr-David G, Codd R, Lay PA, Dixon NE, Hammershoi A, Hendry P: *In vitro* plasmid DNA cleavage by chromium (V) and (IV) 2-hydroxycarboxylato complexes. Chem Res Toxicol 12: 371–381, 1999
3. Shi X, Ding M, Ye J, Wang S, Leonard SS, Zang L, Castranova V, Vallyathan V, Chiu A, Dalal N, Liu KJ: Cr (IV) causes activation of nuclear transcription factor-kappa B, DNA strand breaks and dG hydroxylation via free radical reactions. Inorg Biochem 75: 37–44, 1999
4. Myers CR, Myers JM: Iron stimulates the rate of reduction of hexavalent chromium by human microsomes. Carcinogenesis 19: 1029–1038, 1998
5. Stearns DM, Courtney KD, Giangrande PH, Phieffer LS, Wetterhahn KE: Chromium (VI) reduction by ascorbate: Role of reactive intermediates in DNA damage *in vitro*. Environ Health Perspect 102: 21–25, 1994
6. Wiegand HJ, Ottenwalder H, Bolt HM: The reduction of chromium (VI) to chromium (III) by glutathione: an intracellular redox pathway in the metabolism of the carcinogen chromate. Toxicology 33: 341–348, 1984
7. Sugden KD, Wetterhahn KE: Direct and hydrogen peroxide-induced chromium (V) oxidation of deoxyribose in single-stranded and double-stranded calf thymus DNA. Chem Res Toxicol 10: 1397–1406, 1997
8. Casadevall M, da Cruz Fresco P, Kortenkamp A: Chromium (VI)-mediated DNA damage: Oxidative pathways resulting in the formation of DNA breaks and abasic sites. Chem Biol Interact 123: 117–132, 1999
9. Liu S, Medvedovic M, Dixon K: Environ Mutational specificity in a shuttle vector replicating in chromium (VI)-treated mammalian cells. Mol Mutagen 33: 313–319, 1999
10. Luo H, Lu Y, Mao Y, Shi X, Dalal NS: Role of chromium (IV) in the chromium (VI)-related free radical formation, dG hydroxylation, and DNA damage. J Inorg Biochem 64: 25–35, 1996
11. Petering D, Fowler B: Roles of metallothionein and related proteins in metal metabolism and toxicity: Problems and perspectives. Environ Health Perspect 65: 217–224, 1986
12. Krepkiy D, Petering DH: unpublished information
13. Pattanaik A, Shaw CF III, Petering DH, Garvey J, Kraker AJ: Basal metallothionein in tumors: Widespread presence of apoprotein. J Inorg Biochem 54: 91–105, 1994
14. Sarkar B: Metal replacement in DNA-binding zinc finger proteins and its relevance to mutagenicity and carcinogenicity through free radical generation. Nutrition 11: 646–649, 1995
15. Muñoz A, Laib F, Petering DH, Shaw CF III: Characterization of the cadmium complex of peptide 49–61: A putative nucleation center for cadmium-induced folding in rabbit liver metallothionein IIA. J Biol Inorg Chem 4: 495–507, 1999
16. Sambrook J, Fritsch EF, Maniatis T: In: Molecular Cloning (A Laboratory Manual), 2nd edn. Cold Spring Harbor Laboratory Press, 1989
17. Zhitkovich A, Voitkun V, Costa M: Glutathione and free amino acids form stable complexes with DNA following exposure of intact mammalian cells to chromate. Carcinogenesis 16: 907–913, 1985
18. Mao Q, Fulmer P, Li W, DeRose EF, Petering DH: Different conformations and site selectivity of HO_2-Co (III)-bleomycin A_2 and Co (III)-bleomycin A_2 bound to DNA oligomers. J Biol Chem 271: 6185–6191, 1996
19. Zang WQ, Veldhoen N, Romaniuk PJ: Effects of zinc finger mutations on the nucleic acid binding activities of Xenopus transcription factor IIIA. Biochemistry 34: 15545–15552, 1995
20. Voitkun V, Zhitkovich A, Costa M: Cr (III)-mediated crosslinks of glutathione or amino acids to the DNA phosphate backbone are mutagenic in human cells. Nucleic Acids Res 26: 2024–2030, 1998

Molecular and Cellular Biochemistry **222:** 221–229, 2001.

On the mechanism of Cr (VI)-induced carcinogenesis: Dose dependence of uptake and cellular responses

Kejian Liu,[1] John Husler,[2] Jianping Ye,[3] Stephen S. Leonard,[3] Debbie Cutler,[3] Fei Chen,[3] Suwei Wang,[3] Zhuo Zhang,[3] Min Ding,[3] Liying Wang[3] and Xianglin Shi[3]

[1]*College of Pharmacy;* [2]*Department of Earth and Planetary Sciences, University of New Mexico, Albuquerque, NM;* [3]*Pathology and Physiology Research Branch, National Institute for Occupational Safety and Health, Morgantown, WV, USA*

Abstract

Cr (VI) compounds are widely used industrial chemicals and are recognized human carcinogens. The mechanisms of carcinogenesis associated with these compounds remain to be investigated. The present study focused on dose-dependence of Cr (VI)-induced uptake and cellular responses. The results show that Cr (VI) is able to enter the cells (human lung epithelial cell line A549) at low concentration (< 10 μM) and that the Cr (VI) uptake appears to be a combination of saturable transport and passive diffusion. Electron spin resonance (ESR) trapping measurements showed that upon stimulation with Cr (VI), A549 cells were able to generate reactive oxygen species (ROS). The amount of ROS generated depended on the Cr (VI) concentration. ROS generation involved NADPH-dependent flavoenzymes. Cr (VI) affected the following cellular parameters in a dose-dependent manner, (a) activation of nuclear transcription factors NF-κB, and p53, (b) DNA damage, (c) induction of cell apoptosis, and (d) inhibition of cell proliferation. The activation of transcription factors was assessed by electrophoretic mobility shift assay and western blot analysis, DNA damage by single cell gel electrophoresis assay, cell apoptosis by DNA fragmentation assay, and cell proliferation by a non-radioactive ELISA kit. At the concentration range used in the present study, no thresholds were found in all of these cell responses to Cr (VI). The results may guide further research to better understand and evaluate the risk of Cr (VI)-induced carcinogenesis at low levels of exposure. (Mol Cell Biochem **222**: 221–229, 2001)

Key words: Cr (VI), uptake, cellular responses, reactive oxygen species, carcinogenesis

Introduction

Cr (VI) compounds are widely recognized as human carcinogens [1]. They are potent inducers of tumors in experimental animals and active agents in the induction of DNA damage, such as DNA strand breaks [2–7]. Industrial exposure to these compounds is reported to be associated with a higher incidence of human cancer [8, 9]. The carcinogenic effect of Cr (VI) is generally attributed to cell uptake because Cr (VI), in contrast to Cr (III), actively enters cells by the sulfate transport system [9, 10]. Since Cr (VI) does not react with isolated DNA, the reduction of Cr (VI) by cellular reductants has been considered an important step in the mechanism of Cr (VI)-induced carcinogenesis [2]. Various reductants, such as glutathione, ascorbate and glutathione reductase/NADPH, are able to reduce Cr (VI) to Cr (V) [2]. During the reduction process, molecular oxygen is consumed to produce superoxide radical ($O_2^{\bullet -}$), H_2O_2, and hydroxyl radical ($^{\bullet}OH$), collectively called reactive oxygen species (ROS) [2].

Address for offprints: X. Shi, Pathology and Physiology Research Branch, National Institute for Occupational Safety and Health, 1095 Willowdale Road, Morgantown, WV 26505, USA

ROS-mediated reactions are believed to be involved in various pathological processes. Initiation of carcinogenesis can be associated with an excessive generation of ROS. ROS can cause DNA strand breaks, base modicafication, lipid peroxidation, and protein modification, resulting in oxidative stress. Various signaling pathways are involved in the mechanism of ROS-induced oxidative stress, including activation of nuclear transcription factors, NF-κB [11], AP-1 [12], and p53 [13]. These transcription factors are involved in the cell responses to various stimuli [14, 15]. These responses include expression of oncogenes, apoptosis, and alteration in cell growth [14, 15]. Since Cr (VI) is able to generate ROS from cells, it is expected that Cr (VI) may produce similar cell responses through ROS-mediated reactions. In the past several years our laboratories have investigated the Cr (VI)-induced apoptosis [16], activation of nuclear transcription factors, NF-κB [11, 17], AP-1 [12] and p53 [13], and signal transduction pathways involved. ROS-mediated reactions play an important role in these Cr (VI)-induced cell responses.

The present study focuses on dose-dependence of Cr (VI) uptake by the cells and of Cr (VI)-induced free radical generation, inhibition of cell proliferation, DNA damage, activation of nuclear transcription factors, and apoptosis. It is important to examine cell transport and responses potentially critical to Cr (VI)-mediated carcinogenesis in a cellular test system (i.e. human lung epithelial cell line, A549) related to the human targets. In cells, the carcinogenic potential of Cr (VI) can be modified by defense mechanisms, such as cellular antioxidants that can neutralize the ROS generated by Cr (VI) stimulation. Therefore, understanding the dose-responses relationships in cells stimulated by Cr (VI) can help guide further research to better understand and evaluate the risk of Cr (VI)-induced carcinogenesis at low level exposure.

Materials and methods

Materials

Potassium dichromate (Cr (VI)) and 5,5-diethylenetriamine-pentaacetic acid (DMPO) were purchased from Aldrich (Milwaukee, WI, USA). Glutathione (GSH), β-nicotinamide adenine dinucleotide phosphate (NADPH), and N-ethylmaleimide were purchased from Sigma (St. Louis, MO, USA). The spin trap, DMPO, was purified by charcoal decoloration and vacuum distillation and was free of ESR detectable impurities.

Measurements of Cr (VI) uptake by cells

Human lung epithelial cell line A549 was used. The study was conducted in 6 well culture plate with 1×10^6 cell/well in 1 ml complete culture medium. The complete cell culture medium was prepared from RMPI-1640 supplemented with 5% fetal calf serum, 2 mM glutamine and antibiotics. The cells were cultured in a 5% CO_2 incubator. After 2 h attachment, Cr (VI) was added to the cell culture at various final concentrations and incubation times. Sodium chromate was used as a source of Cr (VI). At the end of cell culture, the cells and cell supernatants were harvested for atomic absorption assay.

Intracellular and extracellular chromium concentration was determined by atomic absorption spectroscopy on a Perkin-Elmer 303 atomic absorption spectrophotometer with linear recorder. Concentrated nitric acid (100 μl) was added to the isolated cells, and samples were heated at 100°C until a homogenous yellowish solution was formed. Twenty μl of H_2O_2 (30%) was added to the mixture to obtain a clear solution, which was then diluted to a total volume of 1.0 ml for flame atomic absorption measurement. Extracellular chromium concentration was measured directly in the supernatant (total volume, 1 ml) without further processing. The instrumental settings were: lamp current, 30 mA; slit width, 0.2 nm; wavelength, 358 nm; flame, air-acetylene. The concentration of chromium is calculated using the calibration curve obtained with various standard chromium samples.

Measurements of the effect of Cr (VI) on cell proliferation

Human lung epithelial cell line A549 was used. The study was conducted in 96 well culture plate with 1×10^5 cells/well in 200 μl complete culture medium. The cells were cultured in a 5% CO_2 incubator. Cr (VI) was added to the cell culture at different concentrations and incubated for 48 h in the presence of Cr (VI). At the end of cell culture, the culture medium was removed and the cells were harvested for proliferation assay. The cell growth was determined using a non-radioactive ELISA kit (Cell proliferation ELISA system version II, code RPN250) which was purchased from Amersham Pharmacia Biotech, Inc. (Piscataway, NJ, USA). The assay result was read with a micro plate reader. The growth of unexposed cells was used as a control. The inhibition of cell growth was derived from a calculation in which growth of the control cell was defined 100%.

Electron spin resonance (ESR) detection of Cr (VI)-induced $^{\bullet}OH$ radical generation in cellular system

Human lung epithelial cell line A549 was used in the study of Cr (VI)-induced $^{\bullet}OH$ radical generation in cellular system. ESR spin trapping was used to detect short-lived free radical intermediates. This technique involves an addition-type reaction of a short-lived radical with a diamagnetic compound (spin trap) to form a relatively long-lived free radical product, the so-called spin adduct, which can be studied by conventional ESR. The intensity of the spin adduct signal corresponds to the amount of short-lived radicals trapped, and the hyperfine splittings of the spin adduct are generally characteristic of the original, short-lived, trapped radicals. This method is specific and sensitive and is considered to be the best for detection and identification of free radical generation. All ESR measurements were made using a Varian E9 spectrometer and a flat cell assembly. Hyperfine splitting were measured (to 0.1 G) directly from magnetic field separations using potassium tetraperoxochromate (K_3CrO_8) and 1,1-diphenyl-2-picrylhydrazyl (DPPH) as standards. The relative radical concentration was estimated by multiplying half of the peak height by $(\Delta H_{pp})^2$, where ΔH_{pp} represents peak-to-peak width. An EPR DAP 2.0 program was used for data acquisition and analysis. Reactants were mixed in test tubes in a total volume of 0.5 ml. The reaction mixture was then transferred to a flat cell for ESR measurement.

DNA damage assay

The human prostate cell line, LNCap (ATCC, CRL-1740) (Ras (–)) was obtained from American Type Culture Collection (ATCC) (Rockville, MD, USA). Ras (+) cells, which overexpress Ras protein, were developed by transfecting LNCap cells with the H-ras expression vector. The cells were cultured at 37°C in a 5% CO_2 incubator with RPMI 1640 cell culture medium supplemented with 5% fetal bovine serum (FBS), 2 mM glutathione, 2 mM HEPES, and 0.075% Na_2HCO_3 (pH 7.4). The cells were cultured in 75 cm^2 cell culture flasks and trypsinized with 0.25% trypsin EDTA for plating and passing.

For single cell gel electrophoresis assay (SCG assay) for Cr (VI)-induced DNA damage, 2.5×10^6 cells were seeded in 35 mm cell culture plates, incubated at 37°C in a 5% CO_2 incubator overnight. The cells were exposed to different concentrations of Cr (VI). After incubation for 24 h, the cells were harvested and re-suspended in the PBS at an approximate density of 2×10^6/ml for SCG assay.

The SCG assay was performed according to the procedure described by Tice *et al.* [18], with minor modification. One hundred µl 0.5% normal melting agarose (Sigma) in Ca^{2+} and Mg^{2+} free PBS was added onto the fully frosted microscope slides, then covered with a 22 × 22 mm coverslip immediately. They were placed on ice for 10 min. One hundred µl cell suspension was mixed with 900 µl low melting point (LMP) agarose (Sigma) at 37°C, then 75 µl of the mixture (approximately $1.5–2.0 \times 10^4$ cells) was pipetted onto the first agarose layer (after gently removing the coverslip). The coverslip was replaced and the slide was maintained at 4°C for solidification. After removal of the coverslip, 75 µl of 0.5% LMP agarose was loaded, the coverslip was replaced, and the gel was allowed to solidify for 10 min. The coverslip was then removed and the slide was immersed in freshly prepared cold lysing solution (2.5 M NaCl, 100 mM Na_2EDTA, 10 mM Tris, and 1% sodium sarcosinate, pH = 10; 1% Triton X-100 and 10% DMSO v/v were added just before use) for at least 1 h at 4°C. All the following steps were conducted under red light to prevent additional DNA damage. The slides were drained and placed in a horizontal gel electrophoresis tray filled with fresh alkaline-EDTA buffer (300 mM NaOH and 1 mM EDTA in distilled water, pH > 13) for 10 min to allow for DNA unwinding and alkaline-labile damage expression. Electrophoresis was carried out for 50 min at room temperature at 20 volts (about 1 V/cm). Then the slides were neutralized by rinsing for 5 min with Tris buffer (0.4 M Tris, pH 7.5) 3 times, stained with 50 µl ethidium bromide (20 µg/ml), and covered with a coverslip. Image analysis was performed at 400× magnification using a fluorescence microscope. The length of DNA migration was determined with an optical micrometer by measuring the tail length (TL, distance between the edge of head and the end of tail) in microns. A minimum of 50 cells per treatment, 25 cells on each of two duplicate slides from a randomly chosen fields, were scored. Statistical analysis was performed by using Kruskal-Wallis One Way Analysis.

NF-κB-dependent luciferase activity

BEAS-2B cells are grown in Dulbecco's modified Eagle's Medium (DMEM) containing 10% fetal bovine serum in 6-well tissue culture plates at a cell concentration of 5×10^6 for 2 days. The cells were transfected by 2 µg/ml of NF-κB-dependent luciferase reporter plasmid using the DEAE-dextran method in the presence of 50 µg/ml chloroquine for 2 h. Then the cells were treated with 10% DMSO for 2 min and washed 3 times. The transfected cells were cultured in complete medium for an additional 48 h. Cells were treated with a various doses of Cr (VI) for an additional 12 h. Cells were harvested at the end of incubation, washed twice with PBS (pH 7.6) and resuspended in lysis buffer (25 mM Tris-HCl, 2 mM EDTA, 2 mM DTT, 10% glycerol, and 1% Triton X-100). Total protein concentration of each extract was quantitated using a Bio-Rad protein assay reagent. Luciferase activity was determined according to protocol recommended by manufacturer. Luciferase activity was expressed as relative luciferase activity and normalized for transfection efficiency on the basis of β-galactosidase expression.

Western blot analysis for p53 protein

The nuclear extract was used for Western blot analysis to determine the p53 protein level. Samples (20 µg protein) denatured with SDS were electrophoretically separated on 10% Tris- Glycine gels (Novex, San Diego, CA, USA) and were transferred to nitrocellulose membrane (Schleicher & Schuell, Keene, NH, USA). The membrane was pre-blocked in milk buffer (Tris-buffered saline Tween (TBS-T) containing 5% non-fat milk) for 20 min and exposed for 1 h to 0.8 µg/ml affinity-purified mouse antibody to p53 protein in fresh milk buffer. The membrane was rinsed and incubated in a 1: 2000 dilution of HRP conjugated goat anti-mouse IgG for 1 h. The membrane was then washed with TBS-T and antibody binding sites were visualized by ECL Western blotting detection reagents (Amersham Pharmacia Biotech, Piscataway, NJ, USA). All the antibodies were from Santa Cruz Biotechnology (Santa Cruz, CA, USA).

Apotosis assay

The cells were plated in a 96 well plates at a density of 2 × 10^4 cells/well in supplemented RPMI-1640 and incubated for 16 h before the cells were subjected to treatment, in triplicate. After treatment, the cells were washed twice in phosphate-buffered solution (PBS) (2.68 mM KCl, 1.47 M KH_2PO_4, 8 mM Na_2KPO_4, 136.75 mM NaCl), and apoptosis was quantitated by measuring the levels of cytosolic histone-bound DNA fragments. This was done by using a cell death ELISA assay kit manufactured by Boringher Mannheim (Indianapolis, IN, USA). The assays were carried out according to the protocol provided by the manufacturer. Briefly, the cells were lysed with 200 µl of lysis buffer at room temperature. The lysate from three identical wells was combined and 20 µl of the resultant lysate was mixed with 80 µl of antibody solution in a well of the ELISA plate. The loaded wells were incubated at room temperature for 2 h. The substrate was added after the wells were washed 3 times in washing buffer. After incubation at 37°C for 10–20 min, the reaction was stopped and optical density was measured using a microplate reader at a wavelength of 405 nm. The readings were used to represent the degree of apoptosis. A mean value of 3 separate experiments was used for analysis.

Results

Cr (VI) uptake by cells

Table 1 shows Cr (VI) uptake by A549 cells. It my be noted from this table that an increase in Cr (VI) concentration in the medium increased Cr (VI) uptake by the cells. The percentage of Cr (VI) uptake decreased when the concentration of Cr (VI) in the medium increased. At a low Cr (VI) concentration of 1 µM in the medium, the Cr (VI) concentration in the cells was 0.29 mM (29 %) after incubation for 2 h while at a concentration of 100 µM, the Cr (VI) concentration in cells was 9.24 mM (9.2 %). An increase in incubation time also increased the Cr (VI) uptake. From the data presented in this Table, it is evident that there was no apparent threshold in Cr (VI)-uptake by the cells.

Effect of Cr (VI) on cell proliferation

As shown in Fig. 1, Cr (VI), even at a concentration as low as 0.1 µM, inhibited the proliferation of A549 cells. The cells were treated for 48 h. An increase in Cr (VI) concentration enhanced the inhibitory effect. Figure 2 shows the percentage of inhibition vs incubation time. An increase in incubation time enhanced the inhibition percentage.

Cr (VI) induced ·OH radical generation

ESR spin trapping was used to detect free radical generation in Cr (VI)-stimulated A549 cells. Figure 3c shows the typical ESR spectrum obtained from a mixture containing 10^6 cells/ml, 2 mM Cr (VI) and 100 mM spin trap (DMPO) in a phosphate buffer solution (pH 7.4). The spectrum exhibits a 1:2:2:1 quartet with hyperfine splittings of $a_H = a_N = 14.9$ G, where a_H and a_N denote hyperfine splittings of β-hydrogen

Table 1. Cr concentrations (µM) in cells

Incubation time	Cr (VI) concentration in medium (µM)				
	1	5	20	100	300
	Cr concentration in cells				
2 h	0.29 ± 0.096	1.35 ± 0.096	4.09 ± 0.91	9.24 ± 2.11	14.43 ± 2.89
4 h	0.38 ± 0.096	1.92 ± 0.19	5.58 ± 1.25	14.96 ± 5.72	24.65 ± 7.1
8 h	0.53 ± 0.53	2.50 ± 0.097	9.57 ± 3.03	29.44 ± 11.93	45.69 ± 15.87
16 h	0.67 ± 0.29	2.98 ± 0.00	10.49 ± 0.58	53.39 ± 15.87	70.71 ± 23.57

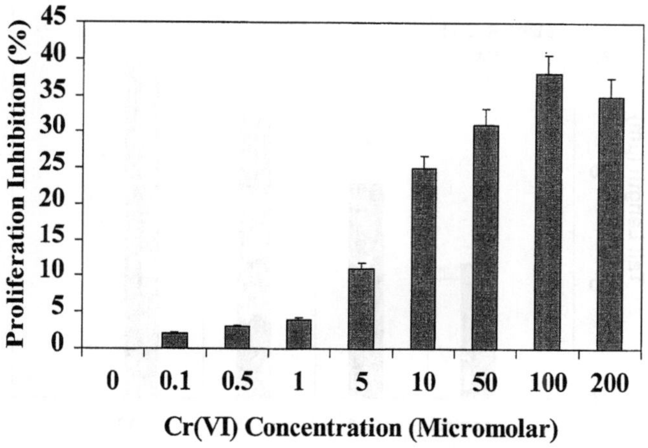

Fig. 1. Inhibition of cell proliferation by Cr (VI). The human lung epithelial cells A549 were plated in a 96 well plate at 3×10^4/well in 200 μl of RPMI-1640 supplemented with 5% FCS. The cells were exposed to Cr (VI) 16 h, and then harvested for cell proliferation assay using the Biotrak cell proliferation ELISA system from Amersham Pharmacia Biotech. The inhibition percentage of unexposed cells was used to express the inhibitory activity of Cr (VI). The inhibition percentage of cell proliferation was calculated from OD readings from a microplate reader. The cells were treated with Cr (VI) at final concentrations from 0.1–200 μM for 48 h.

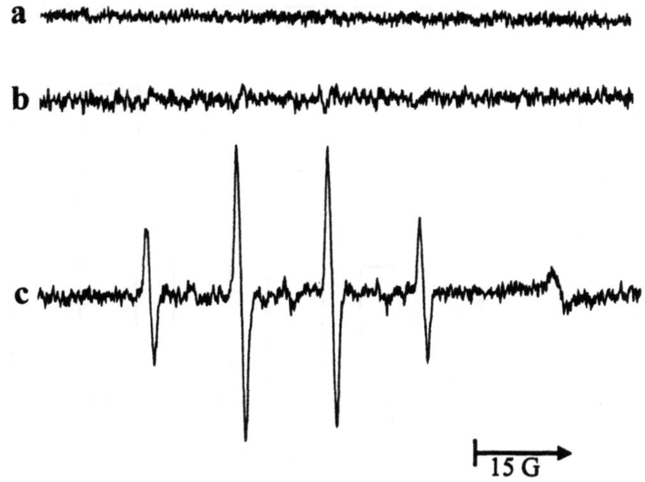

Fig. 3. Generation of free radicals from Cr (VI)-stimulated A549 cells. (a) The ESR spectrum was recorded from a mixture containing 1.0×10^6 cells/ml and 100 mM DMPO in a phosphate-buffered solution (pH 7.4). (b) Same as (a) but the mixture containing 1 mM Cr (VI) and 100 mM DMPO in a phosphate-buffered solution (pH 7.4). (c) same as (a) but the mixture containing 1 mM Cr (VI), 1.0×10^6 cells/ml and 100 mM DMPO in a phosphate-buffered solution (pH 7.4). The spectra were recorded 5 min after reaction initiation.

and nitroxy nitrogen, respectively. Based on these splitting constants [20], this quartet spectrum was assigned to DMPO/˙OH adduct, which is an indirect evidence for ˙OH generation. Either cells alone or Cr (VI) alone did not generate any observable ESR signal (Figs 3a and 3b). The weak signal at the right side is Cr (V), which was generated during the Cr (VI) reduction by the cells. Figure 4 shows spectral intensity (the amount of ˙OH radicals generated) vs time. The radical generation was rapid for the first 10 min and reached its maximum in 20 min. A decrease in Cr (VI) concentration decreased the amounts of ˙OH radicals generated over the concentration range of 0.125–2 mM (Fig. 5).

Effect of antioxidants on free radical generation

Figure 6 shows the effects of various agents on ˙OH radical generation. Addition of N-ethylmaleimide decreased the ˙OH radical generation. N-ethylmaleimide is an inhibitor of a group of flavoenzymes that have capabilities to reduce Cr (VI) using NAD(P)H as a co-factor. Addition of cellular Cr (VI) reductants, glutathione (GSH) and NADPH, enhanced

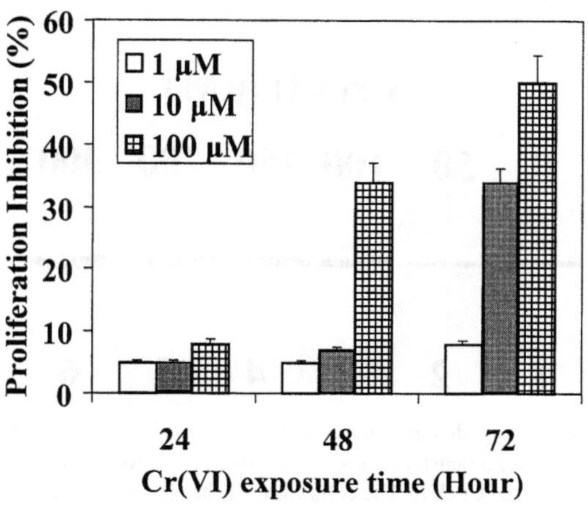

Fig. 2. Time course of Cr (VI) induced inhibition of cell proliferation. The cells were treated with Cr (VI) at 1, 10 and 100 μM final concentrations for different time periods as indicated. Other experimental conditions were the same as those described in Fig. 1.

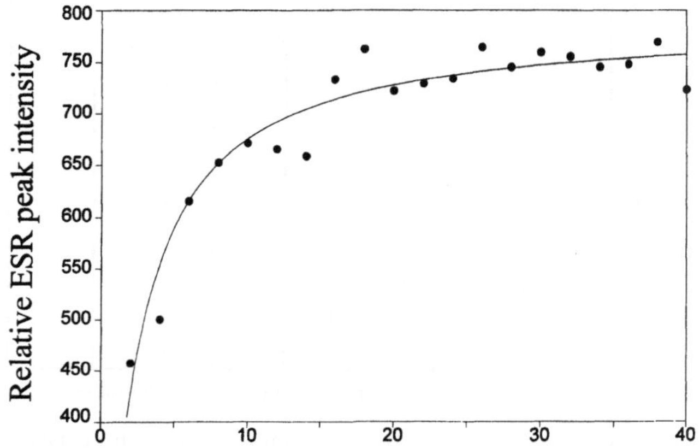

Fig. 4. The time-course of ESR peak intensity recorded from a mixture containing 1 mM Cr (VI), 1.0×10^6 cells/ml and 100 mM DMPO in a phosphate-buffered solution (pH 7.4).

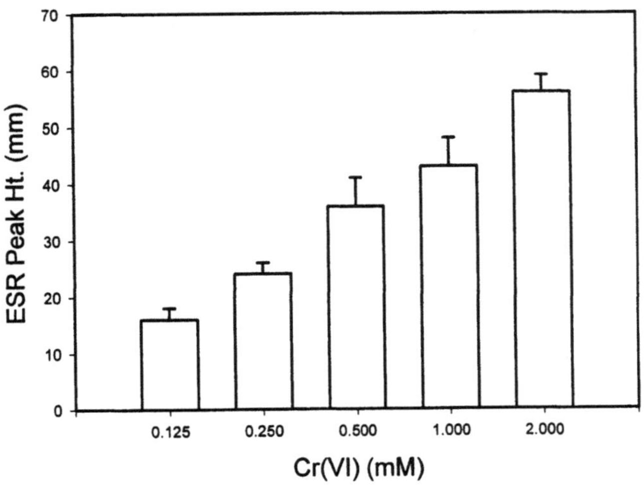

Fig. 5. Concentration dependence of Cr (VI)-stimulated free radical generation from A549 cells. The reaction conditions were the same as those described in Fig. 3.

the ·OH radical generation. These results indicate that reduction of Cr (VI) to its lower oxidation states is required for Cr (VI)-induced ·OH radical generation.

Cr (VI)-induced DNA damage

For the enhancement of detection sensitivity, both Ras(+) cells, which constantly over-express Ras protein, and Ras(–) cells, were used. The results obtained on Cr (VI)-induced DNA damage in whole cells are presented in Fig. 7. As shown by this figure, Cr (VI) caused DNA damage in both Ras(+) and Ras(–) cells but the Ras(+) cells were more susceptible. There was an increase in DNA damage with increasing Cr (VI) concentration in the medium.

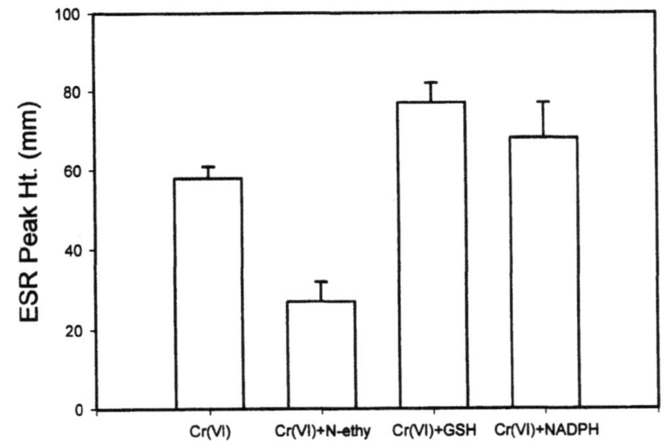

Fig. 6. The effect of a flavoeenzyme inhibitor, flavoenzyme cofactor, NADPH, and antioxidant, glutathione (GSH) on Cr (VI)-induced free radical generation. The experimental conditions were the same as those described in Fig. 1.

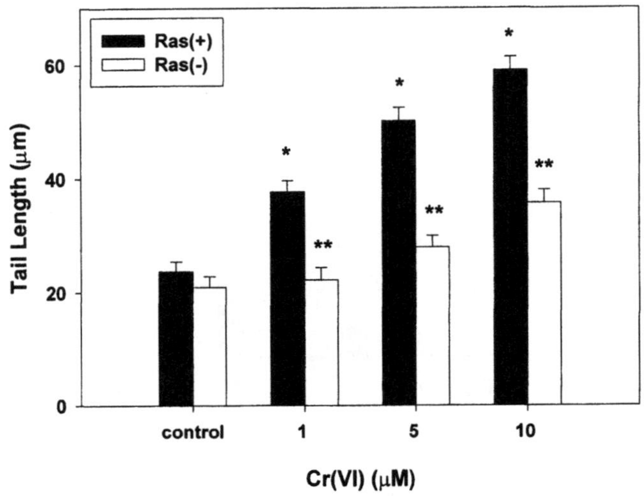

Fig. 7. Cr (VI) induced DNA damage in Ras transformed cells (Ras(+)) and wild type cells (Ras(–)). 2.5 × 10^6 cells were seeded in 35 mm cell culture plates, incubated at 37°C and 5% CO_2 overnight. The cells were exposed to different concentrations of Cr (VI). After incubation for 24 h, the cells were harvested and re-suspended in PBS at an approximate density of 2 × 10^6/ml. DNA damage was determined by single cell gel electrophoresis assay (SCG assay) as described in Materials and methods. The tail length (distance between the edge of head and the end of tail) was utilized for the measurements, the longer the tail, more severe the damage. Values presented are means and S.E. of 3 separate experiments. *indicate a significant increase from the control ($p < 0.05$); **indicate a significant decrease from Ras(+) cells at the same dosage.

Cr (VI)-induced p53 activation

An early study [13] showed that Cr (VI) is able to cause p53 activation. The dose-dependence of Cr (VI)-induced p53 activation at protein level is presented in Fig. 8. Cr (VI) caused a dose-dependent increase in p53 protein over the concentration range of 50–300 μM.

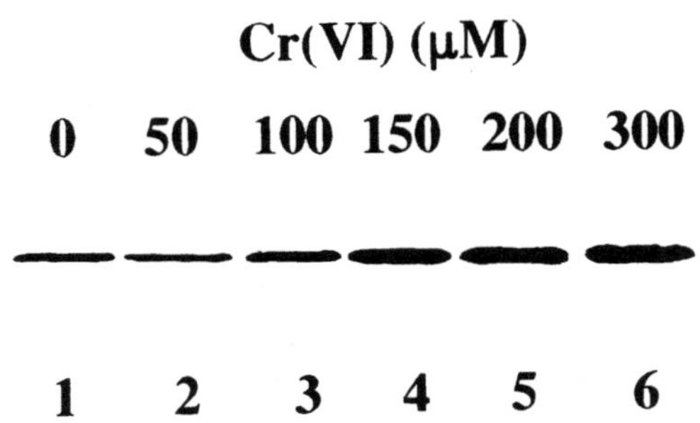

Fig. 8. Dose-dependent p53 activation induced by Cr (VI). The A549 cells were adjusted to a density of 1 × 10^6/ml and treated for 3 h with different concentrations of Cr (VI). The cells were washed, added to fresh medium, and incubated overnight. The nuclear extracts were used for western blot assay. Lane 1: untreated cells; lane 2: cells + 50 μM Cr (VI); lane 3: cells + 100 μM Cr (VI); lane 4: cells + 150 μM Cr (VI); lane 5: cells + 200 μM Cr (VI); lane 6: cells + 300 μM Cr (VI).

Dose-dependence of Cr (VI)-induced NF-κB activation

Cr (VI)-induced NF-κB has also been reported in our laboratories using DNA binding assay [11, 17]. In the present study, we focused on dose-dependence. As shown in Fig. 9, Cr (VI) is able to cause a dose dependent enhancement of NF-κB-dependent luciferase activity over 1–10 μM. Higher concentrations of Cr (VI), (> 10 μM), however, was cytotoxic (data not shown).

Dose dependence of Cr (VI)-induced apoptosis

Cr (VI)-induced apoptosis in A549 cells has also been reported in our laboratories [16]. The mechanism of action involves p53 and free radical reactions. The present study focused on the dose-dependence. A DNA fragmentation assay was carried out by detection of histone-bound DNA fragments in a whole cell lysate with a cell death enzyme-linked immunosorbent assay system. As shown in Fig. 10, Cr (VI)-induced apoptosis was observable at the Cr (VI) concentration of 75 μM. The degree of apoptosis increased with increasing Cr (VI) concentration over the range of 75–300 μM.

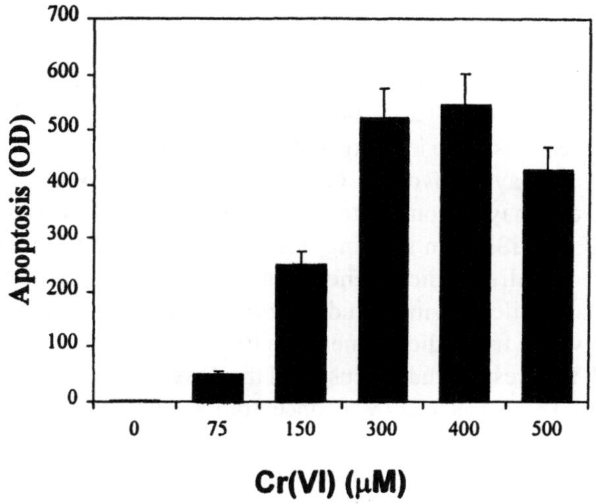

Fig. 10. Cr (VI)-induced apoptosis. A549 cells were seeded in a 96-well plate at a density of 1 × 10^4/well. Cr (VI) was added to the cells culture at the final concentrations as indicated. After exposure of 3 h, Cr (VI) was removed, and the culture was supplemented with the fresh cell medium. Histone protein assay was carried after 24 h. The data presented are mean value ± S.D. from 3 independent experiments. *indicates a significant increase from control ($p < 0.05$).

Discussion

The present study investigated the mechanism of Cr (VI)-induced carcinogenesis with a focus on dose dependence of Cr (VI) cell-uptake, free radical generation, Cr (VI)-induced DNA damage, cell proliferation, free radical generation in whole cells, activation of NF-κB and p53, and apoptosis. It has been established that cellular uptake is an important factor in the carcinogenesis induced by Cr (VI) [9]. Cr (III) exhibits much lower carcinogenesis in both epidemiologic and laboratory studies due to its inability to enter the cells. After entering the cells, Cr (VI) is reduced to Cr (V) and Cr (IV) by cellular reductants, such as glutathione, ascorbate, NADPH, and glutathione reductase/NADPH resulting in redox cycling before Cr (VI) is ultimately converted to Cr (III), the most stable form among several different chromium oxidation states [2]. During the redox cycling, molecular oxygen is reduced to $O_2^{\cdot-}$ radical, which produces H_2O_2 via dismutation. H_2O_2 further reacts with Cr (V) or Cr (IV) to generate ·OH radical. ROS, Cr (V), and Cr (IV) may all be involved in Cr (VI)-induced carcinogenesis [2]. Cr (VI) cell uptake is an essential first step in this model. The results obtained from the present study show that the cellular uptake of Cr (VI) depends on Cr (VI) concentration in the medium. It may be noted that the atomic absorption technique used to measure Cr content of the cells and medium can not analytically distinguish among Cr (VI), Cr (V), Cr (IV), and Cr (III). Because Cr (VI) is known to be actively and rapidly transported into cells, it is believed that the Cr measured in the cells originates from direct cellular uptake of Cr (VI) and not from uptake of Cr (III) following extracellular reduction of Cr (VI). In these studies, the medium did not contain added physiological reductants, such as ascorbate and glutathione. The reductants could facilitate extracellular reduction of Cr (VI) to Cr (III) *in vivo* presenting significant uptake into target cells. This possibility is currently being investigated.

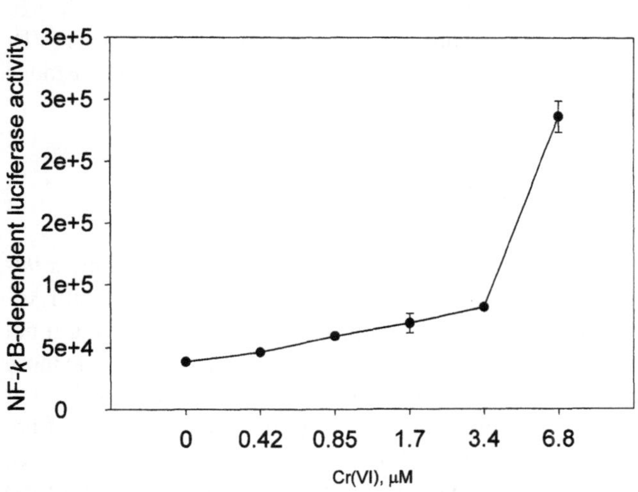

Fig. 9. Activation of NF-κB by Cr (VI). BEAS-2B cells transfected with a NF-κB-dependent luciferase reporter plasmid were treated with various concentrations of Cr (VI) for an additional 12 h. The luciferase activity was determined as described in Materials and methods. The data presented are mean values ± S.D. from 5 independent experiments.

The results obtained from the present study show that the Cr (VI) is able to inhibit cell proliferation. The inhibition is dose-dependent with no apparent response threshold. While the mechanism of Cr (VI)-caused inhibition of cell proliferation remains to be investigated, Cr (VI)-induced generation of ROS may be involved. Cell growth arrest, apoptosis, and necrosis may all contribute to inhibition of cell proliferation.

Using ESR spin trapping, the present study also investigated the free radical generation from Cr (VI)-stimulated cells. While previous studies from our laboratories have shown the free radical generation from a similar system [11, 17], the present study focused on the dose and time dependence. There was a Cr (VI) concentration-related increase in the level of free radical generation above 0.125 mM. Several agents were examined for their abilities to effect free radical generation. N-ethylmaleimide, an inhibitor of a group of flavoenzymes, decreased the radical generation, indicating that flavoenzymes, such as glutathione reductase may play a role in the Cr (VI)-induced free radical generation. GSH, an antioxidant, and NADPH, a co-factor for flavoenzyme, enhanced the radical generation further showing that reduction of Cr (VI) is important in Cr (VI)-induced free radical generation.

While it has been reported previously that Cr (VI) is able to induce DNA damage via free radical reaction [5], the present study examined the dose dependence of Cr (VI)-induced DNA damage in a sensitive human prostate cell line. Using the comet assay, the results obtained from the present study show that Cr (VI) is indeed able to induce DNA damage at a whole cell level. DNA damage is an important step in the cancer development. ROS are typical DNA damaging agents. The free radicals generated by Cr (VI), specially $^{\cdot}OH$ radical, may be the species responsible for Cr (VI)-induced DNA damage. If not repaired or removed, cells with injured DNA can proliferate, leading to cancer development. The dose dependence of Cr (VI)-induced DNA damage is currently being investigated in the human lung epithelial cell line.

The results obtained from the present study show that Cr (VI) is able to cause a dose-dependent p53 activation above 50 μM. In our earlier study [13], we have shown that Cr (VI) caused p53 activation through free radical reactions. This tumor suppressor protein plays an important role in protecting cells from tumorgenic alteration. Mutational inactivation of p53 has been found to be involved in various human cancers, indicating the importance of p53 in human carcinogenesis. The p53 can be activated in response to a variety of stimuli, such as UV, γ-radiation and nucleotide deprivation. The activation of p53 can cause either cell division cycle arrest or apoptosis. The activation of p53 by Cr (VI) may be the cell response to DNA damage or direct response to Cr (VI)-induced oxidative stress.

The results obtained from the present study suggest that Cr (VI) is able to cause a dose dependent NF-κB activation even at concentration below 10 μM. The NF-κB protein is found in many cell types and is a focal point for understanding how extracellular signals induce the expression of specific sets of early-response genes, such as those regulating the secretion of growth promoters [14, 15]. NF-κB is considered to be an intracellular sensor of oxidative stress. This transcription factor plays an important role in the transcriptional regulation of a number of genes including those encoding cytokines, cell adhesion molecules, growth factors, and several anti-apoptotic proteins [14, 15]. The dose-dependent activation of NF-κB by Cr (VI) indicates that this transcription factor may be involved in mechanism of Cr (VI)-induced carcinogenesis.

Cr (VI) is also able to cause a dose-dependent enhancement of cell apoptosis above 75 μM. Apoptosis is a programed cell death mechanism to control cell number in tissues and to eliminate individual cells that may lead to disease states [21, 22]. It is a response to physiologic and pathologic stress that disrupt the balance between the rates of cell division and elimination. In diseases such as cancer, there is an imbalance between the rates of cell proliferation and cell death. Agents that promote or suppress apoptosis can alter the rates of cell division and death, influencing the anomalous accumulation of neoplastic cells. Cr (VI)-induced apopotosis is likely due to the cell response to both Cr (VI)-induced DNA damage and oxidative stress.

From the above discussion, it can be concluded that Cr (VI) is able to enter the cells in a dose dependent manner at low concentrations. Inside the cells, Cr (VI) is reduced to Cr (V) and Cr (IV). During the redox cycling, molecular oxygen is reduced to $O_2^{\cdot-}$, H_2O_2 and $^{\cdot}OH$ radical. The reduced Cr forms and ROS cause DNA damage, activation of p53 and NF-κB, inhibition of cell proliferation, and apoptosis. These Cr (VI)-induced cell responses show dose-dependence with the range of 1–300 μM.

Acknowledgements

Research funded under Interagency Agreement number 98-18-00m2 between the Occupational Safety and Health Administration and the National Institute for Occupational Safety and Health. The views expressed in the paper are those of the authors and are not necessarily the official position of OSHA. We appreciate the suggestions and comments from Dr. Val Schaeffer, Jeff Snyder, Caroline Freeman, William G. Perry, Health Standards Programs, Occupational Safety and Health Administration, Washington, DC, on the present work. We thank Drs. Vince Castranova, Murali Rao, and Val Vallyathan for critical review of the manuscript.

References

1. De Flora S, Bagnasco M, Serra D, Zanacchi P: Genotoxicity of chromium compounds: A review. Mutat Res 238: 99–172, 1990
2. Shi X, Chiu A, Chen CT, Halliwell B, Castranova V, Vallyathan V: Reduction of chromium (VI) and its relationship to carcinogenesis. J Toxicol Environ Health 2: 87–104, 1999
3. Deflora S, Wetterhahn KE: Mechanism of chromium metabolism and genotoxicity. Life Chem Rep 7: 169–244, 1989
4. Sugiyama M, Tsuzuki K, Haramaki N: DNA single-strand breaks and cytotoxicity induced by sodium chromate in hydrogen peroxide-resistant cell lines. Mutat Res 299: 95–102, 1993
5. Leonard S, Wang S, Zang L, Castranova V, Vallyathan V, Shi X: Role of molecular oxygen in the generation of hydroxyl and superoxide anion radicals during enzymatic Cr (VI) reduction and its implication to Cr (VI)-induced carcinogenesis. J Environ Pathol Toxicol Oncol 19: 49–60, 2000
6. Xu J, Wise JP, Patierno SR: DNA damage induced by carcinogenic lead chromate particles in cultured mammalian cells. Mutat Res 280: 129–136, 1992
7. Kortenkamp A, Ozolins Z, Beyersmann D, O'Brien P: Generation of PM2 DNA breaks in the course of reduction of chromium (VI) by glutathione. Mutat Res 216: 19–26, 1989
8. Hayes RB: Review of occupational epidemiology of chromium chemicals and respiratory cancer. Sci Total Environ 71: 331–339, 1988
9. Connett PH, Wetterhahn KE: Metabolism of the carcinogenic chromate by cellular constituents. Struct Bond 54: 93–124, 1983
10. Cohen MD, Kargacin B, Klein CB, Costa M: Mechanism of chromium carcinogenicity and toxicity. Crit Rev Toxicol 23: 255–281, 1983
11. Ye J, Zhang X, Young HA, Mao Y, Shi X: Chromium (VI)-induced nuclear factor-κB activation in intact cells via free radical reactions. Carcinogenesis 16: 2401–2405, 1995
12. Chen F, Ding M, Lu Y, Leonard SS, Vallyathan V, Castranova V, Shi X: Participation of MAP kinase p38 and IκB kinase in chromium (VI)-induced NF-κB and AP-1 activation. J Environ Pathol Toxicol Oncol 19: 231–238, 2000
13. Wang S, Leonard SS, Ye J, Ding M, Shi X: The role of hydroxyl radical as a messenger in Cr (VI)-induced p53 activation. Am J Physiol 279: C868–C875, 2000
14. Baeuerale PA, Henkel T: Function and activation of NF-κB in the immune system. Annu Rev Immunol 12: 141–179, 1994
15. Baldwin AS: The NF-κB and IκB proteins: New discoveries and insights. Annu Rev Immunol 14: 649–681, 1996
16. Ye J, Wang S, Leonard SS, Sun Y, Butterworth L, Antonini J, Ding M, Yongyut Rojanasakul, Vallyathan V, Castranova V, Shi X: Role of reactive oxygen species and p53 in chromium (VI)-induced apoptosis. J Biol Chem 274: 34974–34980, 1999
17. Shi X, Dong Z, Huang C, Ma W, Liu K, Ye P, Chen F, Leonard SS, Ding M, Castranova V, Vallyathan V: The role of hydroxyl radical as a messenger in the activation of nuclear transcription factor NF-κB. Mol Cell Biochem 194: 63–70, 1999
18. Tice RR, Strauss GSH, Peters WP: High dose combination alkylating agents with autologous bone-marrow support in patients with breast cancer: Preliminary assessment of DNA damage in individual peripheral blood lymphocytes using the single cell gel electrophoresis assay. Mutat Res 271: 101–113, 1992
19. Chen F, Sun SC, Kuh DC, Gaydos LJ, Demers LM: Essential role of NF-κB activation in silica-induced inflammatory mediator production in macrophages. Biochem Biophys Res Commun 214: 985–992, 1995
20. Buettner GR: ESR parameters of spin adducts. Free Rad Biol Med 3: 259–303, 1987
21. Seale J, Kerr JFR, Bishop CJ: Necrosis and apoptosis: distinct modes of cell death with fundamentally different significance. Pathol Ann 17: 229–259, 1982
22. Vaux DL: Toward an understanding of the molecular mechanisms of physiological cell death. Proc Natl Acad Sci USA 90: 786–789, 1993

Molecular and Cellular Biochemistry **222:** 231–233, 2001.

Index to Volume 222

Molecular and Cellular Biochemistry **222:** 235–239, 2001.

Molecular and Cellular Biochemistry

INSTRUCTIONS FOR AUTHORS

EDITOR-IN-CHIEF

Naranjan S. Dhalla, Institute of Cardiovascular Sciences, St. Boniface General Hospital Research Centre, Faculty of Medicine, University of Manitoba, 351 Tache Avenue, Winnipeg, Manitoba, Canada R2H 2A6 (Telephone 204-235-3417; Fax 204-233-6723)

AIMS AND SCOPE

Molecular and Cellular Biochemistry: An International Journal for Chemical Biology in Health and Disease will publish original full length research papers and short communications in all areas of biochemical sciences. Emphasis will be given to manuscripts which present novel findings pertinent to the biochemical basis of cellular function and disease processes, as well as mechanisms of action of hormones and chemical agents. Investigations directed towards molecular biology and gene expression in the cell are particularly encouraged. Membrane transport, receptor mechanism, immune responses, secretory processes, and cytoskeletal function are areas of great interest to this journal. Articles in all areas related to biochemical structure-function relationships in the cell will be published. Studies examining adaptation of biochemical processes at the molecular and cellular levels in response to physiological and pathological stimuli are welcomed. In addition to the original articles, this journal will consider the publication of state of the art reviews. Ordinarily, reviews will be invited, however, investigators who wish to publish review articles in the area of their expertise may contact the Editor.

MANUSCRIPT SUBMISSION

Kluwer Academic Publishers prefer the submission of manuscripts and figures in electronic form in addition to a hard-copy printout. The preferred storage medium for your electronic manuscript is a 31/2 inch diskette. Please label your diskette properly, giving exact details on the name(s) of the file(s), the operating system and software used. Always save your electronic manuscript in the word processor format that you use; conversions to other formats and versions tend to be imperfect. In general, use as few formatting codes as possible. For safety's sake, you should always retain a backup copy of your file(s). **After acceptance**, please make absolutely sure that you send the latest (i.e., revised) version of your manuscript, both as hard-copy printout and on diskette.

Kluwer Academic Publishers prefer articles submitted in word processing packages such as MS Word, WordPerfect, etc. running under operating systems MS DOS, Windows and Apple Macintosh, or in the file format LaTeX. Articles submitted in other software programs, as well as articles for conventional typesetting, can also be accepted.

For submission in LaTeX, Kluwer Academic Publishers have developed a Kluwer LaTeX class file, which can be downloaded from: www.wkap.nl/kaphtml.htm/IFAHOME. Use of this class file is highly recommended. Do not use versions downloaded from other sites. Technical support is available at: texhelp@wkap.nl. If you are not familiar with TeX/LaTeX, the class file will be of no use to you. In that case, submit your article in a common word processor format.

For the purpose of reviewing, articles for publication should be submitted as hard-copy printout (three-fold) and on diskette to: Dr. N.S. Dhalla, Editor-in-Chief, Molecular and Cellular Biochemistry, Institute of Cardiovascular Sciences, St. Boniface General Hospital Research Centre, University of Manitoba, 351 Tache Avenue, Winnipeg, Canada R2H 2A6.

Please include the names and addresses of three referees who would be suitable for review

REVIEWING PROCEDURE

Molecular and Cellular Biochemistry follows a double-blind reviewing procedure. Authors are therefore requested to place their name and affiliation on a separate page. Self-identifying citations and references in the article text should either be avoided or left blank when manuscripts are first submitted. Authors are responsible for reinserting self-identifying citations and references when manuscripts are prepared for final submission.

ETHICS

When reporting experiments on human subjects, indicate whether the procedures followed were in accordance with the ethical standards of the responsible committee on human experimentation (institutional or regional) or with the Helsinki Declaration (1964, amended in 1975 and 1983) of the World Medical Association. Do not use patients' names, initials, or hospital numbers, especially in any illustrative material. When reporting experiments on animals, indicate whether the institution's or the National Research Council's guide for, or any national law on, the care and use of laboratory animals was followed.

MANUSCRIPT PRESENTATION

The journal's language is English. British English or American English spelling and terminology may be used, but either one should be followed consistently throughout the article. Manuscripts should be printed or typewritten on A4 or US Letter bond paper, one side only, leaving adequate margins on all sides to allow reviewers' remarks. Please double-space all material, including notes and references. The manuscript should be organized in the following order: Title page, Abstract, Introduction, Materials and Methods, Results, Discussion, Acknowledgements, References, Tables, Figure Legends, and Figures. Quotations of more than 40 words should be set off clearly, either by indenting the left-hand margin or by using a smaller typeface. Use double quotation marks for direct quotations and single quotation marks for quotations within quotations and for words or phrases used in a special sense.

Number the pages consecutively with the first page containing:

- **running head (shortened title)**
- **title**
- **author(s)**
- **affiliation(s)**
- **full address for correspondence, including telephone and fax number and e-mail address**

Abstract

Please provide a short abstract of 100 to 250 words. The abstract should not contain any undefined abbreviations or unspecified references.

Key Words

Please provide 5 to 10 key words or short phrases in alphabetical order.

Abbreviations

Abbreviations should be explained at first occurrence.

Article Types

The Journal welcomes manuscripts in all areas of biochemical sciences under the following categories:

a) **Full length papers** with original new data should provide the background information of the problem with clear indication as to why the study was undertaken.
b) **Short communications** which warrant rapid publication should be identified as such; these communications should not include more than 4 Tables or Figures and be as brief as possible.
c) **State of the art reviews** concerning current concepts and major technological advances in biochemistry and cellular biology are normally invited by the Editor-in-Chief; however, individuals who wish to submit a review article may contact the Editor for possible invitation.

Figures and Tables

Submission of electronic figures

In addition to hard-copy printouts of figures, authors are encouraged to supply the electronic versions of figures in either Encapsulated PostScript (EPS) or TIFF format. Many other formats, e.g., Microsoft Postscript, PiCT (Macintosh) and WMF (Windows), cannot be used and the hard copy will be scanned instead.

Figures should be saved in separate files *without* their captions, which should be included with the text of the article. Files should be named according to DOS conventions, e.g., 'figure1.eps'. For vector graphics, EPS is the preferred format. Lines should not be thinner than 0.25pts and in-fill patterns and screens should have a density of at least 10%. Font-related problems can be avoided by using standard fonts such as Times Roman and Helvetica. For bitmapped graphics, TIFF is the preferred format but EPS is also acceptable. The following resolutions are optimal: black-and-white line figures – 600–1200 dpi; line figures with some grey or coloured lines – 600 dpi; photographs – 300 dpi; screen dumps – leave as is. Higher resolutions will not improve output quality but will only increase file size, which may cause problems with printing; lower resolutions may compromise output quality. Please try to provide artwork that approximately fits within the typeset area of the journal. Especially screened originals, i.e. originals with grey areas, may suffer badly from reduction by more than 10–15%.

AVOIDING PROBLEMS WITH EPS GRAPHICS

Please always check whether the figures print correctly to a PostScript printer in a reasonable amount of time. If they do not, simplify your figures or use a different graphics program.

If EPS export does not produce acceptable output, try to create an EPS file with the printer driver (see below). This option is unavailable with the Microsoft driver for Windows NT, so if you run Windows NT, get the Adobe driver from the Adobe site (www.adobe.com).

If EPS export is not an option, e.g., because you rely on OLE and cannot create separate files for your graphics, it may help us if you simply provide a PostScript dump of the entire document.

HOW TO SET UP FOR EPS AND POSTSCRIPT DUMPS UNDER WINDOWS

Create a printer entry specifically for this purpose: install the printer 'Apple Laserwriter Plus' and specify 'FILE': as printer port. Each time you send something to the 'printer' you will be asked for a filename. This file will be the EPS file or PostScript dump that we can use.

The EPS export option can be found under the PostScript tab. EPS export should be used only for single-page documents. For printing a document of several pages, select 'Optimise for portability' instead. The option 'Download header with each job' should be checked.

Submission of hard-copy figures

If no electronic versions of figures are available, submit only high-quality artwork that can be reproduced as is, i.e., without any part having to be redrawn or re-typeset. The letter size of any text in the figures must be large enough to allow for reduction. Photographs should be in black-and-white on glossy paper. If a figure contains colour, make absolutely clear whether it should be printed in black-and-white or in colour. Figures that are to be printed in black-and-white should not be submitted in colour. Authors will be charged for reproducing figures in colour.

Each figure and table should be numbered and mentioned in the text. The approximate position of figures and tables should be indicated in the margin of the manuscript. On the reverse side of each figure, the name of the (first) author and the figure number should be written in pencil; the top of the figure should be clearly indicated. Figures and tables should be placed at the end of the manuscript following the Reference section. Each figure and table should be accompanied by an explanatory legend. The figure legends should be grouped and placed on a separate page. Figures are not returned to the author unless specifically requested.

In tables, footnotes are preferable to long explanatory material in either the heading or body of the table. Such explanatory footnotes, identified by superscript letters, should be placed immediately below the table.

Section Headings
First-, second-, third-, and fourth-order headings should be clearly distinguishable but not numbered.

Appendices
Supplementary material should be collected in an Appendix and placed before the Notes and Reference sections.

Notes
Please use footnotes rather than endnotes. Notes should be indicated by consecutive superscript numbers in the text. A source reference note should be indicated by means of an asterisk after the title. This note should be placed at the bottom of the first page.

Cross-Referencing
In the text, a reference identified by means of an author's name should be followed by the reference number in square brackets. When there are more than two authors, only the first author's name should be mentioned, followed by *et al.*

Examples:
Winograd [1]
Bullen and Bennett [2]

Acknowledgements
Acknowledgements of people, grants, funds, etc. should be placed in a separate section before the References.

References
References to books, journal articles, articles in collections and conference or workshop proceedings, and technical reports should be listed at the end of the article in numbered order following the MCBI style (see examples below). Please list all authors; do not use "et al."; do *not* add extraneous punctuation. Articles in preparation or articles submitted for publication, unpublished observations, personal communications, etc. should not be included in the reference list but should only be mentioned in the article text (e.g., T. Moore, personal communication). Abbreviate titles of periodicals according to the style of the Index Medicus.

References to books should include the author's name; year of publication; title; page numbers where appropriate; publisher; place of publication, in the order given in the example below.

12. Koelle GB: Cholinesterases and Anticholinesterase Agents. Springer-Verlag, Berlin, 1989, pp 187–294

References to articles in an edited collection should include the author's name; year of publication; article title; editor's name; title of collection; first and last page numbers; publisher; place of publication., in the order given in the example below.

13. Bing RJ: Contribution of myocardial metabolism to the knowledge of myocardial failure and myocardial infarction. In: T. Kobayashi, Y. Ito, G. Rona (eds). Cardiac Adaptation. University Park Press, Baltimore, 1978, pp 3–9

References to articles in conference proceedings should include the author's name; year of publication; article title; editor's name (if any); title of proceedings; first and last page numbers; place and date of conference; publisher and/or organization from which the proceedings can be obtained; place of publication, in the order given in the example below.

14. Corvi R, Savelyeva L, Schwab M: Patterns of oncogene activation in human neuroblastoma cells. In: G. Melina, R.A. Knight, K.-M. Debatim (eds). Biology of Neuroectodermal Tumours. Third workshop on the 'Biochemistry of Neuroectodermal Tumours', Rome, Istituto Dermopatico dell'Immacolata, June 12–14 1995, Kluwer Academic Publishers, Dordrecht, 1995, pp 25–31

References to articles in periodicals should include the author's name; year of publication; article title; full abbreviated title of periodical; volume number (issue number where appropriate); first and last page numbers, in the order given in the example below.

15. Naimark A, Klass BD: The incorporation of palmitate-^{14}C by rat lung *in vitro*. Can J Physiol Pharmacol 45: 52–60, 1981

References to technical reports or doctoral dissertations should include the author's name; year of publication; title of report or dissertation; institution; location of institution, in the order given in the example below.

16. Block KA: Transcriptional Regulation of Multidrug Resistance in Breast Cancer, Erasmus University Rotterdam, Rotterdam, The Netherlands, 1998

PROOFS
Proofs will be sent to the corresponding author. One corrected proof, together with the original, edited manuscript, should be returned to the Publisher within three days of receipt by mail (airmail overseas).

OFFPRINTS
Fifty offprints of each article will be provided free of charge. Additional offprints can be ordered by means of an offprint order form supplied with the proofs.

PAGE CHARGES AND COLOUR FIGURES
No page charges are levied on authors or their institutions. Colour figures are published at the author's expense only.

COPYRIGHT
Authors will be asked, upon acceptance of an article, to transfer copyright of the article to the Publisher. This will ensure the widest possible dissemination of information under copyright laws.

PERMISSIONS
It is the responsibility of the author to obtain written permission for a quotation from unpublished material, or for all quotations in excess of 250 words in one extract or 500 words in total from any work still in copyright, and for the reprinting of figures, tables or poems from unpublished or copyrighted material.

ADDITIONAL INFORMATION
Additional information can be obtained from:

Zachary Rolnik
MCBI Editorial Office
Tel: 1 781 871 6600
Fax: 1 781 871 6528
e-mail: zachary.rolnik@wkap.com

Developments in Molecular and Cellular Biochemistry

Series Editor: Naranjan S. Dhalla, Ph.D., M.D. (Hon.), FACC

1. V.A. Najjar (ed.): *Biological Effects of Glutamic Acid and Its Derivatives.* 1981 ISBN 90-6193-841-4
2. V.A. Najjar (ed.): *Immunologically Active Peptides.* 1981 ISBN 90-6193-842-2
3. V.A. Najjar (ed.): *Enzyme Induction and Modulation.* 1983 ISBN 0-89838-583-0
4. V.A. Najjar and L. Lorand (eds.): *Transglutaminase.* 1984 ISBN 0-89838-593-8
5. G.J. van der Vusse (ed.): *Lipid Metabolism in Normoxic and Ischemic Heart.* 1989 ISBN 0-7923-0479-9
6. J.F.C. Glatz and G.J. van der Vusse (eds.): *Cellular Fatty Acid-Binding Proteins.* 1990 ISBN 0-7923-0896-4
7. H.E. Morgan (ed.): *Molecular Mechanisms of Cellular Growth.* 1991 ISBN 0-7923-1183-3
8. G.J. van der Vusse and H. Stam (eds.): *Lipid Metabolism in the Health and Diseased Heart.* 1992 ISBN 0-7923-1850-1
9. Y. Yazaki and S. Mochizuki (eds.): *Cellular Function and Metabolism.* 1993 ISBN 0-7923-2158-8
10. J.F.C. Glatz and G.J. van der Vusse (eds.): *Cellular Fatty-Acid-Binding Proteins, II.* 1993 ISBN 0-7923-2395-5
11. R.L. Khandelwal and J.H. Wang (eds.): *Reversible Protein Phosphorylation in Cell Regulation.* 1993 ISBN 0-7923-2637-7
12. J. Moss and P. Zahradka (eds.): *ADP-Ribosylation: Metabolic Effects and Regulatory Functions.* 1994 ISBN 0-7923-2951-1
13. V.A. Saks and R. Ventura-Clapier (eds.): *Cellular Bioenergetics: Role of Coupled Creatine Kinases.* 1994 ISBN 0-7923-2952-X
14. J. Slezák and A. Ziegelhöffer (eds.): *Cellular Interactions in Cardiac Pathophysiology.* 1995 ISBN 0-7923-3573-2
15. J.A. Barnes, H.G. Coore, A.H. Mohammed and R.K. Sharma (eds.): *Signal Transduction Mechanisms.* 1995 ISBN 0-7923-3663-1
16. A.K. Srivastava and J.-L. Chiasson (eds.): *Vanadium Compounds: Biochemical and Therapeutic Applications.* 1995 ISBN 0-7923-3763-8
17. J.M.J. Lamers and P.D. Verdouw (eds.): *Biochemistry of Signal Transduction in Myocardium.* 1996 ISBN 0-7923-4067-1
18. E.-G. Krause and R. Vetter (eds.): *Biochemical Mechanisms in Heart Function.* 1996 ISBN 0-7923-4118-X
19. R. Vetter and E.-G. Krause (eds.): *Biochemical Regulation of Myocardium.* 1996 ISBN 0-7923-4259-3
20. G.N. Pierce and W.C. Claycomb (eds.): *Novel Methods in Molecular and Cellular Biochemistry of Muscle.* 1997 ISBN 0-7923-4387-5
21. F.N. Gellerich and S. Zierz (eds.): *Detection of Mitochondrial Diseases.* 1997 ISBN 0-7923-9925-0
22. P.K. Singal, V. Panagia and G.N. Pierce (eds.): *The Cellular Basis of Cardiovascular Function in Health and Disease.* 1997 ISBN 0-7923-9974-9
23. S. Abdel-aleem and J.E. Lowe (eds.): *Cardiac Metabolism in Health and Disease.* 1998 ISBN 0-7923-8104-1
24. A.K. Srivastava and B. Posner (eds.): *Insulin Action.* 1998 ISBN 0-7923-8113-0
25. V.A. Saks, R. Ventura-Clapier, X. Leverve, A. Rossi and M. Rigoulet (eds.): *Bioenergetics of the Cell: Quantitative Aspects.* 1998 ISBN 0-7923-8118-1
26. G.N. Pierce, H. Rupp, T. Izumi and A. Grynberg (eds.): *Molecular and Cellular Effects of Nutrition on Disease Processes.* 1998 ISBN 0-7923-8171-8
27. K. Ahmed, E. Chambaz and O.G. Issinger (eds.): *Molecular and Cellular View of Protein Kinase CK2.* 1998 ISBN 0-7923-8208-0

Developments in Molecular and Cellular Biochemistry

28. M.V. Cohen, J.M. Downey, R.J. Gelpi and J. Slezák (eds.): *Myocardial Ischemia and Reperfusion*. 1998 ISBN 0-7923-8173-4
29. D.A. Bernlohr and L. Banaszak (eds.): *Lipid Binding Proteins within Molecular and Cellular Biochemistry*. 1998 ISBN 0-7923-8223-4
30. R. Albarez-Gonzalez (ed.): *ADP-Ribosylation Reactions: From Bacterial Pathogenesis to Cancer*. 1998 ISBN 0-7923-8235-8
31. S. Imai and M. Endo (eds.): *Muscle Physiology and Biochemistry*. 1998 ISBN 0-7923-8264-1
32. D.K. Das (ed.): *Stress Adaptation, Prophylaxis and Treatment*. 1999 ISBN 0-7923-8406-7
33. H. Rupp and B. Maisch (eds.): *Control of Gene Expression by Catecholamines and the Renin-Angiotensin System* ISBN 0-7923-7981-0
34. X. Shi, V. Castranova, V. Vallyathan and W.G. Perry (eds.): *Molecular Mechanisms of Metal Toxicity and Carcinogenesis*. 2001 ISBN 0-7923-7498-3

KLUWER ACADEMIC PUBLISHERS – DORDRECHT / BOSTON / LONDON

MIX
Papier aus verantwortungsvollen Quellen
Paper from responsible sources
FSC® C105338

If you have any concerns about our products,
you can contact us on
ProductSafety@springernature.com

In case Publisher is established outside the EU,
the EU authorized representative is:
Springer Nature Customer Service Center GmbH
Europaplatz 3, 69115 Heidelberg, Germany

Printed by Libri Plureos GmbH
in Hamburg, Germany

MOLECULAR MECHANISMS OF METAL TOXICITY AND CARCINOGENESIS

Library of Congress Cataloging-in-Publication Data

Molecular mechanisms of metal toxicity and carcinogenesis / edited by Xianglin Shi.
p.cm. –(Developments in molecular and cellular biochemistry; v. 34) "Reprinted from Molecular and cellular biochemistry, Volume 222 (2001)."
ISBN 978-1-4613-5242-6 ISBN 978-1-4615-0793-2 (eBook)
DOI 10.1007/978-1-4615-0793-2
1. Metals–Carcinogenicity–Congresses. 2. Metals–Toxicology–Congress. 3. Molecular toxicology–Congresses. 1. Shi, Xianglin, Ph.D. II. Series.

RC268.7.M45.M65 2001
616.99'4071–dc21 2001046196

Originally published by Kluwer Academic Publishers in 2001
Softcover reprint of the hardcover 1st edition 2001

Printed on acid-free paper

Molecular and Cellular Biochemistry:

An International Journal for Chemical Biology in Health and Disease

CONTENTS VOLUME 222, Nos. 1 & 2, [illegible]

Kluwer

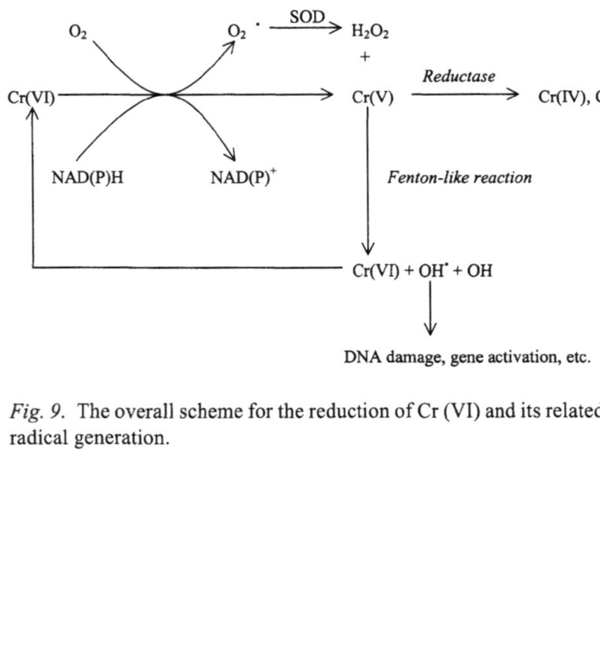

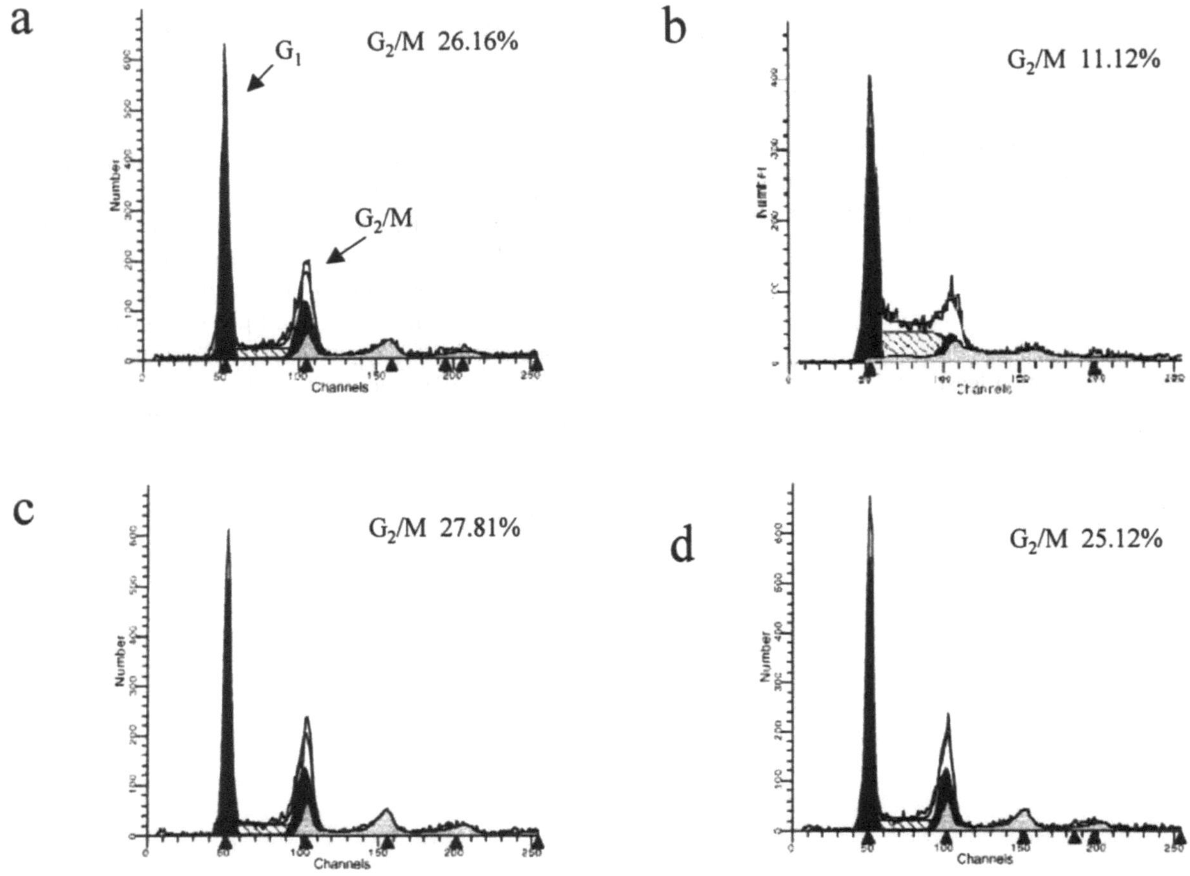

Fig. 2. Effect of antioxidants on Cr (VI)-induced cell growth arrest in human lung epithelial A549 cells. Cells were incubated in a 100 mm dish and pretreated with 50,000 units/ml catalase, 300 μM sodium formate, or 500 units/ml SOD for 0.5 h before Cr (VI) treatment (10 μM) for 24 h. (a) Cr (VI); (b) Cr (VI) + catalase; (c) Cr (VI) + sodium formate; (d) Cr (VI) + SOD.

Effects of antioxidants on Cr (VI)-induced cell growth arrest

Figure 2 shows the effects of antioxidants on Cr (VI)-induced G_2/M phase arrest. SOD, a specific $O_2^{\bullet-}$ radical scavenger, or formate, a specific $^{\bullet}OH$ radical scavenger, did not exhibit significant effect, i.e. the percentage of G_2/M phase was 26.16, 27.81 and 25.12% for Cr (VI), Cr (VI) plus formate and Cr (VI) plus SOD, respectively (Figs 2a, 2c and 2d), indicating that neither $O_2^{\bullet-}$ nor $^{\bullet}OH$ was directly involved in the mechanism of Cr (VI)-induced cell growth arrest. Catalase, a scavenger of H_2O_2, decreased Cr (VI)-induced G_2/M phase arrest from 26.16 to 11.12%, indicating that H_2O_2 was involved in Cr (VI)-induced cell growth arrest (Fig. 2b).

Hydroxyl radical formation from Cr (VI)-stimulated cells

ESR study was used to detect the $^{\bullet}OH$ formation from Cr (VI)-stimulated cells. A549 cells alone did not produce any detectable amount of free radicals (Fig. 3a), whereas addition of 2 mM Cr (VI) generated a 1:2:2:1 quartet ESR spin adduct signal (Fig. 3b). The splittings of this spectrum were $a_H = a_N = 14.9$ G, where a_H and a_N denote hyperfine splittings of the α-hydrogen and the nitroxyl nitrogen, respectively, indicating the DMPO/$^{\bullet}OH$ adduct (Shi and Dalal, 1992a). The detection of this DMPO/$^{\bullet}OH$ spin adduct is evidence for $^{\bullet}OH$ generation. The peak at the right side was assigned to a Cr (V) intermediate based on the lineshape and the g value. Addition of formate (50 mM), an $^{\bullet}OH$ scavenger, reduced the signal intensity by 50% (Fig. 3d). Catalase (2000 units/ml), a specific scavenger of H_2O_2, inhibited the generation of $^{\bullet}OH$ by 90% (Fig. 3c). The inhibition of $^{\bullet}OH$ generation upon addition of catalase indicates that H_2O_2 was generated, and that it was a precursor for $^{\bullet}OH$ generation. SOD (5 μg/ml), a scavenger of $O_2^{\bullet-}$ radical, did not show any significant inhibition, showing that the 1:2:2:1 quartet signal is not due to the $O_2^{\bullet-}$ radical trapping.

Acknowledgements

References

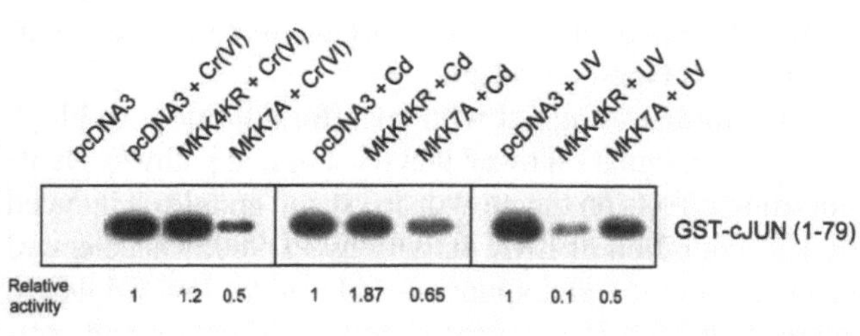

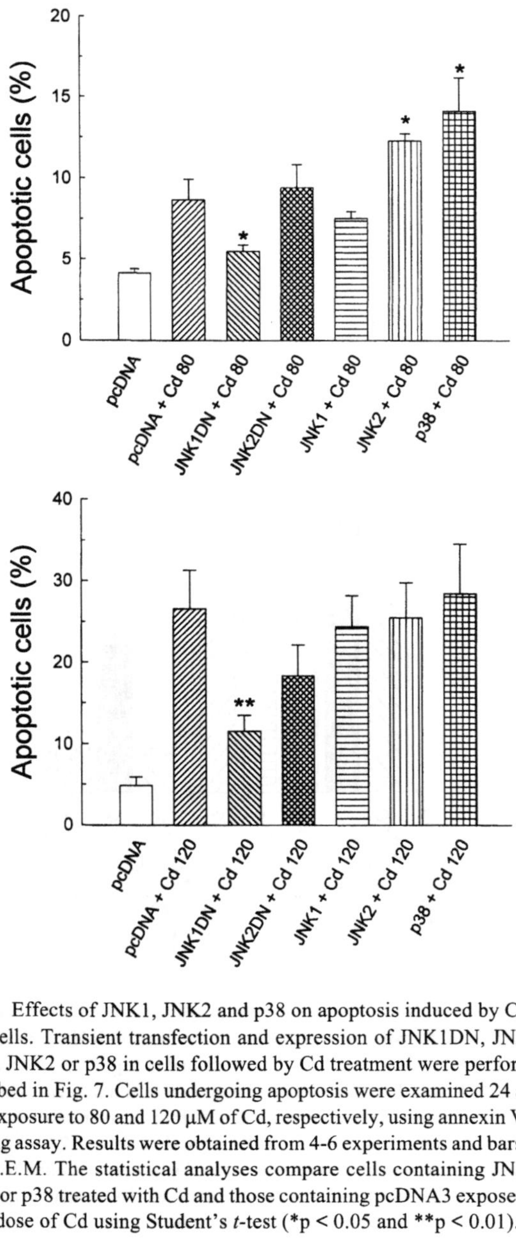

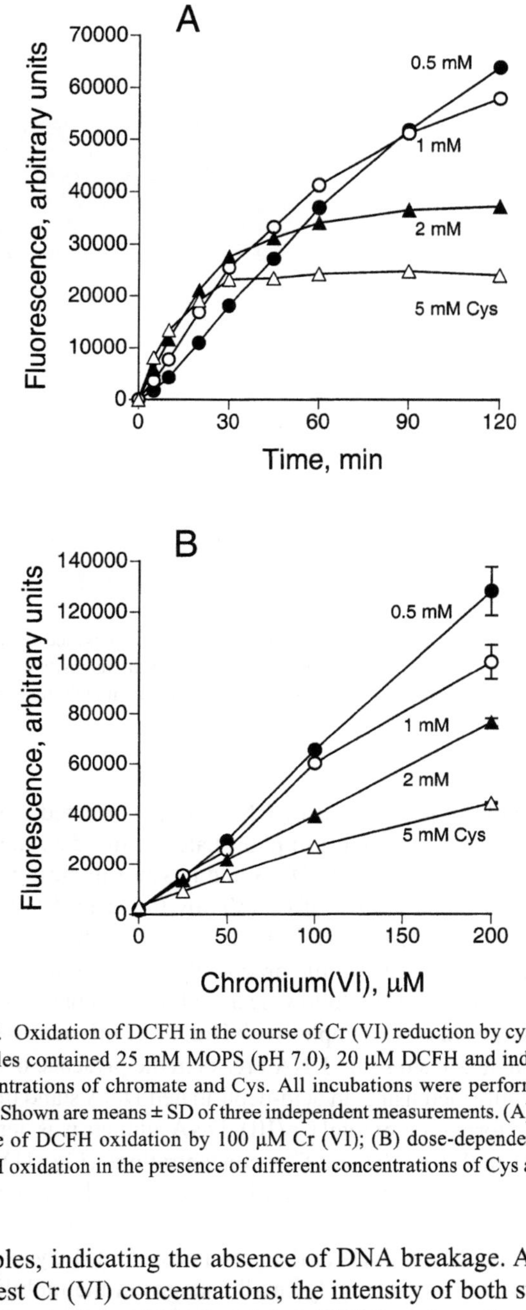

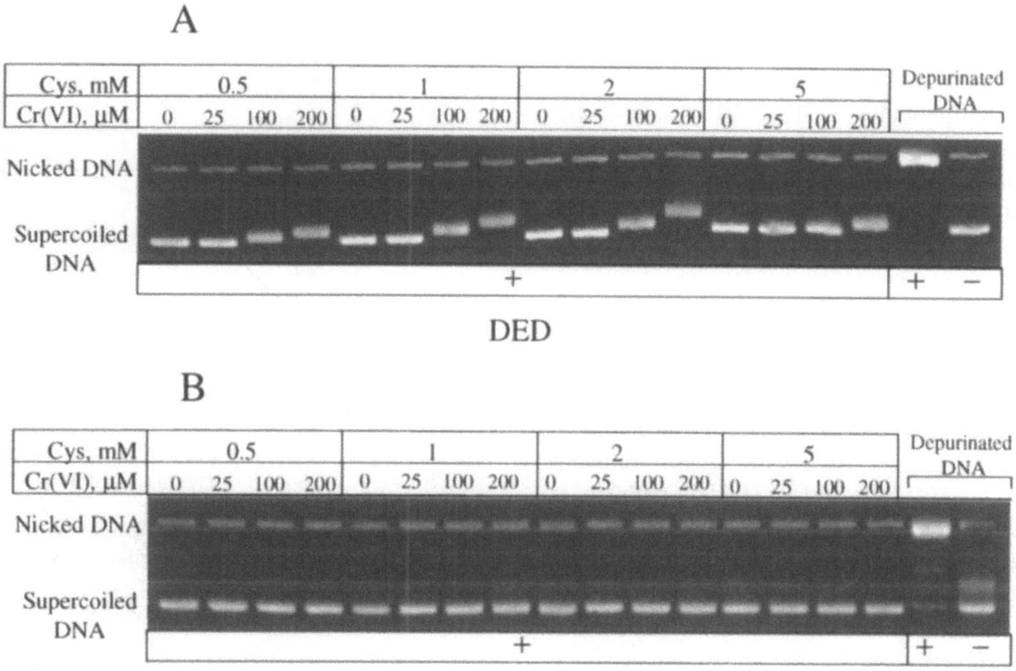

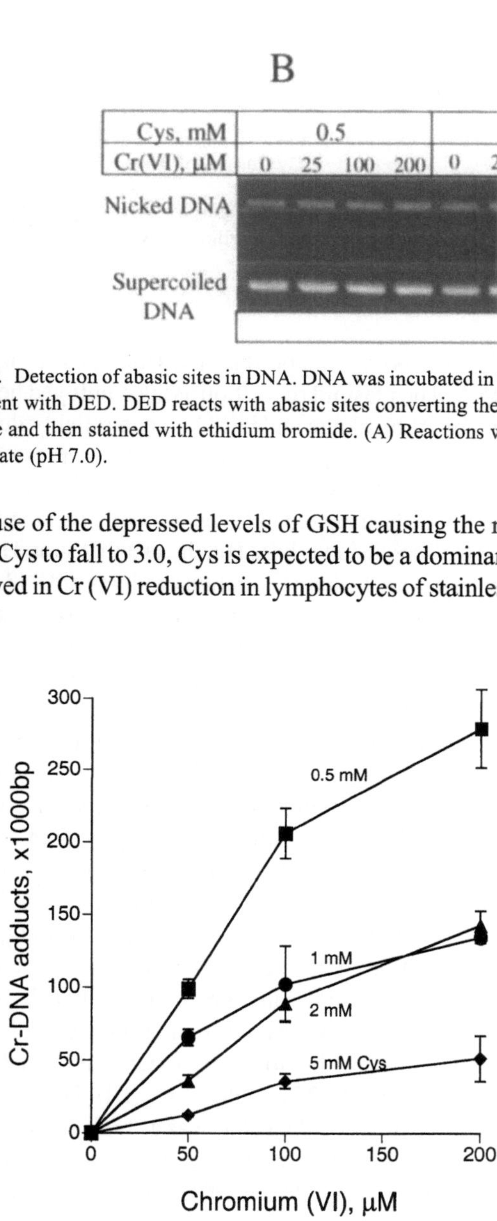

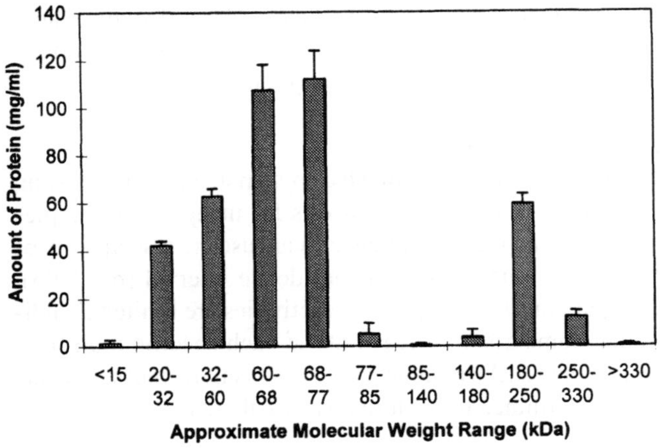

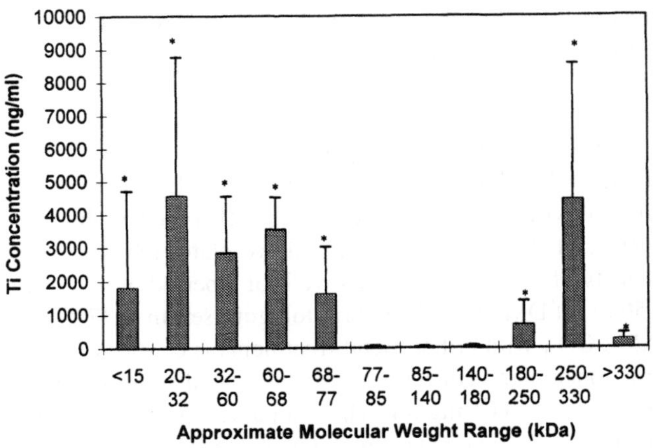

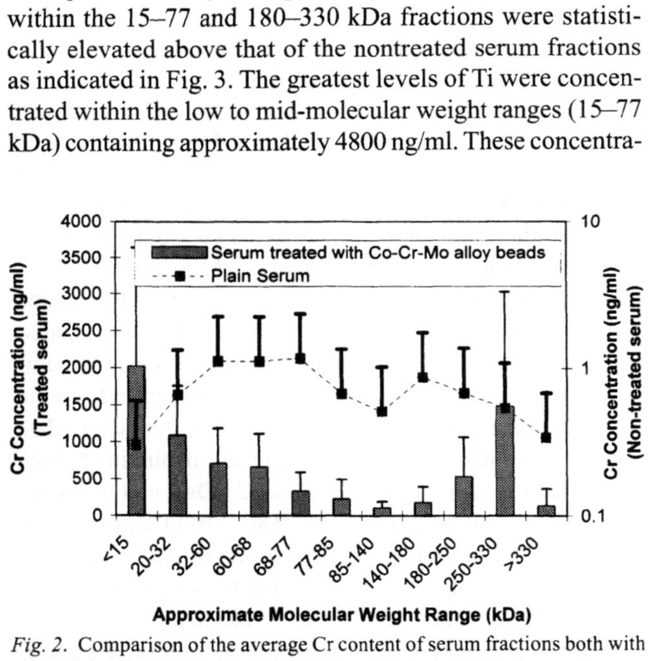

Table 1. Concentration- and time-dependent changes on the enhanced production of superoxide anion and hydroxyl radicals, changes in intracellular oxidized states, DNA fragmentation and apoptotic cell death in human peripheral blood mononuclear (HPBM) cells after treatment with sodium dichromate (chromium (VI) or Cr (VI))

Treatment	Cytochrome *c* reduction (nmoles reduced/15 min/ 1 × 10⁶ cells)	Hydroxyl radical production			Flourescent intensity	DNA fragmentation (% control)	Apoptotic cell death (%)
		2,5-DHBA (nmoles/ml)	2,3-DHBA (nmoles/ml)	Total (nmoles/ml)			
Control 24 h	56.1 ± 8.2^a	0.12 ± 0.02^a	0.22 ± 0.03^a	0.34^a	89 ± 24^a	3.16 ± 0.70^a (100)	0^a
Control 48 h	59.4 ± 4.3^a	0.10 ± 0.01^a	0.25 ± 0.04^a	0.35^a	85 ± 15^a	3.07 ± 0.26^a (100)	ND
Cr (VI) 12.5 μM 24 h	79.4 ± 13.0^b	0.21 ± 0.05^b	0.61 ± 0.05^b	0.82^b	87 ± 22^a	3.19 ± 0.45^a (101)	0^a
Cr (VI) 12.5 μM 48 h	82.0 ± 19.1^b	0.25 ± 0.06^b	0.65 ± 0.09^b	0.90^b	188 ± 56^b	3.45 ± 0.62^a (109)	ND
Cr (VI) 25 μM 24 h	86.9 ± 16.8^b	0.33 ± 0.04^c	0.50 ± 0.06^b	0.83^b	ND	4.04 ± 0.72^b (128)	0^a
Cr (VI) 25 μM 48 h	98.4 ± 22.3^c	0.39 ± 0.06^c	0.68 ± 0.08^b	1.07^c	ND	4.29 ± 0.75^b (136)	ND

Cultured HPBM cells were treated with 12.5 or 25 μM concentration of sodium dichromate (chromium (VI) or Cr (VI)). See Materials and methods section for details. Each value is the mean ± S.D. of 4–6 experiments. Values with non-identical superscripts are significantly different ($p < 0.05$). ND = Not determined.

with 12.5 μM concentration of chromium (VI) for 24 h. However, treatment of HPBM cells with 12.5 μM concentration of chromium (VI) for 48 h resulted in 2.2-fold increase in fluorescent intensity (Table 1). Approximately 5.7-fold increase in fluorescent intensity was observed following treatment of the chronic myelogenous leukemic K562 cells with 12.5 μM concentration of chromium (VI) for 24 h as compared with the control untreated cells. Due to extensive cell necrosis at 48 h, fluorescent intensity could not be measured at 48 h (Table 2).

Concentration-dependent increases in fluorescent intensity were observed following treatment of the J774A.1 macrophage cells with chromium (VI). Approximately 2.8-, 10.1- and 9.2-fold increases in fluorescent intensities were observed following incubation of the cells with 0.20, 0.40 and 0.60 μM concentrations of chromium (VI), respectively, as compared to the control untreated cells (Table 5).

Reduction of MTT

Concentration- and time-dependent effects were studied on the reductions of tetrazolium dye MTT in the cultured J774A.1 macrophages. The cells were incubated with 0.20, 0.40 and 0.60 μM concentrations of chromium (VI) for 24 and 48 h. This assay measures the activity of succinate dehydrogenase, a marker of the mitochondrial electron transport chain. Treatment of the cells with 0.20 and 0.40 μM concentrations of chromium (VI) for 24 h resulted in no significant changes in succinate dehydrogenase activity. When these cells were treated with 0.60 μM concentration of chromium (VI) for 24 h, a small but non-significant increase in succinate dehydrogenase activity was observed. Incubation of the macrophage cells with the three concentrations of chromium (VI) for 48 h resulted in dose-dependent decreases in succinate dehydrogenase activities (Table 5).

Table 2. Concentration-dependent changes in enhanced production of superoxide anion and hydroxyl radicals, changes in intracellular oxidized states, DNA fragmentation and apoptotic cell death in chronic myelogenous leukemic K562 cells following 24 h treatment with sodium dichromate (chromium (VI) or Cr (VI))

Treatment	Cytochrome *c* reduction (nmoles reduced/15 min/ 1 × 10⁶ cells)	Hydroxyl radical production			Flourescent intensity	DNA fragmentation (% control)	Apoptotic cell death (%)
		2,5-DHBA (nmoles/ml)	2,3-DHBA (nmoles/ml)	Total (nmoles/ml)			
Control	99.0 ± 6.7^a	0.18 ± 0.05^a	0.13 ± 0.02^a	0.31	84 ± 14^a	8.17 ± 1.80^a (100)	0^a
Cr (VI) 12.5 μM	262.6 ± 18.7^b	0.80 ± 0.21^b	2.51 ± 0.53^b	3.31^b	475 ± 214^a	18.06 ± 1.12^a (221)	39.5 ± 6.3^a
Cr (VI) 25 μM	479.6 ± 41.6^c	0.89 ± 0.14^c	3.49 ± 0.66^b	4.38^c	ND	24.84 ± 1.86^b (304)	0^a

Cultured K562 cells were treated with 12.5 or 25 μM concentration of sodium dichromate (chromium (VI) or Cr (VI)). See Materials and methods section for details. Each value is the mean ± S.D. of 4–6 experiments. Values with non-identical superscripts are significantly different ($p < 0.05$). ND = Not determined.

Table 3. Hepatic lipid peroxidation, DNA fragmentation, superoxide anion production in sodium dichromate (chromium (VI) or Cr (VI))-treated control and p53-deficient mice

Assay	Control (C57BL/6Ntac) mice		p53-deficient (TSG-p53) mice	
	Control	Cr (VI)-treated	Control	Cr (VI)-treated
Lipid peroxidation (nmoles/mg protein)	2.09 ± 0.25^{a}	6.84 ± 0.91^{b}	1.89 ± 0.26^{a}	20.39 ± 3.16^{a}
DNA fragmentation (% control)	4.57 ± 0.72^{a} (100)	10.06 ± 1.32^{b} (220)	3.03 ± 0.46^{a} (100)	12.26 ± 1.62^{b} (405)
Superoxide anion (nmoles cytochrome *c* reduced/15min/mg/protein)	8.31 ± 1.14^{a}	49.86 ± 6.76^{b}	6.71 ± 0.90^{a}	74.30 ± 10.29^{b}

Control C57BL/6Ntac mice and p53-deficient C57BL/6TSG p53 mice were treated with a single oral 0.50 LD_{50} dose of sodium dichromate (chromium (VI)) (95 mg/kg body wt) and sacrificed after 24 h. Each value is the mean ± S.D. of 4 mice. Assays were performed, as described in the Materials and methods section. Values with non-identical superscripts are significantly different ($p < 0.05$).

Table 4. Brain lipid peroxidation, DNA fragmentation, superoxide anion production in sodium dichromate (chromium (VI) or Cr (VI)-treated control and p53-deficient mice

Assay	Control (C57BL/6Ntac) mice		p53-deficient (TSG-p53) mice	
	Control	Cr (VI)-treated	Control	Cr (VI)-treated
Lipid peroxidation (nmoles/mg protein)	1.69 ± 0.19^{a}	5.88 ± 0.81^{b}	2.02 ± 0.24^{a}	14.71 ± 1.69^{b}
DNA fragmentation (% control)	2.96 ± 0.40^{a} (100)	6.20 ± 0.51^{b} (220)	2.76 ± 0.24^{a} (100)	12.55 ± 1.22^{b} (405)
Superoxide anion (nmoles cytochrome *c* reduced/15min/mg/protein)	5.78 ± 0.41^{a}	23.83 ± 5.56^{b}	7.34 ± 0.65^{a}	56.80 ± 8.94^{b}

Control C57BL/6Ntac mice and p53-deficient C57BL/6TSG p53 mice were treated with a single oral 0.50 LD_{50} dose of sodium dichromate (chromium (VI)) (95 mg/kg body wt) and sacrificed after 24 h. Each value is the mean ± S.D. of 4 mice. Assays were performed, as described in the Materials and methods section. Values with non-identical superscripts are significantly different ($p < 0.05$).

Table 5. Concentration-dependent changes in fluorescent intensity, DNA fragmentation and reduction of tetrazolium dye MTT in cultured J774A.1 murine macrophage cells following treatment with sodium dichromate (chromium (VI) or Cr (VI))

Treatment	Fluorescent intensity	DNA fragmentation	Cell viability (%)	
			24 h	48 h
Control	335 ± 40^{a}	7.25 ± 0.63^{a} (100)	1.07 ± 0.11^{a} (100)	1.02 ± 0.16^{a} (100)
Cr (VI) 0.20 μM	930 ± 155^{b}	13.23 ± 1.89^{b} (183)	0.99 ± 0.03^{a} (93)	0.71 ± 0.03^{b} (70)
Cr (VI) 0.40 μM	3392 ± 617^{c}	20.15 ± 0.71^{c} (278)	1.14 ± 0.17^{a} (107)	$0.65 \pm 0.06^{b,c}$ (64)
Cr (VI) 0.60 μM	3073 ± 341^{c}	21.04 ± 1.08^{c} (290)	1.27 ± 0.10^{a} (119)	0.58 ± 0.03^{c} (57)

Cultured J774A.1 macrophage cells were treated with either 0.20, 0.40, or 0.60 μM sodium dichromate (chromium (VI) or Cr (VI)). See Materials and methods for details. All values are reported as mean ± S.D. from 4–6 replicates. Values with non-identical superscripts are significantly different (*$p < 0.05$).

DNA fragmentation

Following 24 h incubation of the HPBM cells with 12.5 and 25 μM concentrations of chromium (VI), approximately 1.0- and 1.3-fold increases in DNA fragmentation were observed, respectively, and 1.1- and 1.4-fold increases in DNA fragmentation were observed after 48 h, respectively, as compared to the control untreated cells (Table 1). Following treatment of the human K562 cells with 12.5 and 25 μM concentrations of chromium (VI) for 24 h, approximately 2.2- and 3.0-fold increases in DNA fragmentation were observed, respectively (Table 2). Due to extensive cell necrosis at 48 h, DNA fragmentation could not be measured at 48 h. Concentration-dependent increases in DNA fragmentation were observed following incubation of cultured J774A.1 macrophage cells with chromium (VI). Following incubation of J774A.1 macrophage cells with 0.20, 0.40 and 0.60 μM concentrations of chromium (VI) for 24 h approximately 1.8-, 2.8- and 2.9-fold increases in DNA fragmentation were observed, respectively, as compared to control cells (Table 5). The effect of chro-

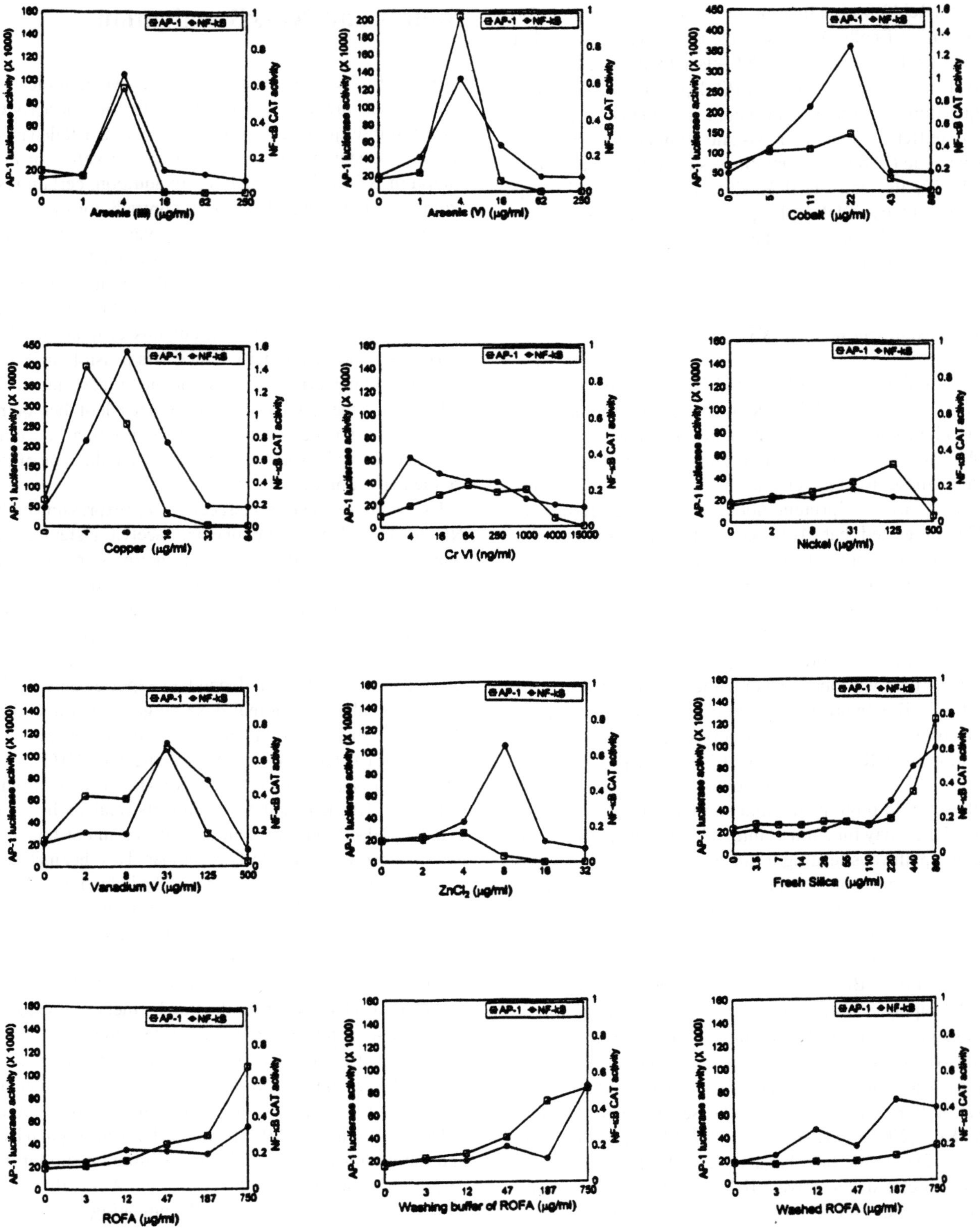

[illegible] of the [illegible] Exch [illegible] 1998.

[illegible] Staal [illegible] repli[cation] [illegible] Acad Sci USA 87 [illegible] 1990.

[illegible] Hay RT. The [illegible] activity of NF-κB [illegible] Nucleic [illegible]

105. [illegible] Hay RT. Integration of the [illegible] signal [illegible] of NF-κB/IκB [illegible] DNA cleavage [illegible] 1993.

106. [illegible] Hay RT. Role [illegible] of NF-κB [illegible] subunit of NF-[illegible] 1997.

107. [illegible] Kaneda K, Kario K. The [illegible] by inhibiting a [illegible] 1996.

108. [illegible] Inhibition of [illegible] Arch [illegible]

[illegible] Mechanisms of [illegible] 1995.

[illegible] the [illegible] site in the [illegible] PALPH [illegible]

[illegible] The role of [illegible] hydroxyl radical. The [illegible] Biochem Biophys [illegible] 1992.

93. [illegible] tissue [illegible]

94. [illegible]

95. [illegible] Perkins [illegible]

96. [illegible] Pathol. Microbiol [illegible]

97. [illegible] [agents] and [illegible] of the [illegible] (1998).

98. [illegible] New Frontiers in Oxygen [illegible]

99. [illegible] The [illegible] of [illegible]

100. [illegible] Kaneda [illegible]

101. [illegible] Med [illegible]

The present studies show that Cr (VI) is able to induce tyrosine phosphorylation in the epithelial cell line A549. H_2O_2 and $^{\bullet}OH$ play a key role in the Cr (VI)-induced tyrosine phosphorylation. Previous studies have demonstrated that reduction of Cr (VI) by cellular reductants, such as glutathione reductase in the presence of NAD(P)H, generates Cr (V) [5, 15]. During the reduction process, molecular oxygen is reduced to $O_2^{\bullet-}$ radical, which produces H_2O_2 via dismutation. Cr (V) can react with H_2O_2 to generate $^{\bullet}OH$ radical via a Fenton-like reaction (Cr (V) + $H_2O_2 \rightarrow$ Cr (VI) + $^{\bullet}OH + OH^-$). Using ESR spin trapping, the present study has demonstrated that Cr (VI)-stimulated cells generate a whole spectrum of ROS and molecular oxygen is the source of these ROS.

Among the ROS generated by Cr (VI)-stimulated cells, both H_2O_2 and $^{\bullet}OH$ appears to be the key species responsible for tyrosine phosphorylation. The following experimental observation support the above conclusion. (a) Sodium formate, an $^{\bullet}OH$ radical scavenger, decreased but did not abolish, tyrosine phosphorylation. (b) SOD, a scavenger of $O_2^{\bullet-}$ only slightly decreased tyrosine phosphorylation. (c) Catalase, a scavenger of H_2O_2, abolished the tyrosine phosphorylation. It may be noted that H_2O_2 is a precursor of $^{\bullet}OH$ radical generation. Elimination of H_2O_2 eliminated $^{\bullet}OH$ generation as shown by ESR spin trapping measurements. As for $O_2^{\bullet-}$, while it is not directly involved in tyrosine phosphorylation, it is an intermediate for the generation of both H_2O_2 and $^{\bullet}OH$.

It has been reported that ROS are involved in various physiological and pathological processes [29]. Biological systems generate ROS naturally. They are also produced in these system from exposure to exogenous substances [30]. Emerging evidence points to the involvement of ROS as signaling intermediates [31]. ROS are involved in the activation of several nuclear transcription factors, such as AP-1, NF-κB, and p53 and appear to regulate apoptosis, cell growth regulation, cytokine production, and oncogene expression [32]. Recently, it has been shown that ROS help Ras transmit growth-stimulating messages to the cell interior [33]. This report adds to the growing evidence that ROS, traditionally viewed as toxins, can also play a role in the cell's normal signaling pathways. Although it is believed that ROS function as signaling intermediates, their mechanism of action remains to be investigated. For example, how ROS can turn certain signals on and off. ROS-mediated tyrosine phosphorylation may be involved in this process. Thus, elucidation of mechanisms involved in Cr (VI)-induced tyrosine phosphorylation may not only increase our understanding of Cr (VI)-induced carcinogenesis, but may also help in understanding ROS-mediated signal pathways in general.

It may also be noted that many other metal ions, mineral particles and chemical carcinogens, such as cobalt, nickel, vanadium, asbestos, and silica, have been reported to be capable of generating ROS. It is possible that these agents may function in the same way as Cr (VI); i.e. they may cause tyrosine phosphorylation through H_2O_2 and $^{\bullet}OH$ mediated common messages. Because tyrosine phosphorylation is important in signal transduction, ROS could play a critical role in various signal pathways in both physiological and pathological processes.

In conclusion, (a) Cr (VI) is able to induce increased tyrosine phosphorylation in human lung epithelial A549 cells; (b) Cr (VI)-stimulated A549 cells generate ROS, including $O_2^{\bullet-}$, H_2O_2 and $^{\bullet}OH$, via reduction of molecular oxygen; and (c) the generated H_2O_2 and $^{\bullet}OH$ are responsible for the increased tyrosine phosphorylation induced by Cr (VI). Therefore, ROS may play a critical role in various signal pathways under both physiological and pathological conditions via their ability to cause tyrosine phosphorylation.

Acknowledgements

The authors are grateful to Drs. Vince Castranova, Murali Rao, and Val Vallyathan for helpful comments and suggestions during the preparation of this manuscript.

References

1. De Flora S, Bagnasco M, Serra D, Zanacchi P: Genotoxicity of chromium compounds: a review. Mutat Res 238: 99–172, 1990
2. Shi X, Chiu A, Chen CT, Halliwell B, Castranova V, Vallyathan V: Reduction of chromium (VI) and its relationship to carcinogenesis. J Toxicol Environ Health 2: 87–104, 1999
3. De Flora S, Wetterhahn KE: Mechanism of chromium metabolism and genotoxicity. Life Chem Rep 7: 169–244, 1989
4. Xu J, Wise JP, Patierno SR: DNA damage induced by carcinogenic lead chromate particles in cultured mammalian cells. Mutat Res 280: 129–136, 1992
5. Leonard S, Wang S, Zang L, Castranova V, Vallyathan V, Shi X: Role of molecular oxygen in the generation of hydroxyl and superoxide anion radicals during enzymatic Cr (VI) reduction and its implication to Cr (VI)-induced carcinogenesis. J Environ Pathol Toxicol Oncol 19: 49–60, 2000
6. Xu J, Wise JP, Patierno SR: DNA damage induced by carcinogenic lead chromate particles in cultured mammalian cells. Mutat Res 280: 129–136, 1992
7. Kortenkamp A, Ozolins Z, Beyersmann D, O'Brien P: Generation of PM2 DNA breaks in the course of reduction of chromium (VI) by glutathione. Mutat Res 216: 19–26, 1989
8. Hayes RB: Review of occupational epidemiology of chromium chemicals and respiratory cancer. Sci Total Environ 71: 331–339, 1988
9. Langard S: One hundred years of chromium and cancer: A review of epidemiological evidence and selected case reports. Am J Ind Med 17: 189–215, 1990
10. Connett PH, Wetterhahn KE: Metabolism of the carcinogenic chromate by cellular constituents. Struct Bond 54: 93–124, 1983
11. Cohen MD, Kargacin B, Klein CB, Costa M: Mechanism of chromium carcinogenicity and toxicity. Crit Rev Toxicol 23: 255–281, 1983
12. Ye J, Wang S, Leonard SS, Sun Y, Butterworth, L, Antonini J, Ding M, Yongyut Rojanasakul, Vallyathan V, Castranova V, Shi X: Role of

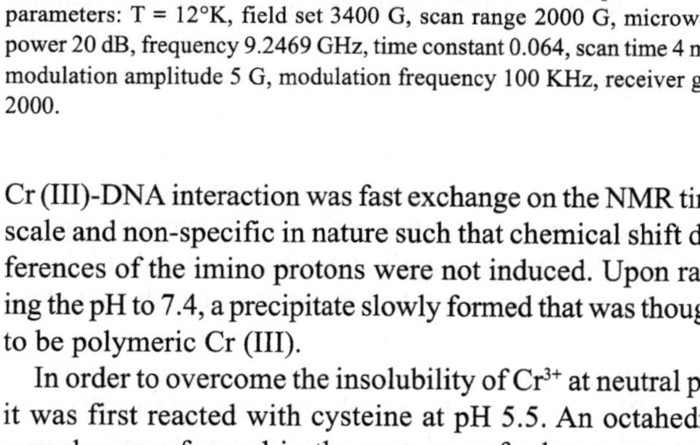